ENZYMATIC FUEL CELLS

ENZYMATIC FUEL CELLS

From Fundamentals to Applications

Edited by

HEATHER R. LUCKARIFT
Air Force Research Laboratory

PLAMEN ATANASSOV
University of New Mexico

GLENN R. JOHNSON
Air Force Research Laboratory

Published by John Wiley & Sons, Inc., Hoboken, New Jersey
Published simultaneously in Canada

For general information on our other products and services or for technical support, please contact our
Customer Care Department within the United States at (800) 762-2974, outside the United States
at (317) 572-3993 or fax (317) 572-4002.

Wiley also publishes its books in a variety of electronic formats. Some content that appears in print may
not be available in electronic formats. For more information about Wiley products, visit our web site at
www.wiley.com.

Library of Congress Cataloging-in-Publication Data:

Enzymatic fuel cells : from fundamentals to applications / edited by Heather
R. Luckarift, Plamen B. Atanasso, Glenn R. Johnson.
    pages cm
  "Published simultaneously in Canada"–Title page verso.
  Includes bibliographical references and index.
  ISBN 978-1-118-36923-4 (cloth)
1. Fuel cells. 2. Fuel cells–Research. 3. Enzymes–Biotechnology. 4.
Electrocatalysis. I. Luckarift, Heather R., 1971- II. Atanasso, Plamen B.,
1962- III. Johnson, Glenn R., 1967-
  TK2931.E59 2014
  621.31'2429–dc23

                                                        2013042736

Printed in the United States of America

10 9 8 7 6 5 4 3 2 1

CONTENTS

**15 *In Situ* X-Ray Spectroscopy of Enzymatic Catalysis:
Laccase-Catalyzed Oxygen Reduction** **304**

*Sanjeev Mukerjee, Joseph Ziegelbauer, Thomas M. Arruda,
Kateryna Artyushkova, and Plamen Atanassov*

16 Enzymatic Fuel Cell Design, Operation, and Application **337**

Vojtech Svoboda and Plamen Atanassov

17 Miniature Enzymatic Fuel Cells 361

Takeo Miyake and Matsuhiko Nishizawa

18 Switchable Electrodes and Biological Fuel Cells 374

Evgeny Katz, Vera Bocharova, and Jan Halámek

19 Biological Fuel Cells for Biomedical Applications   **422**

Magnus Falk, Sergey Shleev, Claudia W. Narváez Villarrubia, Sofia Babanova, and Plamen Atanassov

20 Concluding Remarks and Outlook   **451**

Glenn R. Johnson, Heather R. Luckarift, and Plamen Atanassov

Index   **459**

PREFACE

Substantial research activity is directed to the design and engineering of enzymatic fuel cells in order to optimize their performance, achieve long-term operation, and realize commercial applicability. There has been great interest in the scientific and engineering advances in the area of enzymatic fuel cells and this book covers the state-of-the-art advances of this effort, showing various aspects of the field and demonstrating perspectives. This book will introduce the reader to the scientific aspects of bioelectrochemistry (e.g., electrical wiring of enzymes and charge transfer in enzymatic fuel cell electrodes), the unique engineering problems of enzymatic fuel cells (e.g., design and optimization), and how this understanding lends itself to developing practical applications (such as powering of microdevices, biomedical applications, and autonomous systems). Integration of biocatalysts with devices, however, is a field that is in its infancy in respect to deliverable devices. The book includes an overview of all aspects of enzymatic fuel cell research with a special emphasis on methodology, fabrication, and testing of enzymatic fuel cells. Such a text will be invaluable to investigators within this research field and may encourage a fresh flow of research endeavors.

This book aims to summarize research encompassing all of the aspects required to understand, fabricate, and integrate enzymatic fuel cells. The contribution will span the fields of bioelectrochemistry and biological fuel cell research to inform and instruct any reader interested in this stimulating research area.

The editors thank all the authors for their contributions and the editorial staff at Wiley for their invaluable assistance.

HEATHER R. LUCKARIFT
PLAMEN ATANASSOV
GLENN R. JOHNSON

CONTRIBUTORS

Thomas M. Arruda, Department of Chemistry, Salve Regina University, Newport, RI, USA

Kateryna Artyushkova, Department of Chemical and Nuclear Engineering and Center for Emerging Energy Technologies, University of New Mexico, Albuquerque, NM, USA

Plamen Atanassov, Department of Chemical and Nuclear Engineering and Center for Emerging Energy Technologies, University of New Mexico, Albuquerque, NM, USA

Sofia Babanova, Department of Chemical and Nuclear Engineering and Center for Emerging Energy Technologies, University of New Mexico, Albuquerque, NM, USA

Scott Banta, Department of Chemical Engineering, Columbia University, New York, NY, USA

Lorena Betancor, Laboratorio de Biotecnología, Facultad de Ingeniería, Universidad ORT Uruguay, Montevideo, Uruguay

Vera Bocharova, Department of Chemistry and Biomolecular Science, and Nano-Bio Laboratory (NABLAB), Clarkson University, Potsdam, NY, USA

Elliot Campbell, Department of Chemical Engineering, Columbia University, New York, NY, USA

Michael J. Cooney, Hawai'i Natural Energy Institute, School of Ocean and Earth Science and Technology, Honolulu, HI, USA

Arnab Dutta, Department of Chemistry and Biochemistry and Center for Bio-Inspired Solar Fuel Production, Arizona State University, Tempe, AZ, USA

D. Matthew Eby, Booz Allen Hamilton, Atlanta, GA, USA; Airbase Sciences Branch, Air Force Research Laboratory, Tyndall Air Force Base, FL, USA

Magnus Falk, Department of Biomedical Sciences, Malmö University, Malmö, Sweden

Karen E. Farrington, Universal Technology Corporation, Dayton, OH, USA; Airbase Sciences Branch, Air Force Research Laboratory, Tyndall Air Force Base, FL, USA

Joshua W. Gallaway, The CUNY Energy Institute, The City College of New York, New York, NY, USA

Jan Halámek, Department of Chemistry and Biomolecular Science, and NanoBio Laboratory (NABLAB), Clarkson University, Potsdam, NY, USA

Dmitri M. Ivnitski, Department of Chemical and Nuclear Engineering and Center for Emerging Energy Technologies, University of New Mexico, Albuquerque, NM, USA

Glenn R. Johnson, Airbase Sciences Branch, Air Force Research Laboratory, Tyndall Air Force Base, FL, USA

Anne K. Jones, Department of Chemistry and Biochemistry and Center for Bio-Inspired Solar Fuel Production, Arizona State University, Tempe, AZ, USA

Evgeny Katz, Department of Chemistry and Biomolecular Science, and NanoBio Laboratory (NABLAB), Clarkson University, Potsdam, NY, USA

Patrick Kwan, Department of Chemistry and Biochemistry and Center for Bio-Inspired Solar Fuel Production, Arizona State University, Tempe, AZ, USA

Carolin Lau, Department of Chemical and Nuclear Engineering and Center for Emerging Energy Technologies, University of New Mexico, Albuquerque, NM, USA

Heather R. Luckarift, Universal Technology Corporation, Dayton, OH, USA; Airbase Sciences Branch, Air Force Research Laboratory, Tyndall Air Force Base, FL, USA

Chelsea L. McIntosh, Department of Chemistry and Biochemistry and Center for Bio-Inspired Solar Fuel Production, Arizona State University, Tempe, AZ, USA

Shelley D. Minteer, Departments of Chemistry and Materials Science and Engineering, University of Utah, Salt Lake City, UT, USA

Takeo Miyake, Department of Bioengineering and Robotics, Tohoku University, Sendai, Japan; Core Research for Evolutional Science and Technology (CREST), Japan Science and Technology Agency, Tokyo, Japan

Sanjeev Mukerjee, Department of Chemistry and Chemical Biology, Northeastern University, Boston, MA, USA

Claudia W. Narváez Villarrubia, Department of Chemical and Nuclear Engineering and Center for Emerging Energy Technologies, University of New Mexico, Albuquerque, NM, USA

Matsuhiko Nishizawa, Department of Bioengineering and Robotics, Tohoku University, Sendai, Japan; Core Research for Evolutional Science and Technology (CREST), Japan Science and Technology Agency, Tokyo, Japan

Lindsey N. Pelster, Departments of Chemistry and Materials Science and Engineering, University of Utah, Salt Lake City, UT, USA

Ramaraja P. Ramasamy, Nano Electrochemistry Laboratory, College of Engineering, University of Georgia, Athens, GA, USA

Michelle Rasmussen, Departments of Chemistry and Materials Science and Engineering, University of Utah, Salt Lake City, UT, USA

Rosalba A. Rincón, Department of Chemical and Nuclear Engineering and Center for Emerging Energy Technologies, University of New Mexico, Albuquerque, NM, USA

Souvik Roy, Department of Chemistry and Biochemistry and Center for Bio-Inspired Solar Fuel Production, Arizona State University, Tempe, AZ, USA

Sergey Shleev, Department of Biomedical Sciences, Malmö University, Malmö, Sweden

Guinevere Strack, Oak Ridge Institute of Science and Engineering, Oak Ridge, TN, USA; Airbase Sciences Branch, Air Force Research Laboratory, Tyndall Air Force Base, FL, USA

Vojtech Svoboda, School of Materials Science and Engineering, Georgia Institute of Technology, Atlanta, GA, USA

Shuai Xu, Departments of Chemistry and Materials Science and Engineering, University of Utah, Salt Lake City, UT, USA

Sijie Yang, Department of Chemistry and Biochemistry and Center for Bio-Inspired Solar Fuel Production, Arizona State University, Tempe, AZ, USA

Joseph Ziegelbauer, Department of Chemistry and Chemical Biology, Northeastern University, Boston, MA, USA

1

INTRODUCTION

HEATHER R. LUCKARIFT

Universal Technology Corporation, Dayton, OH, USA; Airbase Sciences Branch, Air Force Research Laboratory, Tyndall Air Force Base, FL, USA

PLAMEN ATANASSOV

Department of Chemical and Nuclear Engineering and Center for Emerging Energy Technologies, University of New Mexico, Albuquerque, NM, USA

GLENN R. JOHNSON

Airbase Sciences Branch, Air Force Research Laboratory, Tyndall Air Force Base, FL, USA

During the past century, energy consumption has increased so dramatically that an unbalanced energy management scenario now exists. We are now aware of the transience of nonrenewable resources and the irreversible damage caused to the environment through their collection and use. There is no sign that this growth in demand will abate as global industrialization and the consumer culture expand. The call for resources to satisfy economies and individual requirements will outpace supply of classic resources. In addition, there is a trend toward the miniaturization and portability of computing and communications devices. These energy-intense applications require small, lightweight power sources that will sustain operation over long periods of time, even in remote locations separated from "civilization," such as military engagement and exploration. Biological fuel cells (BFCs) may provide a solution to provide small, lightweight, sustainable sources of power using simple renewable fuels (sugars, alcohols). BFCs have two primary operational advantages over alternative technologies for generating power from organic substrates: high

Enzymatic Fuel Cells: From Fundamentals to Applications, First Edition. Edited by Heather R. Luckarift, Plamen Atanassov, and Glenn R. Johnson.
© 2014 John Wiley & Sons, Inc. Published 2014 by John Wiley & Sons, Inc.

conversion efficiency and operation at ambient temperatures. Furthermore, BFCs will scale down more effectively than conventional batteries. That scaling complements advances in the medical sciences, which are leading to an increased prevalence of implantable electrically operated devices (e.g., pacemakers). These items need power supplies that can operate for long period of time, so as to avoid issues with maintenance or replacement that would require surgery. Ideally, implanted devices would take advantage of the natural substances found in the body, thus deriving power from a continuous and renewable fuel source. BFCs potentially offer solutions to all these challenges, by applying Nature's solutions to energy conversion and tailoring them to meet specific power requirements. Because BFCs use concentrated sources of chemical energy, and will operate at high conversion efficiency, these devices can be small and lightweight, particularly when the fuel is derived directly from a living organism (e.g., glucose from the bloodstream), or harvested from the environment ("fuel scavenging").

Conventional fuel cells use inorganic catalysts and precious metals in anodic and cathodic half-reactions separated by a barrier that selectively allows passage of positively charged ions. BFCs follow the same basic principle, but biological molecules (redox enzymes or whole microbial cells) catalyze the electrochemical processes. Microbial BFCs require continuous maintenance of whole living cells to sustain physiological processes, and as a result, dictate stringent working conditions to maintain output. To overcome this constraint, the redox enzymes responsible for desired processes may be separated and purified from living organisms and directly applied as biocatalysts in BFCs. The resulting systems, termed enzymatic fuel cells (EFCs) use specific enzymes that when electrically contacted with an electrode will oxidize energy-rich abundant organic raw materials, such as alcohols, organic acids, or sugars, in the anode. In the cathode, substrates such as molecular oxygen (O_2) or hydrogen peroxide (H_2O_2) are reduced. In combination with the anodic reaction, this process generates electrical power (voltage and current). Enzymes are environmentally sensitive, however, and may degrade over time when exposed to the environment, and as a result, special ways for stabilization and utilization must be established. Advances in characterizing and manipulating nanomaterial architectures, for example, provide enhanced connectivity at the biomaterial interface.

Undeniably, EFCs are a rapidly developing area of scientific research and technological development. Within the last decade, publications concerning optimization and characterization of anodic and cathodic biocatalysts have risen by an order of magnitude (e.g., 3 manuscripts in 2002 compared with >30 in 2010).

This book will address BFC technology in five distinct sections:

1. Introduction to EFC.
2. Fundamentals of EFC.
3. Optimizing and characterizing biological catalysis.
4. System design and integration.
5. Outlook to future development and emerging applications.

The utility of BFCs is determined largely by service life and power density. Accordingly, maximizing the lifetime of catalysts by effective electrocatalytic association with electrodes is essential to EFC development. There are numerous biochemical and biophysical facets of EFCs that can be optimized to enhance their efficiency and power, and many of those factors will be discussed and presented herein.

BFCs have evolved exponentially over the past decade and are emerging as a simple and robust technology for generating power that is limited only by device engineering. Provided the biological understanding increases, the electrochemical technology advances, and the overall electrode prices decrease, this concept may soon qualify as a core technology for conversion of carbohydrates to electricity.

LIST OF ABBREVIATIONS

BFC   biological fuel cell
EFC   enzymatic fuel cell

2

ELECTROCHEMICAL EVALUATION OF ENZYMATIC FUEL CELLS AND FIGURES OF MERIT[*]

SHELLEY D. MINTEER

Departments of Chemistry and Materials Science and Engineering, University of Utah, Salt Lake City, UT, USA

HEATHER R. LUCKARIFT

Universal Technology Corporation, Dayton, OH, USA; Airbase Sciences Branch, Air Force Research Laboratory, Tyndall Air Force Base, FL, USA

PLAMEN ATANASSOV

Department of Chemical and Nuclear Engineering and Center for Emerging Energy Technologies, University of New Mexico, Albuquerque, NM, USA

2.1 INTRODUCTION

Over the last decade, biological fuel cells (BFCs) have become an increasingly popular area of research. As a new technology, many of the analytical techniques that have been previously used for conventional electrochemical devices have been adapted and tailored for use in studying biological systems.

Many reports evaluate energy conversion on the basis of energy density ($Wh\,l^{-1}$) and specific energy ($Wh\,kg^{-1}$), that is, energy converted per unit volume or per unit

[*]This chapter is reproduced in part with permission from Moehlenbrock MJ, Arechederra RL, Sjöholm KH, Minteer SD. Analytical techniques for characterizing enzymatic biofuel cells. *Anal Chem* 2009;81:9538–9545. Copyright 2009, American Chemical Society.

Enzymatic Fuel Cells: From Fundamentals to Applications, First Edition. Edited by Heather R. Luckarift, Plamen Atanassov, and Glenn R. Johnson.

mass. In enzymatic fuel cells (EFCs), enzymes convert energy from a remarkable range of chemical substrates. Using these types of catalysts offers great benefits in terms of their catalytic activity, specificity, and cost. However, development and characterization of these devices require a diverse range of expertise to optimize and fully understand the activity of biologically functionalized electrodes and BFCs.

In this chapter, the terminology and methodology used for characterizing bio-electrocatalytic processes (as they are manifested in the electrochemical power source of question) are discussed and BFC performance is presented in terms of relevant figures of merit.

2.2 ELECTROCHEMICAL CHARACTERIZATION

There are basic electroanalytical characterization techniques that are consistently used to evaluate performance characteristics of BFCs. Standard electroanalytical techniques include linear sweep voltammetry, cyclic voltammetry, amperometry, and both galvanostatic and potentiostatic coulometry [1–5].

2.2.1 Open-Circuit Measurements

The power generated by a BFC is quantified in terms of power output, P (W) $= V$ (V) $\times I$ (A) as a derivation of Ohm's law ($I = V/R$). One of the most common and simplest evaluation tools is the measurement of the electrochemical potential at open circuit ($I = 0$). This is done in a similar method to measuring potentials with a voltmeter and provides information about the thermodynamics of the cell, but it does not provide information about the kinetics. Early on, this was considered an important tool in evaluating BFCs, but as research has progressed, researchers have realized that kinetics and transport are typically the limiting issues. When the potential at zero current (or rest potential) is recorded with respect to a reference electrode (in a three-electrode measurement), it is denoted as "open circuit potential" (OCP). When the measurement is made in a two-electrode configuration (with respect to another fuel cell electrode), it is reported as "open-circuit voltage" (OCV), signifying that the electrode against which the measurement is made is subject to polarization as well.

2.2.2 Cyclic Voltammetry

Voltammetry is a common electroanalytical technique for characterization of enzyme-modified electrodes. In cyclic voltammetry (CV), a potential window is scanned in the forward and reverse directions while the resulting current is measured. This technique is useful for determining the reduction potential of the enzyme or coenzyme and for determining the overpotential for the system, which, in turn, corresponds to efficiency. Using this technique, detailed information about the catalytic cycle of the system can be determined including electron transfer kinetics, reaction mechanisms, current densities, and reduction potentials [6,7].

2.2.3 Electron Transfer

Specific redox characteristics of a catalyst derived from CV scans are also used to confirm an enzyme's ability for bioelectrocatalysis by either direct electron transfer (DET) or mediated electron transfer (MET) to the electrode. DET and MET are two distinct mechanisms of bioelectrocatalysis. MET has the advantage of being compatible with almost all naturally occurring oxidoreductase enzymes and coenzymes, but it requires additional components (either small-molecule redox mediators or redox polymers) because the enzymes cannot efficiently transfer electrons to the electrode. These additional components make the system more complex and less stable [8]. The vast majority of oxidoreductase enzymes that require MET to an electrode are nicotinamide adenine dinucleotide (NAD^+) dependent. Two of the most commonly encountered NAD^+-dependent enzymes in BFC anodes are glucose dehydrogenase (GDH) and alcohol dehydrogenase (ADH). These enzymes have been thoroughly characterized in respect to half-cell electrochemistry and have been demonstrated for operation in BFC. More information about MET can be found in Chapter 9.

DET is a more straightforward process resulting in a simpler system, but it is limited to specific enzymes capable of transferring electrons directly from the enzyme to the current collector. Pyrroloquinoline quinone (PQQ)-dependent, heme-containing, and flavin adenine dinucleotide (FAD)-dependent enzymes have demonstrated the ability to undergo DET. Examples include PQQ-dependent ADH and FAD-dependent glucose oxidase (GOx). Both enzymes demonstrate DET as characterized by CV and can be used as anodic catalysts in BFCs. One disadvantage of some FAD-dependent enzymes, such as GOx, is that they produce hydrogen peroxide during operation; hydrogen peroxide is highly reactive and can damage BFC components.

Researchers have also used voltammetry to evaluate enzyme inhibition mechanisms. One of the most commonly studied enzyme inhibition systems is hydrogenases (see Chapter 6), primarily because these enzymes are inhibited by a wide variety of species, and experimental difficulty comes from the inherent instability of the enzymes in low oxygen concentrations [9].

2.2.4 Polarization Curves

Polarization curves are a common method for evaluation of cell performance and are developed by applying a variable load and plotting potential as a function of current density [10]. Figure 2.1 is an example of a polarization curve showing deviations from the ideal behavior of the electrochemical device. The contributions can result from many factors in the system, including concentration, activation, transfer, resistance, and earlier or simultaneous reactions. Polarization curves can be used to generate power curves by plotting the power produced by the system versus the current density or potential. This allows for the determination of the maximum power density that the BFC can generate at the optimal potential. One method of obtaining a polarization curve is very slow ($<1\,mV\,s^{-1}$) linear scan voltammetry of a complete BFC. Other methods for determining polarization curves and power curves include employing amperometry at a variety of potentials and measuring the resulting current, or employing a series of external electrical

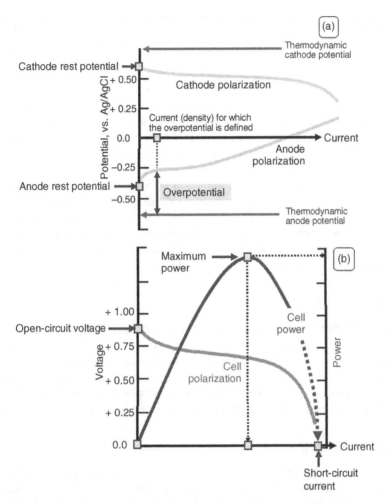

FIGURE 2.1 (a) Examples of polarization characteristics of a fuel cell cathode and anode outlining the definition of the overpotential as deviation from the equilibrium (theoretical) potential at a given current density (reaction rate). (b) Overall cell polarization as a sum of anode and cathode polarization and the power of a fuel cell as an integral of the voltage by the current. Usual figures of merit are indicated on both polarization and power curves.

loads and measuring current and potential at each external resistance. Chronoamperometry is an electrochemical technique where the potential of the electrode is held constant while the resulting current is measured as a function of time. In contrast, chronopotentiometry is the study of potential as a function a time at an electrode operating with a constant current.

Obtaining electrode polarization as a function of the current (and even better, of the current density) is the main method of analysis of the fuel cell performance. The closer the data are collected to a steady-state operation mode, the more adequate is the performance represented. In an ideal case, "hydrodynamic polarization" is the preferred method that eliminates all hydrodynamic and transport effects and is obtained by steady-state measurements of current at a given potential, with substantial time allowed to achieve a constant potential response (chronopotentiometry), or, vice versa, a current measured after allowing for the current to reach a steady-state response after a potential has been applied (chronoamperometry). If that is deemed unpractical, for a well-defined cell configuration, one can use slow scans of the potential to record polarization. Figure 2.1 presents an illustration of such anodic or cathodic polarization independently obtained in a three-electrode configuration with the anode and the cathode being interrogated as "working electrodes" in connection with a potentiostat and independent reference electrode. The OCP and the over-potential are shown (Figure 2.1a). If the cell is well designed, the formal algebraic sum of the anodic and cathodic polarization will coincide with the cell polarization as recorded when the two-electrode scheme (i.e., full cell) is used (Figure 2.1b). The deviation from this represents the internal cell resistance and can be subtracted for analysis purposes after being independently measured as high-frequency impedance. A full cell assembly is usually characterized by the OCV, which should be the sum of the OCPs of the corresponding cathode and anode.

2.2.5   Power Curves

Often researchers use power curves, which are a derivative of the polarization curve as the power axis is an integral expression of both the current and the voltage. The power curve is usually characterized by the short-circuit current (this popular figure of merit has little engineering consequence) and maximum derived power or power density. The last is often used as a measure of success, but it is important to note that there is hardly any practical device that operates at the maximum power density as a design point.

2.2.6   Electrochemical Impedance Spectroscopy

Electrochemical impedance spectroscopy (EIS) is also commonly employed for analysis of enzymatic electrode systems [11]. EIS is performed by overlaying a range of alternating current (AC) perturbation signals to an electrode that is under direct current (DC) bias. A Nyquist plot is then generated, and variations of the frequency response can then be used to interpret limiting mechanisms associated with charge transfer.

2.2.7   Multienzyme Cascades

In some cases, multienzyme cascades are used where an artificial metabolic pathway has been created on an electrode to take a specific substrate and electrocatalytically convert it to produce energy [5,12–14]. With these systems, coulometry is useful for determining coulombic efficiency, as well as combining coulometry with nuclear magnetic resonance (NMR) analysis to determine reaction intermediates and prod-ucts, and elucidate bottlenecks in the artificial metabolic pathway, or to determine the energy density of a fuel cell. Please refer to Chapter 5 for further information regarding multienzyme cascades.

2.2.8 Rotating Disk Electrode Voltammetry

Rotating disk electrode (RDE) voltammetry is a powerful technique for probing catalyst kinetics and substrate/fuel diffusion. It is similar to CV, but the working electrode is rotated at several different rates. This technique allows for a thorough characterization of the kinetics of an enzyme on an electrode. Immobilization polymers demonstrate typical rates of diffusion two to four orders of magnitude slower than the substrate's diffusion coefficient in solution. This makes it important to characterize both the diffusion-limiting aspect of an immobilization polymer and the kinetics of the enzyme at the electrode surface without interference from each other, and RDE voltammetry allows for this. If the enzyme is immobilized in a polymer matrix on the RDE, the substrate diffusion coefficient through the polymer membrane can be determined [15,16]. Other types of immobilization materials can also be examined with RDE. Rotating ring–disk electrodes (RRDEs) have an added ring on the outer perimeter of the disk, allowing for products formed at the disk to be immediately detected at the ring. A good example of this would be a disk with immobilized GOx with a platinum ring to detect the hydrogen peroxide produced during glucose oxidation [17]. Advances in immobilization of enzymes on carbon nanotubes, and making those protein–nanomaterial adducts in a form of an ink, have allowed for precise RRDE electrode measurements of oxygen reduction reaction catalyzed by bilirubin oxidase and laccase [18,19] (Figure 2.2).

2.3 OUTLOOK

Development and characterization of BFCs require an extensive range of expertise and ever-developing characterization techniques in order to fully evaluate the nuances of

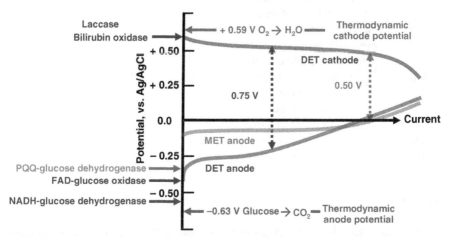

FIGURE 2.2 Conceptual drawing of polarization curves that illustrates deviations from ideality for a bioanode that is catalyzing oxidation of glucose by enzymes with several different cofactors and a biocathode that is reducing molecular oxygen by a copper oxidase. (Please see the color version of this figure in Color Plates section.)

bioelectrochemical systems. The techniques described throughout this book span not only the obvious electroanalytical characterization techniques needed to evaluate a power source or analytical device but also biological and materials characterization techniques, spectroscopy, and biological imaging. As EFCs continue to develop, new analytical techniques will be extended to provide a better understanding of the advantages and limitations of this technology, along with understanding the engineering design envelope for applications.

Readers are referred to a number of sources for further information [20–25].

ACKNOWLEDGMENT

This research was prepared as an account of work sponsored by the United States Air Force Research Laboratory, Materials and Manufacturing Directorate, Airbase Technologies Division (AFRL/RXQ), but the views of authors expressed herein do not necessarily reflect those of the United States Air Force.

LIST OF ABBREVIATIONS

AC	alternating current
ADH	alcohol dehydrogenase
BFC	biological fuel cell
CV	cyclic voltammetry
DC	direct current
DET	direct electron transfer
EFC	enzymatic fuel cell
EIS	electrochemical impedance spectroscopy
FAD	flavin adenine dinucleotide
GDH	glucose dehydrogenase
GOx	glucose oxidase
MET	mediated electron transfer
NAD^+	nicotinamide adenine dinucleotide
NMR	nuclear magnetic resonance
OCP	open-circuit potential
OCV	open-circuit voltage
PQQ	pyrroloquinoline quinone
RDE	rotating disk electrode
RRDE	rotating ring–disk electrode

REFERENCES

1. Arechederra RL, Minteer SD. Organelle-based biofuel cells: immobilized mitochondria on carbon paper electrodes. *Electrochim Acta* 2008;53:6698–6703.
2. Calabrese Barton S, Kim H, Binyamin G, Zhang Y, Heller A. Electroreduction of O_2 to water on the "wired" laccase cathode. *J Phys Chem B* 2001;105:11917–11921.
3. Atanassov P, Apblett C, Banta S, Brozik S, Calabrese Barton S, Cooney MJ, Liaw BY, Mukerjee S, Minteer SD. Enzymatic biofuel cells. *Electrochem Soc Interface* 2007;16:28–31.

4. Svoboda V, Cooney M, Liaw BY, Minteer S, Piles E, Lehnert D, Calabrese Barton S, Rincón R, Atanassov P. Standard characterization of electrocatalytic electrodes. *Electroanalysis* 2008;20:1099–1109.

5. Arechederra RL, Treu BL, Minteer SD. Development of glycerol/O_2 biofuel cell. *J Power Sources* 2007;173:156–161.

6. Treu BL, Minteer SD. Isolation and purification of PQQ-dependent lactate dehydrogenase from *Gluconobacter* and use for direct electron transfer at carbon and gold electrodes. *Bioelectrochemistry* 2008;74:73–77.

7. Armstrong FA, Hill HAO, Walton NJ. Direct electrochemistry of redox proteins. *Acc Chem Res* 1988;21:407–413.

8. Bullen RA, Arnot TC, Lakeman JB, Walsh FC. Biofuel cells and their development. *Biosens Bioelectron* 2006;21:2015–2045.

9. Vincent KA, Parkin A, Armstrong FA. Investigating and exploiting the electrocatalytic properties of hydrogenases. *Chem Rev* 2007;107:4366–4413.

10. Mano N, Fernandez JL, Kim Y, Shin W, Bard AJ, Heller A. Oxygen is electroreduced to water on a "wired" enzyme electrode at a lesser overpotential than on platinum. *J Am Chem Soc* 2003;125:15290–15291.

11. Katz E, Willner I. A biofuel cell with electrochemically switchable and tunable power output. *J Am Chem Soc* 2003;125:6803–6813.

12. Arechederra RL, Minteer SD. Complete oxidation of glycerol in an enzymatic biofuel cell. *Fuel Cells* 2009;9:63–69.

13. Sokic-Lazic D, Minteer SD. Carbon nanotube/gold nanoparticles/polyethyleneimine-functionalized ionic liquid thin film composites for glucose biosensing. *Biosens Bioelectron* 2008;24:945–950.

14. Kjeang E, Sinton D, Harrington DA. Strategic enzyme patterning for microfluidic cells. *J Power Sources* 2006;158:1–12.

15. Pardo-Yissar V, Katz E, Willner I, Kotylar AB, Sanders C, Lill H. Biomaterial engineered electrodes for bioelectronics. *Faraday Discuss* 2000;116:119–134.

16. Mano N, Kim H-H, Zhang Y, Heller A. An oxygen cathode operating in a physiological solution. *J Am Chem Soc* 2002;124:6480–6486.

17. Castner JF, Wingard LB. Mass transport and reaction kinetic parameters determined electrochemically for immobilized glucose oxidase. *Biochemistry* 1984;23:2203–2210.

18. Brocato S, Lau C, Atanassov P. Mechanistic study of direct electron transfer in bilirubin oxidase. *Electrochim Acta* 2012;61:44–49.

19. Parimi NS, Umasankar Y, Atanassov P, Ramasamy RP. Electrochemical kinetic and mechanistic parameters of laccase-catalyzed oxygen reduction reaction. *ACS Catal* 2012;2:38–44.

20. Bartlett PN (ed.), *Bioelectrochemistry: Fundamentals, Experimental Techniques and Applications*. John Wiley & Sons, Ltd, Chichester, UK, 2008.

21. Bard AJ, Faulkner LR. *Electrochemical Methods: Fundamentals and Applications*, 2nd edition. John Wiley & Sons, Ltd, Chichester, UK, 2001.

22. Kano K, Ikeda T. Fundamentals and practices of mediated bioelectrocatalysis. *Anal Sci* 2000;16:1013–1021.

23. Meredith MT, Minteer SD. Biofuel cells: enhanced enzymatic bioelectrocatalysis. *Annu Rev Anal Chem* 2012;5:157–179.

24. Alkire RC, Kolb DM, Lipkowski J (eds), *Bioelectrochemistry: Fundamentals, Applications and Recent Developments*. Wiley-VCH Verlag GmbH, Weinheim, 2012.

25. Ghindilis AL, Atanassov P, Wilkins E. Enzyme-catalyzed direct electron transfer: fundamentals and analytical applications. *Electroanalysis* 1997;9:661–674.

3

DIRECT BIOELECTROCATALYSIS: OXYGEN REDUCTION FOR BIOLOGICAL FUEL CELLS

DMITRI M. IVNITSKI AND PLAMEN ATANASSOV

Department of Chemical and Nuclear Engineering and Center for Emerging Energy Technologies, University of New Mexico, Albuquerque, NM, USA

HEATHER R. LUCKARIFT

Universal Technology Corporation, Dayton, OH, USA; Airbase Sciences Branch, Air Force Research Laboratory, Tyndall Air Force Base, FL, USA

3.1 INTRODUCTION

Effective redox bioelectrocatalysts are the crux to design and development of the next generation of microscale electrochemical biosensors, biomedical devices, and biological fuel cells (BFCs) [1–5]. BFCs, for example, have garnered interest as promising alternative power sources for long-term *in vivo* medical devices [6]. Miniaturized power sources have the potential for integration into implantable devices such as cardiac pacemakers or sensor–transmitters to monitor glucose concentration, body temperature, or blood pressure differences [1,6–8]. For efficient operation of enzyme-based BFCs, however, operational parameters must be optimized and addressed. First, the enzymes should have high catalytic activity, stability, and be inexpensive to produce. Second, the process of bioelectrocatalysis necessitates specific methods for mediation and enzyme immobilization that optimize electron transfer (ET) kinetics between the enzyme and the electrode. Third, electrode fabrication that allows for an open-circuit potential (OCP) that operates close to

Enzymatic Fuel Cells: From Fundamentals to Applications, First Edition. Edited by Heather R. Luckarift, Plamen Atanassov, and Glenn R. Johnson.
© 2014 John Wiley & Sons, Inc. Published 2014 by John Wiley & Sons, Inc.

the theoretical redox potential of the enzyme provides the opportunity to maximize the potential difference between anode and cathode and hence maximize the output of the fuel cell [1,9–11].

There are several strategies for establishing effective electrical communication between enzymes and an electrode. The most common approach is to supply a soluble or an immobilized redox mediator that can shuttle electrons [1,9,10]. The mediators are able to engage a large number of proteins in a 3D architecture, which typically allows higher enzyme loading. Mediated ET, however, has inherent disadvantages, particularly in respect to thermodynamic losses resulting from a negative change in Gibbs energy that arises during ET between enzymes and mediators [1,12]. Another disadvantage of mediated systems is the risk of mediator leakage from electrode half-cells, which can lead to crossover between anode and cathode and inhibition of reaction kinetics. A full discussion of mediated ET is provided in Chapter 9. The alternative to mediated systems is for bioelectrocatalysis to occur directly, thereby simplifying the coupling of redox molecules with a transducer [2,3,10,13–15]. A BFC based on direct electron transfer (DET) between the redox centers of the enzyme and an electrode can operate without exogenous redox mediators—and at a potential approaching that of the redox enzyme itself.

This chapter will summarize state-of-the-art concepts, strategies, and methodologies specifically related to the design and development of biological cathodes that demonstrate bioelectrocatalysis via DET during enzyme-catalyzed reduction of oxygen to water. The core of this research area is a fundamental understanding of multicopper oxidases (MCOs), elucidating specific mechanisms of interfacial and intramolecular ET and characterizing the redox properties of MCOs. Laccase from *Trametes versicolor*, bilirubin oxidase (BOx) from the fungus *Myrothecium verrucaria*, and ascorbate oxidase (AOx) from *Cucurbita* sp. are discussed as representative MCOs originating from very different species. The diversity of MCOs has been well studied, not least because the enzymes exhibit excellent catalytic activity, which is beneficial to a number of commercial technologies [1,8,11,13–17].

3.2 MECHANISTIC STUDIES OF INTRAMOLECULAR ELECTRON TRANSFER

3.2.1 Determining the Redox Potential of MCO

The MCO family of enzymes has unique redox-active centers that include four copper ions: a type 1 (T1) copper ion and a trinuclear cluster including one type 2 (T2) and two type 3 (T3) copper ions (Figure 3.1) [18–21]. The copper ions contained within the protein structure facilitate intramolecular ET by switching their oxidation states between Cu(II) and Cu(I). The accepted mechanism of intramolecular ET for MCO suggests that electrons are transferred from the donor group to the redox center of MCO via a hopping mechanism [18–20]. The T1 copper center acts as an electron acceptor for the substrate and then provides long-range molecular ET to the trinuclear (T2/T3) redox center, which in turn plays a key role in the reduction of oxygen to

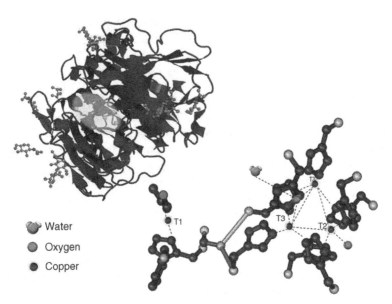

FIGURE 3.1 Ball-and-stick representation of the T1 and T2/T3 copper binding sites of laccase from *T. versicolor* (NCBI Protein Data Bank: 1GYC from *T. versicolor*) viewed using Cn3D ver. 4.1 (NCBI) showing coordination of oxygen. (Adapted with permission from Ref. [27]. Copyright 2010, Elsevier.) Protein structure of *T. versicolor* laccase is shown (RCSB Protein Data Bank: 1GYB); copper atoms in the active site are highlighted in yellow. (Please see the color version of this figure in Color Plates section.)

water [19,20,22,23]. Between the two T3 coppers there is an oxygen ligand, either OH^- or O_2, that coordinates with the T2 and T3 copper ions. The solvent and oxygen have access to the T2/T3 center through two channels [24]. The fully reduced trinuclear copper center reacts with dioxygen to generate a peroxide-level intermediate, and finally, dioxygen is reduced to water [19,25,26].

Analysis of the intramolecular ET from T1 to the trinuclear cluster is a crucial step in understanding the kinetics and mechanism of enzymatic oxygen reduction [22]. An important thermodynamic parameter of the MCO is the redox potential of the enzymes' copper centers. In a MCO, each copper type exhibits a different redox potential, which corresponds to different intermediate redox states [19,22,25,28,29]. The majority of electrochemical studies of MCOs, however, show a single redox response under all conditions in the cathodic and anodic waves [26,30]. Attempts to pinpoint individual redox events are certainly more complex, but improvements in technology and methodologies have begun to offer the opportunity to characterize MCO redox events on a much more fundamental level. There is still a challenge, however, to obtain individual reproducible redox signatures for all MCOs [20,25,26,29]. As a consequence, typical electrochemical studies of MCO show variable values for redox responses [19,25,26,28,29,31]. Cyclic voltammograms of *Trametes hirsuta* laccase adsorbed on a gold electrode, for example, exhibited two distinct low and high redox potentials, at approximately 0.195 and 0.595 V (vs. standard hydrogen electrode

(SHE)) [19,26,28]. It has been suggested that these two values correspond to the two formal redox potentials of the T2/T3 and T1 sites, respectively. The redox potential of the T1 copper center, however, can range considerably (0.23–0.59 V vs. Ag/AgCl) for various laccases [19,26,28,29]. Variation in redox potentials may also be affected by the purity of protein preparations, and electrochemical data may reflect single or multiple isozymes [32].

Similarly, ET processes, in the low- and high-potential range, 0.2 and 0.47 V, respectively, were seen for BOx from *M. verrucaria*—formal redox potential of T1 at 0.26 V (vs. Ag/AgCl at pH 5.3)—yet other studies report potentials more positive than 0.47 V (vs. Ag/AgCl at pH 7.0) [30,31,33]. The evidence of DET for AOx has also been reported recently, but only a single redox response was detected in the cathodic and anodic waves of AOx, although several redox centers are known to be present in the protein [19,25,34]. The difficulties in using electrochemistry to determine accurate redox potentials of MCO are associated with the fact that different conditions—electrode material and treatment, method of enzyme immobilization, ionic strength, and pH of the buffer solution—all contribute to the relative potential values of the redox events. In addition, variations in redox potentials may be attributed to non-covalent binding of copper ions in the active site of the MCO, whereby changes in hydrogen bonding around the copper ions may affect the bond lengths between the copper atoms and coordinating histidine residues [35].

Protein film voltammetry (PFV) has been demonstrated in order to evaluate ET and catalytic properties of MCO directly on the electrode surface. In this approach, the electrode and redox centers of the enzyme are considered to be a donor–acceptor pair. The electrode donates or accepts electrons over a continuous range of potentials, and current is used as an indicator of the movement of electrons within the enzyme [36–39].

Specific methods for enzyme immobilization influence the formation of a homogeneous enzyme monolayer on a carbon electrode surface (as observed by atomic force microscopy (AFM)), but the orientation of MCOs is inherently random [27,40]. This indicates that MCOs are positioned in a distribution of conformational orientations, for example, some having the T1 copper center closest to the electrode surface and some having the trinuclear copper center (T2/T3) in close alignment with the electrode. The PFV method therefore provides an overview of the redox events of all three redox centers, rather than of events specifically related to any single molecular orientation. Typical cyclic voltammograms of laccase, BOx, and AOx immobilized on carbon electrodes, for example, show discrete redox events under anaerobic conditions (Figure 3.2) [27,40,41].

Using the PFV approach, distinct pairs of redox peaks corresponding to reduction and oxidation of three redox copper centers of MCO are distinguishable in three separate potential areas: low (0.0–0.3 V), mid (0.3–0.5 V), and high (0.5–0.8 V) versus Ag/AgCl. Evidence for three distinct redox events for distinct MCOs suggests that the enzymes are immobilized in such a way that the T1, T2, and T3 copper atoms can all accept or pass electrons to the electrode at appropriate potentials. Reports from different research groups examining various representative MCOs consistently indicate that the redox event at a high potential (0.5–0.8 V vs. Ag/AgCl) corresponds to

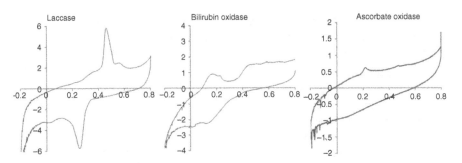

FIGURE 3.2 Cyclic voltammograms of MCOs on screen-printed carbon electrodes (0.1 M phosphate buffer, pH 6.8 for AOx, pH 5.8 for BOx and laccase, $10 \, mV \, s^{-1}$ under N_2). (Adapted with permission from Ref. [27]. Copyright 2010, Elsevier.)

the T1 redox copper center in MCO [15,19,24,27–30,40,41]. In all cases, the anodic peak of the T1 site is observed, but the corresponding cathodic peak is not. This is probably caused by rapid oxidation of copper ions in the redox centers that is countered by a corresponding reduction event whose rate is limited by the slow kinetics of intramolecular ET [14,19,22]. Consistent and repeatable observations revealed that the redox potential of the T1 copper center correlates well with both the onset potential of oxygen reduction and the OCP under aerobic conditions [27]. Laccase-modified electrodes exhibit an OCP of 0.56 ± 0.04 V, which corresponds precisely with the onset of oxygen reduction. Similarly, for BOx, an electrode OCP of 0.55 V correlates with the onset potential for oxygen reduction [27]. It is clear from this correlation that (i) the redox potential of the T1 copper center, (ii) the onset potential of oxygen reduction, and (iii) the OCPs under aerobic conditions are all intimately related and, as such, can be used as a definitive measurement of the true redox potential of the T1 copper center. Redox peaks in the potential area below 0.5 V versus Ag/AgCl are attributed to the redox potential of the trinuclear copper center of MCO [19,29].

Questions remain, however, about how to discriminate the individual redox potentials of the T2 and T3 copper sites of MCO. Cyclic voltammograms in the presence of nitrogen show strong redox events at ~0.2 V and can probably be attributed to the T2 copper center, in agreement with previous studies [14,19]. The assignment of the T2 copper redox potential is also supported by studies of BOx and AOx in the presence of an inhibitor, bathocuproine disulfonate (BCS), under anaerobic conditions [27,40]. BCS acts as a chelating agent that creates complexes exclusively with the T2 copper ion in MCO [42]. Based on electron paramagnetic resonance and resonance Raman spectroscopy [42], the removal of T2 copper as a result of the complex formation [(BCS)$_2$Cu(I)] is accompanied by structural changes of the enzyme that also indirectly affect the T1 copper site. In the presence of BCS, the anodic and cathodic peaks attributed to the T2/T3 copper center (in the region from 0 to 0.3 V vs. Ag/AgCl) completely disappear. Concurrently, the anodic peak related to the T1 copper center (potential region 0.5–0.6 V) shifts to a more positive potential and the current associated with the anodic peak increases more than threefold [27,40].

TABLE 3.1   Redox Potentials and Putative Assignment of Copper Centers in MCOs

Redox Copper Center	Laccase, $E^{\circ\prime}$ (V)	BOx, $E^{\circ\prime}$ (V)	AOx, $E^{\circ\prime}$ (V)
T1	0.520 ± 0.04	0.546 ± 0.02	0.420 ± 0.04
T2	0.197 ± 0.03	0.217 ± 0.02	0.153 ± 0.03
T3	0.360 ± 0.03	0.385 ± 0.02	n/d

The prior assignment of T2 at ~ 0.2 V (vs. Ag/AgCl) [19,28–30] leads us to speculate that, in all cases, anodic and cathodic redox peaks at potentials between 0.3 and 0.5 V (vs. Ag/AgCl) belong to the T3 copper center. Based on these assumptions, the assigned values of formal redox potentials for T1, T2, and T3 redox copper centers of laccase, BOx, and AOx can be surmised (Table 3.1) [27].

Bioelectrocatalytic activity of MCO in the presence of oxygen is dominated by a large sigmoidal catalytic wave, attributed to oxygen reduction. The current density and potential for the onset of oxygen reduction provide an indication of the effective coupling between the intramolecular ET and catalytic reduction of oxygen to water. The corresponding ET rate constant and transfer coefficient for MCO immobilized on the electrode surface can be determined from this observation using Laviron's theory at different scan rates [43]. It was found, for example, that the ET rate constant and the transfer coefficient for electrode-immobilized laccase are $k = 3.4\,\text{s}^{-1}$ and $\alpha = 0.5$, respectively [27]. Further in-depth characterization of DET by studies employing rotating disk electrodes have been used to identify the mechanisms of DET and to confirm a four-electron transfer process [44,45].

3.2.2   Effect of pH and Inhibitors on the Electrochemistry of MCO

Electrolyte pH inherently affects the redox characteristics of MCO [46–48]. The value of anodic and cathodic peaks for the assigned T1 copper center of AOx, for example, is highest (0.66 ± 0.02 V vs. Ag/AgCl) at pH 6.8, which corresponds with the pH optimum of the enzyme. At lower pH (4.5 and 5.8), the potential shifts to ~ 0.48 V versus Ag/AgCl. As a consequence, protein dynamics in response to physiological variations is a critical parameter controlling interfacial and intramolecular ET reactions [46]. The protein pocket containing the T1 copper center is the flexible structural component of the enzyme that allows for variations in substrate specificity. Small internal conformational changes in the protein pocket and around the T1 copper center may therefore contribute significantly to the rate-limiting process of ET [46]. In contrast, the redox potential of the T2 copper center is more conserved, stable, and varies little with changes in environmental conditions. The effect of redox mediators (such as 1,4-hydroquinone and 2,2-azinobis(3-ethylbenzothiazoline-6-sulfonate)) or inhibitors (fluoride ion, BCS, and sodium azide) can dramatically affect the kinetics and mechanism of intramolecular ET [27,40,41]. It is interesting to note that the anodic current peak of laccase increases more than eightfold in the potential area related to T1 (~ 0.6 V vs. Ag/AgCl), when inactivated by hydroquinone [41]. From the crystal structure of *T. versicolor* laccase, it is proposed that oxygen has access to the T2/T3 copper center through two solvent channels [24]. The solvent channels allow fast access of oxygen and water molecules to the T1 and trinuclear copper centers, but water-soluble molecules in general, including redox mediators and inhibitors, can also access the copper centers directly. Because of this accessibility, partial enzyme inactivation has been observed in the presence of relatively high concentrations of redox mediators, which is attributed to blockage of the reduction of oxygen by accumulation of redox molecules inside the channels [22,23,41].

3.3 ACHIEVING DET OF MCO BY RATIONAL DESIGN

Demonstration of DET between MCO and an electrode has progressed steadily over the last 30 years, but the efficiency of energy conversion has not increased in parallel [18–20,24]. Optimal conditions for efficient DET depend on a combination of factors, including electrode material, enzyme orientation and proximity to the electrode surface, and the location of the enzyme redox centers within the enzyme structure. The structure and biochemistry of MCO are reasonably well understood from a fundamental standpoint but are less well documented in respect to standardized electrochemical characteristics [37,49]. In particular, mechanisms of interfacial and intramolecular ET vary depending on the enzyme connection with the electrode, which in turn is dictated by the functionality and conductivity of the electrode architecture. Immobilization of laccase to screen-printed electrodes, for example, varies significantly when the electrode architecture is carbon, planar gold, or carbon modified with gold nanoparticles (Figure 3.3) [50]. A recent study revealed that the interfacial ET kinetics between an immobilized enzyme and an electrode surface can

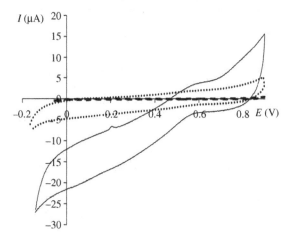

FIGURE 3.3 Oxygen reduction activity of laccase following immobilization on screen-printed gold electrodes (dashed line), screen-printed carbon electrodes (dotted line), and screen-printed carbon electrodes modified with gold nanoparticles (solid line). (Adapted with permission from Ref. [50]. Copyright 2012, Wiley-VCH Verlag GmbH.)

be significantly improved by incorporation of gold nanoparticles into a carbon-based electrode by increasing the available surface area and by preferential binding of MCO to gold. Retention of catalytic activity of the immobilized enzyme is evidenced by the presence of a reduction current during cyclic voltammetry (Figure 3.3). The electrocatalytic current at the electrode begins at a potential of approximately 0.6 V (vs. Ag/AgCl) and provides direct evidence of efficient DET in response to substrate (oxygen).

The redox centers of most enzyme molecules are located deep inside the protein structure, leading to a long electron tunneling distance and thus inefficient ET. Ideally, the electron tunneling distance of the redox centers should be minimized, so the biocatalytic reaction is the limiting process. In addition, there are difficulties associated with engineering the enzyme–electrode interface to establish DET. As such, a detailed understanding of the enzyme–electrode interface, the inter- and intramolecular ET mechanisms, and methodological principles to reduce the electron tunneling distance between enzyme and electrode are all critical components to enhance activity of an enzymatic cathode.

Conditions for achieving efficient DET via enzyme immobilization are dictated partly by materials architecture. Enzyme immobilization techniques may include nonspecific adsorption, covalent linkage, entrapment in conductive polymeric films, association with metal colloids, and encapsulation within porous matrices (see Chapter 11). The simplest method is nonspecific adsorption, but control is limited; various noncovalent interactions will yield different orientations of the redox center with respect to the electrode interface and, as a result, inefficient DET.

Conductive nanostructures (e.g., Au, ZnO, carbon nanotubes (CNTs)) provide a ready means to create a 3D, porous, conductive catalytic matrix on an electrode surface, to which proteins can then be specifically immobilized through careful selection of surface chemistry [51]. By selective attachment of biomolecules via specific protein functional groups, redox-active centers can be positioned close to the electrode surface, thereby reducing the tunneling distance. It was recently demonstrated that substrate mimics, particularly aromatic compounds such as anthracene or 4-(2-aminoethyl)benzoic acid hydrochloride, can be used to functionalize carbon surfaces—for example, pyrolytic graphite or multiwalled carbon nanotubes (MWCNTs)—and enhance DET to laccase [45,49,52–54]. Similarly, the cross-linking agents, 1-pyrenebutyric acid N-hydroxysuccinimide ester (PBSE) and structural homologs (e.g., 4,4′-[(8,16-dihydro-8,16-dioxodibenzo[a,j]perylene-2,10-diyl) dioxy]dibutyric acid di(N-succinimidyl ester) (DDPSE)), can interact specifically with CNTs to create a molecular tether to covalently attach MCOs and juxtapose the enzyme with the CNTs (Figure 3.4). The resulting composite appears to favor specific protein interaction that ensures highly effective DET, with OCP approaching the theoretical maximum (0.6 V vs. Ag/AgCl) and current densities of 0.1–0.5 mA cm^{-2} [52,55]. CNTs provide a highly versatile surface for electrode fabrication, particularly when formed as flexible "buckypaper"-type architectures [55,56]. Materials fabrication and the influence on electrode design are topics further covered in Chapter 10.

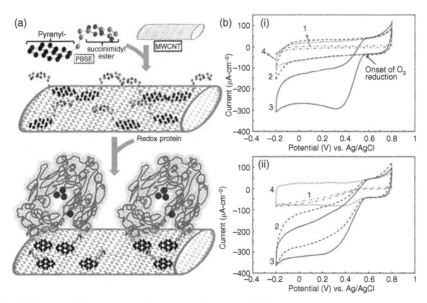

FIGURE 3.4 Illustration of MCO immobilized onto PBSE-modified carbon nanotubes (a) and cyclic voltammograms (b) of laccase (i) and BOx (ii) modified electrodes. MCO physisorbed to carbon (1) and carbon/ MWCNT (2). MCO immobilized via PBSE to carbon/MWCNT; with oxygen (3) and nitrogen (4). CV scans in phosphate buffer (pH 5.8), scan rate $10\,\mathrm{mV\,s^{-1}}$. (Adapted with permission from Ref. [52]. Copyright 2010, Royal Society of Chemistry.)

3.3.1 Surface Analysis of Enzyme-Modified Electrodes

Precise design of biointerfacial properties is a key factor in the field of bioengineering. As the interactions between enzyme and transducer matrix may significantly affect enzyme conformation and, hence, catalytic activity, it is crucial to evaluate and understand enzyme organization on specific surfaces, such as a fuel cell electrode. The physical architecture of ultrathin MCO films on the electrode surface, including elemental and chemical composition, and relative thickness and assembly of layers, has been investigated in detail by using techniques such as angle-resolved X-ray photoelectron spectroscopy (ARXPS) and AFM (Chapter 14) [57]. ARXPS is a surface-sensitive, nondestructive technique that can provide elemental and chemical information for various multilayered systems, such as Langmuir–Blodgett films and self-assembled monolayers (SAMs), by providing unique spectral peaks that can discriminate between individual components [40]. In fact, during layer-by-layer enzyme deposition, ARXPS spectra reveal that the layers are intermixed rather than discrete. BOx immobilized to a carbon electrode, for example, exhibits unique spectral signals for both BOx and the carbon electrode, which can be used in a substrate–overlayer model to estimate the relative thickness of the enzyme monolayer on the surface [40]. The thickness of the BOx coating increases approximately twofold, as expected, when enzyme loading is increased from one to two sequential layers. Adding additional layers of BOx shows no further increase in loading thickness, due to partial interpenetration of the composite layers [40].

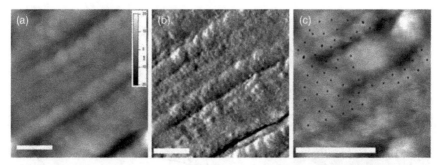

FIGURE 3.5 AFM images in liquid (tapping mode) of laccase on a screen-printed carbon electrode: (a) height, (b) amplitude, and (c) expanded section showing enzyme molecules marked by black dots. Scale bar = 50 nm. (Adapted with permission from Ref. [27]. Copyright 2010, Elsevier.)

AFM can provide complementary visible characterization of the bio–nano interface via phase contrast and topography, from which dimensions and surface coverage of biomolecules can be calculated [19,24,27,58]. Laccase immobilized to carbon electrodes functionalized with glutaraldehyde or dithiobis(succinimidyl propionate), for example, will appear as enzyme molecules adsorbed to the carbon surface as an organized monolayer of tightly packed protein particles. Isolated areas of protein are discernible with dimensions $(8 \pm 1 \text{ nm})$ consistent with the size reported for laccase (Figure 3.5) [27,59]. In addition, from AFM topography and force curve measurements, the area per enzyme $(66 \pm 14 \text{ nm}^2)$ can be calculated and used to estimate percent surface coverage on the electrode compared with theoretical predictions [24]. Based on force curve measurements, an unmodified carbon electrode behaves as an ideal rigid surface, but the AFM tip encounters resistance on initial contact of protein-modified carbon electrodes, which requires a force on the order of 0.5 nN to overcome. Further compression contacts the rigid surface and confirms that a soft layer of several nanometers of protein is absorbed to the hard carbon electrode [5,16].

3.3.2 Design of MCO-Modified Biocathodes Based on Direct Bioelectrocatalysis

With an understanding of MCO catalysis in hand, bioelectrocatalytic activity can be integrated into effective electrode design. The assembly of conductive nanoparticles functionalized with protein provides optimal DET when combining a nanostructured biofunctional surface that is controllable and robust [60–62]. CNTs and gold nanoparticles, for example, provide a functional surface in the nanometer size range that demonstrate high chemical stability and electrical conductivity, essentially serving as "nanowires" for efficient DET between the redox center of an enzyme and the electrode surface [4,6,8,16,22,23,26–28,30,38–41,63]. By studying the examples of DET reported for MCO on carbon and gold electrodes, a toolbox for standardized electrode design can be envisioned [17]. First, preparation of SAMs of enzyme on an electrode surface is typically critical for efficient DET. Second, tether or linker

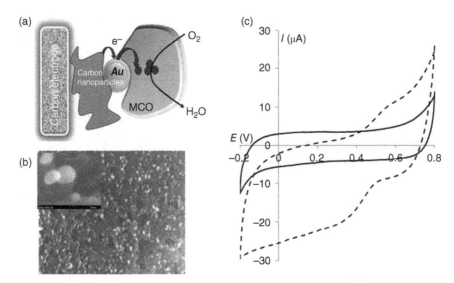

FIGURE 3.6 (a) Schematic of MCO immobilized to carbon electrodes modified with gold nanoparticles. (b) Surface as viewed by SEM and (c) enhancement of electrocatalytic activity as observed by oxygen reduction, with (dashed line) and without (solid line) gold nanoparticles. (Adapted with permission from Ref. [50]. Copyright 2012, Wiley-VCH Verlag GmbH.)

molecules play a significant role in conformational flexibility of immobilized enzyme molecules. Third, conductive nanoparticles play an active role as a conductive connection between the active site of the enzyme and the surface of the electrode. Size-specific and functionalized gold nanoparticles, for example, can bind to laccase in such a way that electrocatalytic activity is maximized and DET is enhanced (Figure 3.6) [50,64–66]. Coupling to carbon can be achieved using coupling reactions that generate aromatic functional groups that bind to the hydrophobic residues of laccase that define the T1 Cu site and thereby facilitate fast intermolecular ET [54,67].

3.3.3   Design of MCO-Modified "Air-Breathing" Biocathodes

One of the limiting factors of DET for oxygen reduction is oxygen transport across an unstirred layer (Nernst diffusion layer) on the electrode surface [1,10,68–70]. In addition, low solubility and small diffusion coefficients for oxygen dissolved in aqueous solutions present a challenge for efficient biocatalytic oxygen reduction. To address the limitations, significant advances in cathode design have been achieved by moving away from diffusion-limited bulk solution environments to gas diffusion electrodes (GDEs), the essentially "air-breathing" cathode designs that are inherently more feasible in practical applications of BFCs (Figure 3.7) [55,68]. The advantage of a GDE is that molecular oxygen can be supplied through the gaseous phase, reducing the solubility and diffusion limitations of liquid electrolytes. By increasing the availability of oxygen, the current density of GDE cathodes is similarly enhanced.

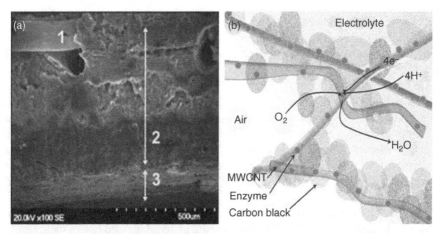

FIGURE 3.7 (a) Cross-sectional scanning electron microscope image of a gas diffusion electrode, showing (1) the nickel mesh current collector, (2) the gas diffusion layer (carbon black/35 wt% PTFE), and (3) the MWCNT catalytic layer (3.5 wt% PTFE). (b) Schematic representation of the three-phase interphase in a gas diffusion biocathode. (Reproduced with permission from Ref. [55]. Copyright 2011, Wiley-VCH Verlag GmbH.)

GDEs are widely used in various types of fuel cells and in metal–air batteries in which oxygen serves as the cathode oxidant, supplied from ambient air [71].

Typically, a GDE consists of a layered carbon–polymer composite (e.g., carbon black mixed with Teflon (polytetrafluoroethylene (PTFE)) or polyaniline as a binder) formed onto a backing material, usually a metal mesh, that serves as an electron conductor and provides mechanical support. The key component of GDE design is the gas diffusion layer (GDL), which allows gas (typically oxygen) to flow through the electrode into the electrolyte. As such, the GDL should be highly gas permeable and retain electronic conductivity. Ideally, the GDL will be completely hydrophobic (water repellant), but hydrophilicity must increase across the electrode and interface with electrolyte. The balance of hydrophilic and hydrophobic properties of the carbon–air and carbon–electrolyte interfaces determines the performance of such GDE [72]. With the introduction of polymers such as PTFE, it became possible to fabricate GDE by controlling the relative hydrophobicity and hydrophilicity of carbon within a composite electrode. In short, the aim for GDEs is to achieve a high surface area material with a triphase interface (of gaseous, liquid, and solid phases) that maximizes exposure of the immobilized catalysts to the gas (air) and the liquid (electrolyte) phases [55,68].

GDEs, however, have been poorly explored for integrating biocatalytic reactions, such as oxygen reduction as catalyzed by MCO. Integration of enzymes in GDE presents an additional materials challenge as the GDE must contain a catalytic layer that is hydrophilic enough for enzyme immobilization from aqueous solution. Yet, the material should be adaptable to treatments that yield transition to the superhydrophobicity required in the GDL. To date, only a handful of enzyme-based GDEs have been reported. For example, in 1995 Iliev et al. demonstrated an amperometric glucose sensor fabricated from an enzyme GDE consisting of glucose oxidase immobilized in Nafion [73]. Recently, attempts have been made to integrate various carbon-based GDEs with MCOs including laccase, BOx, and the MCO CueO [68,74–77].

With the introduction of CNTs, it was possible to integrate CNT-based catalytic layers with a GDE. By modifying MWCNTs with PTFE, for example, a GDL is created that provides microstructured gas channels for efficient oxygen supply and provides a materials architecture with electric conductivity, mechanical stability, and water repellency. The resulting triphase interface is formed on the outside of the teflonized CNT particles. The aqueous electrolyte will primarily occupy the larger pore spaces, whereas the gas transport is ensured by the hydrophobic PTFE microchannels.

In 2012, Lau et al., used the PTFE-based approach to fabricate a gas diffusion biocathode using a MWCNT–PTFE composite material pressed onto a nickel mesh as the current collector. A cross-sectional scanning electron microscope (SEM) image (Figure 3.7) of the electrode shows the nickel wire embedded within the gas diffusion layer ($\sim$500 μm total thickness) that is in direct contact with the MWCNT–PTFE catalytic layer of $\sim$100 μm thickness. With laccase immobilized directly to the catalytic layer, the resulting electrode yielded a 5–10-fold increase in catalytic performance compared with dissolved oxygen in the electrolyte. In the presence of oxygen or ambient air, the laccase-functionalized GDE exhibits an OCP of 0.53 V (vs. Ag/AgCl), which is close to the theoretical redox potential of the T1 copper center [55].

By replacing the carbon black with MWCNT, higher surface area is introduced that can be specifically functionalized with PBSE for covalent attachment of enzyme. Substituting PBSE with DDPSE, however, provides a molecular tether that encourages stronger interaction through π–π stacking and results in electrodes with current densities of threefold higher than for PBSE. DDPSE attaches to the protein molecule via two amine groups that position the protein in a defined orientation and significantly reduce the electron transfer distance ($\sim$1 Å) [55]. The method resulted in laccase-modified GDE with OCP approaching that of the theoretical potential for laccase (0.55 V vs. Ag/AgCl) and current densities of up to 0.5 mA cm^{-2} (at zero potential in air-breathing mode). Such air-breathing cathodes are further integrated into enzymatic and hybrid microbial/enzymatic BFCs [78,79].

In an air-breathing design, the enzyme is exposed to a triphase interface: the current-collecting solid phase for DET, the liquid phase for efficient proton transfer, and the gas phase for effective oxygen transport. The biocathode includes two layers: (i) a gas diffusion layer prepared from a hydrophobic carbon, and (ii) a catalytic layer made of untreated hydrophilic carbon. The hydrophilic layer contains the physically adsorbed MCO and provides a porous, 3D, conductive matrix, which is electrochemically accessible for the electrolyte. Thus, both liquid electrolyte and a gaseous reactant are brought together within the porous GDE. Because the magnitude of the diffusion-limited current is inversely proportional to the thickness of the diffusion layer, it is important to optimize conditions that will facilitate diffusion of oxygen from the bulk of the solution to the electrode surface [32]. An OCP of 0.6 V (vs. Ag/AgCl) was observed for a cathode fabricated in this way using laccase from *T. hirsuta*, and a similar OCP of 0.65 V (vs. Ag/AgCl) was observed for BOx. The resulting current density was ~350 μA cm^{-2}, approximately threefold higher in air-breathing mode than in oxygenated electrolyte. The porous materials provide high surface area to volume ratio architectures in which microchannels are reduced to a cross-sectional dimension of tens or hundreds of micrometers, which significantly decreases the Nernst diffusion layer and the associated transport limitations.

3.4 OUTLOOK

Extending the utility of MCO to legitimate applications in fuel cell development will depend primarily on protein availability at a large, yet cost-effective scale and the ability to manipulate isolated enzymes—for example, by tuning protein structure—to achieve defined or higher redox potentials [32,80]. Changes in the axial ligand at the T1 Cu site inherently influences the redox potential of MCO [81]. Site-directed mutagenesis of the substrate-binding pocket of MCO could theoretically alter the redox potential for electrocatalytic activity, but, to date, with the exception of the single-copper azurin [73,82], improvements in the catalytic activity of MCO have been incremental and typically limited to shifts in the substrate specificity for xenobiotic compounds [83] and physiological changes such as a shift in pH optimum [47]. Molecular engineering of fungal laccases is typically complicated by the presence of protein isomers and posttranslational glycosylation of the proteins, which limit their synthesis in heterologous hosts (Chapter 7) [84,85]. Isolation and characterization of laccases from bacteria may provide fortuitous catalysts for electrocatalysis, but it is a relatively recent area of endeavor [86]. The use of a bacterially expressed laccase, such as the small laccase of *Streptomyces coelicolor* (SLAC), can simplify protein expression and genetic modification. SLAC-functionalized electrodes demonstrate effective DET and a redox potential of ~0.4 V (vs. Ag/AgCl). Redox potentials of the T1 Cu site in four different MCOs were recently modeled and compared with SLAC using density functional theory and molecular dynamics simulations [87]. The model included laccase from *T. versicolor*, a laccase-like enzyme from *Bacillus subtilis*, and the MCO CueO from *Escherichia coli*. The modeling highlighted how the protein backbone and side chain amino acids influence the T1 redox potential, in good agreement with experimental data. With the model in hand, predictions could be made of substitutions in the amino acid sequence that theoretically may increase the T1 redox potential in SLAC (Figure 3.8).

An alternative strategy to extend the toolbox of catalysts available for electrocatalytic oxygen reduction is to identify MCOs from diverse environments. A thermostable laccase, for example, was recently isolated from a marine fungus and shows optimum growth at 70 °C [88]. Similarly, marine isolates demonstrate catalytic activity in high-salinity environments due to inherent salt-tolerant characteristics [89]. Identifying and isolating such catalysts may extend the potential working range for future BFC applications.

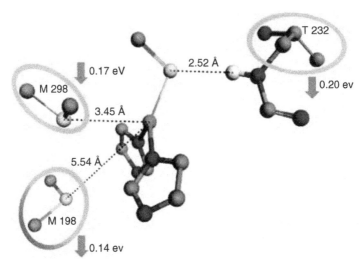

FIGURE 3.8 Comparison of the T1 Cu sites from *T. versicolor* (TvL), *B. subtilis* (CotA), and *E. coli* (CueO) by molecular modeling can be used to identify axial amino acid ligands and proximal residues that may influence the redox potential of the T1 Cu center in SLAC. (Reproduced with permission from Ref. [87]. Copyright 2011, American Chemical Society.)

ACKNOWLEDGMENTS

This research was prepared as an account of work sponsored by the United States Air Force Research Laboratory, Materials and Manufacturing Directorate, Airbase Technologies Division (AFRL/RXQ), but the views of authors expressed herein do not necessarily reflect those of the United States Air Force. This work (DMI) was also supported in part by a grant from DoD/Air Force Office of Scientific Research (MURI on Fundamentals and Bioengineering of Enzymatic Fuel Cells).

LIST OF ABBREVIATIONS

AFM	atomic force microscopy
AOx	ascorbate oxidase
ARXPS	angle-resolved X-ray photoelectron spectroscopy
BCS	bathocuproine disulfonate
BFC	biological fuel cell
BOx	bilirubin oxidase
CNT	carbon nanotube
DDPSE	4,4′-[(8,16-dihydro-8,16-dioxodibenzo[*a,j*]perylene-2,10-diyl) dioxy]dibutyric acid di(*N*-succinimidyl ester)

DET	direct electron transfer
ET	electron transfer
GDE	gas diffusion electrode
GDL	gas diffusion layer
MCO	multicopper oxidase
MWCNT	multiwalled carbon nanotube
OCP	open-circuit potential
PBSE	1-pyrenebutyric acid N-hydroxysuccinimide ester
PFV	protein film voltammetry
PTFE	polytetrafluoroethylene (Teflon)
SAM	self-assembled monolayer
SEM	scanning electron microscope
SHE	standard hydrogen electrode
SLAC	small laccase (*Streptomyces coelicolor*)
T1	type 1
T2	type 2
T3	type 3

REFERENCES

1. Calabrese Barton S, Gallaway J, Atanassov P. Enzymatic biofuel cells for implantable and microscale devices. *Chem Rev* 2004;104:4867–4886.

2. Carrara S. Nano-bio-technology and sensing chips: new systems for detection in personalized therapies and cell biology. *Sensors* 2010;10:526–543.

3. Ghindilis AL, Atanassov P, Wilkins E. Enzyme catalyzed direct electron transfer: fundamentals and analytical applications. *Electroanalysis* 1997;9:661–674.

4. Kano K, Ikeda T. Bioelectrocatalysis, powerful means of connecting electrochemistry to biochemistry and biotechnology. *Electrochemistry* 2003;71:86–99.

5. Vaddiraju S, Tomazos I, Burgess DJ, Jain FC, Papadimitrakopoulos F. Emerging synergy between nanotechnology and implantable biosensors: a review. *Biosens Bioelectron* 2010;25:1553–1565.

6. Heller A. Miniature biofuel cells. *Phys Chem Chem Phys* 2004;6:209–216.

7. Heller A. Potentially implantable miniature batteries. *Anal Bioanal Chem* 2006;385:469–473.

8. Mano N, Mao F, Heller A. A miniature biofuel cell operating in a physiological buffer. *J Am Chem Soc* 2002;124:12962–12963.

9. Bullenn RA, Arnot TC, Lakeman JB, Walsh FC. Biofuel cells and their development. *Biosens Bioelectron* 2006;21:2015–2045.

10. Cooney MJ, Svoboda V, Lau C, Martin G, Minteer SD. Enzyme catalyzed biofuel cells. *Energy Environ Sci* 2008;1:320–337.

11. Willner I, Yan YM, Willner B, Tel-Vered R. Integrated enzyme-based biofuel cells. *Fuel Cells* 2009;9:7–24.

12. Kim J, Jia H, Wang P. Challenges in biocatalysis for enzyme-based biofuel cells. *Biotechnol Adv* 2006;24:296–308.

13. Ramanavicius A, Kausaite A, Ramanaviciene A. Biofuel cell based on direct bioelectrocatalysis. *Biosens Bioelectron* 2005;20:1962–1967.

14. Shleev S, Christenson A, Serezhenkov V, Burbaev D, Yaropolov A, Gorton L, Ruzgas T. Electrochemical redox transformations of T1 and T2 copper sites in native *Trametes hirsuta* laccase at a gold electrode. *Biochem J* 2005;385:745–754.

15. Shleev S, Tkac J, Christenson A, Ruzgas T, Yaropolov AI, Whittaker JW, Gorton L. Direct electron transfer between copper-containing proteins and electrodes. *Biosens Bioelectron* 2005;20:2517–2554.

16. Calabrese Barton S, Kim H-H, Binyamin G, Zhang Y, Heller A. The "wired" laccase cathode: high current density electroreduction of O_2 to water at +0.7 V (NHE) at pH 5. *J Am Chem Soc* 2001;123:5802–5803.

17. Minteer SD, Liaw BY, Cooney MJ. Enzyme-based biofuel cells. *Curr Opin Biotechnol* 2007,18:228–234.

18. Messerschmidt A, Huber R. The blue oxidases, ascorbate oxidase, laccase and ceruloplasmin. Modeling and structural relationships. *Eur J Biochem* 1990;187:341–352.

19. Solomon EI, Sundaram UM, Machonkin TE. Multicopper oxidases and oxygenases. *Chem Rev* 1996:2563–2605.

20. Solomon EI, Szilagyi RK, DeBeer George S, Basumallick L. Electronic structures of metal sites in proteins and models: contributions to function in blue copper proteins. *Chem Rev* 2004;104:419–458.

21. Solomon EI, Chen P, Metz M, Lee S-K, Palmer AE. Oxygen binding, activation, and reduction to water by copper proteins. *Angew Chem Int Ed* 2001;40:4570–4590.

22. Farver O, Wherland S, Koroleva O, Loginov DS, Pecht I. Intramolecular electron transfer in laccases. *FEBS J* 2011;278:3463–3471.

23. Leger C, Lederer F, Guigliarelli B, Bertrand P. Electron flow in multicenter enzymes: theory, applications, and consequences on the natural design of redox chains. *J Am Chem Soc* 2006;128:180–187.

24. Piontek K, Antorini M, Choinowski T. Crystal structure of a laccase from the fungus *Trametes versicolor* at 1.90-Å resolution containing a full complement of coppers. *J Biol Chem* 2002;277:37663–37669.

25. Rulisek L, Solomon EI, Ryde U. A combined quantum and molecular mechanical study of the O_2 reductive cleavage in the catalytic cycle of multicopper oxidases. *Inorg Chem* 2005;44:5612–5628.

26. Shleev S, Pita M, Yaropolov AI, Ruzgas T, Gorton L. Direct heterogeneous electron transfer reactions of *Trametes hirsuta* laccase at bare and thiol-modified gold electrodes. *Electroanalysis* 2006;18:1901–1908.

27. Ivnitski D, Khripin C, Luckarift HR, Johnson GR, Atanassov P. Surface characterization and direct bioelectrocatalysis of multicopper oxidases. *Electrochim Acta* 2010;55:7385–7393.

28. Frasconi M, Boer H, Koivula A, Mazzei F. Electrochemical evaluation of electron transfer kinetics of high and low redox potential laccases on gold electrode surface. *Electrochim Acta* 2010;56:817–827.

29. Reinhammar BRM. Oxidation–reduction potentials of the electron acceptors in laccases and stellacyanin. *Biochim Biophys Acta: Bioenerg* 1972;275:245–259.

30. Christenson A, Shleev S, Mano N, Heller A, Gorton L. Redox potentials of the blue copper sites of bilirubin oxidases. *Biochim Biophys Acta: Bioenerg* 2006;1757:1634–1641.

31. Tsujimura S, Tatsumi H, Ogawa J, Shimizu S, Kano K, Ikeda T. Bioelectrocatalytic reduction of dioxygen to water at neutral pH using bilirubin oxidase as an enzyme and 2,2-azinobis(3-ethylbenzothiazolin-6-sulfonate) as an electron transfer mediator. *J Electroanal Chem* 2001;496:69–75.

32. Rodgers CJ, Blanford CF, Giddens SR, Skamnioti P, Armstrong FA, Gurr SJ. Designer laccases: a vogue for high-potential fungal enzymes? *Trends Biotechnol* 2009;28:63–72.

33. Schubert K, Göbel G, Lisdat F. Bilirubin oxidase bound to multi-walled carbon nanotube-modified gold. *Electrochim Acta* 2009;54:3033–3038.

34. Murata K, Nakamura N, Ohno H. Direct electron transfer reaction of ascorbate oxidase immobilized by a self-assembled monolayer and polymer membrane combined system. *Electroanalysis* 2007;19:530–534.

35. Kranich A, Ly HK, Hildebrandt P, Murgida DH. Direct observation of the gating step in protein electron transfer: electric-field-controlled protein dynamics. *J Am Chem Soc* 2008;130:9844–9848.

36. Armstrong FA. Insights from protein film voltammetry into mechanism of complex biological electron-transfer reactions. *J Chem Soc, Dalton Trans* 2002:661–671.

37. Armstrong FA. Recent developments in dynamic electrochemical studies of adsorbed enzymes and their active sites. *Curr Opin Chem Biol* 2005;9:110–117.

38. Hirst J. Elucidating the mechanisms of coupled electron transfer and catalytic reactions by protein film voltammetry. *Biochim Biophys Acta* 2006;1757:225–239.

39. Rusling JF, Forster RJ. Electrochemical catalysis with redox polymer and polyion–protein films. *J Colloid Interface Sci* 2003;262:1–15.

40. Ivnitski D, Artyushkova K, Atanassov P. Surface characterization and direct electro-chemistry of redox copper centers of bilirubin oxidase from the fungi *Myrothecium verrucaria*. *Bioelectrochemistry* 2008;74:101–110.

41. Ivnitski D, Atanassov P. Electrochemical studies of intramolecular electron transfer in laccase from *Trametes versicolor*. *Electroanalysis* 2007;19:2307–2313.

42. Malkin R, Malmström BG, Vänngard T. The reversible removal of one specific copper(II) from fungal laccase. *Eur J Biochem* 1969;7:253–259.

43. Laviron E. General expression of the linear potential sweep voltammogram in the case of diffusionless electrochemical systems. *J Electroanal Chem* 1979;101:19–28.

44. Brocato S, Lau C, Atanassov P. Mechanistic study of direct electron transfer in bilirubin oxidase. *Electrochim Acta* 2012;61:44–49.

45. Parimi NS, Umasankar Y, Atanassov PB, Ramasamy RP. Kinetic and mechanistic parameters of laccase catalyzed direct electrochemical oxygen reduction reaction. *ACS Catal* 2012;2:38–44.

46. Augustine AJ, Kragh ME, Sarangi R, Fujii S, Liboiron BD, Stoj CS, Kosman DJ, Hodgson KO, Hedman B, Solomon EI. Spectroscopic studies of perturbed T1 Cu sites in the multicopper oxidases *Saccharomyces cerevisiae* Fet3p and *Rhus vernicifera* laccase: allosteric coupling between the T1 and trinuclear Cu sites. *Biochemistry* 2008;47: 2036–2045.

47. Madzak C, Mimmi MC, Caminade E, Brault A, Baumberger S, Briozzo P, Mougin C, Jolivalt C. Shifting the optimal pH of activity for a laccase from the fungus *Trametes versicolor* by structure-based mutagenesis. *Protein Eng Des Sel* 2006;19: 77–84.

48. Xu F. Dioxygen reactivity of laccase: dependence on laccase source, pH, and anion inhibition. *Appl Biochem Biotechnol* 2001;95:125–133.

49. Meredith MT, Minson M, Hickey D, Artyushkova K, Glatzhofer DT, Minteer SD. Anthracene-modified multi-walled carbon nanotubes as direct electron transfer scaffolds for enzymatic oxygen reduction. *ACS Catal* 2011;1:1683–1690.

50. Luckarift HR, Ivnitski D, Lau C, Khripin C, Atanassov P, Johnson GR. Gold-decorated carbon composite electrodes for enzymatic oxygen reduction. *Electroanalysis* 2012;24: 931–937.

51. Jensen UB, Vagin M, Koroleva O, Sutherland DS, Besenbacher F, Ferapontova EE. Activation of laccase bioelectrocatalysis of O_2 reduction to H_2O by carbon nanoparticles. *J Electroanal Chem* 2012;667:11–18.

52. Ramasamy RP, Luckarift HR, Ivnitski DM, Atanassov PB, Johnson GR. High electrocatalytic activity of tethered multicopper oxidase–carbon nanotube conjugates. *Chem Commun* 2010;46:6045–6047.

53. Stolarczyk K, Sepelowska M, Lyp D, Zelechowska K, Biernat JF, Rogalski J, Farmer KD, Roberts KN, Bilewicz R. Hybrid biobattery based on arylated carbon nanotubes and laccase. *Bioelectrochemistry* 2012;87:154–163.

54. Martinez-Ortiz J, Flores R, Vazquez-Duhalt R. Molecular design of laccase cathode for direct electron transfer in a biofuel cell. *Biosens Bioelectron* 2011;26: 2626–2631.

55. Lau C, Adkins ER, Ramasamy RP, Luckarift HR, Johnson GR, Atanassov P. Design of carbon nanotube-based gas-diffusion cathode for O_2 reduction by multicopper oxidases. *Adv Energy Mater* 2011;2:162–168.

56. Hussein L, RubenWolf S, von Stetten F, Urban G, Zengerle R, Krueger M, Kerzenmacher S. A highly efficient buckypaper-based electrode material for mediatorless laccase-catalyzed dioxygen reduction. *Biosens Bioelectron* 2011;26:4133–4138.

57. Artyushkova K, Fulghum JE, Reznikov Y. Orientation of 5CB molecules on aligning substrates studied by angle resolved X-ray photoelectron spectroscopy. *Mol Cryst Liq Cryst* 2005;438:1769–1777.

58. Giessibl FJ. Advances in atomic force microscopy. *Rev Mod Phys* 2003;75:949–983.

59. Vigil RD, Ziff RM. Random sequential adsorption of unoriented rectangles onto a plane. *J Chem Phys* 1989;91:2599–2602.

60. Ivnitski D, Branch B, Atanassov P, Apblett C. Glucose oxidase anode for biofuel cell based on direct electron transfer. *Electrochem Commun* 2006;8:1204–1210.

61. Li J, Cassell A, Delzeit L, Han J, Meyyappan M. Novel three-dimensional electrodes: electrochemical properties of carbon nanotube ensembles. *J Phys Chem B* 2002;106: 9299–9305.

62. Wang J, Liu G, Jan MR. Ultrasensitive electrical biosensing of proteins and DNA: carbon nanotube derived amplification of the recognition and transduction events. *J Am Chem Soc* 2004;126:3010–3011.

63. Gupta G, Rajendran V, Atanassov P. Bioelectrocatalysis of oxygen reduction reaction by laccase on gold electrode. *Electroanalysis* 2004;16:1182–1185.

64. Scanlon MD, Salaj-Kosla U, Belochapkine S, MacAodha D, Leech D, Ding L, Magner E. Characterization of nanoporous gold electrodes for bioelectrochemical applications. *Langmuir* 2011;28:2251–2261.

65. Thorum MS, Anderson CA, Hatch JJ, Campbell AS, Marshall NM, Zimmerman SC, Lu Y, Gewirth AA. Direct, electrocatalytic oxygen reduction by laccase on anthracene-2-methanethiol-modified gold. *J Phys Chem Lett* 2010;1:2251–2254.

66. Dagys M, Haberska K, Shleev S, Arnebrant T, Kulys J, Ruzgas T. Laccase–gold nanoparticle assisted bioelectrocatalytic reduction of oxygen. *Electrochem Commun* 2010;12:933–935.

67. Blanford CF, Foster CE, Heath RS, Armstrong FA. Efficient electrocatalytic oxygen reduction by the 'blue' copper oxidase, laccase, directly attached to chemically modified carbons. *Faraday Discuss* 2008;140:319–335.

68. Gupta G, Lau C, Branch B, Rajendran V, Ivnitski D, Atanassov P. Direct bio-electro-catalysis by multi-copper oxidases: gas-diffusion laccase-catalyzed cathodes for biofuel cells. *Electrochim Acta* 2011;56(28):10767–10771.

69. Rincón RA, Lau C, Luckarift HR, Garcia KE, Adkins E, Johnson GR, Atanassov P. Enzymatic fuel cells: integrating flow-through anode and air-breathing cathode into a membrane-less biofuel cell design. *Biosens Bioelectron* 2011;27:132–136.

70. Tsujimura S, Kamitaka Y, Kano K. Diffusion-controlled oxygen reduction on multi-copper oxidase-adsorbed carbon aerogel electrodes without mediator. *Fuel Cells* 2007:463–469.

71. Barsukov IV, Johnson CS, Doninger JE, Barsukov VZ. Carbon materials for gas diffusion electrodes, metal air cells and batteries. In: *New Carbon Based Materials for Electrochemical Energy Storage Systems*, Vol. 229. Springer, Dordrecht, The Netherlands, 2007, pp. 85–88.

72. Shteinberg GV, Dribinsky AV, Kukushkina IA, Musilova M, Mrha J. Influence of structure and hydrophobic properties on the characteristics of carbon–air electrodes. *J Power Sources* 1982;8:17–33.

73. Iliev I, Kaisheva A, Scheller F, Pfeiffer D. Amperometric gas-diffusion/enzyme electrode. *Electroanalysis* 1995;7:542–546.

74. Tarasevich MR, Bogdanovskaya VA, Kapustin AV. Nanocomposite material laccase/dispersed carbon carrier for oxygen electrode. *Electrochem Commun* 2003;5:491–496.

75. Shleev S, Shumakovich G, Morozova O, Yaropolov A. Stable 'floating' air diffusion biocathode based on direct electron transfer reactions between carbon particles and high redox potential laccase. *Fuel Cells* 2010;10:726–733.

76. Kontani R, Tsujimura S, Kano K. Air diffusion biocathode with CueO as electrocatalyst adsorbed on carbon particle-modified electrodes. *Bioelectrochemistry* 2009;76:10–13.

77. Gupta G, Lau C, Ranjendran V, Colon F, Branch B, Ivnitski D, Atanassov P. Direct electron transfer catalyzed by bilirubin oxidase for air breathing gas-diffusion electrodes. *Electrochem Commun* 2011;13:247–249.

78. Rincón RA, Lau C, Luckarift HR, Garcia KE, Adkins E, Johnson GR, Atanassov P. Enzymatic fuel cells: integrating flow-through anode and air-breathing cathode into a membrane-less biofuel cell design. *Biosens Bioelectron* 2011;27:132–136.

79. Higgins S, Lau C, Minteer SD, Atanassov P, Cooney M. Hybrid biofuel cell: microbial fuel cell with an enzymatic air-breathing cathode. *ACS Catal* 2011;1:994–997.

80. Kataoka K, Komori H, Ueki Y, Konno Y, Kamitaka Y, Kurose S, Tsujimura S, Higuchi Y, Kano K, Seo D, Sakurai T. Structure and function of the engineered multicopper oxidase CueO from *Escherichia coli*: deletion of the methionine-rich helical region covering the substrate-binding site. *J Mol Biol* 2007;373:141–152.

81. Quintanar L, Stoj CS, Taylor AB, Hart PJ, Kosman DJ, Solomon EI. Shall we dance? How a multicopper oxidase chooses its electron transfer partner. *Acc Chem Res* 2007;40:445–452.

82. Marshall NM, Garner DK, Wilson TD, Gao YG, Robinson H, Nilges MJ, Lu Y. Rationally tuning the reduction potential of a single cupredoxin beyond the natural range. *Nature* 2009;462:113–116.

83. Galli C, Gentili P, Jolivalt C, Madzak C, Vadala R. How is the reactivity of laccase affected by single-point mutations? Engineering laccase for improved activity towards sterically demanding substrates. *Appl Microbiol Biotechnol* 2011;91(1):123–131.

84. Bohlin C, Jonsson LJ, Roth R, van Zyl WH. Heterologous expression of *Trametes versicolor* laccase in *Pichia pastoris* and *Aspergillus niger*. *Appl Biochem Biotechnol* 2006;129(1–3):195–214.

85. Mate D, Garcia-Burgos C, Garcia-Ruiz E, Ballesteros AO, Camarero S, Alcalde M. Laboratory evolution of high-redox potential laccases. *Chem Biol* 2010;17:1030–1041.

86. Gallaway J, Wheeldon I, Rincón R, Atanassov P, Banta S, Calabrese Barton S. Oxygen-reducing enzyme cathodes produced from SLAC, a small laccase from *Streptomyces coelicolor*. *Biosens Bioelectron* 2008;23:1229–1235.

87. Hong G, Ivnitski DM, Johnson GR, Atanassov P, Pachter R. Design parameters for tuning the type 1 Cu multicopper oxidase redox potential: insight from a combination of first principles and empirical molecular dynamics simulations. *J Am Chem Soc* 2011;133:4802–4809.

88. D'Souza-Ticlo D, Sharma D, Raghukumar C. A thermostable metal-tolerant laccase with bioremediation potential from a marine-derived fungus. *Mar Biotechnol* 2009;11:725–727.

89. Mtui G, Nakamura Y. Lignocellulosic enzymes from *Flavodon flavus*, a fungus isolated from Western Indian Ocean off the coast of Dar es Salaam, Tanzania. *Afr J Biotechnol* 2008;7:3066–3072.

4

ANODIC CATALYSTS FOR OXIDATION OF CARBON-CONTAINING FUELS

ROSALBA A. RINCÓN, CAROLIN LAU, AND PLAMEN ATANASSOV

Department of Chemical and Nuclear Engineering and Center for Emerging Energy Technologies, University of New Mexico, Albuquerque, NM, USA

HEATHER R. LUCKARIFT

Universal Technology Corporation, Dayton, OH, USA; Airbase Sciences Branch, Air Force Research Laboratory, Tyndall Air Force Base, FL, USA

4.1 INTRODUCTION

One of the main advantages of using enzymes as catalysts for biological fuel cell (BFC) development is their selectivity and substrate specificity, thereby eliminating the need for separation of fuel and oxidant. Because of this inherent selectivity, enzymatic anodes can be designed and developed based on entirely different classes of oxidoreductase enzymes, depending on the fuel of choice. Ideally, the optimal biocatalyst can achieve oxidation of fuels at low redox potentials and through direct electron transfer (DET), in order to obtain the maximum possible difference between the anode's and the cathode's operating potentials, that is, the driving force of the fuel cell.

Oxidoreductase enzymes include dehydrogenases, reductases, and oxidases that typically contain flavin adenine dinucleotide (FAD), nicotinamide adenine dinucleotide phosphate ($NAD(P)^+$), or pyrroloquinoline quinone (PQQ) as redox cofactors. Although some specific enzymes possess cofactors with theoretically low redox potentials, this does not always translate to efficient performance when immobilized onto electrodes for BFCs. The caveat provides a need to study

Enzymatic Fuel Cells: From Fundamentals to Applications, First Edition. Edited by Heather R. Luckarift, Plamen Atanassov, and Glenn R. Johnson.

oxidoreductase enzymes from different sources and evaluate their relative advantages and disadvantages. In the following sections, we will discuss why some anodic catalysts may perform better for certain fuels, and why compromise is often needed to maximize performance from a BFC.

4.2   OXIDASES

4.2.1   Electron Transfer Mechanisms of Glucose Oxidase

Oxidase enzymes catalyze a redox reaction in which molecular oxygen (O_2) is the electron acceptor. Because of its abundance in Nature and high activity as a substrate, glucose is the focus of many BFC studies [1–11]. As such, one of the most studied anodic oxidase enzymes for BFC applications is glucose oxidase (GOx). The electron transfer mechanism during oxidation of glucose by GOx is now well understood and has found application in development of both biosensors and BFC applications. GOx catalyzes the oxidation of glucose to D-glucono-δ-lactone and hydrogen peroxide. The required cofactor for GOx to catalyze this reaction is FAD, which acts as the initial electron acceptor and is reduced to $FADH_2$. Efficient electrical contact is imperative between enzymes and electrodes, and in GOx (as well as in many other redox proteins), the redox active site (FAD) is buried deeply within the protein core, impeding the accessibility for direct communication with electrode surfaces (Figure 4.1). Therefore, great effort has been invested in searching for methods for

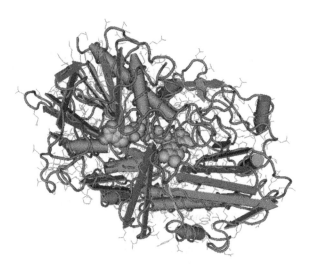

FIGURE 4.1   Glucose oxidase monomer from *Aspergillus niger*. (Reproduced with permission from Ref. [12]. Copyright 2011, American Chemical Society.) Structure retrieved from NCBI (PDB ID: 3QVR), viewed in Cn3D 4.3. Protein backbone viewed as tube worm structure with protein side chains shown as wire frames. FAD cofactor is represented as space fill model.

achieving DET with GOx. The biggest challenge in attaining DET, however, is to overcome the electron tunneling distance.

Typically, the addition of redox mediators is used as a way to shuttle electrons from the active site to electrode surfaces [13]. This mechanism is referred to as *mediated electron transfer* (MET). Redox mediators are usually small, mobile molecules containing redox moieties that assist in transferring electrons between redox enzymes and electrodes by diffusing in and out of the enzyme active site. Despite helping overcome the tunneling distance between the active site and electrode surface, the use of mediators also has certain disadvantages; these include high costs, potential toxicity, and instability of the systems, since mediators can diffuse over time.

Typical mediator systems for GOx include the use of redox polymers and chemically modified electrodes (CMEs) [14–16]. In certain cases, attachment of mediators directly to the enzyme surface or to a surrounding hydrogel is practiced in order to avoid diffusion problems [17–20]. MET can be useful in the development of biosensors, but in BFCs, DET is preferred in order to operate at the lowest possible anodic potential.

Achieving DET with GOx has therefore been a major field of study in the development of BFCs. Over the past decade, there have been various reports on DET with GOx occurring on the surfaces of carbonaceous materials such as graphene and carbon nanotubes (CNTs) [8,21–25]. The incorporation of CNTs into the electrode architecture is thought to allow electrical contact between the active site of GOx and the electrode surface. The diameter of a CNT can be as small as 1 nm for a single-walled CNT (SWCNT); in contrast, a GOx molecule is ~8 nm. Their comparable sizes theoretically allow CNTs to be positioned within proximity of the cofactor and reduce the electron tunneling distance [8,22].

In spite of the recent reports of DET of GOx, there are still questions of how the contact occurs between CNT and the FAD site. Understanding such mechanisms is fundamental for optimization and application of this system. Furthermore, overcoming the electron transfer obstacle of GOx is not enough. GOx only catalyzes a two-electron oxidation of glucose, and complete oxidation of fuel remains a major impediment for the development of efficient BFCs.

4.3 DEHYDROGENASES

4.3.1 The NADH Reoxidation Issue

Dehydrogenase enzymes play an essential role in the breakdown of sugars and oxidation of alcohols. There are more than 300 dehydrogenases that are known to be dependent on the nicotinamide adenine dinucleotide cofactor (NAD^+/NADH), comprising the largest group of redox enzymes known today [26]. Despite their abundance in Nature, their utility in biosensor and BFC applications has been limited. Unlike other redox enzymes (e.g., oxidases), dehydrogenase activity relies on the presence of a soluble cofactor, such as NAD^+/NADH [26], that must be incorporated into the fuel cell configuration. In the presence of cofactor, dehydrogenases catalyze

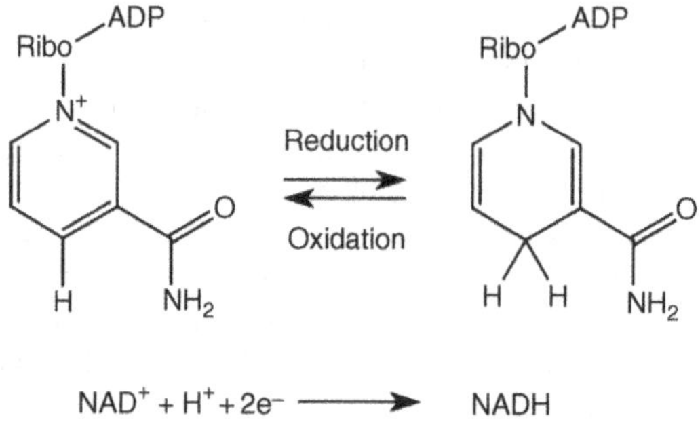

$$NAD^+ + H^+ + 2e^- \longrightarrow NADH$$

FIGURE 4.2   Redox reaction of the $NAD^+/NADH$ cofactor.

the oxidation of a specific substrate and the cofactor NAD^+ is concurrently reduced to NADH (Figure 4.2).

The electrochemistry of this cofactor in both oxidized and reduced forms is irreversible and conveys an additional complexity in fuel cell design. Electrochemical reduction of NAD^+ requires specific conditions in order to limit or avoid adsorption of NAD^+ that causes enzyme inhibition, reportedly due to the adenine moiety of the cofactor [27–32].

The mechanism for electrochemical oxidation of NADH has been proposed in a number of studies as an electrochemical–chemical–electrochemical mechanism characterized by the following reaction scheme (Equation 4.1) [32–34]:

$$NADH \xrightarrow{-e^-} NADH^{\bullet+} \overset{-H^+}{\longleftrightarrow} NAD^{\bullet} \xrightarrow{-e^-} NAD^+ \qquad (4.1)$$

The high overpotential of the direct electrochemical oxidation of NADH described in Equation 4.1 is caused by the very high potential of the $NAD^+/NADH$ redox couple (the first step in the reaction) [35], which results in an initial rate-limiting electron transfer step. It has also been suggested that the intermediate radicals resulting from the first and second steps of this reaction might participate in alternative reaction routes for electrochemical oxidation of NADH [34].

To summarize, the direct electrochemical reactions of the $NAD^+/NADH$ cofactors at metallic or carbon electrodes are highly irreversible, occur at large overpotentials, and can be affected by side reactions and fouling (adsorption) of cofactor-related products [26]. The development of BFCs based on NAD^+-dependent dehydrogenases, therefore, cannot rely on direct electrochemical reactions of the cofactor. As a result, a major focus for this field of research has been directed at finding suitable methods for the electrochemical oxidation of NADH.

As previously stated, in an enzyme-catalyzed reaction involving the $NAD^+/NADH$ cofactor, NAD^+ is reduced to NADH whereas the substrate is concurrently oxidized. In a BFC, any NAD^+ that is reduced to NADH must be reoxidized in order to perpetuate the reaction cycle and provide a continuous oxidation of fuel. Direct oxidation of NADH at the electrodes, such as gold or glassy carbon, however, requires high overpotentials exceeding 1 V [36–38]. By overcoming the high overpotential, it may be possible to use any NAD^+-dependent enzyme in a BFC anode and create entire cascades of enzymes that could achieve deep oxidation of complex fuels and support a wider variety of fuel choices. Hence, finding ways to reoxidize NADH is a crucial task that will broaden the opportunities for enzymatic BFCs.

4.3.2 Mediators for Electrochemical Oxidation of NADH

The use of mediators with CMEs to reduce the overpotential of NADH oxidation and improve its kinetics started in the late 1970s and has been extensively reported on and summarized in several reviews since then [39–44]. An appropriate mediator will meet the following demands: (i) it must substantially decrease the overpotential of NADH oxidation while preserving good reaction rates that approach diffusion-controlled regimes; in BFCs, the $E^{\circ\prime}$ (formal redox potential) value of the mediator should approach that of the enzyme cofactor NAD^+/NADH so that the energy losses are minimized [37,45]; (ii) it must possess high chemical and electrochemical stability in the presence of NADH, as well as long-term stability (weeks to months) [44]; and (iii) it should be selective for NADH oxidation and yield enzymatically active NAD^+ while maintaining high reaction rates of the mediator-modified system [44].

The first researchers to report a significant decrease in the overpotential of electro-chemical oxidation of NADH on CMEs were Tse and Kuwana [46], who in 1978 activated glassy carbon electrodes with cyanuric chloride, followed by modification with immobilized primary amines (containing o-quinone derivatives, dopamine, or 3,4-dihydroxybenzylamine) to form a monolayer film on the electrode. The resulting electrodes decreased the NADH oxidation overpotential to ~0.4 V. This report was followed by many others that investigated similar structures for the mediation of electrochemical oxidation of NADH, all containing a basic catalytic functionality that can be characterized by being o- or p-quinone [46–50], phenothiazine and phenoxazine derivatives that contain positively charged p-phenylenediimine structures (Figure 4.3) [51–54]. It has also been reported that organic two-electron acceptors (also proton acceptors) are more efficient in decreasing the overpotential of NADH electrochemical oxidation than one-electron acceptors [26].

Formerly, the mediator was typically immobilized by physical adsorption [50,51,53–60], or covalent binding [46,48,61], onto the surface of the electrodes to create the CMEs, which limited their long-term stability because of leaking or desorption of the mediator from the electrode surface [62]. Physical attachment of the mediator to the electrode also limits the operational mode of the device; CMEs with adsorbed mediators could only work in quiescent conditions (i.e., a biobattery) [63] and not under a flow-through regime (BFC) where desorption will be more likely. This issue was addressed by depositing films of mediators, such as azine monomers, through electrochemical polymerization directly onto the surface of electrodes [64–72], resulting in a polymeric form of the azine with improved stability and reaction rates that can then be considered an electrocatalyst. This method has proven to be the most reliable immobilization strategy to address stability [73].

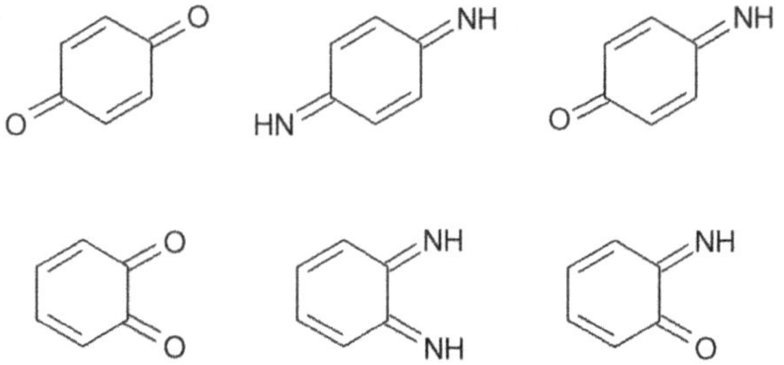

FIGURE 4.3   Basic catalytic functionalities of many organic two-electron proton acceptors efficient for catalytic NADH oxidation. (Reprinted with permission from Ref. [44]. Copyright 2002, Elsevier.)

There have been numerous reports on electrochemical polymerization of different monomers for applications in biosensors [65,66,74–82] and BFCs [71,83–85]. For example, in 1993, Karyakin et al. reported that methylene blue can be successfully electropolymerized in weakly acidic solutions and that basic media are actually optimal for the polymer growth [86]. They later reported that the main advantage of electropolymerized methylene blue as a catalyst for NADH oxidation was its improved long-term stability compared with the monomeric form [64]. The work of Karyakin's group has continued over the years, with investigations extending to a variety of different azine monomers that can be electropolymerized, in search of the optimal catalyst for NADH oxidation [65,66]. In 1999, poly(methylene green) (poly (MG)) was reported as the optimal catalyst among the many studied [65]. More recently, Dai et al. confirmed in 2008 that a poly(MG)-modified electrode can oxidize NADH by a simultaneous two-electron reaction and reduce the overpotential of NADH oxidation by approximately 0.65 V [81].

4.3.3   Electropolymerization of Azines

When a suitable catalyst for NADH oxidation is identified, the next stage is associating that activity with an electrode surface. From the viewpoint of applications, the electrochemical polymerization of cheap, simple aromatic compounds (such as benzoids, nonbenzoids, and heterocyclics) is of extreme interest. The reaction is usually an oxidative polymerization, although reductive polymerization is also possible. Chemical polymerization can also be used in certain cases, but electro-polymerization is preferred if the resulting polymer is intended for use as a polymer film electrode or thin-layer sensor. Because of the potential control imparted during synthesis, a good-quality polymer film produced at the desired location is attractive for practical applications. In general, electropolymerization consists of polymerizing a monomer by an electrochemical technique, by potentiostatic (constant potential), galvanostatic (constant current), or cyclic voltammetry (multiscan). Figure 4.4, for example, shows electropolymerization of methylene green onto a glassy carbon electrode via consecutive cyclic voltammetry.

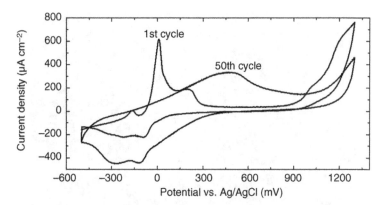

FIGURE 4.4 Electrochemical polymerization of methylene green by cyclic voltammetry (50 cycles) from −0.5 to 1.3 V versus Ag/AgCl.

Polymer-modified electrodes that are prepared by electropolymerization of mono-mers, with mediating properties toward NADH oxidation, are attractive because the synthesis is irreversible, the long-term stability of the electrode is enhanced, and the catalytic properties of the monomer are retained. A diversity of reports on electro-polymerization of *o*-quinone derivatives [87–89], and phenothiazine and phenoxazine derivatives with different levels of utility for practical applications, can be found in the literature [62,65,66,68,72,90–92]. Electropolymerization of azine derivatives has mainly been reported on gold [66,91], platinum [93], glassy carbon electro-des [65,68,71,81,92,94–97], and other carbonaceous electrode materials (Toray paper [98–100], graphite [101], and screen-printed carbon [70]).

In 2007, Svoboda et al. studied the morphology of poly(MG) films grown on platinum electrodes and observed that a conformal coating of the electropolymerized film is deposited on the two-dimensional structure of the electrode [93]. This conformal coating on two-dimensional materials has also been observed on glassy carbon electrodes [102], but the integrity of the coating is lost when depositing poly (MG) onto a three-dimensional surface, such as reticulated vitreous carbon (RVC) (Figure 4.5) [103]. The coated RVC will still oxidize NADH [103].

Figure 4.6 shows the electrochemical activity of deposited poly(MG) onto RVC, and in each case, a nonlinear dependence that resembles Michaelis–Menten-type kinetics is observed. This observation agrees with the model proposed in 1985 by Gorton et al. [51] for mediator-modified electrodes for NADH oxidation, and it agrees with similar studies in 1990 and 2001 [53,105]. This model postulates the formation of a charge transfer (CT) complex in the reaction sequence between NADH and the mediator, because the observed reaction rate starts to decrease with the increase in NADH concentration, analogous to the Michaelis–Menten kinetics of enzymatic reactions. Catalytic activity of poly(MG) is inversely proportional to the thickness of the polymer, and the number of deposition cycles is consistent with observations in the literature for other NADH mediators [26,44,47,49]. This is attributed to the low partition coefficient of NADH and the diffusion coefficient of NADH within the

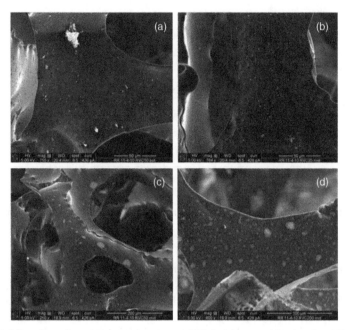

FIGURE 4.5   Scanning electron microscope (SEM) micrographs of poly(MG)-modified RVC by various deposition cycles: (a) 10 cycles, (b) 25 cycles, (c) 50 cycles, and (d) 200 cycles. (Reprinted with permission from Ref. [104]. Copyright 2011, Elsevier).

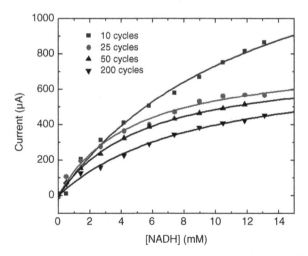

FIGURE 4.6   Amperometric response to consecutive additions of NADH to poly(MG)-modified RVC. (Reprinted with permission from Ref. [104]. Copyright 2011, Elsevier).

polymer [44]. This experimental observation has helped build an electrode with a layer of polymer particles thin enough to reach their best electrocatalytic activity toward NADH oxidation for the development of a BFC anode (see Section 4.3.4).

4.3.4 Alcohol Dehydrogenase as a Model System

In 2011, alcohol dehydrogenase (ADH) was used as a model enzyme coupled with poly(MG) for NADH reoxidation in the construction of a three-dimensional BFC with ethanol as fuel [103]. In combination with an air-breathing/gas diffusion cathode (using laccase as an oxygen reduction enzyme), a BFC was fabricated that was able to successfully exploit ethanol oxidation by an NAD^+-dependent ADH, immobilized by entrapment in a multiwalled CNT (MWCNT)/chitosan matrix [106]. The feasibility and reproducibility of the resulting BFC were demonstrated in 2008 with a series of standardized multilaboratory experiments [96].

The result was a fully enzymatic, membrane-free BFC operating near neutral pH using ethanol as fuel, with an open-circuit voltage of 0.618 V and a power density of $22\,\mu W\,cm^{-3}$. Figure 4.7 shows the polarization curves of the BFC as described, with polarization characteristics demonstrating the regions of kinetic, ohmic, and transport limitations. The cathode shows the typical behavior of an air-breathing electrode with kinetic limitations at low currents and ohmic losses between 30 and $150\,\mu A$; the significant drop in potential at high currents is due to transport limitations associated with the diffusion of air.

Large ohmic losses are observed with the ADH anode revealed by a polarization curve that depicts resistive behavior and a BFC that is anode limited. This ohmic

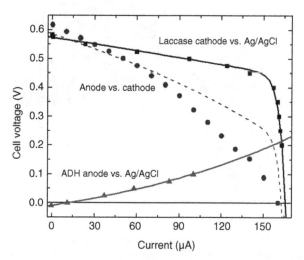

FIGURE 4.7 Polarization curves of (▲) ADH anode, (■) laccase cathode, and (●) the fully assembled biological fuel cell. Dashed line represents the theoretical full cell polarization curve. Ethanol concentration: 475 mM; open-circuit voltage = 0.618 V. (Reprinted with permission from Ref. [103]. Copyright 2011, Elsevier.)

effect is a consequence of the low conductivity of the supporting electrolytes as well as the macroscopic separation (~ 1 cm) between the anode and the cathode. These are key design factors for further improving BFC design.

4.4   PQQ-DEPENDENT ENZYMES

Many PQQ-dependent enzymes have received interest as electrocatalysts because of their ability to oxidize a wide range of organic substrates [107–110]. PQQ-dependent glucose dehydrogenase (GDH), in particular, has been a workhorse in the development of blood glucose sensors for diabetes screening [111]. *Gluconobacter* sp. are acetic acid bacteria that have been the focus of many studies as they contain numerous PQQ-dependent dehydrogenases, including ADH, aldehyde dehydrogenase (ALDH), GDH, and glycerol dehydrogenases [112]. All PQQ-containing enzymes belong to a group of quinohemoenzymes that contain the cofactor PQQ and one (or more) heme *c* moieties. The factor group PQQ is bound to the protein via electrostatic interactions between the carboxyl groups of the protein in the presence of Ca^{2+} ions (Figure 4.8) [113]. The heme *c* serves as a shuttle that facilitates internal electron transfer, and evidence suggests that such electron transfer proceeds through the formation of a semiquinone free radical intermediate (Equation 4.2) in which PQQ_{OX}, PQQ_{SEM}, and

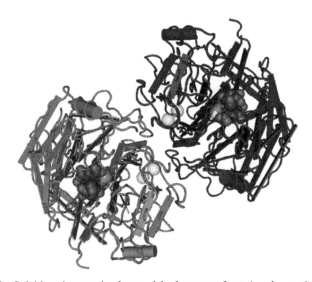

FIGURE 4.8   Soluble quinoprotein glucose dehydrogenase from *A. calcoaceticus* in complex with PQQH$_2$ and glucose. (Reprinted with permission from Ref. [110]. Copyright 2003, Elsevier.) Structure retrieved from NCBI (PDB ID: 1CQ1), viewed in Cn3D 4.3. Protein homodimer colored in blue/purple tube worm structure with calcium ions colored yellow for clarity; PQQ cofactor represented as space fill model. (Please see the color version of this figure in Color Plates section.)

PQQ$_{RED}$ are the oxidized, semiquinone, and reduced forms of PQQ, respectively [114,115]:

$$\begin{aligned} \text{(Step 1)} \quad & PQQ_{OX} + e^- \leftrightarrow PQQ_{SEM} \\ \text{(Step 2)} \quad & PQQ_{SEM} + e^- \leftrightarrow PQQ_{RED} \end{aligned} \qquad (4.2)$$

PQQ enzymes can catalyze redox reactions through the use of soluble mediators or through conductive polymers [116]. The ability to shuttle electrons within these proteins, however, allows for DET of many PQQ-dependent dehydrogenases, increasing their utility as bioelectrocatalysts. A PQQ-dependent lactate dehydrogenase, for example, was isolated in 2008 from *Gluconobacter* sp. 33 and *Gluconobacter suboxydans*, and DET was demonstrated following immobilization to gold and carbon electrode surfaces [117]. Carbon electrodes exhibited a lower anodic potential during cyclic voltammetry (0.16 V vs. Ag/AgCl compared with 0.32 V vs. Ag/AgCl for gold electrodes) and was attributed to the relative positioning of the heme *c* on the electrode surface. The immobilized protein was further used in a BFC with lactate/air and exhibited a high open-circuit potential (0.85 V) and an average power density approaching 20 μW cm^{-2}, which was sustained for more than 45 days.

Similarly, in 2005, Ferapontova and Gorton demonstrated DET of D-fructose dehydrogenase from *Gluconobacter industrius* by immobilization on gold electrodes modified with an alkanethiol self-assembled monolayer (SAM). A formal redox potential of −0.158 and −0.089 V was observed at pH 5.0 and 4.0, respectively. It was found that a positively charged SAM was essential for DET and the potential for fructose oxidation could be shifted from −0.080 to −0.120 V by simply varying the functional groups of the SAM [118].

In 2007, Ivnitski et al. reported the association of PQQ-dependent GDH with SWCNTs (Figure 4.9) by formation of ester groups on the SWCNTs that considerably enhanced hydrophilicity and, therefore, improved the immobilization of the enzymes [114]. As a result, a linear response in electrocatalytic current was observed in response to an increase

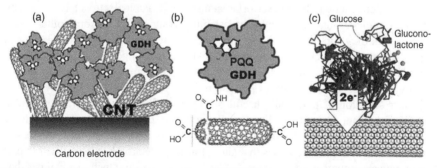

FIGURE 4.9 (a) Schematic representation of SWCNT enzyme electrode; (b) PQQ-dependent glucose dehydrogenase is coupled to SWCNT on a carbon electrode via chemical attachment of the enzyme to carboxyl-functionalized SWCNT (c) that allows for direct electron transfer during bioelectrocatalysis of glucose oxidation. (Reproduced with permission from Ref. [114]. Copyright 2007, Wiley-VCH Verlag GmbH.)

in glucose concentration. Similarly, in 2009, Treu et al. combined two PQQ-dependent enzymes, ADH (oxidizes ethanol to aldehyde) and ALDH (oxidizes aldehyde to acetate), by immobilization within a hydrophobically modified Nafion polymer that incorporated carboxyl-modified SWCNTs [116]. The technique was used to fabricate a BFC that used ethanol and air and produced power densities of $\sim 0.25\,\mathrm{mW\,cm^{-2}}$. Changing the dimensions and functionality of the CNTs was shown to dramatically affect the power density of the resulting electrode composites, and this observation was attributed to preferential binding of the enzymes to the CNT matrix.

Incremental improvements in power density can be attained through protein engineering [119]. PQQ-dependent GDH from *Acinetobacter calcoaceticus*, for example, exhibits high catalytic efficiency, wide substrate specificity, and oxygen tolerance, but it is unstable. A single amino acid substitution (Ser415Cys), however, was sufficient to improve stability sixfold and retain activity for 6.5 days. When the variant PQQ-dependent GDH was used as the anodic catalyst and bilirubin oxidase as the cathodic catalyst, the maximum power density was $\sim 17\,\mathrm{\mu W\,cm^{-2}}$ (under $200\,\mathrm{k\Omega}$ external load) in the presence of 2,2′-azinobis(3-ethylbenzothiazoline-6-sulfonate) as a mediator. The substitution also imparted enhanced protein thermostability up to $70\,°C$ without affecting the catalytic activity. In a similar study, another single amino acid substitution in PQQ-dependent GDH from *A. calcoaceticus* was sufficient to double the catalytic and electrocatalytic activity of the protein. Following immobilization to highly porous carbon cryogel electrodes, catalytic currents of $>1\,\mathrm{mA\,cm^{-2}}$ for the variant enzymes were observed [115].

4.5 OUTLOOK

The integration of cofactor-dependent enzymes as anodic catalysis clearly faces a number of technical hurdles that must be considered and addressed. To some extent, the stability of the NAD$^+$/NADH couple has been addressed by introduction of electropolymerized polymers that has directly led to extended bioanode lifetimes.

PQQ-dependent dehydrogenases offer an alternative as bioelectrocatalysts, as they are oxygen tolerant and show enhanced stability compared with some other redox enzymes, but they still require effective methods to increase enzyme loading that can impact power density. One of the current limitations of using PQQ-dependent dehydrogenases in BFCs is commercial availability. As such, methods to rapidly screen, identify, and purify dehydrogenases with low redox potentials will undoubtedly provide a broader selection of potentially useful biocatalysts. Protein engineering may also provide significant advances in direct application to enzymatic anodes by selectively engineering the redox potential of enzymes. Alternatively, the operation of enzymes in concert to create enzyme cascades that allow for deep electron harvesting will impart higher power outputs (see Chapter 5). Stability of oxidoreductase enzymes has historically been a challenge, and the means to extend lifetime to months or years will be a key aspect of bioanode development. This may be addressed by rational enzyme immobilization and specific development of hierarchical architectures that optimize enzyme loading, retain cofactor stability, and maintain efficient mass transport.

ACKNOWLEDGMENT

This research was prepared as an account of work sponsored by the United States Air Force Research Laboratory, Materials and Manufacturing Directorate, Airbase Technologies Division (AFRL/RXQ), but the views of authors expressed herein do not necessarily reflect those of the United States Air Force.

LIST OF ABBREVIATIONS

ADH	alcohol dehydrogenase
ALDH	aldehyde dehydrogenase
BFC	biological fuel cell
CME	chemically modified electrode
CNT	carbon nanotube
CT	charge transfer
DET	direct electron transfer
FAD	flavin adenine dinucleotide
$FADH_2$	flavin adenine dinucleotide, reduced form
GDH	glucose dehydrogenase
GOx	glucose oxidase
MET	mediated electron transfer
MWCNT	multiwalled carbon nanotube
NADH	nicotinamide adenine dinucleotide, reduced form
$NAD^+/NADH$	nicotinamide adenine dinucleotide cofactor
$NAD(P)^+$	nicotinamide adenine dinucleotide phosphate
poly(MG)	poly(methylene green)
PQQ	pyrroloquinoline quinone
RVC	reticulated vitreous carbon
SAM	self-assembled monolayer
SEM	scanning electron microscope
SWCNT	single-walled carbon nanotube

REFERENCES

1. Ivnitski D, Artyushkova K, Rincón RA, Atanassov P, Luckarift HR, Johnson GR. Entrapment of enzymes and carbon nanotubes in biologically synthesized silica: glucose oxidase-catalyzed direct electron transfer. *Small* 2008;4:357–364.

2. Calabrese Barton S, Gallaway J, Atanassov P. Enzymatic biofuel cells for implantable and microscale devices. *Chem Rev* 2004;104:4867–4886.

3. Bullen RA, Arnot TC, Lakeman JB, Walsh FC. Biofuel cells and their development. *Biosens Bioelectron* 2006;21:2015–2045.

4. Heller A. Miniature biofuel cells. *Phys Chem Chem Phys* 2004;6:209–216.

5. Mano N, Mao F, Heller A. Characteristics of a miniature compartment-less glucose–O_2 biofuel cell and its operation in a living plant. *J Am Chem Soc* 2003;125:6588–6594.

6. Katz E, Willner I, Kotlyar AB. A non-compartmentalized glucose/O_2 biofuel cell by bioengineered electrode surfaces. *J Electroanal Chem* 1999;479:64–68.

7. Katz E, Willner I. A biofuel cell with electrochemically switchable and tunable power output. *J Am Chem Soc* 2003;125:6803–6813.

8. Ivnitski D, Branch B, Atanassov P, Apblett C. Glucose oxidase anode for biofuel cell based on direct electron transfer. *Electrochem Commun* 2006;8:1204–1210.

9. Ramanavicius A, Kausaite A, Ramanaviciene A. Biofuel cell based on direct bioelectrocatalysis. *Biosens Bioelectron* 2005;20:1962–1967.

10. Barriere F, Kavanagh P, Leech D. A laccase–glucose oxidase biofuel cell prototype operating in a physiological buffer. *Electrochim Acta* 2006;51:5187–5192.

11. Tsujimura S, Kano K, Ikeda T. Glucose/O_2 biofuel cell operating at physiological conditions. *Electrochemistry* 2002;70:940–942.

12. Kommoju PR, Chen ZW, Brucknet RC, Mathews FS, Jorns MS. Probing oxygen activation sites in two flavoprotein oxidases using chloride as an oxygen surrogate. *Biochemistry* 2011;50:5521–5534.

13. Frew JE, Hill HAO. Direct and indirect electron transfer between electrodes and redox proteins. *Eur J Biochem* 1988;172:261–269.

14. Hale PD, Boguslavsky LI, Inagaki T, Karan HI, Lee HS, Skotheim TA, Okamoto Y. Amperometric glucose biosensors based on redox polymer-mediated electron transfer. *Anal Chem* 1991;63:677–682.

15. Antiochia R, Gorton L. Development of a carbon nanotube paste electrode osmium polymer-mediated biosensor for determination of glucose in alcoholic beverages. *Biosens Bioelectron* 2007;22:2611–2617.

16. Jonsson G, Gorton L, Pettersson L. Mediated electron transfer from glucose oxidase at a ferrocene-modified graphite electrode. *Electroanalysis* 1989;1:49–55.

17. Degani Y, Heller A. Direct electrical communication between chemically modified enzymes and metal electrodes. 1. Electron transfer from glucose oxidase to metal electrodes via electron relays, bound covalently to the enzyme. *J Phys Chem* 1987;91:1285–1289.

18. Battaglini F, Bartlett PN, Wang JH. Covalent attachment of osmium complexes to glucose oxidase and the application of the resulting modified enzyme in an enzyme switch responsive to glucose. *Anal Chem* 2000;72:502–509.

19. Schuhmann W. Electron-transfer pathways in amperometric biosensors. Ferrocene-modified enzymes entrapped in conducting-polymer layers. *Biosens Bioelectron* 1995;10:181–193.

20. Mao F, Mano N, Heller A. Long tethers binding redox centers to polymer backbones enhance electron transport in enzyme "wiring" hydrogels. *J Am Chem Soc* 2003;125:4951–4957.

21. Shan CS, Yang HF, Song JF, Han DX, Ivaska A, Niu L. Direct electrochemistry of glucose oxidase and biosensing for glucose based on graphene. *Anal Chem* 2009;81:2378–2382.

22. Guiseppi-Elie A, Lei C, Baughman RH. Direct electron transfer of glucose oxidase on carbon nanotubes. *Nanotechnology* 2002;13:559–564.

23. Liang W, Zhuobin Y. Direct electrochemistry of glucose oxidase at a gold electrode modified with single-wall carbon nanotubes. *Sensors* 2003;3:544–554.

24. Cai C, Chen J. Direct electron transfer of glucose oxidase promoted by carbon nanotubes. *Anal Biochem* 2004;332:75–83.

25. Liu Y, Wang M, Zhao F, Xu Z, Dong S. The direct electron transfer of glucose oxidase and glucose biosensor based on carbon nanotubes/chitosan matrix. *Biosens Bioelectron* 2005;21:984–988.

26. Gorton L, Hale PD, Persson B, Boguslavsky LI, Karan HI, Lee HS, Skotheim TA, Lan HL, Okamoto Y. Electrocatalytic oxidation of nicotinamide adenine dinucleotide cofactor at chemically modified electrodes. *ACS Symp Ser* 1992;487:56–83.

27. Bresnahan WT, Elving PJ. The role of adsorption in the initial one-electron electrochemical reduction of nicotinamide adenine dinucleotide (NAD^+). *J Am Chem Soc* 1981;103:2379–2386.

28. Underwood AL, Burnett RW. Electrochemistry of biological compounds. In: Bard AJ (ed.), *Electroanalytical Chemistry: A Series of Advances*. Marcel Dekker, New York, 1973, pp. 1–85.

29. Schmakel CO, Santhanam KSV, Elving PJ. Nicotinamide adenine dinucleotide (NAD^+) and related compounds. Electrochemical redox pattern and allied chemical behavior. *J Am Chem Soc* 1975;97:5083–5092.

30. Jensen MA, Bresnahan WT, Elving PJ. Comparative adsorption of adenine and nicotinamide adenine dinucleotide (NAD^+) at an aqueous solution/mercury interface. *Bioelectrochem Bioenerg* 1983;11:299–306.

31. Moiroux J, Deycard S, Malinski T. Electrochemical reduction of NAD^+ and pyridinium cations adsorbed at the mercury/water interface. Electrochemical behavior of adsorbed pyridinyl radicals. *J Electroanal Chem* 1985;194:99–108.

32. Blankespoor RL, Miller LL. Electrochemical oxidation of NADH: kinetic control by product inhibition and surface coating. *J Electroanal Chem* 1984;171:231–241.

33. Moiroux J, Elving PJ. Mechanistic aspects of the electrochemical oxidation of dihydronicotinamide adenine dinucleotide (NADH). *J Am Chem Soc* 1980;102:6533–6538.

34. Jaegfeldt H. Adsorption and electrochemical oxidation behavior of NADH at a clean platinum electrode. *J Electroanal Chem* 1980;110:295–302.

35. Gorton L, Bartlett PN. NAD(P)-based biosensors. In: Bartlett PN (ed.), *Bioelectrochemistry: Fundamentals, Experimental Techniques, and Applications*. John Wiley & Sons, Inc., Chichester, UK, 2008, pp. 157–198.

36. Blaedel WJ, Jenkins RA. Study of electrochemical oxidation of reduced nicotinamide adenine dinucleotide. *Anal Chem* 1975;47:1337–1343.

37. Chenault HK, Whitesides GM. Regeneration of nicotinamide cofactors for use in organic synthesis. *Appl Biochem Biotechnol* 1987;14:147–197.

38. Karyakin AA, Ivanova YN, Revunova KV, Karyakina EE. Electropolymerized flavin adenine dinucleotide as an advanced NADH transducer. *Anal Chem* 2004;76:2004–2009.

39. Bartlett PN, Tebbutt P, Whitaker RG. Kinetic aspects of the use of modified electrodes and mediators in bioelectrochemistry. *Prog React Kinet* 1991;16:55–155.

40. Katakis I, Domínguez E. Catalytic electrooxidation of NADH for dehydrogenase amperometric biosensors. *Microchim Acta* 1997;126:11–32.

41. Lobo MJ, Miranda AJ, Tuñón P. Amperometric biosensors based on NAD(P)-dependent dehydrogenase enzymes. *Electroanalysis* 1997;9:191–202.

42. Armstrong FA, Wilson GS. Recent developments in faradaic bioelectrochemistry. *Electrochim Acta* 2000;45:2623–2645.

43. Habermuller K, Mosbach M, Schuhmann W. Electron-transfer mechanisms in amperometric biosensors. *Fresenius J Anal Chem* 2000;366:560–568.

44. Gorton L, Domínguez E. Electrocatalytic oxidation of NAD(P)H at mediator-modified electrodes. *Rev Mol Biotechnol* 2002;82:371–392.

45. Palmore GTR, Bertschy H, Bergens SH, Whitesides GM. A methanol/dioxygen biofuel cell that uses NAD^+-dependent dehydrogenases as catalysts: application of an electroenzymatic method to regenerate nicotinamide adenine dinucleotide at low overpotentials. *J Electroanal Chem* 1998;443:155–161.

46. Tse DCS, Kuwana T. Electrocatalysis of dihydronicotinamide adenosine diphosphate with quinones and modified quinones electrodes. *Anal Chem* 1978;50:1315–1318.

47. Degrand C, Miller LL. An electrode modified with polymer-bound dopamine which catalyzes NADH oxidation. *J Am Chem Soc* 1980;102:5728–5732.

48. Ueda C, Tse DCS, Kuwana T. Stability of catechol modified carbon electrodes for electrocatalysis of dihydronicotinamide adenine dinucleotide and ascorbic acid. *Anal Chem* 1982;54:850–856.

49. Fukui M, Kitani A, Degrand C, Miller LL. Propagation of a redox reaction through a quinoid polymer film on an electrode. *J Am Chem Soc* 1982;104:28–33.

50. Jaegfeldt H, Kuwana T, Johansson G. Electrochemical stability of catechols with a pyrene side chain strongly adsorbed on graphite electrodes for catalytic oxidation of dihydronicotinamide adenine dinucleotide. *J Am Chem Soc* 1983;105: 1805–1814.

51. Gorton L, Johansson G, Torstensson A. A kinetic study of the reaction between dihydronicotinamide adenine dinucleotide (NADH) and electrode modified by adsorption of 1,2-benzophenoxazine-7-one. *J Electroanal Chem* 1985;196:81–92.

52. Gorton L. Chemically modified electrodes for the electrocatalytic oxidation of nicotinamide coenzymes. *J Chem Soc, Faraday Trans 1* 1986;82:1245–1258.

53. Persson B, Gorton L. A comparative study of some 3,7-diaminophenoxazine derivatives and related compounds for electrocatalytic oxidation of NADH. *J Electroanal Chem* 1990;292:115–138.

54. Polasek M, Gorton L, Appelqvist R, Markovarga G, Johansson G. Amperometric glucose sensor based on glucose dehydrogenase immobilized on a graphite electrode modified with an *N,N'*-bis(benzophenoxazinyl) derivative of benzene-1,4-dicarboxamide. *Anal Chim Acta* 1991;246:283–292.

55. Torstensson A, Gorton L. Catalytic oxidation of NADH by surface-modified graphite electrodes. *J Electroanal Chem* 1981;130:199–207.

56. Huck H. Catalytic oxidation of NADH on graphite electrodes with adsorbed phenoxazine derivatives. *Fresenius Z Anal Chem* 1982;313:548–552.

57. Appelqvist R, Markovarga G, Gorton L, Torstensson A, Johansson G. Enzymatic determination of glucose in a flow system by catalytic oxidation of the nicotinamide coenzyme at a modified electrode. *Anal Chim Acta* 1985;169:237–247.

58. Munteanu FD, Mano N, Kuhn A, Gorton L. Mediator-modified electrodes for catalytic NADH oxidation: high rate constants at interesting overpotentials. *Bioelectrochemistry* 2002;56:67–72.

59. de Lucca AR, Santos AD, Pereira AC, Kubota LT. Electrochemical behavior and electrocatalytic study of the methylene green coated on modified silica gel. *J Colloid Interface Sci* 2002;254:113–119.

60. Gligor D, Muresan LM, Dumitru A, Popescu IC. Electrochemical behavior of carbon paste electrodes modified with methylene green immobilized on two different X type zeolites. *J Appl Electrochem* 2007;37:261–267.

61. Miyawaki O, Wingard LB. Electrochemical and glucose oxidase coenzyme activity of flavin adenine dinucleotide covalently attached to glassy carbon at the adenine amino group. *Biochim Biophys Acta* 1985;838:60–68.

62. Lawrence NS, Wang J. Chemical adsorption of phenothiazine dyes onto carbon nanotubes: toward the low potential detection of NADH. *Electrochem Commun* 2006;8:71–76.

63. Sakai H, Nakagawa T, Tokita Y, Hatazawa T, Ikeda T, Tsujimura S, Kano K. A high-power glucose/oxygen biofuel cell operating under quiescent conditions. *Energy Environ Sci* 2009;2:133–138.

64. Karyakin AA, Karyakina EE, Schuhmann W, Schmidt HL, Varfolomeyev SD. New amperometric dehydrogenase electrodes based on electrocatalytic NADH-oxidation at poly(methylene blue)-modified electrodes. *Electroanalysis* 1994;6:821–829.

65. Karyakin AA, Karyakina EE, Schmidt HL. Electropolymerized azines: a new group of electroactive polymers. *Electroanalysis* 1999;11:149–155.

66. Karyakin AA, Karyakina EE, Schuhmann W, Schmidt HL. Electropolymerized azines. Part II. In a search of the best electrocatalyst of NADH oxidation. *Electroanalysis* 1999;11:553–557.

67. Mao LQ, Yamamoto K. Glucose and choline on-line biosensors based on electro-polymerized Meldola's blue. *Talanta* 2000;51:187–195.

68. Kertesz V, Van Berkel GJ. Electropolymerization of methylene blue investigated using on-line electrochemistry/electrospray mass spectrometry. *Electroanalysis* 2001;13:1425–1430.

69. Shan D, Mousty C, Cosnier S, Mu SL. A composite poly azure B–clay–enzyme sensor for the mediated electrochemical determination of phenols. *J Electroanal Chem* 2002;537:103–109.

70. Sha YF, Gao Q, Qi B, Yang XR. Electropolymerization of azure B on a screen-printed carbon electrode and its application to the determination of NADH in a flow injection analysis system. *Microchim Acta* 2004;148:335–341.

71. Blackwell AE, Moehlenbrock MJ, Worsham JR, Minteer SD. Comparison of electro-polymerized thiazine dyes as an electrocatalyst in enzymatic biofuel cells and self powered sensors. *J Nanosci Nanotechnol* 2009;9:1714–1721.

72. Gligor D, Dilgin Y, Popescu IC, Gorton L. Poly-phenothiazine derivative-modified glassy carbon electrode for NADH electrocatalytic oxidation. *Electrochim Acta* 2009;54:3124–3128.

73. Prieto-Simon B, Fabregas E. Comparative study of electron mediators used in the electrochemical oxidation of NADH. *Biosens Bioelectron* 2004;19:1131–1138.

74. Schuhmann W, Huber J, Mirlach A, Daub J. Covalent binding of glucose oxidase to functionalized polyazulenes. The first application of polyazulenes in amperometric biosensors. *Adv Mater* 1993;5:124–126.

75. Schuhmann W. Conducting polymers and their application in amperometric biosensors. In: Usmani AM, Akmal N (eds.), *Diagnostic Biosensor Polymers*. American Chemical Society, Washington, DC, 1994, pp. 110–123.

76. Yamato H, Ohwa M, Wernet W. Stability of polypyrrole and poly(3,4-ethylenediox-ythiophene) for biosensor application. *J Electroanal Chem* 1995;397:163–170.

77. Hiller M, Kranz C, Huber J, Bauerle P, Schuhmann W. Amperometric biosensors produced by immobilization of redox enzymes at polythiophene-modified electrode surfaces. *Adv Mater* 1996;8:219–222.

78. Chen J, Burrell AK, Collis GE, Officer DL, Swiegers GF, Too CO, Wallace GG. Preparation, characterization and biosensor application of conducting polymers based on ferrocene substituted thiophene and terthiophene. *Electrochim Acta* 2002;47:2715–2724.

79. Wang J, Musameh M. Carbon-nanotubes doped polypyrrole glucose biosensor. *Anal Chim Acta* 2005;539:209–213.

80. Mariotti MP, Riccardi CDS, Fertonani FL, Yamanaka H. Strategies for developing NADH detector based on Meldola blue in different immobilization methods: a comparative study. *J Braz Chem Soc* 2006;17:689–696.

81. Dai ZH, Liu FX, Lu GF, Bao JC. Electrocatalytic detection of NADH and ethanol at glassy carbon electrode modified with electropolymerized films from methylene green. *J Solid State Electrochem* 2008;12:175–180.

82. Ates M, Sarac AS. Conducting polymer coated carbon surfaces and biosensor applications. *Prog Org Coat* 2009;66:337–358.

83. Raitman OA, Katz E, Buckmann AF, Willner I. Integration of polyaniline/poly(acrylic acid) films and redox enzymes on electrode supports: an *in situ* electrochemical/surface plasmon resonance study of the bioelectrocatalyzed oxidation of glucose or lactate in the integrated bioelectrocatalytic systems. *J Am Chem Soc* 2002;124:6487–6496.

84. Lee S, Choi B, Tsutsumi A. Polyaniline/carboxydextran–gold hybrid nanomaterials as a biofuel cell electrode platform. *J Chem Eng Jpn* 2009;42:596–599.

85. Wen D, Deng L, Zhou M, Guo SJ, Shang L, Xu GB, Dong SJ. A biofuel cell with a single-walled carbon nanohorn-based bioanode operating at physiological condition. *Biosens Bioelectron* 2010;25:1544–1547.

86. Karyakin AA, Strakhova AK, Karyakina EE, Varfolomeyev SD, Yatsimirsky AK. The electrochemical polymerization of methylene blue and bioelectrochemical activity of the resulting film. *Bioelectrochem Bioenerg* 1993;32:35–43.

87. Arai G, Matsushita M, Yasumori I. Electrochemical oxidation of nicotinamide adenine dinucleotide (NADH) with quinonoid polymer modified electrode. *Nippon Kagaku Kaishi* 1985;5:894–897.

88. Pariente F, Lorenzo E, Tobalina F, Abruna HD. Aldehyde biosensor based on the determination of NADH enzymically generated by aldehyde dehydrogenase. *Anal Chem* 1995;67:3936–3944.

89. Pariente F, Tobalina F, Moreno G, Hernandez L, Lorenzo E, Abruna HD. Mechanistic studies of the electrocatalytic oxidation of NADH and ascorbate at glassy carbon modified with electrodeposited films derived from 3,4-dihydroxybenzaldehyde. *Anal Chem* 1997;69:4065–4075.

90. Zhou DM, Fang HQ, Chen HY, Ju HX, Wang Y. The electrochemical polymerization of methylene green and its electrocatalysis for the oxidation of NADH. *Anal Chim Acta* 1996;329:41–48.

91. Silber A, Hampp N, Schuhmann W. Poly(methylene blue)-modified thick-film gold electrodes for the electrocatalytic oxidation of NADH and their application in glucose biosensors. *Biosens Bioelectron* 1996;11:215–223.

92. Cai CX, Xue KH. The effects of concentration and solution pH on the kinetic parameters for the electrocatalytic oxidation of dihydronicotinamide adenine dinucleotide (NADH) at

glassy carbon electrode modified with electropolymerized film of toluidine blue O. *Microchem J* 2000;64:131–139.

93. Svoboda V, Cooney MJ, Rippolz C, Liaw BY. *In situ* characterization of electrochemical polymerization of methylene green on platinum electrodes. *J Electrochem Soc* 2007;154: D113–D116.

94. Li NB, Duan JP, Chen GN. Electrochemical polymerization of azure blue II and its electrocatalytic activity toward NADH oxidation. *Chin J Chem* 2003;21:1191–1197.

95. Klotzbach T, Watt M, Ansari Y, Minteer SD. Effects of hydrophobic modification of chitosan and Nafion on transport properties, ion exchange capacities, and enzyme immobilization. *J Membr Sci* 2006;282(1–2):276–283.

96. Svoboda V, Cooney M, Liaw BY, Minteer S, Piles E, Lehnert D, Calabrese Barton S, Rincón R, Atanassov P. Standardized characterization of electrocatalytic electrodes. *Electroanalysis* 2008;20:1099–1109.

97. Cooney MJ, Petermann J, Lau C, Minteer SD. Characterization and evaluation of hydrophobically modified chitosan scaffolds: towards design of enzyme immobilized flow-through electrodes. *Carbohydr Polym* 2009;75:428–435.

98. Sokic-Lazic D, Minteer SD. Citric acid cycle biomimic on a carbon electrode. *Biosens Bioelectron* 2008;24:939–944.

99. Sokic-Lazic D, Minteer SD. Pyruvate/air enzymatic biofuel cell capable of complete oxidation. *Electrochem Solid State Lett* 2009;12:F26–F28.

100. Addo PK, Arechederra RL, Minteer SD. Evaluating enzyme cascades for methanol/air biofuel cells based on NAD$^+$-dependent enzymes. *Electroanalysis* 2010;22: 807–812.

101. Malinauskas A, Niaura G, Bloxham S, Ruzgas T, Gorton L. Electropolymerization of preadsorbed layers of some azine redox dyes on graphite. *J Colloid Interface Sci* 2000;230:122–127.

102. Rincón RA, Artyushkova K, Mojica M, Germain MN, Minteer SD, Atanassov P. Structure and electrochemical properties of electrocatalysts for NADH oxidation. *Electroanalysis* 2010;22:799–806.

103. Rincón RA, Lau C, Luckarift HR, Garcia KE, Adkins ER, Johnson GR, Atanassov P. Enzymatic fuel cells: integrating flow-through anode and air-breathing cathode into a membrane-less biofuel cell design. *Biosens Bioelectron* 2011;27:132–136.

104. Rincón RA, Lau C, Garcia KE, Atanassov P. Flow-through 3D biofuel cell anode for NAD$^+$-dependent enzymes. *Electrochim Acta* 2011;56:2503–2509.

105. de los Santos Álvarez N, Ortea PM, Pañeda AM, Castañón MJL, Ordieres AJM, Blanco PT. A comparative study of different adenine derivatives for the electrocatalytic oxidation of β-nicotinamide adenine dinucleotide. *J Electroanal Chem* 2001;502:109–117.

106. Lau C, Cooney MJ, Atanassov P. Conductive macroporous composite chitosan–carbon nanotube scaffolds. *Langmuir* 2008;24:7004–7010.

107. Anthony C. The pyrroloquinoline quinone (PQQ)-containing quinoprotein dehydrogenases. *Biochem Soc Trans* 1998;26:413–417.

108. Anthony C. Pyrroloquinoline quinone (PQQ) and quinoprotein enzymes. *Antioxid Redox Signal* 2001;3:757–774.

109. Duine JA. PQQ and quinoproteins: an important novel field in enzymology. *Antonie van Leeuwenhoek* 1989;56:3–12.

110. Oubrie A. Structure and mechanism of soluble glucose dehydrogenase and other PQQ-dependent enzymes. *Biochim Biophys Acta* 2003;1647:143–151.

111. Heller A, Feldman B. Electrochemical glucose sensors and their applications in diabetes management. *Chem Rev* 2008;108:2482–2505.

112. Prust C, Hoffmeister M, Liesegang H, Wiezer A, Fricke WF, Ehrenreich A, Gottschalk G, Deppenmeier U. Complete genome sequence of the acetic acid bacterium *Gluconobacter oxydans*. *Nat Biotechnol* 2005;23:195–200.

113. Oubrie A, Dijkstra BW. Structural requirements of pyrroloquinoline quinone dependent enzymatic reactions. *Protein Sci* 2000;9:1265–1273.

114. Ivnitski D, Atanassov P, Apblett C. Direct bioelectrocatalysis of PQQ-dependent glucose dehydrogenase. *Electroanalysis* 2007;19:1562–1568.

115. Flexer V, Durand F, Tsujimura S, Mano N. Efficient direct electron transfer of PQQ–glucose dehydrogenase on carbon cryogel electrodes at neutral pH. *Anal Chem* 2011;83:5721–5727.

116. Treu BL, Arechederra RL, Minteer SD. Bioelectrocatalysis of ethanol via PQQ-dependent dehydrogenases utilizing carbon nanomaterial supports. *Nanosci Nanotechnol* 2009;9:2374–2380.

117. Treu BL, Minteer SD. Isolation and purification of PQQ-dependent lactate dehydrogenase from *Gluconobacter* and use for direct electron transfer at carbon and gold electrodes. *Bioelectrochemistry* 2008;74:73–77.

118. Ferapontova EE, Gorton L. Direct electrochemistry of heme multicofactor-containing enzymes on alkanethiol-modified gold electrodes. *Bioelectrochemistry* 2005;66:55–63.

119. Yuhashi N, Tomiyama M, Okuda J, Igarashi S, Ikebukuro K, Sode K. Development of a novel glucose enzyme fuel cell system employing protein engineered PQQ glucose dehydrogenase. *Biosens Bioelectron* 2005;20:2145–2150.

5

ANODIC BIOELECTROCATALYSIS: FROM METABOLIC PATHWAYS TO METABOLONS

SHUAI XU, LINDSEY N. PELSTER, MICHELLE RASMUSSEN, AND SHELLEY D. MINTEER

Departments of Chemistry and Materials Science and Engineering, University of Utah, Salt Lake City, UT, USA

5.1 INTRODUCTION

Enzymatic biological fuel cells (BFCs) have expanded the applications that previously used conventional fuel cells. BFCs use more diverse potential fuels and offer the opportunity to use more complex molecules as energy sources at low operation temperatures, which are great advantages over conventional fuel cells. Anodic bioelectrocatalysis is the study of enzymes responsible for oxidation of fuels in BFCs, where chemical energy is converted to electrical energy upon the oxidation of the fuel. This makes bioanode design an important part of enzymatic BFC development. This chapter will give an overview of the research on anodic bioelectrocatalysis as it has expanded over the last decade and will also discuss the development of enzymatic BFCs using a variety of fuels.

5.2 BIOLOGICAL FUELS

Although traditional fuel cells have long been studied using fuels consisting of simple molecules such as hydrogen gas, methane, and methanol [1], studies on enzymatic

Enzymatic Fuel Cells: From Fundamentals to Applications, First Edition. Edited by Heather R. Luckarift, Plamen Atanassov, and Glenn R. Johnson.

BFCs have opened a window to a much larger selection of fuels. Oxidoreductase enzymes in living cells play a key role in breaking down a variety of substrates in metabolic pathways, such as the citric acid cycle (Krebs cycle), glycolysis, and fatty acid degradation. Use of these enzymes in BFCs allows for the metabolic substrates to be considered as possible fuel choices. In the past several years, the diversity of fuels that have been used in enzymatic BFCs has expanded from glucose and ethanol to fructose, glycerol, pyruvate, lactate, and many others, as shown in Figure 5.1.

When choosing the appropriate fuel for a BFC, researchers first consider the theoretical energy density of the fuel, because this provides a first approximation of the maximum energy density of a BFC or biobattery. However, it is important to keep in mind that most reported BFCs only use one enzyme to cause partial oxidation of each of the fuels, so most of the energy density is not converted to electricity. For example, most reported glucose BFCs employ only one enzyme, glucose oxidase (GOx) or glucose dehydrogenase (GDH), to oxidize glucose to gluconolactone [2,3]. Although glucose has a high energy density itself, it has a relatively low energy density when used in an enzymatic BFC with a single enzyme catalyst on the anode. In 2012, the Minteer research group reported a glucose BFC

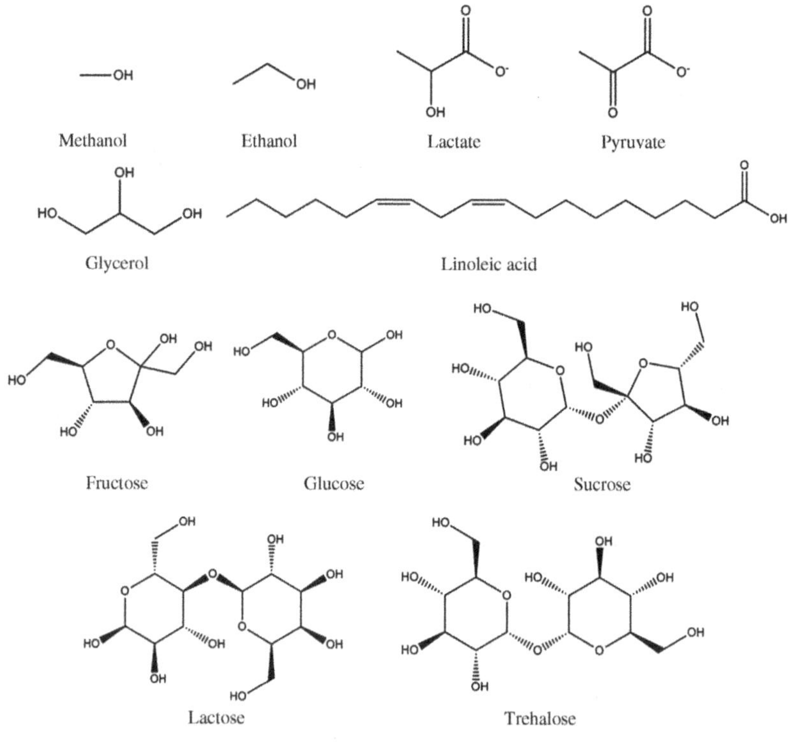

FIGURE 5.1   Structures of fuels that have been studied with biological fuel cells.

that employs a six-enzyme cascade to completely oxidize glucose to carbon dioxide and water, thereby increasing the amount of energy that can be converted per molecule of glucose fuel [4].

Application is another important factor when choosing a fuel. Glucose and lactate are logical choices for implantable devices, since these fuels are found in relatively high concentrations in the bloodstream [5]. Portable electronic devices require fuels that are easy to package and can be produced in high quantities, so glycerol, ethanol, and methanol are good choices for these types of applications.

Toxicity of the fuel and by-products along the pathway are also factors to consider. For instance, methanol is more toxic than ethanol, which is more toxic than glycerol. When considering by-products of these fuels, it would be important to compare the relative toxicities of formaldehyde (methanol oxidation product) with those of acetaldehyde (ethanol oxidation product) and glyceraldehyde (glycerol oxidation product). Glycerol might seem to be the best fuel choice because it is the least toxic of the three; nevertheless, the toxicity of glyceraldehyde must be considered. It is important to examine this issue when developing a realistic BFC.

Further considerations include the physical and chemical properties of the fuel, its by-products, and any potential ability to interfere with the cathodic chemistry. When thinking about the physical and chemical properties, researchers are usually interested in (1) whether the separator between the anode and the cathode is stable in the fuel and by-product solution, (2) the solubility of the fuel in water, (3) whether fuel evaporation will be an issue, and (4) the viscosity of the fuel. The first three points affect the actual amount of usable fuel, which further defines the energy density of the BFC, and the viscosity has an important effect on the mass transfer of the fuel to the electrode surface. Fuel tolerance is a special issue for BFC designs where fuel may enter the cathode half-cell. The cathodic enzyme, for example, could be denatured or inhibited by the presence of the fuel or its by-products.

5.3 PROMISCUOUS ENZYMES VERSUS MULTIENZYME CASCADES VERSUS METABOLONS

5.3.1 Promiscuous Enzymes

One of the practical attributes of enzymes is the specificity to individual molecular substrates. However, not all enzymes are specific to a single substrate. Enzymes that are able to catalyze reactions with different substrates or different transition states are given the label "promiscuous." It has been discussed that enzymes come from a common evolutionary starting point and through time have evolved in such a way as to preserve catalytic mechanisms that lead to increased selectivity in the present day [6]. By using enzymes with the ability to catalyze multiple substrates, it may be possible to use fewer enzymes in a biological pathway or use a higher energy density fuel mixture with the enzyme [4]. Alcohol dehydrogenase (ADH) catalyzes reversible reactions with primary and secondary alcohols, so a variety of fuels are options for this promiscuous enzyme. Evolutionary changes in proteins occur over a long time period, but protein engineering allows for more rapid changes to protein sequence or structure. Single changes near or in the active site have been used to increase activity or add new activity in an enzyme [7]. Such changes may lead to possibilities for improved and/or complete oxidation of fuels to produce the highest power densities.

5.3.2 Multienzyme Cascades

Enzymes typically catalyze two-electron oxidation reactions. Because complex fuels typically require more than two electrons to be completely oxidized, and because most enzymes are not promiscuous, multiple sequential enzymes (commonly termed *enzyme cascades*) are needed to completely oxidize a fuel to a small-molecule final product. An example of this was first demonstrated using methanol as fuel by Palmore et al. in 1998 [8]. They showed that ADH, aldehyde dehydrogenase (ALDH), and formate dehydrogenase could be used together as an enzyme cascade to completely oxidize methanol to carbon dioxide. Each of the three enzymes catalyzed a single two-electron oxidation step to form the substrate for the next enzyme in the cascade. More recently, enzyme cascades have surfaced in BFCs due to the fact that important metabolic enzymes, such as those involved in the citric acid cycle, glycolysis, and fatty acid degradation, are used by living cells for complete oxidation of substrates/fuels, making them attractive for use in bioanodes. Each of these pathways involves cofactors and substrates that must be in a working concentration and regenerated to avoid bottlenecks that decrease current densities.

5.3.3 Metabolons

Organisms have a seemingly disordered design but, in fact, are organized to make the most efficient use of substrates and produce the most energy for the cell in the entire organism. Metabolons were first named by Srere in 1987 and are defined as structures containing enzymes that participate in sequential catalyzed reactions in close proximity to each other, that is, multienzyme complexes, multifunctional proteins, and large groups of complex enzymes together [9]. Evidence has been shown that citric acid cycle enzymes have protein–protein interactions close to the inner membrane of the mitochondria, providing substrate to the electron transport chain [10]. The close proximity of these enzymes leads to better substrate channeling between catalytic centers of the enzymes [11], promoting an efficiency that is valued for BFCs. In 2010, Moehlenbrock et al. cross-linked citric acid cycle enzymes to form a metabolon complex that showed an increase in current and power density on an anode with pyruvate as fuel [12]. Advances in studying proteins *in vivo* and *in vitro* allow for development of enzymatic BFCs that contain one or more enzymes with multiple catalytic sites, making possible the most efficient conversion of substrate fuel to final oxidized product.

5.4 DIRECT AND MEDIATED ELECTRON TRANSFER

When discussing the transfer of electrons from the enzyme active site to the electrode surface, thus generating catalytic current, there are two types of electron transfer mechanisms: mediated electron transfer (MET) and direct electron transfer (DET) [13]. Most oxidoreductase enzymes that have been commonly used in BFC development are unable to promote the transfer of electrons themselves because of the long electron transfer distance between the enzyme active site and the electrode surface; as a result, DET is slow. In such a case, a redox-active compound is incorporated to allow for MET. In this approach, a small molecule or redox-active polymer participates directly in the catalytic reaction by reacting with the enzyme or its cofactor to become oxidized or reduced and diffuses to the electrode surface, where rapid electron transfer takes place [14]. Frequently, this redox molecule is a diffusible coenzyme or cofactor for the enzyme. Characteristic requirements for mediator species include stability and selectivity of both the oxidized and reduced forms of the species. The redox chemistry for the chosen mediator is to be reversible and with minimal overpotential [15].

Mediators can be polymerized on the electrode surface prior to enzyme immobilization, co-immobilized with enzyme, or simply added to the fuel solution. Common mediators used in BFC applications include low molecular weight, polymerizable, organic dyes such as methylene green, phenazines, and azure dyes, along with other redox-active compounds such as ferrocene, ferrocene derivatives, and conductive salts [14]. These mediators are often required for nicotinamide adenine dinucleotide (NAD^+)- and flavin adenine dinucleotide (FAD)-dependent enzymes, such as ADH, ALDH, and GOx. MET has been achieved at both cathodic and anodic interfaces through solution-phase mediators and mediators immobilized in various ways with or near the enzymes themselves [16,17]. However, these mediated systems do have drawbacks in that the species used to assist electron transfer are often not biocompatible, have short lifetimes themselves, or cause large potential losses. Table 5.1 lists common enzyme cofactors that can mediate or undergo DET with an enzyme on the electrode.

A mediator would not be needed if the enzymes used in BFCs are capable of DET via the active site of the enzyme. Several enzymes capable of DET have been reported [18,19]. Many of these enzymes contain redox-active metal centers, such as iron–sulfur groups, heme groups, and metallic centers, that perform the catalytic transfer of electrons. These enzymes convert the chemical signal directly to an electrical signal through the transfer of charge to the redox center, which is in turn

TABLE 5.1 List of Cofactor/Coenzyme Names and Abbreviations

Cofactor/Coenzyme Name	Abbreviation
Nicotinamide adenine dinucleotide	NAD^+ (NADH)
Nicotinamide adenine dinucleotide phosphate	$NADP^+$ (NADPH)
Nicotinamide mononucleotide	NMN (NMNH)
Flavin adenine dinucleotide	FAD ($FADH_2$)
Pyrroloquinoline quinone	PQQ

capable of transferring the charge to another molecule or directly to the electrode surface [20]. Many pyrroloquinoline quinone (PQQ)-dependent enzymes, as an example, contain one or more heme groups that are capable of existing in several redox states and accept resultant electrons generated through the oxidation of substrates, such as alcohol, aldehyde, and glucose. DET eliminates the need for mediator molecules that can be nonselective and add fuel cell resistance, which limits the optimal performance of the fuel cell.

5.5 FUELS

5.5.1 Hydrogen

Hydrogen oxidation has been widely studied for fuel cells [21]. Most conventional hydrogen fuel cells use platinum anodes for the catalysis of hydrogen oxidation, but platinum is quite expensive and finding an alternative catalyst is a large area of current research. One alternative that has been studied is the use of hydrogenases. Hydrogenases catalyze the oxidation of hydrogen with an electron acceptor, such as O_2, NAD^+, nicotinamide adenine dinucleotide phosphate ($NADP^+$), cytochrome c, and quinones, being reduced in the process. These enzymes are divided into several categories based on their catalytic sites: Ni–Fe, Fe–Fe, and Fe. Most hydrogenases are sensitive to oxygen and carbon dioxide content in the fuel, which can be an issue when developing this type of fuel cell [22].

In 2001, Ikeda and coworkers developed the first hydrogenase fuel cell using a bacterial cell with high hydrogenase activity as an anode with a bilirubin oxidase (BOx) cathode [23]. Several groups have studied multiple anaerobic and aerobic bacterial isolates to find an enzyme before discovering a thermophilic bacterial hydrogenase that is stable and tolerant to inhibitors, with a Ni–Fe catalytic center and a Fe–S site [21,24]. In 2006, Vincent et al. used purified membrane-bound Ni–Fe hydrogenase from *Ralstonia metallidurans* to make an anode which, when coupled with a laccase cathode, was able to power a wristwatch for 24 h [25]. This enzyme was active with just 3% H_2 in the chamber with sufficient O_2 tolerance. The group also used a hydrogenase from *Ralstonia eutropha* for a membraneless BFC with a laccase cathode; this cell was operational with ambient oxygen and CO-contaminated fuel [26]. Luo et al. found that the addition of carbon nanotubes (CNTs) with bacterial hydrogenase increased the electrochemical connection and led to DET to the electrode surface [27]. In 2010, Wait et al. studied a BFC with and without a Nafion membrane between the hydrogenase anode and the BOx cathode, showing the tolerance of the enzyme to oxygen [28]. Although hydrogenase catalytic activity can match platinum with hydrogen fuel, enzyme stability and tolerance to oxygen and carbon dioxide in air are problems that still need to be solved for efficient hydrogen oxidation with enzymes.

5.5.2 Ethanol

Ethanol is oxidized by NAD^+-dependent ADH or alcohol oxidase into acetaldehyde. One of the first ethanol BFCs was produced by Yahiro et al. in 1964, although it did not produce any power [29]. In 2004, Akers et al. immobilized the enzyme in hydrophobically modified Nafion to prolong the lifetime of the enzyme on an anode [30]. The electropolymerized methylene green enzymatic anode, coupled with a cathode (ELAT electrode with 20% platinum on Vulcan XC-72), produced a power density of $1\,mW\,cm^{-2}$ for ADH and $2.04\,mW\,cm^{-2}$ for two enzymes, ADH and ALDH. The anode was active for more than 1 week with a slight decrease in activity. Later, in 2006, Topcagic and Minteer developed a membraneless fuel cell with an anode consisting of ADH and ALDH in hydrophobically modified Nafion and a BOx cathode with a $Ru(bpy)_3{}^{2+}$ mediator [31]. The cell produced a power density of $0.4\,mW\,cm^{-2}$ with an open-circuit potential (OCP) of 0.51 V. This cell design also illustrates the use of a membraneless BFC to reduce the cost by eliminating the need for a proton exchange membrane.

In 2002, Ramanavicius and coworkers investigated the use and stability of PQQ-dependent enzymes for DET processes on biosensor electrodes. The PQQ-dependent enzymes eliminate the need for NAD^+ in the fuel and at the electrode [32]. Three years later, they developed an anode with PQQ-dependent ADH and GOx capable of producing $2.6\,\mu A\,cm^{-2}$ [33]. When coupled with an alcohol oxidase/microperoxidase cathode, this fuel cell produced a maximum power density of $1.5\,\mu W\,cm^{-2}$. Although the power density is low, this system eliminated the challenge of substrate contamination by requiring only ethanol as fuel [34]. In 2008, Sokic-Lazic and Minteer fabricated a multienzyme cascade based on the citric acid cycle, using ADH and ALDH to consume ethanol as the fuel [35]. With the addition of ADH and ALDH and S-acetyl coenzyme A (CoA) synthetase, the six dehydrogenases immobilized on the electrode show a dramatic ninefold increase in power and current densities to $1\,mW\,cm^{-2}$ and $3.60\,mA\,cm^{-2}$, respectively. Even though not all the enzymes employed are dehydrogenases, they continue the oxidation pathway for the fuel, preventing inhibition of enzymes by the build up of product that would normally occur in one- and two-enzyme systems.

In 2009, Treu et al. purified PQQ-dependent ADH from *Gluconobacter* and immobilized the enzyme with PQQ-dependent ALDII and high surface area nanotube supports on Toray paper with a hydrophobically modified Nafion layer to bind the mixture to the surface [36]. With this type of electrode fabrication, the PQQ-dependent enzymes were able to achieve better electrical contact on the electrodes. This biological fuel anode with carboxylated single-walled CNTs (SWCNTs) achieved the best result with a power density of $1.25\,mA\,cm^{-2}$, a current density of $0.252\,mA\,cm^{-2}$, and an OCP of 1.15 V. Short and long carboxylated multiwalled nanotubes were also studied and all nanotube anodes showed increased performance compared with carbon paper anodes without nanotubes.

Another demonstration of an ethanol BFC was shown in 2010 with the development of an ADH BFC that could work with wine as fuel [37]. This fuel cell had a power density of $1.56\,mW\,cm^{-2}$ and a current density of $2.07\,mA\,cm^{-2}$ in ethanol. The fuel cell displayed an even higher activity in wine with a power density of $1.78\,mW\,cm^{-2}$ and a current density of $3.21\,mA\,cm^{-2}$. Aquino Neto et al. developed a layer-by-layer technique in 2011 with polyamidoamine (PAMAM) dendrimers for one- and two-enzyme systems for the oxidation of ethanol, as shown in Figure 5.2 [38]. The

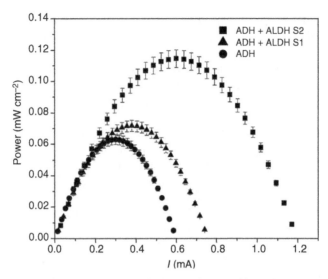

FIGURE 5.2   Representative power density curves for both bioanode samples containing a double enzymatic system prepared with 36 bilayers. (Reproduced with permission from Ref. [38]. Copyright 2011, Elsevier.)

intercalation of ADH and ALDH between PAMAM layers led to an increase in power density of $0.12\,mW\,cm^{-2}$, believed to be caused by the improvement of diffusion of substrates between the layers. Also in 2011, Rincón et al. used a flow cell with ADH to characterize the system [39]. This research into fuel utilization may prolong the usable lifetime of the cell because of the continuous supply of fuel and could possibly lead to integration into other areas for BFC applications.

5.5.2.1 Protein Engineering   With advanced biotechnology, it is possible to carefully design enzymes and cofactors that will give the most promising result. In 2010, Campbell et al. used rational design to change the specificity of an ADH to have a higher affinity for nicotinamide mononucleotide (NMNH) rather than NADH, broadening the specificity for the cofactor [40]. By engineering the active site of an NADH-dependent ADH from *Pyrococcus furiosus*, Campbell et al. were able to substitute NMN^+ in place of NAD^+ and improve BFC performance [41]. Although the OCP dropped from 0.642 to 0.593 V, the current density increased from 16.1 to $22.8\,\mu A\,cm^{-2}$, suggesting increased mass transfer for the minimal cofactor. With continuing advances in proteomics, protein design and engineering could lead to great increases in enzymatic BFC performance.

5.5.3   Methanol

Similar techniques have been used for BFCs using methanol as the fuel. Davis et al. used ADH in solution to oxidize methanol in a BFC in 1983 [42]. This BFC had a low power output but was used to compare the performance with different cation membranes for separation of anodic and cathodic compartments and to investigate the use of other primary alcohols. In 1998, Palmore et al. developed an anode with NAD^+-dependent ADH, ALDH, and formate dehydrogenase using diaphorase and benzyl viologen for NADH regeneration at the electrode [8]. The three-enzyme fuel cell produced a maximum power density of 0.68 mW cm^{-2} at 0.49 V. With diaphorase and three dehydrogenases, methanol was oxidized completely to carbon dioxide to produce more power than just one enzyme. In 2010, Addo et al. immobilized the enzymes on the electrode to prolong the lifetime of the enzymes in the fuel cell [16]. The enzymes were immobilized in modified Nafion at carbon paper electrodes for the BFC assembly. The concentrations of NAD^+ and methanol, as well as the pH, were optimized to produce the best results. The cell produced a maximum current density of 845 μA cm^{-2} and a power density of 261 μW cm^{-2} with an OCP of 1.21 V. This anode setup was modified with the addition of carbonic anhydrase IV to reverse the oxidation, reducing carbon dioxide to methanol at the electrode. The addition of carbonic anhydrase increased the rate at which carbon dioxide was reduced back to formate and eventually to methanol as a proof of concept for the reversal of catalytic enzyme systems [43].

5.5.4 Methane

Methane is regularly used as a fuel in microbial fuel cells and traditional metal-catalyzed fuel cells, but it has yet to be studied with enzymatic BFC systems. Methane monooxygenase is a membrane-bound enzyme that reduces methane to methanol with low redox potentials (<100 mV) [44]. It is well characterized, and one source of the enzyme has a copper center that would enable DET at the electrode [45]. This enzyme could be added to a methanol anode, thus allowing the use of methane as a fuel in a gas-permeable fuel cell.

5.5.5 Glucose

An overwhelming portion of BFC research involves glucose as the anodic fuel. There are many reasons why glucose draws so much interest. The first reason is its high energy density (2805 kJ mol^{-1} if oxidized to carbon dioxide), which indicates a potential for designing a biological fuel cell with energy density in excess of lithium ion batteries. The second reason is its abundance in Nature (as a main product of photosynthesis) and its ubiquity as a fuel in biology. Glucose is used as an energy source by most organisms ranging from bacteria to humans, which makes it a possible fuel for implantable BFCs in humans. These properties, coupled with its low volatility, nontoxicity, and low cost, support glucose as a promising fuel.

5.5.5.1 Early Work The first application of enzyme-based BFCs using glucose as fuel was reported in 1962 when Davis and Yarbrough employed the enzyme GOx to oxidize glucose at the anode, which generated small current densities [46]. Similarly, in 1964, Yahiro et al. reported a glucose/O_2 fuel cell that also used GOx as the catalyst [29]. The OCP of that BFC was 0.3–0.5 V, and the introduction of elemental iron to the anode compartment increased the potential to 0.75 V and generated a current density of $\sim 17 \, \mu A \, cm^{-2}$.

Exciting advances have been made since that time. In 1975, Weibel and Dodge used dichloroindophenol as a mediator to aid the electron transfer to the electrode surface and successfully reached 100% faradaic efficiency [47]. In 1985, Persson et al. reported a GOx-catalyzed glucose/O_2 biological fuel cell using N,N-dimethyl-7-amino-1,2-benzophenoxazinium ion (Meldola's blue or MB^+) as mediator [48]. MB^+ was chosen over other mediators because of its higher stability and irreversible adsorption to graphite. The fuel cell delivered a current density of $0.2 \, mA \, cm^{-2}$ at 0.8 V for more than 8 h with a simulated oxygen cathode.

5.5.5.2 Recent Work

Poor power density and short lifetimes are two key problems in the application of enzymatic BFCs. To address these issues, many efforts and significant improvements have been made during the last 15 years. In 1999, Katz et al. reported a BFC with GOx and microperoxidase-11 monolayer assembled on gold electrodes (0.4 cm diameter) and applied in a glucose/cumene fuel mixture [49]. A power output of $520 \, \mu W$, which corresponded to $4.1 \, mW \, cm^{-2}$, was observed. Five years later, Heller reported a miniature glucose/O_2 BFC with improved power density per unit volume [50]. In Heller's work, GOx was immobilized on the surface of two $7 \, \mu m$ carbon fibers by being entrapped in an osmium redox polymer. A power output of $4.3 \, \mu W \, cm^{-2}$ was achieved with a total fiber volume of $0.0026 \, mm^3$, representing $1.65 \, mW \, mm^{-3}$.

From the beginning, the short lifetimes of enzymatic BFCs have been a concern for their practical application. Several factors regulate the lifetime of BFCs; however, in most cases, the stability of the biocatalysts themselves is the determining factor. Most enzymatic BFCs are operable for only a few days [49]. Researchers have been exploring methods to improve the immobilization techniques to extend the lifetime of enzymes. In 2002, Mano et al. reported a miniature BFC with GOx and BOx immobilized in an osmium redox polymer that functioned for 20 days at 37 °C (estimated by extrapolating the power decay curve reported in the reference) [51]. In 2004, promising results were reported by Moore et al. using tetrabutylammonium bromide (TBAB)-modified Nafion membranes to entrap dehydrogenases such as GDH. The half-life of the native parent enzyme is only 7–8 h in solution, whereas an active lifetime of more than 45 days was achieved after immobilization [52].

In recent years of glucose enzymatic biological fuel cell research, one of the most significant advances is electrode modification by employing CNT. Figure 5.3 shows the modifications by Saleh et al. with CNTs [53,54]. Several research groups have addressed the application of a SWCNT hybrid system [55,56]. Surface-assembled GOx is in good electric contact with the electrode due to the application of SWCNTs, which act as conductive nanoneedles that electrically wire the enzyme's active site to the transducer surface [55,56]. Other studies have been reported on improving electrochemical and electrocatalytic behavior and fast electron transfer kinetics of

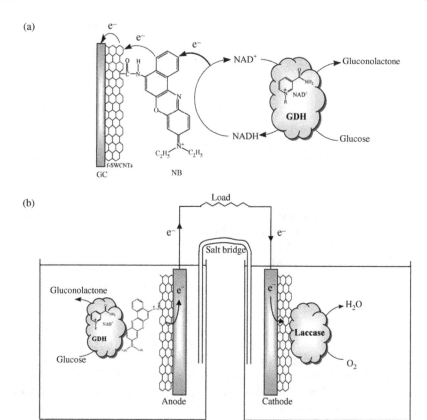

FIGURE 5.3 Scheme of (a) the GDH-based composite electrode for the electrocatalytic oxidation of glucose and (b) the setup of a glucose/O_2 biological fuel cell composed of the GDH/NB/f-SWCNTs/GC bioanode and the laccase/MG-SWCNTs/GC biocathode. (Reproduced with permission from Ref. [54]. Copyright 2011, Elsevier.)

CNT [53]. Table 5.2 summarizes the work done between 2001 and 2011 on MET and DET performed with glucose anodes.

5.5.5.3 Miniature and Implantable Glucose BFCs
Because blood glucose concentration is high in humans, the development of miniature glucose BFCs offers a great opportunity to serve as long-term power sources for implantable devices where frequent battery replacement is not practical. The first miniature glucose BFC was reported by Chen et al. in 2001 [57]. In their research, a glucose/O_2 BFC consisted of two $7\,\mu m \times 2\,cm$ electrocatalyst-coated carbon fibers operating at ambient temperature in a pH 5 solution. The power density of the cell was $64\,\mu W\,cm^{-2}$ at $23\,°C$ and $137\,\mu W\,cm^{-2}$ at $37\,°C$. In 2002, Mano et al. reported a miniature glucose BFC with the same carbon fibers operating in a physiological buffer [58]. In a week of operation, the cell generated $1.5\,\mu W$ of electrical energy while passing $1.7\,C$ of charge. In 2003, the same research group reported a miniature compartmentless glucose/O_2 BFC

TABLE 5.2 Summary of Single-Enzyme Glucose Biological Fuel Cells Developed After 2000

Enzymes	Electrode	Electron Transfer	OCV (V)	Current Density ($\mu A\,cm^{-2}$)	Reference
GOx/laccase	Carbon fiber	MET	0.8	64	[57]
GOx/BOx	Carbon fiber	MET	0.84	432	[58]
GOx/BOx	Glassy carbon	MET	0.44	58	[59]
GOx/BOx	Carbon fiber	MET	0.68	50	[60]
GOx/BOx	Carbon fiber	MET	0.8	440	[61]
GOx/BOx	Carbon fiber	MET	0.63	244	[62]
GDH/PDMS	Platinum	DET	0.8	11 000	[63]
GOx/laccase	Silicon/SWCNTs	DET	NA	30	[55]
GOx/laccase	Gold/SWCNTs	DET	0.46	960	[56]
PQQ-GDH/BOD	Gold/SWCNTs	DET	0.60	200	[53]
GDH/NB	Glassy carbon/ SWCNTs	DET	0.35	100	[54]

operating in a grape, which produced 2.4 μW at 0.52 V [62]. The performance of this cell was later improved to 0.78 V when operating at 37 °C in buffered fuel at pH 5 [61]. The most recent report on miniature glucose/O_2 BFCs used carbon fiber microelectrodes modified with SWCNTs. The power density for this cell reached 58 $\mu W\,cm^{-2}$ at 0.40 V [64].

The first glucose BFC implanted in a live animal was reported by Cinquin et al. in 2010 [65]. They designed a glucose bioelectrode based on composite graphite disks containing GOx and ubiquinone at the anode and polyphenol oxidase (PPO) and quinone at the cathode. Their BFC was implanted in a rat and produced a peak specific power of 24.4 $\mu W\,cm^{-3}$, reportedly greater than the power required to operate a pacemaker.

5.5.5.4 Promiscuous Enzymes for Glucose Anodes
GOx is the most commonly used glucose-oxidizing enzyme in enzymatic bioanode design; however, one of the major drawbacks with this enzyme is its anomeric selectivity. GOx only demonstrates activity on β-glucose, so in many applications GOx is employed together with mutarotase, an isomerase that catalyzes the interconversion of the α- and β-forms of D-glucose. Pyranose oxidase (PyOx) produced by white rot fungi has shown the ability to oxidize glucose and several aldopyranoses at the C-2 position to their corresponding 2-keto sugars with concomitant generation of H_2O_2 [66]. Pyranose dehydrogenase (PDH) from *Agaricus meleagris* can also oxidize glucose at the C-2 and C-3 positions [67]. In 1997, Lidén et al. reported PyOx-modified carbon paste electrodes that served as a biosensor interface to test a series of monosaccharides [68]. Since that time, several studies on the use of PyOx as an anodic BFC catalyst have been reported [69,70]. In 2010, Tasca et al. reported a glucose-fueled bioanode by

using a combination of PDH and cellobiose dehydrogenase (CDH) [71]. PDH and CDH oxidize glucose at different carbons, and electrochemical measurements revealed that the product of one enzyme can serve as a substrate for the other. The kinetic pathway analysis showed that up to six electrons could be gained from one glucose molecule through the combination of CDH and PDH, thus increasing the coulombic efficiency of the bioanode.

To reach the goal of oxidation of glucose to carbon dioxide, an enzyme cascade was used to perform the multistep oxidation. In 2011, Xu and Minteer designed a BFC that used a six-enzyme cascade to completely oxidize glucose to carbon dioxide, demonstrated in Figure 5.4 [4]. The enzyme cascade included PQQ-GDH, PQQ-gluconate dehydrogenase, PQQ-ADH, PQQ-ALDH, aldolase, and oxalate oxidase. This glucose enzymatic bioanode, coupled with an air-breathing cathode, yielded a maximum power density of $6.74\,\mu W\,cm^{-2}$. This bioanode is also capable of undergoing DET to the carbon electrode surface, eliminating the need for a mediator.

5.5.6 Sucrose

Sucrose is a disaccharide that is composed of one molecule of glucose and one molecule of fructose. It is also known as "table sugar" and is a common sweetener in the food industry. Early sucrose bioelectrochemistry research was focused on biosensors. In the 1980s, several groups reported biosensors containing invertase (INV) and GOx to obtain currents that demonstrated good linearity in the range of 10^{-6} to $10^{-3}\,M$ sucrose [72–74]. These sensors showed fairly long lifetimes (10–20 s) and good storage stability.

A large portion of sucrose BFC studies are based on microbial-catalyzed bioanodes; however, a few enzymatic sucrose BFCs were reported in recent years [75–77]. In 2009, Tam et al. reported a BFC with an "enzyme logic network" [78]. In their design, INV, GDH, ADH, and amyloglucosidase (AGS) were immobilized on an osmium-containing indium tin oxide electrode, as shown in Figure 5.5. These enzymes were activated by different combinations of chemical input signals: NADH, acetaldehyde, maltose, and sucrose. When "successful" patterns of the chemical input signals were applied, gluconic acid was produced, thus lowering the pH in the solution and activating the pH-sensitive redox polymer-modified cathode, turning the whole cell on. "Unsuccessful" patterns of the input signals resulted in the cell being switched off. In other words, the "smart" BFC is able to release power on demand, depending on the specific patterns of chemical signals. The system can switch the BFC on or off by logical processing of the chemical input signals.

5.5.7 Trehalose

Trehalose is a disaccharide most commonly found in insects and fungi. It can be broken down to two molecules of glucose by the enzyme trehalase [79]. A BFC for implantation in insects is one area of current research using trehalose. The goal is to use the power generated by such a cell to power small devices in or on the insect. In

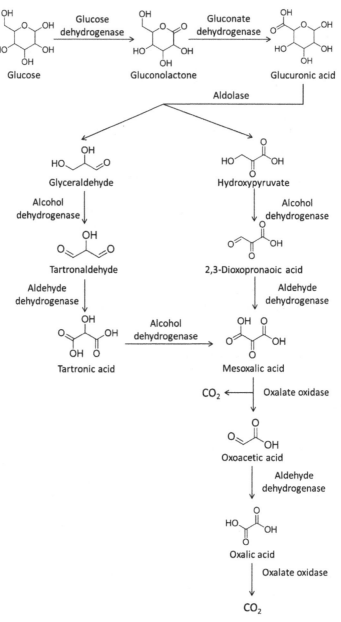

FIGURE 5.4 Schematic of the six-enzyme oxidation of glucose to carbon dioxide. (Reproduced with permission from Ref. [4]. Copyright 2012, American Chemical Society.)

2006, Heller's group developed a trehalose anode using a genetically engineered glucose 3-dehydrogenase that was able to catalytically convert trehalose to glucose, but this electrode was not tested in a BFC [80]. Recently, the Scherson group designed a trehalose–oxygen BFC that was implanted into a cockroach [81]. The anode

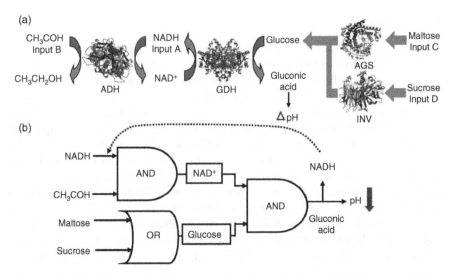

FIGURE 5.5 (a) The cascade of reactions biocatalyzed by ADH, AGS, INV, and GDH and triggered by chemical input signals: NADH, acetaldehyde, maltose, and sucrose added in different combinations. (b) The logic network composed of three concatenated gates and equivalent to the cascade of enzymatic reactions outlined in (a). (Reproduced with permission from Ref. [78]. Copyright 2009, Elsevier.)

consisted of two enzymes, GOx and trehalase, along with an osmium redox polymer to allow for MET to the carbon electrode surface. The cathode used BOx along with a second osmium redox polymer for the reduction of oxygen. This BFC was able to produce $55 \, \mu W \, cm^{-2}$ at 0.2 V and a maximum current density of $65 \, \mu A \, cm^{-2}$.

5.5.8 Fructose

As one of the three most important blood sugars along with glucose and galactose, fructose has been studied as a potential anodic fuel in the last decade. Fructose dehydrogenase (FDH) is the most common biocatalyst to be employed on these bioanodes. In 2006, Kamitaka et al. reported an FDH-adsorbed bioanode with high current density [82]. FDH was slowly adsorbed to Ketjenblack-modified glassy carbon electrodes, which, in the absence of mediator, produced an enzyme-functionalized electrode with catalytic oxidation currents as high as $10 \, mA \, cm^{-2}$ via DET processes. Soon after this, the same research group used this same enzyme adsorption technique to immobilize FDH on a carbon paper electrode and coupled the FDH anode with a laccase biocathode to construct an enzymatic BFC [83]. A maximum catalytic current density of $4 \, mA \, cm^{-2}$ was observed. In 2009, Murata et al. reported a mediator-free fructose/O_2 BFC with a BOx biocathode [84]. A maximum current density of $2.6 \, mA \, cm^{-2}$ and a maximum power density of $0.66 \, mW \, cm^{-2}$ were achieved at 360 mV. A maximum current density of $4.9 \, mA \, cm^{-2}$ and a maximum power density of $0.87 \, mW \, cm^{-2}$ were achieved at 300 mV while rotating at 1000 rpm.

5.5.9 Lactose

Lactose is a disaccharide and is most commonly found in mammals' milk, making it a practical anodic fuel for implantable BFCs or self-powered sensors for milk analysis. Most of the research on lactose BFCs employs CDH, whose natural substrate is cellobiose. However, CDH isolated from ascomycete fungi has also shown activity on other disaccharides, such as lactose, as well as some monosaccharides, including glucose. CDH exhibits both DET, through heme domains, and MET, through FAD cofactors. In 2008, Tasca et al. reported a CDH-based bioanode for BFCs [85]. By using an osmium-containing redox polymer to mediate the electron transfer, the electrocatalytic current was much higher than that obtained using DET for CDH. Later that year, Stoica et al. reported a study on the influence of osmium polymer modification and the enzyme loading on the performance of a CDH electrode for lactose oxidation [86]. Several different types of osmium redox hydrogels were tested and a laccase-based biocathode was incorporated. At pH 4.3, in the presence of 7.4 mM lactose, a maximum power density of $1.9\,\mu\text{W}\,\text{cm}^{-2}$ was obtained at 0.28 V.

5.5.10 Lactate

Lactate is a small biological molecule that functions as metabolite in the mitochondria and a precursor to pyruvate in the citric acid cycle [87]. In 1997, Bardea et al. developed a lactate BFC using NAD^+-dependent lactate dehydrogenase (LDH) [88]. They introduced a new method for enzyme immobilization that enabled better oxidation of the substrate and allowed the enzyme to have electrical contact with the electrode. Covalently linked PQQ and native NAD^+ form a monolayer on gold electrodes to induce affinity interactions with cross-linked NAD^+-dependent LDH. In 2001, Katz et al. further improved on the concept of this anode, coupling it with a cytochrome c oxidase cathode to produce a self-powered biosensor that is active only in the presence of the anode's substrate, lactate [89]. With this addition, the cell is completely dependent on substrate for voltage and current output, which is ideal for BFCs.

Lee et al. introduced a novel technique for covalently immobilizing LDH to PQQ with N-ethyl-3-(3-dimethylaminopropyl)carbodiimide hydrochloride (EDC) chemistry for a more stable connection to the electrode with NAD^+ and $CaCl_2$ in the fuel solution. The enzyme was immobilized in the presence of lactate, NAD^+, and $CaCl_2$ to avoid cross-linking the active site, which led to an increase in the power density of 26%, up to $142\,\mu\text{W}\,\text{cm}^{-2}$ [90]. In 2005, Katz further characterized the LDH anode in the presence of a magnetic field to study the magnetohydrodynamic effect on the voltage and current output of the anode. The fuel cell with a cytochrome c oxidase cathode produced higher current and power densities based on the exposure to a magnetic strength of 0.92 T [91].

In 2008, Treu and Minteer purified and characterized a PQQ-LDH from *Gluconobacter* sp. 33 that was able to undergo DET adsorbed on gold and immobilized in polymer on carbon paper electrodes [92]. This membrane-bound enzyme was cast in hydrophobically modified Nafion, allowing for the flux of lactate through the

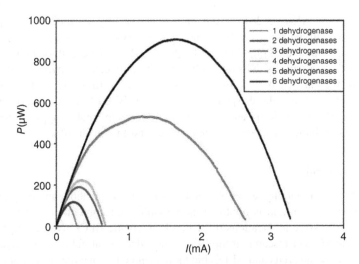

FIGURE 5.6 Representative power curves for lactate/air biological fuel cells in 500 mM sodium lactate at room temperature with different degrees of oxidation of the fuel. (Reproduced with permission from Ref. [93]. Copyright 2011, Elsevier.) (Please see the color version of this figure in Color Plates section.)

membrane and direct contact to the electrode. This anode, coupled with a cathode (ELAT electrode with 20% Pt Vulcan) and Nafion 117 membrane, produced optimal current and power densities of 0.138 mA cm^{-2} and 22.0 µW cm^{-2}. In 2011, Sokic-Lazic and Minteer immobilized all of the enzymes and cofactors involved in the citric acid cycle in hydrophobically modified Nafion as a biomimic on a carbon paper electrode [93]. This anode is able to completely oxidize lactate to carbon dioxide, as a modification of previous work done for pyruvate oxidation [35]. With six dehydrogenases, the fuel cell produces a current density of 2.7 mA cm^{-2} and a power density of 800 µW cm^{-2}, well above what a single dehydrogenase would produce on an electrode, as shown in Figure 5.6. Immobilization of enzymes with directed assembly and encapsulation provide an effort to orient the enzymes close to the electrode, enabling the chance for the highest activity and current density from an enzymatic anode.

5.5.11 Pyruvate

Pyruvate is the natural metabolic substrate in the mitochondria for the citric acid cycle. In 2009, Sokic-Lazic and Minteer incorporated the enzymes for the entire citric acid cycle onto an electrode to completely oxidize pyruvate to carbon dioxide (detected by ^{13}C nuclear magnetic resonance (NMR)), thereby releasing more electrons [94]. The fuel cell had a current density of 3.92 mA cm^{-2} and a power density of 0.93 mW cm^{-2}. Later, in 2010, Moehlenbrock et al. used citric acid cycle enzymes, cross-linked *in vivo*, to produce a metabolon formation for greater efficiency of the oxidation pathway [12]. The power density reached 24.14 µW cm^{-2} and had a current density of

92.40 $\mu A\,cm^{-2}$. Also in 2010, Treu et al. successfully purified PQQ-dependent pyruvate dehydrogenase to study its bioelectrocatalysis and use in BFCs [95]. The enzyme is similar to the PQQ-ADH and PQQ-ALDH enzymes, although this enzyme has a broad specificity for substrate. It has the ability to catalyze pyruvate along with the rest of the citric acid cycle intermediates, including fumarate, malate, and even succinate. This allows the enzyme to be useful in BFCs having a matrix of fuels. Pyruvate BFCs have the potential to be used for implantable applications, since pyruvate is an indicator of energy metabolism and the health of cells [96].

5.5.12 Glycerol

Glycerol is a potentially excellent fuel source; it is an abundant by-product of biodiesel production and is nontoxic, nonvolatile, and nonflammable [97]. One of the most impressive applications of glycerol enzymatic BFCs was accomplished in 2007 by the Minteer research group [98]. Arechederra et al. fabricated a glycerol/O_2 BFC employing an ADH and ALDH bioanode, where the enzymes were immobilized onto the electrode by co-casting with a modified Nafion polymer. This BFC was able to oxidize glycerol to mesoxalic acid, yielding power densities of up to 1.21 mW cm^{-2}. By incorporation of a third enzyme, oxalate oxidase, the three-enzyme cascade was able to oxidize glycerol all the way to carbon dioxide and the power density increased to 1.32 mW cm^{-2} [99]. The enzyme pathway for this fuel cell is shown in Figure 5.7. The dehydrogenases that Arechederra et al. employed are PQQ-dependent enzymes, containing heme groups capable of existing in several redox states and accepting resultant electrons that are generated through the oxidation of substrate, thus performing DET. In 2005, Lapenaite et al. extracted PQQ-dependent glycerol dehydrogenase (GlyDH) from *Gluconobacter* sp. 33 and immobilized the enzyme onto a graphite electrode by absorption [100]. They constructed a biosensor with this enzyme for the determination of glycerol in wines and obtained good correlations for different types of wines. These results indicate the applicability of this type of biosensor for food quality control. However, low stability of the enzymes is the main restriction in the application of this biosensor for the wine fermentation monitoring online.

5.5.13 Fatty Acids

Fatty acids are present in abundance in all living organisms and have high energy densities ideal for use in BFCs. Although other fuels have been widely studied, fatty acid pathways have not been fully incorporated into BFCs. In 2006, lipoxygenase was studied by Kerr and Minteer in a BFC with soybean oil [101]. This enzyme has broad specificity for multiple unsaturated fatty acids, making it ideal for plant oils composed of several unsaturated acids. The enzyme was immobilized within a quaternary ammonium bromide-modified Nafion membrane, which has enlarged pores to enable encapsulation of enzymes and flux of substrate or highly hydrophobic oils [102]. Lipoxygenase is also inhibited by product concentrations, so the concentration of the fuel oil is low (at 1 mM), causing low energy densities and low power densities [101].

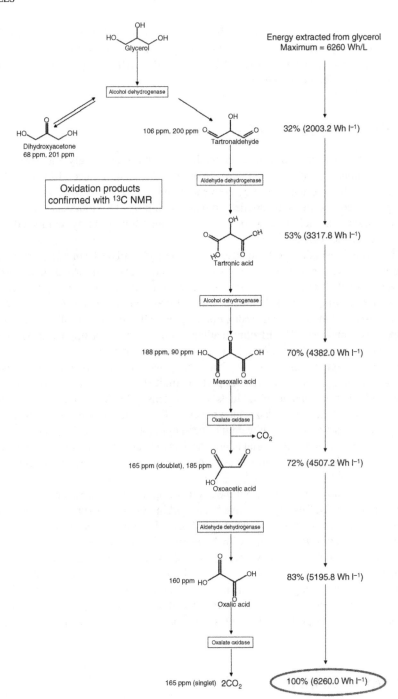

FIGURE 5.7 Enzyme cascade for the complete oxidation of glycerol to carbon dioxide. (Reproduced with permission from Ref. [99]. Copyright 2009, Wiley-VCH Verlag GmbH.)

Monooxygenase is another enzyme that requires an NADP(H) cofactor to oxidize unsaturated fatty acid bonds to a hydroxyl group. Immobilization of the flavin-containing monooxygenase-3 has not been studied for BFC applications, but it has been used in a trimethylamine biosensor [103].

5.6   OUTLOOK

In the past two decades, enzymatic bioanode design has increasingly drawn researchers' attention and expedited the improvement of BFCs. Numerous efforts have built promise into this alternative energy conversion device. Several limitations, however, have prevented enzymatic bioanodes from being employed for practical applications. These drawbacks include low power density and low stability compared with conventional fuel cells.

Improving bioanodes' performances and efficiencies will be the most important task in future studies of enzymatic anodic catalysis. Based on research carried out in the past few years, trends for improving performance rely on better electron transport methods and higher enzyme loading. Electron transport could be improved, for example, by developing novel mediators and redox polymers for MET or by controlling orientation of enzymes to improve DET. Enzyme loading techniques could be improved to increase active enzyme concentration per unit of electrode area or volume.

Future research on improving BFC efficiency and energy density will most likely be based on using enzyme cascades to perform multistep oxidation of fuels in order to convert a higher percentage of the fuels' energy density to electrical energy. Another interesting direction of research is enzyme cross-linking techniques, which will yield higher enzyme activity per unit weight of protein immobilized on electrodes. The use of carbon nanomaterials and metal nanomaterials may also play a more important role in increasing bioanodes' performances, as development in this research area continues to expand quickly.

Increasing bioanode stability will be another important direction for future studies of enzymatic BFCs. Researchers have been working on new enzyme immobilization materials and novel casting techniques [86,104]. Much effort has been put in increasing biocatalysts' stability with bioengineering techniques [18,105]. Several studies on miniature BFCs have been reported during the past few years, and it is expected that more endeavors in this area will be seen in the future as well. Anodic catalysis is an interesting and promising area of research with a large number of issues to study and address. With the current fast-paced development of bioanode design, it is possible that high-performance BFCs may soon take a role in the dynamic energy market.

ACKNOWLEDGMENT

The authors acknowledge the Air Force Research Laboratory, Air Force Office of Scientific Research, and the National Science Foundation.

LIST OF ABBREVIATIONS

ADH	alcohol dehydrogenase
AGS	amyloglucosidase
ALDH	aldehyde dehydrogenase
BFC	biological fuel cell
BOx	bilirubin oxidase
CDH	cellobiose dehydrogenase
CNT	carbon nanotube
CoA	coenzyme A
DET	direct electron transfer
EDC	N-ethyl-3-(3-dimethylaminopropyl)carbodiimide hydrochloride
FAD	flavin adenine dinucleotide
$FADH_2$	flavin adenine dinucleotide (reduced form)
FDH	fructose dehydrogenase
GDH	glucose dehydrogenase
GlyDH	glycerol dehydrogenase
GOx	glucose oxidase
INV	invertase
LDH	lactate dehydrogenase
MB^+	Meldola's blue (N,N-dimethyl-7-amino-1,2-benzophenoxazinium ion)
MET	mediated electron transfer
NAD^+	nicotinamide adenine dinucleotide
NADH	nicotinamide adenine dinucleotide (reduced form)
$NADP^+$	nicotinamide adenine dinucleotide phosphate
NADPH	nicotinamide adenine dinucleotide phosphate (reduced form)
NMN (NMNH)	nicotinamide mononucleotide
NMR	nuclear magnetic resonance
OCP	open-circuit potential
PAMAM	polyamidoamine
PDH	pyranose dehydrogenase
PPO	polyphenol oxidase
PQQ	pyrroloquinoline quinone
PyOx	pyranose oxidase
SWCNT	single-walled carbon nanotube
TBAB	tetrabutylammonium bromide

REFERENCES

1. Topcagic S, Treu BL, Minteer SD. Alcohol-based biofuel cells. In: Minteer SD (ed.), *Alcoholic Fuels*. CRC Press, Boca Raton, FL, 2006, pp. 215–231.
2. Atanassov P, Colon F, Rajendran V. *Glucose–air enzymatic bio-fuel cell*. ACS National Meeting, Philadelphia, PA, 2004, COLL-207.

3. Barriere F, Kavanaugh P, Leech D. A laccase–glucose oxidase biofuel cell prototype operating in a physiological buffer. *Electrochim Acta* 2006;51:5187–5192.

4. Xu S, Minteer SD. Enzymatic biofuel cell for oxidation of glucose to CO_2. *ACS Catal* 2012;2:91–94.

5. Calabrese Barton S, Atanassov P. Enzymatic biofuel cells for implantable and micro-scale devices. *Abstr Pap Am Chem Soc* 2004;228:U653–U653.

6. Gerlt JA, Babbitt PC. Mechanistically diverse enzyme superfamilies: the importance of chemistry in the evolution of catalysis. *Curr Opin Chem Biol* 1998;2:607–12.

7. Morley KL, Kazlauskas RJ. Improving enzyme properties: when are closer mutations better? *Trends Biotechnol* 2005;23:231–237.

8. Palmore G, Bertschy H, Bergens SH, Whitesides GM. A methanol/dioxygen biofuel cell that uses NAD^+-dependent dehydrogenases as catalysts: application of an electro-enzymic method to regenerate nicotinamide adenine dinucleotide at low overpotentials. *J Electroanal Chem* 1998;443:155–161.

9. Srere PA. Complexes of sequential metabolic enzymes. *Annu Rev Biochem* 1987;56:89–124.

10. Morgunov I, Srere P. Interaction between citrate synthase and malate dehydrogenase. *J Biol Chem* 1998;273:29540–29544.

11. Moehlenbrock MJ, Toby TK, Pelster LN, Minteer SD. Metabolon catalysts: an efficient model for multi-enzyme cascades at electrode surfaces. *ChemCatChem* 2011;3:561–570.

12. Moehlenbrock MJ, Toby TK, Waheed A, Minteer SD. Metabolon catalyzed pyruvate/air biofuel cell. *J Am Chem Soc* 2010;132:6288–6289.

13. Moehlenbrock MJ, Minteer SD. Extended lifetime biofuel cells. *Chem Soc Rev* 2008;37:1188–1196.

14. Chaubey A, Malhorta BD. Mediated biosensors. *Biosens Bioelectron* 2002;17:441–456.

15. Ghindilis AL, Atanasov P, Wilkins E. Enzyme catalyzed direct electron transfer: fundamentals and analytical applications. *Electroanalysis* 1997;9:661–674.

16. Addo PK, Arechederra RL, Minteer SD. Evaluating enzyme cascades for methanol/air biofuel cells based on NAD^+-dependent enzymes. *Electroanalysis* 2010;22:807–812.

17. Arechederra RL, Waheed A, Sly WS, Minteer SD. Electrically wired mitochondrial electrodes for measuring mitochondrial function for drug screening. *Analyst* 2011;136:3747–3752.

18. Yuhashi N, Tomiyama M, Okuda J, Igarashi S, Ikebukuro K, Sode K. Development of a novel glucose enzyme fuel cell system employing protein engineered PQQ glucose dehydrogenase. *Biosens Bioelectron* 2005;20:2145–2150.

19. Calabrese Barton S, Gallaway J, Atanassov P. Enzymatic biofuel cells for implantable and microscale devices. *Chem Rev* 2004;104:4867–4886.

20. Yow Tsong T. Fluctuation-driven directional flow of energy in biochemical cycle: electric activation of Na,K-ATPase. *Am Phys Soc Abstr* 1998; X12.05.

21. Lojou E. Hydrogenases as catalysts for fuel cells: strategies for efficient immobilization at electrode interfaces. *Electrochim Acta* 2011;56:10385–10397.

22. Karyakin AA, Morozov SV, Karyakina EE, Varfolomeyev SD, Zorin NA, Cosnier S. Hydrogen fuel electrode based on bioelectrocatalysis by the enzyme hydrogenase. *Electrochem Commun* 2002;4:417–420.

23. Tsujimura S, Fujita M, Tatsumi H, Kano K, Ikeda T. Bioelectrocatalysis-based dihydrogen/dioxygen fuel cell operating at physiological pH. *Phys Chem Chem Phys* 2001;3:1331–1335.

24. Guiral M, Tron P, Belle V, Aubert C, Leger C, Guigliarelli B, Giudiciorticoni M. Hyperthermostable and oxygen resistant hydrogenases from a hyperthermophilic bacterium *Aquifex aeolicus*: physicochemical properties. *Int J Hydrogen Energy* 2006;31:1424–1431.

25. Vincent KA, Cracknell JA, Clark JR, Ludwig M, Lenz O, Friedrich BR, Armstrong FA. Electricity from low-level H_2 in still air? An ultimate test for an oxygen tolerant hydrogenase. *Chem Commun* 2006; 5033–5035.

26. Vincent KA. From the cover: electrocatalytic hydrogen oxidation by an enzyme at high carbon monoxide or oxygen levels. *Proc Natl Acad Sci USA* 2005;102: 16951–16954.

27. Luo X, Brugna M, Tron-Infossi P, Giudici-Orticoni MT, Lojou É. Immobilization of the hyperthermophilic hydrogenase from *Aquifex aeolicus* bacterium onto gold and carbon nanotube electrodes for efficient H_2 oxidation. *J Biol Inorg Chem* 2009;14:1275–1288.

28. Wait AF, Parkin A, Morley GM, dos Santos L, Armstrong FA. Characteristics of enzyme-based hydrogen fuel cells using an oxygen-tolerant hydrogenase as the anodic catalyst. *J Phys Chem C* 2010;114:12003–12009.

29. Yahiro AT, Lee SM, Kimble DO. Bioelectrochemistry. Enzyme utilizing biofuel cell studies. *Biochim Biophys Acta* 1964;88:375–383.

30. Akers NL, Moore CM, Minteer SD. Development of alcohol/O_2 biofuel cells using salt-extracted tetrabutylammonium bromide/Nafion membranes to immobilize dehydrogenases enzymes. *Electrochim Acta* 2005;50:2521–2525.

31. Topcagic S, Minteer SD. Development of a membraneless ethanol/oxygen biofuel cell. *Electrochim Acta* 2006;51:2168–2172.

32. Laurinavicius V, Razumiene J, Kurtinaitiene B, Lapenaite I, Bachmatova I, Marcinkeviciene L, Meskys R, Ramanavicius A. Bioelectrochemical application of some PQQ-dependent enzymes. *Bioelectrochemistry* 2002;55:29–32.

33. Ramanavicius A, Kausaite A, Ramanaviciene A. Biofuel cell based on direct bioelectrocatalysis. *Biosens Bioelectron* 2005;20:1962–1967.

34. Ramanavicius A, Kausaite A, Ramanaviciene A. Enzymatic biofuel cell based on anode and cathode powered by ethanol. *Biosens Bioelectron* 2008;24:761–766.

35. Sokic-Lazic D, Minteer SD. Citric acid cycle biomimic on a carbon electrode. *Biosens Bioelectron* 2008;24:945–950.

36. Treu BL, Arechederra RL, Minteer SD. Bioelectrocatalysis of ethanol via PQQ-dependent dehydrogenases utilizing carbon nanomaterial supports. *J Nanosci Nanotechnol* 2009;9:2374–2380.

37. Deng L, Shang L, Wen D, Zhai J, Dong S. A membraneless biofuel cell powered by ethanol and alcoholic beverage. *Biosens Bioelectron* 2010;26:70–73.

38. Aquino Neto S, Forti JC, Zucolotto V, Ciancaglini P, de Andrade AR. Development of nanostructured bioanodes containing dendrimers and dehydrogenases enzymes for application in ethanol biofuel cells. *Biosens Bioelectron* 2011;26:2922–2926.

39. Rincón RA, Lau C, Luckarift HR, Garcia KE, Adkins E, Johnson GR, Atanassov P. Enzymatic fuel cells: integrating flow-through anode and air-breathing cathode into a membrane-less biofuel cell design. *Biosens Bioelectron* 2011;27:132–136.

40. Campbell E, Wheeldon IR, Banta S. Broadening the cofactor specificity of a thermostable alcohol dehydrogenase using rational protein design introduces novel kinetic transient behavior. *Biotechnol Bioeng* 2010;107:763–774.

41. Campbell E, Meredith M, Minteer SD, Banta S. Enzymatic biofuel cells utilizing a biomimetic cofactor. *Chem Commun* 2012;48:1898–1900.

42. Davis G, Hill HAO, Aston WJ, Higgins IJ, Turner APF. Bioelectrochemical fuel cell and sensor based on a quinoprotein, alcohol dehydrogenase. *Enzyme Microb Technol* 1983;5:383–388.

43. Addo PK, Arechederra R, Waheed A, Shoemaker JD, Sly WS, Minteer SD. Methanol production via bioelectrocatalytic reduction of carbon dioxide: role of carbonic anhydrase in improving electrode performance. *Electrochem Solid State Lett* 2011;14:E9–E13.

44. Wallar BJ, Lipscomb JD. Dioxygen activation by enzymes containing binuclear non-heme iron clusters. *Chem Rev* 1996;96:2625–2658.

45. Xin J-Y, Cui J-R, Hu X-X, Li S-B, Xia C-G, Zhu L-M, Wang Y-Q. Particulate methane monooxygenase from *Methylosinus trichosporium* is a copper-containing enzyme. *Biochem Biophys Res Commun* 2002;295:182–186.

46. Davis JB, Yarbrough HF. Preliminary experiments on a microbial fuel cell. *Science* 1962;137:615–616.

47. Weibel M, Dodge C. Biochemical fuel cells: demonstration of an obligatory pathway involving an external circuit for the enzymatically catalyzed aerobic oxidation of glucose. *Arch Biochem Biophys* 1975;169:146–151.

48. Persson B, Gorton L, Johansson G, Torstensson A. Biofuel anode based on D-glucose dehydrogenase, nicotinamide adenine dinucleotide and a modified electrode. *Enzyme Microb Technol* 1985;7:549–552.

49. Katz E, Filanovsky B, Willner I. A biofuel cell based on two immiscible solvents and glucose oxidase and microperoxidase-11 monolayer-functionalized electrodes. *New J Chem* 1999;23:481–487.

50. Heller A. Miniature biofuel cells. *Phys Chem Chem Phys* 2004;6:209–216.

51. Mano N, Kim H, Zhang Y, Heller A. An oxygen cathode operating in a physiological solution. *J Am Chem Soc* 2002;124:6480–6486.

52. Moore CM, Akers NL, Hill AD, Johnson ZC, Minteer SD. Improving the environment for immobilized dehydrogenase enzymes by modifying Nafion with tetraalkylammonium bromides. *Biomacromolecules* 2004;5:1241–1247.

53. Tanne C, Goebel G, Lisdat F. Development of a (PQQ)-GDH-anode based on MWCNT-modified gold and its application in a glucose/O_2-biofuel cell. *Biosens Bioelectron* 2010;26:530–535.

54. Saleh FS, Mao L, Ohsaka T. Development of a dehydrogenase-based glucose anode using a molecular assembly composed of Nile blue and functionalized SWCNTs and its applications to a glucose sensor and glucose/O_2 biofuel cell. *Sens Actuators B: Chem* 2011;B152:130–135.

55. Wang SC, Patlolla A, Iqbal Z. Carbon nanotube-based, membrane-less and mediator-free enzymatic biofuel cells. *ECS Trans* 2009;19:55–60.

56. Lee JY, Shin HY, Kang SW, Park C, Kim SW. Application of an enzyme-based biofuel cell containing a bioelectrode modified with deoxyribonucleic acid-wrapped single-walled carbon nanotubes to serum. *Enzyme Microb Technol* 2010;48:80–84.

57. Chen T, Barton SC, Binyamin G, Gao Z, Zhang Y, Kim H-H, Heller A. A miniature biofuel cell. *J Am Chem Soc* 2001;123:8630–8613.

58. Mano N, Mao F, Heller A. A miniature biofuel cell operating in a physiological buffer. *J Am Chem Soc* 2002;124:12962–12963.

59. Tsujimura S, Kano K, Ikeda T. Glucose/O_2 biofuel cell operating at physiological conditions. *Electrochemistry (Tokyo)* 2002;70:940–942.

60. Kim HH, Mano N, Zhang Y, Heller A. A miniature membrane-less biofuel cell operating under physiological conditions at 0.5 V. *J Electrochem Soc* 2003;150:A209–A213.

61. Mano N, Mao F, Shin W, Chen T, Heller A. A miniature biofuel cell operating at 0.78 V. *Chem Commun* 2003;4:518–519.

62. Mano N, Mao F, Heller A. Characteristics of a miniature compartment-less glucose–O_2 biofuel cell and its operation in a living plant. *J Am Chem Soc* 2003;125:6588–6594.

63. Sakai H, Nakagawa T, Tokita Y, Hatazawa T, Ikeda T, Tsujimura S, Kano K. A high–power glucose/oxygen biofuel cell operating under quiescent conditions. *Energy Environ Sci* 2009;2:133–138.

64. Li X, Zhou H, Yu P, Su L, Ohsaka T, Mao L. A miniature glucose/O_2 biofuel cell with single-walled carbon nanotubes-modified carbon fiber microelectrodes as the substrate. *Electrochem Commun* 2008;10:851–854.

65. Cinquin P, Gondran C, Giroud F, Mazabrard S, Pellissier A, Boucher F, Alcaraz J-P, Gorgy K, Lenouvel F, Mathe S, Porcu P, Cosnier S. A glucose biofuel cell implanted in rats. *PLoS One* 2010;5(5):e10476.

66. Daniel G, Volc J, Filonova L, Plihal O, Kubatova E, Halada P. Characteristics of *Gloeophyllum trabeum* alcohol oxidase, an extracellular source of H_2O_2 in brown rot decay of wood. *Appl Environ Microbiol* 2007;73:6241–53.

67. Sygmund C, Kittl R, Volc J, Halada P, Kubátová E, Haltrich D, Peterbauer CK. Characterization of pyranose dehydrogenase from *Agaricus meleagris* and its application in the C-2 specific conversion of δ-galactose. *J Biotechnol* 2008;133:334–342.

68. Lidén H, Volc J, Marko-Varga G, Gorton L. Pyranose oxidase modified carbon paste electrodes for monosaccharide determination. *Electroanalysis* 1998;10:223–230.

69. Tasca F, Timur S, Ludwig R, Haltrich D, Volc J, Antiochia R, Gorton L. Amperometric biosensors for detection of sugars based on the electrical wiring of different pyranose oxidases and pyranose dehydrogenases with osmium redox polymer on graphite electrodes. *Electroanalysis* 2007;19:294–302.

70. Timur S, Yigzaw Y, Gorton L. Electrical wiring of pyranose oxidase with osmium redox polymers. *Sens Actuators B: Chem* 2006;113:684–691.

71. Tasca F, Gorton L, Kujawa M, Patel I, Harreither W, Peterbauer CK, Ludwig R, Noell G. Increasing the coulombic efficiency of glucose biofuel cell anodes by combination of redox enzymes. *Biosens Bioelectron* 2010;25:1710–1716.

72. Scheller F, Karsten C. A combination of invertase reactor and glucose oxidase electrode for the successive determination of glucose and sucrose. *Anal Chim Acta* 1983;155:29–36.

73. Hamid JA, Moody GJ, Thomas JDR. Chemically immobilised tri-enzyme electrode for the determination of sucrose using flow injection analysis. *Analyst* 1988;113:81.

74. Matsumoto K, Kamikado H, Matsubara H, Osajima Y. Simultaneous determination of glucose, fructose, and sucrose in mixtures by amperometric flow injection analysis with immobilized enzyme reactors. *Anal Chem* 1988;60:147–151.

75. Katz E, Pita M. Biofuel cells controlled by logically processed biochemical signals: towards physiologically regulated bioelectronic devices. *Chem Eur J* 2009;15:12554–12564.

76. Katz E, Willner I. Enzyme-based biofuel cells with switchable and tunable power output. *Prepr Pap Am Chem Soc Div Fuel Chem* 2005;50:623–624.

77. Hickey D, Giroud F, Schmidke D, Glatzhofer D, Minteer SD. Enzyme cascade for catalyzing sucrose oxidation. *ACS Catal*, 2013;3(12):2729–2737.

78. Tam TT, Pita M, Ornatska M, Katz E. Biofuel cell controlled by enzyme logic network—approaching physiologically regulated devices. *Bioelectrochemistry* 2009;76(1–2):4–9.

79. Van Beers EH, Büller HA, Grand RJ, Einerhand AWC, Dekker J. Intestinal brush border glycohydrolases: structure, function, and development. *Crit Rev Biochem Mol Biol* 1995;30:197–262.

80. Pothukuchy A, Mano N, Georgiou G, Heller A. A potentially insect-implantable trehalose electrooxidizing anode. *Biosens Bioelectron* 2006;22:678–684.

81. Rasmussen M, Ritzmann RE, Lee I, Pollack AJ, Scherson D. An implantable biofuel cell for a live insect. *J Am Chem Soc* 2012;134:1458–1460.

82. Kamitaka Y, Tsujimura S, Kano K. High current density bioelectrolysis of D-fructose at fructose dehydrogenase-adsorbed and Ketjen black-modified electrodes without a mediator. *Chem Lett* 2007;36:218–219.

83. Kamitaka Y, Tsujimura S, Setoyama N, Kajino T, Kano K. Fructose/dioxygen biofuel cell based on direct electron transfer-type bioelectrocatalysis. *Phys Chem Chem Phys* 2007;9:1793–1801.

84. Murata K, Suzuki M, Kajiya K, Nakamura N, Ohno H. High performance bioanode based on direct electron transfer of fructose dehydrogenase at gold nanoparticle-modified electrodes. *Electrochem Commun* 2009;11:668–671.

85. Tasca F, Gorton L, Harreither W, Haltrich D, Ludwig R, Noll G. Direct electron transfer at cellobiose dehydrogenase modified anodes for biofuel cells. *J Phys Chem C* 2008;112:9956–9961.

86. Stoica L, Dimcheva N, Ackermann Y, Karnicka K, Guschin DA, Kulesza PJ, Rogalski J, Haltrich D, Ludwig R, Gorton L, Schuhmann W. Membrane-less biofuel cell based on cellobiose dehydrogenase (anode)/laccase (cathode) wired via specific Os-redox polymers. *Fuel Cells* 2009;9:53–62.

87. Brooks GA. Role of mitochondrial lactate dehydrogenase and lactate oxidation in the intracellular lactate shuttle. *Proc Natl Acad Sci USA* 1999;96:1129–1134.

88. Bardea A, Katz E, Bueckmann AF, Willner I. NAD^+-dependent enzyme electrodes: electrical contact of cofactor-dependent enzymes and electrodes. *J Am Chem Soc* 1997;119:9114–9119.

89. Katz E, Bueckmann AF, Willner I. Self-powered enzyme-based biosensors. *J Am Chem Soc* 2001;123:10752–10753.

90. Lee JY, Shin HY, Lee JH, Song YS, Kang SW, Park C, Kim JB, Kim SW. A novel enzyme-immobilization method for a biofuel cell. *J Mol Catal B: Enzym* 2009;59 (4):274–278.

91. Katz E, Lioubashevski O, Willner I. Magnetic field effects on bioelectrocatalytic reactions of surface-confined enzyme systems: enhanced performance of biofuel cells. *J Am Chem Soc* 2005;127:3979–3988.

92. Treu BL, Minteer SD. Isolation and purification of PQQ-dependent lactate dehydrogenase from *Gluconobacter* and use for direct electron transfer at carbon and gold electrodes. *Bioelectrochemistry* 2008;74:73–77.

93. Sokic-Lazic D, Minteer SD. Utilization of enzyme for complete oxidation of lactate in an enzymatic biofuel cell. *Electrochim Acta* 2011;56:10772–10775.

94. Sokic-Lazic D, Minteer SD. Pyruvate/air enzymatic biofuel cell capable of complete oxidation. *Electrochem Solid State Lett* 2009;12:F26–F28.

95. Treu BL, Sokic-Lazic D, Minteer SD. Bioelectrocatalysis of pyruvate with PQQ-dependent pyruvate dehydrogenase. *ECS Trans* 2010;25:1–11.

96. Eldridge F. Blood lactate and pyruvate in pulmonary insufficiency. *New Engl J Med* 1966;274:878–883.

97. Pagliaro M, Ciriminna R, Kimura H, Rossi M, Della Pina C. From glycerol to value-added products. *Angew Chem Int Ed* 2007;46:4434–4440.

98. Arechederra RL, Treu BL, Minteer SD. Development of glycerol/O$_2$ biofuel cell. *J Power Sources* 2007;173:156–161.

99. Arechederra RL, Minteer SD. Complete oxidation of glycerol in an enzymatic biofuel cell. *Fuel Cells* 2009;9:63–69.

100. Lapenaite I, Kurtinaitene B, Razumiene J, Laurinavicius V, Marcinkevicene L, Bachmatova I, Meskys R, Ramanavicius A. Properties and analytical application of PQQ-dependent glycerol dehydrogenase from *Gluconobacter* sp. 33. *Anal Chim Acta* 2005;249:140–150.

101. Kerr J, Minteer SD. Development of lipoxygenase bioanodes for biofuel cells. *Polym Mater Sci Eng* 2006;94:594–595.

102. Moore CM, Akers NL, Hill AD, Johnson ZC, Minteer SD. Immobilizing oxidoreductase enzymes in surfactant and symmetrical ammonium treated Nafion. *Polym Prepr* 2004;45:15–16.

103. Fillit C, Jaffrazic-Renault N, Bessueille F, Mitsubayashi K, Tardy J. Development of a microconductometric biosniffer for detection of trimethylamines. *Mater Sci Eng C* 2008;28:781–786.

104. Kang C, Shin H, Heller A. On the stability of the "wired" bilirubin oxidase oxygen cathode in serum. *Bioelectrochemistry* 2006;68:22–26.

105. Okuda J, Yamazaki T, Fukasawa M, Kakehi N, Sode K. The application of engineered glucose dehydrogenase to a direct electron-transfer-type continuous glucose monitoring system and a compartmentless biofuel cell. *Anal Lett* 2007;40:431–440.

6

BIOELECTROCATALYSIS OF HYDROGEN OXIDATION/REDUCTION BY HYDROGENASES

ANNE K. JONES, ARNAB DUTTA, PATRICK KWAN, CHELSEA L. MCINTOSH, SOUVIK ROY, AND SIJIE YANG

Department of Chemistry and Biochemistry and Center for Bio-Inspired Solar Fuel Production, Arizona State University, Tempe, AZ, USA

6.1 INTRODUCTION

Hydrogen is both an important energy storage molecule in the context of the human economy and an essential chemical in the metabolism of many prokaryotes and eukaryotes. Some microorganisms, such as knallgas bacteria, use hydrogen as a fuel to power metabolism; others, such as many fermentative bacteria, use protons as terminal electron acceptors (producing hydrogen gas) to dispose of excess reducing equivalents and recycle cellular redox carriers [1,2]. As the consequences of long-term fossil fuel exploitation become more apparent and pressing, development of technologies to exploit hydrogen, an energy rich, carbon-neutral fuel, in the human energy economy has become more urgent [3–5]. Both production and oxidation of hydrogen require efficient catalysts, and the current industry standards are based primarily on platinum or palladium [6,7]. Hydrogenases are the enzymatic biocatalysts that catalyze the reversible interconversion of hydrogen to protons and electrons ($H_2 \leftrightarrow 2H^+ + 2e^-$). They are extremely efficient catalysts that can operate with turnover frequencies (TOFs) in excess of $1000\,s^{-1}$ at low electrochemical overpotentials using organometallic active sites based only on the abundant metals, nickel and iron [8,9]. A number of hydrogenases have been immobilized at electrode

Enzymatic Fuel Cells: From Fundamentals to Applications, First Edition. Edited by Heather R. Luckarift, Plamen Atanassov, and Glenn R. Johnson.
© 2014 John Wiley & Sons, Inc. Published 2014 by John Wiley & Sons, Inc.

surfaces, and their electrocatalytic properties investigated, both to understand the fundamental mechanisms of these enzymes and to optimize either fuel production or power output in a hydrogen fuel cell. This chapter will discuss the basic enzymology of hydrogenases, the advantages and limitations of their application in electro-catalysis, and bio-inspired small inorganic catalyst mimics that have been designed, with the hope of being cheaper and more robust, to reproduce the functionality of these enzymes in technological applications.

6.2 HYDROGENASES

Three types of hydrogenases, defined by the metal content and structure of the hydrogen-activating (or hydrogen-producing) active site, have been discovered in Nature: [NiFe]-, [FeFe]-, and [Fe]-hydrogenases, and their structures are shown in Figure 6.1. Both the [NiFe]- and [FeFe]-hydrogenases consist of a bimetallic active site, at which the reaction with hydrogen (or protons) is catalyzed, and a series of [FeS] clusters linking the buried active site to the protein surface to allow fast exchange of electrons with the physiological electron transfer partner [10–23]. The [NiFe] active site consists of a redox-active nickel ion coordinated by four cysteine thiolates, two of which also bridge to a low-spin Fe(II) ion. The remainder of the first coordination sphere of the iron is comprised of three diatomic ligands: two cyanides and a carbon monoxide (CO) [24–27]. The active site of the [FeFe]-hydrogenase,

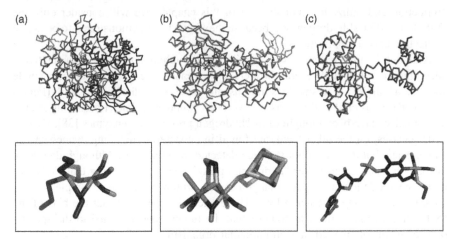

FIGURE 6.1 Representative structures of hydrogenases. Crystal structures of (a) the [NiFe]-hydrogenase from *D. vulgaris* Miyazaki F (PDB ID: 1H2R, 1.40 Å), (b) the [FeFe]-hydrogenase from *C. pasteurianum* (PDB ID: 3C8Y, 1.39 Å), and (c) the [Fe]-hydrogenase from *Methanocaldococcus jannaschii* (PDB ID: 3F47, 1.75 Å). The top row shows the holoprotein structure and the bottom row shows the hydrogen-activating site in stick represen-tation. Atoms are labeled as carbon (black), oxygen (red), nitrogen (blue), sulfur (orange), and iron (rust). All protein structure figures were created in PyMOL [144]. (Please see the color version of this figure in Color Plates section.)

commonly referred to as the H-cluster, is a unique six-iron assembly consisting of a [4Fe4S] cluster bridged via a cysteine thiolate to a diiron subsite. The diiron site is coordinated by a combination of five diatomic CN^- and CO ligands as well as a bridging, nonproteinaceous, dithiolate ligand. The iron adjacent to the cubane, like the iron in [NiFe]-hydrogenases, is thought to be a low-spin Fe(II) ion, and the iron distal to the cubane is believed to be the site of catalysis [28–33]. The [Fe]-hydrogenase, sometimes referred to as the *H₂-forming methylenetetrahydromethanopterin*, has been found only in some hydrogenotrophic methanogenic archaea and is structurally distinct from the other two groups in that it possesses a mononuclear active site and no [FeS] clusters [34]. It is also functionally distinct from the other groups of hydrogenases, with its requirement of a redox-active partner, methylenetetrahydromethanopterin, to produce hydrogen gas from protons. However, despite the initial differences, there are also striking similarities between the [Fe]-hydrogenases and the bimetallic enzymes. The mononuclear iron center is coordinated by two CO ligands [35,36], one sulfur and one or two N/O ligands, and Mössbauer spectroscopy has shown that it is a low-spin center in either the Fe(0) or Fe(II) oxidation state [37]. It is remarkable that all three hydrogenases feature organometallic active sites with intrinsic diatomic ligands, some combination of CO and CN^-, in the first coordination sphere; these ligands are otherwise unprecedented in biology. Because the three types of hydrogenases are not evolutionarily related, it suggests that these ligands play a mechanistic role that is essential for hydrogen activation under biological conditions: aqueous solution, circumneutral pH, and temperatures below $100\,°C$. We will return to this idea later when considering small-molecule electrocatalysts for hydrogen production and utilization. For the rest of this chapter, we will consider only the [NiFe]- and [FeFe]-hydrogenases, because they have been demonstrated to be active in electrocatalytic applications.

Although hydrogenases are well distributed throughout anaerobic or facultatively anaerobic microbes, and several different hydrogenases are often present in a single species, a single organism does not usually possess both [NiFe]- and [FeFe]-hydrogenases. [FeFe]-hydrogenases have been found in eubacteria and algae and, traditionally, have been thought of as "hydrogen production" enzymes [38]. [NiFe]-hydrogenases, on the other hand, are found in eubacteria, including cyanobacteria, and archaea, but they have not yet been detected in eukaryotes. Although they have historically been described as "hydrogen-uptake" enzymes, meaning that TONs for hydrogen oxidation are better than those for proton reduction, examples to the contrary are starting to emerge [39]. These exceptions suggest that the bias of the [NiFe]-hydrogenases may, to some extent, be tunable and matched to the specific physiological role played in each particular organism.

At the quaternary structural level, hydrogenases can also be extremely diverse, and this variation can be both advantageous and challenging when considering applications in electrocatalysis. [FeFe]-hydrogenases are generally smaller than [NiFe]-hydrogenases, and algal [FeFe]-hydrogenases are among the smallest. These approximately 48 kDa proteins consist of a single monomeric protein housing only the H-cluster [40,41]. Bacterial [FeFe]-hydrogenases, such as those from *Clostridium pasteurianum* and *Desulfovibrio desulfuricans*, possess an additional N-terminal ferredoxin-like domain that coordinates a variable number of [FeS] clusters, depending on species, believed to be essential for transferring electrons between the active site and the physiological redox partner *in vivo* or a redox dye or electrode *in vitro* [22]. [NiFe]-hydrogenases tend to have more diverse subunit and cofactor composition, reflecting the wider range of physiological roles they can play. Phylogenetically, [NiFe]-hydrogenases have been divided into four distinct groups: membrane-bound uptake hydrogenases, H_2-sensing hydrogenases, bidirectional heteromultimeric cytoplasmic hydrogenases, and hydrogen-evolving, energy-conserving, membrane-associated hydrogenases [2]. Although additional subunits may be present, the hydrogenase subcomplex of each of these enzymes consists of a heterodimer, and although the topic has not been thoroughly explored, there are preliminary studies indicating that an active-site-containing subunit alone may not be sufficient for catalytic activity [42]. The large subunit, a protein of approximately 60 kDa, coordinates the [NiFe] active site. The primary sequence of this protein is relatively highly conserved among all types of [NiFe]-hydrogenases. The small subunit is an approximately 30 kDa protein that contains one or more [FeS] clusters electronically connecting the active site to the protein surface [10]. The number and composition of [FeS] clusters in the small subunit is highly variable and has been hypothesized to be an important determinant of catalytic properties [43]. In this regard, it is interesting to note that the membrane-bound uptake enzymes usually possess a chain of three [FeS] clusters in the small subunit. Although the chain is flanked by [4Fe4S] clusters on both ends, the medial cluster is a [3Fe4S]. This cluster is remarkable for its higher reduction potential, −30 mV compared with approximately −300 mV versus standard hydrogen electrode (SHE) for the [4Fe4S] clusters, and might be expected to be a barrier to fast electron transfer through the protein [44–47]. On the other hand, the bidirectional soluble [NiFe]-hydrogenases, thought to be less dramatically biased toward hydrogen oxidation than their uptake cousins, do not possess a [3Fe4S] cluster in their small subunit [39,48,49].

Although hydrogenases might appear to be ideal catalysts for biological fuel cell (BFC) applications, their technological use has been hindered by their sensitivity to oxygen. Molecular oxygen is an effective inhibitor because, like the substrate hydrogen, it can serve as a π-accepting ligand to low-valent organometallic sites, such as the hydrogenase active sites [50]. [FeFe]-hydrogenases are irreversibly inactivated on binding oxygen to the active site, and it has been suggested that it is the [4Fe4S] cluster portion of the active site that is irreparably destroyed [51]. [NiFe]-hydrogenases, on the other hand, are reversibly inactivated by aerobic or anaerobic but oxidizing conditions [52–55]. However, as shown in Figure 6.2, a complicated array of states are produced by changes in redox potential. The distinct states of the [NiFe]-hydrogenase active site have been characterized using primarily Fourier transform infrared (FTIR) and electron paramagnetic resonance (EPR) spectroscopies, and the kinetics of the reactivation of inactive states have been evaluated using both solution assays and protein film electrochemistry (PFE) [27,56–58]. Under anaerobic but oxidizing conditions, the so-called standard [NiFe]-hydrogenases, those membrane-bound uptake hydrogenases (MBHs) of the first phylogenetic group that are oxygen-sensitive, are slowly inactivated to the Ni-B or "ready"

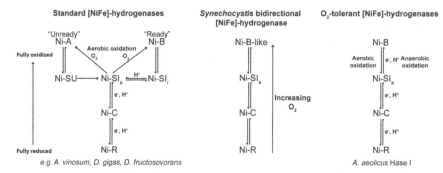

FIGURE 6.2   Schematic overview of the spectroscopically identified redox intermediates in standard, oxygen-sensitive [NiFe]-hydrogenases (left), HoxEFUYH from *Synechocystis* sp. PCC6803 (middle), and O$_2$-tolerant uptake [NiFe]-hydrogenases. The most oxidized states appear at the top of the diagram and the most reduced at the bottom. States believed to be analogous and/or at the same redox level are horizontally aligned across the figure. (Reproduced with permission from Ref. [39]. Copyright 2011, American Chemical Society.)

state. This state is characterized not only by a distinct Ni-based EPR signal but also by its ability to be reactivated in seconds to minutes on a return to reducing conditions. In the presence of oxygen, these [NiFe]-hydrogenases are nearly instantaneously deactivated to a mixture of the Ni-B state and the Ni-A or "unready" state: a state characterized by the requirement for prolonged incubation (minutes to hours) at reducing potentials to regain catalytic activity. It is the formation of Ni-A that makes practical use of these enzymes in fuel cell applications problematic, because crossover of oxygen from the cathode is virtually unavoidable. However, there exists a subgroup of uptake [NiFe]-hydrogenases, the so-called oxygen-tolerant enzymes, which maintain some hydrogen oxidation activity for extended periods of time in the presence of oxygen. This group includes the relatively well-characterized enzymes from *Ralstonia eutropha* [53,59], *Aquifex aeolicus* [60,61], and *Escherichia coli* [62,63]. As shown in Figure 6.2, these enzymes are believed to maintain some activity in the presence of oxygen by inactivating to a single state: the Ni-B state. Thus, although they are slowly converted to Ni-B at high potentials, because Ni-B can also be quickly reactivated at high potentials, a fraction of the enzyme remains active under aerobic conditions. In short, oxygen tolerance appears to be achieved by preventing formation of Ni-A. Two recent crystal structures of oxygen-tolerant [NiFe]-hydrogenases demonstrated the presence of a [4Fe3S] cluster in the small subunit immediately adjacent to the [NiFe] site, and it has been hypothesized that this unprecedented cluster is responsible for the oxygen tolerance [13,63–65]. In this line of argument, resistance to Ni-A formation is thought to be a result of the ability to reduce the attacking oxygen completely with four electrons, thus preventing formation of any blocking species.

The soluble bidirectional [NiFe]-hydrogenases of group 3 soluble hydrogenases (SHs) couple hydrogen oxidation/production with reduction/oxidation of nicotinamide adenine dinucleotide phosphate (NAD(P)) and have thus garnered attention not only as hydrogenases but also as electrocatalysts for recycling of NAD(P) [66–68]. In the second context, they could be used as a diaphorase in concert with other NAD(P)-dependent oxidoreductases in BFC applications [69]. However, few SHs have been electrochemically characterized, and their reactions with oxygen have still not been completely characterized. As shown in Figure 6.2, the inactive states are different from those of both the oxygen-sensitive and the oxygen-tolerant MBHs. Although only a single inactive state was detected through FTIR spectroscopy for the SHs from *Synechocystis* sp. PCC6803 and *R. eutropha* [70,71], two distinct inactive states were detected for the first enzyme using PFE and for the hydrogenase subdimer of the second enzyme. Perhaps surprisingly, neither enzyme was an oxygen-tolerant electrocatalyst, but none of the inactive states was a classical "unready" state [39]. Thus, it is clear that the mechanisms controlling oxygen tolerance of SHs and MBHs must have subtle differences, and elucidating these principles may prove essential in the development of [NiFe]-hydrogenases as electrocatalysts for either hydrogen production or oxidation.

6.3 BIOLOGICAL FUEL CELLS USING HYDROGENASES: ELECTROCATALYSIS

TOFs for hydrogenases have been reported to be as high as thousands per second, and catalysis in a number of PFE studies has been shown to be diffusion limited [8,45,72]. Thus, it is perhaps not surprising that several researchers have reported comparisons between the electrocatalytic rates of hydrogenases and platinum catalysts, the industry standard in fuel cells [8,73–75]. Studies at modest pH values, temperatures, and hydrogen pressures have suggested that, on a per catalytic site basis, hydrogenases may even be considerably more catalytically active than platinum. Although biological conditions are not typically employed in fuel cells, a number of niche applications could operate in these regimes. Particular examples might include cells that need to operate on limited fuel (hydrogen) in an atmosphere contaminated by other small molecules, cells operating on a mixed biologically derived fuel source, or cells functioning without a membrane between compartments. Such conditions would be extremely unfavorable for platinum, but hydrogenases, with relatively high substrate specificity, tend to thrive in such complicated environments. This has inspired several groups to construct prototype devices using hydrogenases as electrocatalysts in fuel cells. This section will survey the accomplishments in this area identifying both the current successes and the future challenges.

The first construction of a complete fuel cell oxidizing hydrogen as fuel with concomitant reduction of oxygen to water using only biological catalysts at both the cathode and anode was reported by Tsujimura et al. in 2001 [76]. This cell was a hybrid of an enzymatic fuel cell and a microbial fuel cell. Bilirubin oxidase (BOx) was used as the cathodic electrocatalyst for the reduction of oxygen, and hydrogen oxidation at the anode was actually catalyzed by whole *Desulfovibrio vulgaris* cells. The oxidative catalysis presumably relied on the periplasmically localized

[NiFe]-hydrogenase in *D. vulgaris*. Interestingly, the catalysis in both chambers was indirect. In the anode, methyl viologen was used as a mediator between the cells and a carbon felt electrode, and in the cathode, 2,2'-azinobis(3-ethylbenzothiazoline-6-sulfonate) played an analogous role. The open-circuit voltage (OCV) of 1.17 V obtained is very close to the theoretical limit of 1.23 V. However, the current was severely limited, and it was hypothesized that diffusion of substrate and perhaps mediator played a significant role in this limitation.

The first use of a purified hydrogenase in a functional H_2/O_2 fuel cell was reported 4 years later by Armstrong and coworkers [53]. In addition to using purified enzyme as opposed to whole cells, their fuel cell was built on a number of studies of hydrogenases using PFE to *directly* interface the hydrogenase with a graphite electrode; that is, a chemical mediator was not used to facilitate electron transfer between the enzyme and the electrode. The enzyme chosen was the MBH from *R. eutropha* [77,78]. This organism is a well-known knallgas bacterium, organisms that grow using hydrogen as the sole energy source and oxygen as the terminal electron acceptor for respiration. The MBH from *R. eutropha* is generally considered a prototypical example of an oxygen-tolerant [NiFe]-hydrogenase [79,80], and the electrocatalytic characteristics of the adsorbed MBH were first carefully defined using PFE. Although oxygen did inhibit hydrogen oxidation, catalytic activity could be observed even at some oxygen concentrations in excess of those in air, setting the stage for construction of a BFC. A simple fuel cell was constructed merely by placing a graphite electrode coated with *Trametes versicolor* laccase in the same beaker with a hydrogenase-covered electrode. The two reactants, hydrogen and oxygen, were then introduced through gas tubes near their respective electrodes. The beauty of this experiment lies in its simplicity in that the design does not require a membrane separating the cathodic and anodic chambers. This is made possible by the ability of MBH to sustain some of its activity in the presence of the oxygen required for the cathodic reaction and the specificity of the two enzymes for their respective substrates. However, we note that the design of this particular cell helps the catalysts somewhat by actively introducing hydrogen immediately adjacent to the hydrogenase, so that its concentration is relatively high and that of oxygen is relatively low. Nonetheless, a cell of a similar design cannot be constructed using platinum due to its promiscuous reactivity with small molecules. This has led to the need to develop better proton exchange membranes (PEMs) to facilitate ionic conductivity between half-cells but prevent gas diffusion. The OCV for the BFC was 970 mV, and a maximum power of approximately 5 µW was reported. The initial power of the cell dropped rapidly on the timescale of 2 min but thereafter remained stable at considerably lower values for 15 min. Although the cell would admittedly not be practical for any applications, it served as proof of principle that purified hydrogenases could function under relatively adverse conditions. In fact, it was further shown that the adsorbed enzyme continued to oxidize hydrogen unhindered by the presence of 0.9 bar CO. Thus, it is imaginable that [NiFe]-hydrogenases, unlike platinum, could be used in applications with hydrogen produced from synthesis gas, a product contaminated with significant CO.

Michaelis–Menten binding constants (K_m) for hydrogen by uptake [NiFe]-hydrogenases have been estimated to be in the low micromolar range [81]. This is consistent with the fact that hydrogen is not present at high concentrations in the atmosphere, and hydrogen-oxidizing microbes must exploit substrate produced nearby by hydrogen-producing microorganisms [82]. With this in mind, in 2006, Armstrong and coworkers modified their initial BFC design to construct a cell capable of operation in a fuel-poor, nonexplosive mixture of 3% hydrogen in air [83]. In this second cell design, the gases were no longer bubbled directly into the half-cells; instead a still, single-gas atmosphere above the cell supplied both the oxidant and the reductant. The *R. eutropha* MBH was replaced with that from *Ralstonia metallidurans* CH34 because the latter had been described as being more active and stable than its cousin. An OCV of 950 mV and a maximum power density of 5.2 μW cm^{-2} at 47 kΩ (11 μA cm^{-2}, 500 mV) were observed, making the performance of this cell, under more stringent conditions, relatively comparable to the previous generation. Additionally, in a display of applicability, the authors connected three of these fuel cells in series to power a wristwatch for 24 h. Considering that each cell consisted of a 12 L tank, this may initially seem unremarkable. However, there are several reasons to give this a second thought. First, although oxygen-tolerant, *R. metallidurans* [NiFe]-hydrogenase can be directly inhibited by oxygen, and one might anticipate that on the timescale of hours this inhibition would become complete and cell operation would halt. Second, reactive oxygen species produced by incomplete reduction of oxygen could also be expected to steadily decrease enzymatic activity. Third, because both enzymes were associated with their respective electrodes simply through adsorption from dilute solution, there was ample opportunity for them to dissociate from the surface during fuel cell operation. Thus, this work served as the first demonstration that, with further engineering, hydrogenase-based fuel cells could see use in niche applications.

Although both *R. eutropha* membrane-bound hydrogenase (*Re*MBH) and *R. metallidurans* membrane-bound hydrogenase (*Rm*MBH) are oxygen-tolerant [NiFe]-hydrogenases, exposure to oxygen in the membraneless fuel cells described above is likely to have negatively impacted the performance of the enzymatic anodes and, by extension, the fuel cells. Wait et al. directly addressed this issue in 2010 by comparing the performance of an analogous fuel cell operating with and without a membrane [84]. They also probed the question of how the voltammetric characteristics determined using PFE for an isolated enzyme electrode, a situation in which the electrons are provided by an external power source at the desired electrochemical potential, correlate with the performance of that same electrode in a fuel cell, a self-powering device. This second question is particularly interesting, because electrochemical characterization of a number of isolated hydrogenases has been reported, and it would be convenient if knowledge derived from these studies could be directly translated into improved device development.

The fuel cell constructed by Wait et al. used the oxygen-tolerant [NiFe]-hydrogenase from *E. coli* (*Ec*Hyd1) in the anode and BOx from *Myrothecium verrucaria*(*Mv*BOx) for oxygen reduction at the cathode. The novelty of this work is that a direct comparison was made between membraneless operation and the configuration in which cathode and anode were separated by a Nafion membrane. As shown in Figure 6.3, three distinct operating modes were characterized:

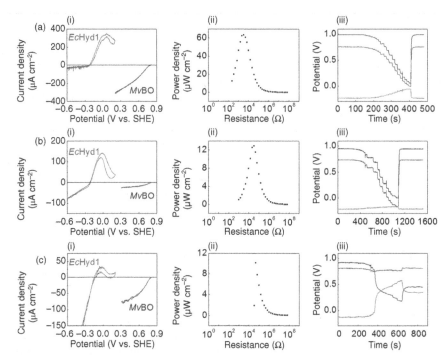

FIGURE 6.3 Three different configurations of EcHyd1 and MvBOx H_2/O_2 fuel cells operating at 25 °C using stationary electrodes of 1.25 cm^2: (a) PEM, (b) 96% H_2/4% O_2, and (c) 4% H_2/96% air. In each case, (i) shows cyclic voltammograms for the EcHyd1 (red) and MvBOx (blue) recorded in the fuel cell at 10 mV s^{-1}, (ii) shows the power of the fuel cell as a function of resistance, and (iii) shows the changes in voltage of the fuel cell (black) and each of the electrodes with respect to the reference electrode (anode vs. reference shown in red, cathode vs. reference shown in blue, quoted vs. SHE) as a function of time, over the course of the power curve being recorded. (Reprinted with permission from Ref. [84]. Copyright 2010, American Chemical Society.) (Please see the color version of this figure in Color Plates section.)

Figure 6.3a shows a fuel- and oxidant-rich cell (100% hydrogen and oxygen, respectively, introduced directly into the appropriate half-cell) separated by a Nafion PEM operating on the benchtop; Figure 6.3b shows a membraneless cell fed a fuel-rich mixture of 96% H_2/4% O_2; and Figure 6.3c shows a membraneless cell using a fuel-poor mixture of 4% H_2 in air. The traditional cell configuration, using a membrane to isolate the two reactions, with an OCV of 990 mV, clearly produced the largest power output at 63 µW cm^{-2} at a cell voltage of 511 mV. Furthermore, the power versus load curve for this cell was symmetric (Figure 6.3a(ii)). Nonetheless, voltammetric experiments probing only the hydrogen oxidation reaction at the anode showed that the activity of the hydrogenase at high potentials was noticeably impacted by oxygen crossover from the cathode even in the presence of a PEM. The second condition, an oxygen-poor gas mixture without a membrane between the half-cells, produced much less power at 12.9 µW cm^{-2} due to the limited cathodic reaction of MvBOx, which, under these conditions, lacked substrate. Characterization of the isolated cathode confirmed that decreasing the oxygen concentration from 100 to 4% caused a drop in the reductive current by approximately an order of magnitude. However, a symmetric power versus resistance curve was still observed (Figure 6.3b(ii)). The third condition, a membraneless configuration fed 4% hydrogen in air, had an OCV of 930 mV, noticeably lower than that for the other two configurations. This is explained by the direct reduction of oxygen at the anode, a reaction that tends to decrease the hydrogen oxidation current and subsequently the operating voltage of the device. The maximum power output of 10.1 μW cm^{-2} at 772 mV was also considerably lower than that in the other two cell configurations. However, the most interesting result of operation in this third configuration was that the power versus resistance curve was asymmetric (Figure 6.3c(ii)). In particular, under conditions of low load, the power dropped dramatically. EcHyd1 was rapidly, oxidatively inactivated due to the combination of both high potential and high oxygen concentration. By analogy to the other studies of [NiFe]-hydrogenases, it is likely that these conditions favor formation of the Ni-ready state. As expected, application of a large resistance alone, that is, a greater load, was not sufficient to reactivate an inactivated hydrogenase anode. Instead, as demonstrated for many [NiFe]-hydrogenases using PFE, the inactive enzyme could only be reactivated by a brief return to reducing conditions. These extra electrons could be supplied either potentiostatically or via connection of another active hydrogenase anode. The take-home message was that additional electrons were required to reactivate the hydrogenase active site; the cell could not spontaneously reactivate on its own.

Although the fuel cell described by Wait et al. was by no means optimized, it had the best performance characteristics (63 μW cm^{-2}) of any reported hydrogenase fuel cell. By way of comparison, it is worth noting that the best glucose-based enzymatic BFCs can now exceed power densities of 1 mW cm^{-2} [85]. This suggests that there is likely room for substantial improvement through engineering such properties as high surface area electrodes and robust enzyme–electrode connections [86–91]. Nonetheless, these early pioneering studies have now made it imaginable that fuel cells using hydrogenases at the anode could actually be exploited in a few niche applications. However, it is also clear that hydrogenases only function optimally in a narrow range of potentials. This means that the applicability of even currently known oxygen-tolerant [NiFe]-hydrogenases is severely limited by the reactions of the enzymes with oxygen. Future research will need to focus not only on improving the fuel cell designs themselves but also on discovering or engineering hydrogenases with increased oxygen tolerance. Alternatively, it may prove more fruitful to design bio-inspired catalysts based on abundant metals that can be directly tethered to electrode surfaces to efficiently catalyze hydrogen oxidation without concomitant reactions with oxygen [92]. We will turn to this topic below.

We pause here to consider a final reported application of [NiFe]-hydrogenases as oxidative catalysts. Although the devices are not always technically "fuel cells," the electrocatalytic principles are closely related. In 2011, Shastik et al. created a unique combination of a bioreactor together with a hydrogenase electrode [93]. The bioreactor was stocked with a microbial consortium adept at fermentation of starch.

Fermentation of glucose (in the reported experiments) produces, among other things, hydrogen as a by-product. The hydrogen could then be oxidized by a carbon filament electrode modified with [NiFe]-hydrogenase from *Thiocapsa roseopersicina*, producing electrical current. The system generated 200 μA of current and maintained 81% of the initial activity after 220 h of operation demonstrating unexpected longevity. Unfortunately, the yield of hydrogen was relatively low at 1.42 mol per mol glucose. Complete conversion of a mole of glucose to hydrogen should yield 12 mol. Nonetheless, the system of Shastik et al. is a particularly interesting idea since the hydrogenase electrode can successfully use unpurified hydrogen contaminated by a number of small molecules (H_2S, CO, etc.) that ordinarily poison nonbiological catalysts. Finally, it is also interesting to note parallels between this device and microbial fuel cells or other biological systems that have also been reported for hydrogen production [94]. There should be numerous opportunities for combining developments from these diverse research areas to produce better processes and devices.

In 2008, using an [FeFe]-hydrogenase, Hambourger et al. described the first bioelectrochemical device employing a hydrogenase in the cathode [72]. This cell is a unique combination of an enzymatic BFC together with a dye-sensitized solar cell, creating a device that uses solar energy to oxidize biofuel while producing both photocurrent and hydrogen gas (Figure 6.4). The anode of the cell featured a porphyrin-sensitized electrode that on photoexcitation injects electrons into the conduction band of a TiO_2 (titanium dioxide) electrode and is then re-reduced by NADH (reduced form of nicotinamide adenine dinucleotide). The NADH, a ubiquitous biological redox cofactor, is cogenerated from oxidation of a biofuel by an

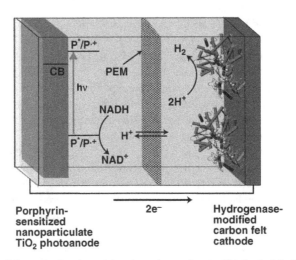

FIGURE 6.4 Schematic drawing of the photoelectrochemical biological fuel cell developed by Hambourger et al. for the production of hydrogen from biomass using solar energy. (Reprinted with permission from Ref. [72]. Copyright 2008, American Chemical Society.) (Please see the color version of this figure in Color Plates section.)

appropriate nicotinamide adenine dinucleotide (NAD)-dependent oxidoreductase. In the cathodic compartment, the electrons were then used to produce hydrogen by *Clostridium acetobutylicum* [FeFe]-hydrogenase (*Ca*HydA) adsorbed to a carbon felt electrode. Electrochemical investigation of *Ca*HydA in the same work showed that it is an extremely efficient catalyst for proton reduction with current densities approximately 40% of those of platinum under the same conditions. Over the course of an hour of illumination, hydrogen could be produced at a rate of 23.4 nmol H_2 min^{-1} in the BFC, as opposed to 19.8 nmol H_2 min^{-1} when the hydrogenase anode was replaced by a platinum electrode; other results suggested that the photoanode was the rate-limiting factor in these experiments. The cell serves as an exciting demonstration of the ability to produce hydrogen using an artificial photosynthetic approach.

In 2009, Reisner et al. described a system to convert solar energy into hydrogen relying on the [NiFeSe]-hydrogenase from *Desulfomicrobium baculatum* (*Db*) as the hydrogen production catalyst. The enzyme was adsorbed to TiO_2 nanoparticles modified with a range of well-known platinum- and ruthenium-based photosensitizers (RuPs) [95,96]. As for the photoelectrochemical cell described above, on excitation, the RuP injects an electron into the conduction band of the TiO_2. Accumulation of electrons in the nanoparticle allows the adsorbed enzyme to produce hydrogen from protons. Holes produced in the RuP were refilled by electrons from a sacrificial electron donor present in the solution: triethanolamine. The *Db*[NiFeSe]-hydrogenase was chosen in this study because of its enhanced H_2 production activity relative to other [NiFe]-hydrogenases, its relative lack of H_2 inhibition, and its ability to continuously produce hydrogen in the presence of at least 1% oxygen [97]. Perhaps surprisingly, the enzyme also exhibited extreme affinity for the TiO_2 electrode, a property dubbed "titaniaphilicity." TiO_2 electrodes with *Db*[NiFeSe]-hydrogenase adsorbed retained 80% of electrocatalytic activity after 48 h and 50% of activity after a month stored in electrolyte at room temperature under N_2. In the most optimized system, the authors demonstrated a TOF of 50 s^{-1} and generated 2070 μmol H_2 h^{-1} (mg enzyme)$^{-1}$ or 712 μmol H_2 h^{-1} (g TiO_2)$^{-1}$. The system was even able to produce H_2 under natural sunlight. Although the enzyme itself is stable on a TiO_2 electrode in the presence of air, the electrocatalytic system quickly decomposed in the presence of air and light. In the complete photochemical system, it is likely that reduction of oxygen at the nanoparticle surface produces reactive oxygen species that attack both the photosensitizer and the enzyme.

Enzymatic fuel cells have been plagued by the notion that current output is too low to be practically useful, both because the catalysts are large and because their TOFs can be lower than those of precious metals. In the case of hydrogenases, with TOFs on the order of 9000 s^{-1}, inherent catalytic rate is probably not a limitation [45]. Nonetheless, the catalysts are large, and, in many cases, increasing the surface area of the electrode has allowed creation of bioelectrodes with relatively high enzymatic loadings and activity. Inspired by a number of mechanistic studies of hydrogenases and other redox enzymes adsorbed to edge-plane pyrolytic graphite, in 2011, Healy et al. incorporated hydrogenase immobilized on small graphite particles into a matrix of the polymer electrolyte Nafion to create a three-dimensional electrode. They achieved hydrogen oxidation currents under optimal conditions of 1.6 mA per geometric cm^2 of electrode at approximately 0.2 V versus SHE (the height dimension is unknown so that volumes cannot be considered) [89]. By comparison, in 2007, Alonso-Lomillo et al. constructed a hydrogenase electrode via covalent immobilization of the [NiFe]-hydrogenase from *Desulfovibrio gigas* at multiwalled carbon nanotubes grown on an underlying gold electrode. These nanotube electrodes yielded hydrogen oxidation current of 4.1 mA cm^{-2} at −0.058 V versus SHE (the data are originally reported as −0.3 V vs. saturated calomel electrode and have been converted using $E_{SHE} = E_{SCE} + 0.242$ V [98]), approximately 33 times the current density measured for the enzyme immobilized at a flat pyrolytic graphite electrode using the same procedure [86].

6.4   ELECTROCATALYSIS BY FUNCTIONAL MIMICS OF HYDROGENASES

Although hydrogenases have exceptional activities, they present a number of challenges to successful use in enzymatic fuel cells. First, their isolation from native sources can be difficult and time consuming, and development of overexpression systems has proven challenging because of the necessity for a number of multiprotein complexes to ensure correct active site incorporation [12,99–102]. Second, enzymes are considerably larger than atoms or small molecules, so that achieving high loadings of functional protein to obtain catalytic rates comparable to precious metal catalysts may prove an insurmountable obstacle. Third, functional and robust connections between the enzymes and electrodes must be constructed. Fourth, enzymes must be prevented from either inactivating or denaturing during the lifetime of the device. The combination of these limitations has led many researchers to conclude that, rather than relying on enzymes, practical devices will use functional molecular mimics of hydrogenase active sites inspired by the mechanistic under-pinnings of the enzymes and built from nonprecious metals. Although no molecular catalyst functional at low overpotentials with rates similar to the enzyme has been described to date, the uncanny similarities of the hydrogenase active sites to well-known classes of organometallic compounds have ensured that a considerable number of structural models has been synthesized. In this section, we will briefly consider some of the most catalytically promising compounds that have been synthesized to date. Before beginning, it is worth noting that although hydrogenases are reversible catalysts that require very little overpotential to observe catalysis, such bidirectionality is almost completely absent in synthetic models.

6.4.1   [FeFe]-Hydrogenase Models

The active site of [FeFe]-hydrogenases is remarkably similar to well-known organo-iron compounds such as $[Fe_2^I(\mu\text{-SR})_2(CO)_6]$, where R is an organic group. Perhaps the best known of this class of compounds, complex **1** in Figure 6.5, is the complex featuring propane dithiol as the bridging ligand. Derivatives of this class of com-pounds, readily formed by variation of the organic thiol and/or substitution of one or

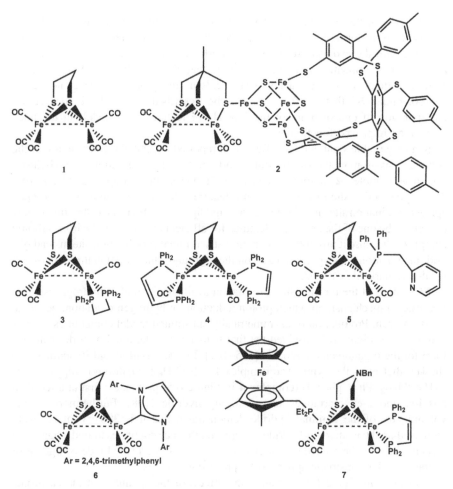

FIGURE 6.5 Schematic representations of selected models of [FeFe]-hydrogenases.

more of the carbonyls, have proven interesting both in their own right as proton reduction catalysts and as models for understanding the mechanism of hydrogenases. As testament to the fruitfulness of this research area, more than 250 related complexes have been reported in the literature as of 2009 [103].

There are four features of the [FeFe]-hydrogenase active site, shown in Figure 6.1, that differ from the hexacarbonyl complexes described above. First, the iron distal to the [4Fe4S] cluster possesses a so-called rotated geometry, resulting in an open apical coordination site for substrate binding. Second, the atom in the bridgehead position of the dithiolate ligand of the enzyme is thought to be nitrogen, providing a built-in base to facilitate proton transfer. Third, in addition to carbonyls, the active site of [FeFe]-hydrogenases features two cyanide ligands, strong σ-donors, in the first coordination sphere. Fourth, the enzyme includes additional reducing equivalents immediately accessible in the form of the covalently connected [4Fe4S] cluster. To date, no model complex with the exquisite reactivity of the [FeFe]-hydrogenases has been characterized. These enzymes are remarkable in that they catalyze proton reduction extremely quickly in water using weak acids as the proton source at negligible overpotentials. Synthetic efforts, as will be described below, have focused on incorporation of one or more of the four above features into models in the hope of creating improved functionality.

In a remarkable synthetic feat, Tard et al. reported in 2005 the construction of the complete six-iron metallosulfur framework of the [FeFe]-hydrogenase H-cluster shown as complex **2** in Figure 6.5 [104]. However, this compound differs from the enzyme active site in that it does not feature a "rotated structure" with an open apical coordination site and a bridging carbonyl ligand. Such a rotated structure is now thought to be one of the necessary features for efficient catalysis by the iron carbonyl complexes because it promotes protonation at a terminal coordination site instead of a bridging position. As expected, based on these grounds, the model in this experiment is an unremarkable catalyst.

Evaluation of the mechanistic information available for [FeFe]-hydrogenases has suggested a mechanism in which proton reduction, or hydrogen oxidation, occurs at the distal iron. Protonation of asymmetrically substituted model compounds such as complex **3** in Figure 6.5 provided the first evidence that terminal hydrides can also form for the organometallic models [105–107]. In 2008, Barton and Rauchfuss also showed that even the symmetric complex $Fe_2(S_2C_3H_6)(CO)_2(dppv)_2$ (dppv = *cis*-$C_2H_2(PPh_2)_2$; Ph = phenyl) (complex **4** in Figure 6.5) protonates first at a terminal position and then rearranges to a bridging hydride [108]. Furthermore, cyclic voltammetry of the terminal hydride demonstrated that it is 200 mV more easily reduced than the isomeric μ-hydride. At present, the race continues to find a ligand set that can stabilize the rotated geometry during catalysis. Sterically hindered ligands or supramolecular constructs appear to be promising options.

As shown in Figure 6.5, to mimic the effects of the cyanides on the electronic environment and enhance the rather poor proton reduction catalysis of the hexacarbonyl complexes, a range of ligands that function simultaneously as σ-donors and π-acceptors such as phosphine, isocyanide, cyanide, nitrosyl, and N-heterocyclic carbenes (NHCs) have been substituted for one or more of the carbonyls to form $Fe_2(SR)_2(CO)_{6-x}L_x$ complexes. Strong σ-donors, such as phosphines, increase electron density at the metallocenter facilitating protonation. However, they can also make it more difficult to reduce the iron. Thus, it is possible to create (μ-SRS)[Fe(CO)$_2$L]$_2$ complexes (where L is a phosphine), which can catalyze proton reduction at the Fe(I)Fe(0) redox state, in contrast to the Fe(0)Fe(0) state that is required for the hexacarbonyls. In short, hydrogen evolution can occur over a range of different oxidation levels through various electrochemical mechanisms, depending on the degree of substitution, the nature of the ligands, and the strength of the acid used in catalysis. One particularly interesting complex, complex **5** in Figure 6.5, features a pyridyl-phosphine ligand. This phosphine includes not only the ligating phosphorous but also a pendant base that can serve as a proton relay much like the bridging azadithiolate unit in the enzyme [109]. Electrochemical studies showed that this complex requires approximately 360–490 mV less overpotential than other phosphine complexes for catalysis. Remarkable among these complexes, the NHC-containing complex, $(\mu\text{-}S(CH_2)_3S)[Fe(CO)_3][Fe(CO)_2IMes]$, where IMes is 1,3-bis(2,4,6-trimethylphenyl)imidazol-2-ylidene (complex **6** in Figure 6.5), has unusual electrochemical properties. It undergoes a concerted two-electron reduction that involves reduction of both the iron and the carbene, resulting in electrocatalytic proton reduction via an electrochemical–electrochemical–chemical–chemical mechanism [110].

Combining successful features from a number of different model complexes, in 2011, Camara and Rauchfuss reported a complex, $Fe_2[(SCH_2)_2NBn](CO)_3(FcP^*)(dppv)$ $(Bn = CH_2Ph; FcP^* = Cp^*Fe(C_5Me_4CH_2PEt_2); Cp^* = $ pentamethylcyclopentadiene; Me = methyl; Et = ethyl) (complex **7** in Figure 6.5), designed to simultaneously incorporate a diiron catalytic center, an acid/base functionality in the bridging thiolate ligand, and a redox-active functionality to provide an additional reduction equivalent [111]. This is the first example of a compound that includes a redox-active component intended to mimic the [4Fe4S] cluster in which that moiety also plays a functional role in catalysis by the model compound. Chemical oxidation of [**7**] to [**7**]$^{2+}$ resulted in a complex active in catalytic oxidation of hydrogen at a modest rate of 0.4 turnover h^{-1}. Although this rate is not comparable to that of a hydrogenase, this is the first example of catalytic hydrogen oxidation by an organosulfur iron carbonyl. All other reported transformations have been stoichiometric [111].

6.4.2 [NiFe]-Hydrogenase Models

A number of close structural mimics of [NiFe]-hydrogenases that include tetrathiolate coordination of the nickel, strongly π-accepting ligands to the iron, and bridging sulfur coordination between the two metals have been reported [112–116]. However, despite structural similarity to the enzyme, none of these complexes were found to have either hydrogen oxidation or proton reduction activity. Only three [Ni–Fe] organometallic complexes, **8**, **9**, and **10** (Figure 6.6), have demonstrated nascent proton reduction activity. Complex **8**, a binuclear Ni–Fe complex in which the terminal thiolates of the enzyme were replaced by a chelating phosphine ligand, reduced protons from trifluoroacetic acid in acetonitrile at a rate of 20–50 s^{-1} with 300–400 mV of overpotential [117,118]. The trimetallic complexes **9** and **10** also have catalytic activity under similar conditions, albeit at much lower rates: 0.1–6 h^{-1} [119–121]. Incorporation of organoruthenium fragments into complexes has been a more successful route to production of catalysts. In electrocatalytic experiments, several Ni–Ru complexes have been shown to be competent proton reduction catalysts for close to 100 turnovers with 95–98% coulombic efficiency of electrons ultimately appearing in hydrogen. However, relatively large overpotentials, 0.5–1.0 V, were required to observe appreciable catalytic rates, and early models were only functional in organic solvents [122–127].

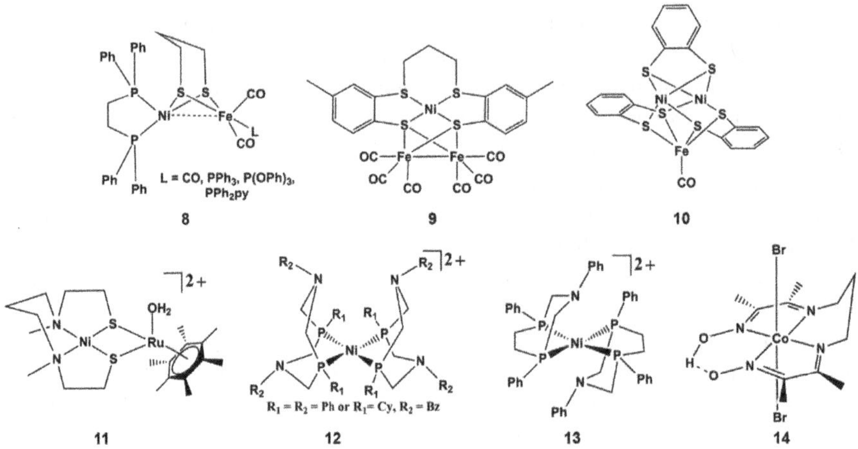

FIGURE 6.6   Schematic representations of selected models of [NiFe]-hydrogenases.

Complex **11** (Figure 6.6) is perhaps the most functionally similar model complex to [NiFe]-hydrogenases, although it is not capable of hydrogenase catalysis. It is the first water-soluble Ni–Ru complex that is able to heterolytically cleave hydrogen to form a hydride [128–130]. Interestingly, at low pH, the hydride ligands behave more like protons and undergo H^+/D^+ exchange with D^+. This is a property shared with hydrogenases but not possessed by most model compounds. Although not a hydrogen oxidation catalyst, at high pH (7–10), this complex catalyzes catalytic hydrogenation of carbonyls to the corresponding alcohols, indicating that it may be a promising route for development of other catalysts.

A number of mononuclear organometallic hydrogen production/oxidation catalysts based on nickel and cobalt are also known, and incorporation of ideas gleaned from enzymological studies has helped to improve these compounds [131–133]. In an approach pioneered by DuBois and coworkers, complex **12** (Figure 6.6) uses phosphine ligands with pendant amines to coordinate nickel. These amines are believed to serve as proton relay sites, much like the azadithiolate ligand of [FeFe]-hydrogenases, leading to proton reduction [134,135] turnover frequencies of $130–350 \, s^{-1}$ at an overpotential of approximately $300 \, mV$ [129,130]. In 2011, Helm et al. reported a variation, complex **13** (Figure 6.6), in which only a single amine is present in each phosphine ligand. This complex was designed to prevent formation of a stable *exo*-protonated complex that is catalytically inactive [136]. Although the TOF was somewhat increased by this change, to about $500 \, s^{-1}$ at $300 \, mV$ of overpotential, it is interesting to note that the rate-determining steps for these two compounds are quite different. Whereas complex **12** is limited by formation of the catalytically competent *endo–endo* isomer, complex **13** is limited by hydrogen elimination. This difference arises from significant structural differences between complexes **12** and **13**. Although **12** can be described as tetrahedral, **13** is much more planar. The result is that the Ni(II)/Ni(I) reduction potential is much lower for **13** and more overpotential is required for catalysis. It is thus perhaps worth reflecting on these compounds as examples that a relatively subtle change in ligand can have a significant impact on metallocenter properties and catalysis.

Finally, in surveying the mechanisms of [NiFe]-hydrogenase biomimetic complexes, the family of mononuclear cobalt and nickel diimine–dioxime catalysts reported by Fontecave and coworkers is also remarkable; see, for example, complex **14** (Figure 6.6). These compounds were designed to create more stable catalysts than previously characterized cobaloxime compounds, and that goal was achieved [137]. Interestingly, detailed electrochemical studies led to a hypothesized catalytic pathway for proton reduction by these compounds that is a homolytic process, that is, combination of two hydrogen atoms, on formation of a transient bimetallic complex. Once again, it can be difficult to predict the reaction pathway simply from the starting state of the catalyst, and even the nuclearity of the metallocatalyst may be, in some cases, suspect.

6.4.3   Incorporation of Outer Coordination Sphere Features

Fast catalysis by hydrogenases relies not only on efficient active sites but also on careful incorporation of these active sites into a protein matrix that contributes to stability, proton and electron transfers, and avoidance of side reactions with products or inhibitors. Small organometallic complexes largely ignore the myriad of possible roles played by interactions between the protein and the metallocenter, but this theme has been gaining traction with a handful of recent publications. In 2011, Shaw and coworkers incorporated mono- and dipeptides into the outer sphere of complex **12** (Figure 6.6) to explore the roles of secondary coordination sphere interactions in modulating catalysis [138]. Starting in 2007, our own group has paved the way for developing synthetic routes to incorporate diiron carbonyls into larger protein and peptide scaffolds [139,140]. Also in 2011, Hayashi and coworkers used this methodology to incorporate a diiron complex into cytochrome c, creating a catalyst capable of light-driven hydrogen production in the presence of a photosensitizer and a sacrificial electron donor [141]. Evidence is mounting that simply mixing such components in solution does not create ideal functional systems, and we anticipate that more reports will soon describe covalent connections designed to stabilize and enhance catalytic activity. For example, in a mimic of the iron–sulfur conduit of hydrogenases, compound **12** (Figure 6.6) was grafted to carbon nanotubes to provide a plentiful source of electrons for electrocatalysis resulting in a system with a TON on the order of 10^5 at overpotentials of $<20\,\text{mV}$ [142,143].

6.5   OUTLOOK

As described herein, electrochemical studies of hydrogenases have been crucial in developing a mechanistic understanding of both the catalytic cycle of the enzyme and its oxidative inactivation. In turn, that understanding has been exploited in the construction of enzyme-based first-generation fuel cells and hydrogen production devices. It is likely that these devices can and will be substantially improved through device engineering for use in niche applications. However, hydrogenases are likely to prove more valuable as templates for construction of nonprecious metal-based organometallic electrocatalysts. Hydrogenases can generally be thought of as models of biological multielectron catalysis with earth-abundant transition metals, and, on that front, there is still much to be learned.

ACKNOWLEDGMENTS

This research was supported by the Center for Bio-Inspired Solar Fuel Production, an Energy Frontier Research Center funded by the U.S. Department of Energy, Office of Science, Office of Basic Energy Sciences, under Award No. DE-SC0001016.

LIST OF ABBREVIATIONS

BFC	biological fuel cell
Bn	CH_2Ph
BOx	bilirubin oxidase
*Ca*HydA	*Clostridium acetobutylicum* [FeFe]-hydrogenase
Cp^*	pentamethylcyclopentadiene
Db	*Desulfomicrobium baculatum*
dppv	*cis*-$C_2H_2(PPh_2)_2$
*Ec*Hyd1	*Escherichia coli* hydrogenase I (oxygen-tolerant)
EPR	electron paramagnetic resonance
Et	ethyl
FcP^*	$Cp^*Fe(C_5Me_4CH_2PEt_2)$
FTIR	Fourier transform infrared spectroscopy
MBH	membrane-bound hydrogenase
Me	methyl
*Mv*BOx	*Myrothecium verrucaria* bilirubin oxidase
NAD	nicotinamide adenine dinucleotide
NADH	nicotinamide adenine dinucleotide (reduced form)
NAD(P)	nicotinamide adenine dinucleotide phosphate
NHC	N-heterocyclic carbene
OCV	open-circuit voltage
PEM	proton exchange membrane
PFE	protein film electrochemistry
Ph	phenyl
*Re*MBH	*Ralstonia eutropha* membrane-bound hydrogenase
*Rm*MBH	*Ralstonia metallidurans* membrane-bound hydrogenase
RuP	ruthenium-based photosensitizer
SH	soluble hydrogenase
SHE	standard hydrogen electrode
TOF	turnover frequency

REFERENCES

1. Cammack R, Frey M, Robson R. *Hydrogen as a Fuel: Learning from Nature.* Taylor & Francis, London, 2001.

2. Vignais P, Billoud B. Occurrence, classification and biological function of hydrogenases: an overview. *Chem Rev* 2007;107:4206–4272.

3. Murray J, King D. Oil's tipping point has passed. *Nature* 2012;481:433–435.

4. U.S. DOE Office of Basic Energy Sciences. *Basic research needs for the hydrogen economy.* Report of the Basic Energy Sciences Workshop on Hydrogen Production, Storage, and Use, May 13–15, 2003 (http://science.energy.gov/~/media/bes/pdf/reports/files/nhe_rpt.pdf).

5. *Global warming & climate change.* The New York Times, 2011.

6. Antolini E. Palladium in fuel cell catalysis. *Energy Environ Sci* 2009;2:915–931.

7. Serov A, Kwak C. Review of non-platinum anode catalysts for DMFC and PEMFC application. *Appl Catal B: Environ* 2009;90:313–320.

8. Jones AK, Sillery E, Albracht SPJ, Armstrong FA. Direct comparison of the electro-catalytic oxidation of hydrogen by an enzyme and a platinum catalyst. *Chem Commun* 2002; 866–867.

9. Madden C, Vaughn MD, Díez-Pérez I, Brown KA, King PW, Gust D, Moore AL, Moore TA. Catalytic turnover of [FeFe]-hydrogenase based on single-molecule imaging. *J Am Chem Soc* 2012;134:1577–1582.

10. Volbeda A, Charon M-H, Piras C, Hatchikian EC, Frey M, Fontecilla-Camps JC. Crystal structure of the nickel–iron hydrogenase from *Desulfovibrio gigas. Nature* 1995;373:580–587.

11. Volbeda A, Garcin E, Piras C, De Lacey AL, Fernandez VM, Hatchikian EC, Frey M, Fontecilla-Camps JC. Structure of the [NiFe] hydrogenase active site: evidence for biologically uncommon Fe ligands. *J Am Chem Soc* 1996;118:12989–12996.

12. Mulder DW, Shepard EM, Meuser JE, Joshi N, King PW, Posewitz MC, Broderick JB, Peters JW. Insights into [FeFe]-hydrogenase structure, mechanism, and maturation. *Structure* 2011;19:1038–1052.

13. Shomura Y, Yoon K-S, Nishihara H, Higuchi Y. Structural basis for a [4Fe–4S] cluster in the oxygen-tolerant membrane-bound [NiFe]-hydrogenase. *Nature* 2011;479:253–256.

14. Higuchi Y, Ogata H, Miki K, Yasuoka N, Yagi T. Removal of the bridging ligand atom at the Ni–Fe active site of [NiFe] hydrogenase upon reduction with H_2, as revealed by X-ray structure analysis at 1.4 Angstrom resolution. *Struct Fold Des* 1999;7:549–556.

15. Higuchi Y, Yagi T, Yasuoka N. Unusual ligand structure in [Ni–Fe] active center and an additional Mg site in hydrogenase revealed by high resolution X-ray structure analysis. *Structure* 1997;5:1671–1680.

16. Volbeda A, Martin L, Cavazza C, Matho M, Faber BW, Roseboom W, Albracht SPJ, Garcin E, Rouset M, Fontecilla-Camps JC. Structural differences between the ready and unready oxidized states of [NiFe] hydrogenases. *J Biol Inorg Chem* 2005;10: 239–249.

17. Montet Y, Amara P, Volbeda A, Vernede X, Hatchikian EC, Field MJ, Frey M, Fontecilla-Camps JC. Gas access to the active site of [Ni–Fe] hydrogenases probed by X-ray crystallography and molecular dynamics. *Nat Struct Biol* 1997;4:523–526.

18. Volbeda A, Montet Y, Vernede X, Hatchikian EC, Fontecilla-Camps JC. High-resolution crystallographic analysis of *Desulfovibrio fructosovorans* [NiFe] hydrogenase. *Int J Hydrogen Energy* 2002;27:1449–1461.

19. Ogata H, Kellers P, Lubitz W. The crystal structure of the [NiFe] hydrogenase from the photosynthetic bacterium *Allochromatium vinosum*: characterization of the oxidized enzyme (Ni-A state). *J Mol Biol* 2010;402:428–444.

20. Marques MC, Coelho R, De Lacey AL, Pereira IAC, Matias PM. The three-dimensional structure of [NiFeSe] hydrogenase from *Desulfovibrio vulgaris* Hildenborough: a hydro-genase without a bridging ligand in the active site in its oxidised, "as-isolated" state. *J Mol Biol* 2010;396:893–907.

21. Garcin E, Vernede X, Hatchikian EC, Volbeda A, Frey M, Fontecilla-Camps JC. The crystal structure of a reduced [NiFeSe] hydrogenase provides an image of the activated catalytic center. *Struct Fold Des* 1999;7:557–566.

22. Peters JW, Lanzilotta WN, Lemon BJ, Seefeldt LC. X-ray crystal structure of the Fe-only hydrogenase (CpI) from *Clostridium pasteurianum* to 1.8 Angstrom resolution. *Science* 1998;282:1853–1858.

23. Nicolet Y, Piras C, Legrand P, Hatchikian CE, Fontecilla-Camps JC. *Desulfovibrio desulfuricans* iron hydrogenase: the structure shows unusual coordination to an active site Fe binuclear center. *Struct Fold Des* 1999;7:13–23.

24. Pierik AJ, Roseboom W, Happe RP, Bagley KA, Albracht SPJ. Carbon monoxide and cyanide as intrinsic ligands to iron in the active site of [NiFe]-hydrogenases. NiFe $(CN)_2CO$, biology's way to activate H_2. *J Biol Chem* 1999;274:3331–3337.

25. van der Spek TM, Arendsen AF, Happe RP, Yun S, Bagley KA, Stufkens DJ, Hagen WR, Albracht SPJ. Similarities in the architecture of the active sites of Ni-hydrogenases and Fe-hydrogenases detected by means of infrared spectroscopy. *Eur J Biochem* 1996;237:629–634.

26. Happe RP, Roseboom W, Pierik AJ, Albracht SPJ, Bagley KA. Biological activation of hydrogen. *Nature* 1997;385:126–126.

27. Bagley KA, Duin EC, Roseboom W, Albracht SPJ, Woodruff WH. Infrared-detectable groups sense changes in charge density on the nickel center in hydrogenase from *Chromatium vinosum*. *Biochemistry* 1995;34:5527–5535.

28. Bennett B, Lemon BJ, Peters JW. Reversible carbon monoxide binding and inhibition at the active site of the Fe-only hydrogenase. *Biochemistry* 2000;39:7455–7460.

29. Lemon BJ, Peters JW. Binding of exogenously added carbon monoxide at the active site of the iron-only hydrogenase (CpI) from *Clostridium pasteurianum*. *Biochemistry* 1999;38:12969–12973.

30. Silakov A, Kamp C, Reijerse E, Happe T, Lubitz W. Spectroelectrochemical characteri-zation of the active site of the [FeFe]-hydrogenase HydA1 from *Chlamydomonas reinhardtii*. *Biochemistry* 2009;48:7780–7786.

31. Silakov A, Reijerse EJ, Albracht SPJ, Hatchikian EC, Lubitz W. The electronic structure of the H-cluster in the [FeFe]-hydrogenase from *Desulfovibrio desulfuricans*: a Q-band Fe-57-ENDOR and HYSCORE study. *J Am Chem Soc* 2007;129:11447–11458.

32. Popescu CV, Münck E. Electronic structure of the H cluster in Fe-hydrogenases. *J Am Chem Soc* 1999;121:7877–7884.

33. Cao ZX, Hall MB. Modeling the active sites in metalloenzymes. 3. Density functional calculations on models for Fe-hydrogenase: structures and vibrational frequencies of the

observed redox forms and the reaction mechanism at the diiron active center. *J Am Chem Soc* 2001;123:3734–3742.

34. Shima S, Pilak O, Vogt S, Schick M, Stagni MS, Meyer-Klaucke W, Warkentin E, Thauer RK, Ermler U. The crystal structure of Fe-hydrogenase reveals the geometry of the active site. *Science* 2008;321:572–575.

35. Lyon EJ, Shima S, Boecher R, Thauer RK, Grevels F-W, Bill E, Roseboom W, Albracht SPJ. Carbon monoxide as an intrinsic ligand to iron in the active site of the iron–sulfur-cluster-free hydrogenase H_2-forming methylenetetrahydromethanopterin dehydrogenase as revealed by infrared spectroscopy. *J Am Chem Soc* 2004;126:14239–14248.

36. Korbas M, Vogt S, Meyer-Klaucke W, Bill E, Lyon EJ, Thauer RK, Shima S. The iron–sulfur cluster-free hydrogenase (Hmd) is a metalloenzyme with a novel iron binding motif. *J Biol Chem* 2006;281:30804–30813.

37. Shima S, Lyon EJ, Thauer RK, Mienert B, Bill E. Mössbauer studies of the iron–sulfur cluster-free hydrogenase: the electronic state of the mononuclear Fe active site. *J Am Chem Soc* 2005;127:10430–10435.

38. Frey M. Hydrogenases: hydrogen-activating enzymes. *Chem Biol Chem* 2002;3:153–160.

39. McIntosh CL, Germer F, Schulz R, Appel J, Jones AK. The [NiFe]-hydrogenase of the cyanobacterium *Synechocystis* sp. PCC 6803 works bidirectionally with a bias to H_2 production. *J Am Chem Soc* 2011;133:11308–11319.

40. Ghirardi ML, Dubini A, Yu J, Maness P-C. Photobiological hydrogen-producing systems. *Chem Soc Rev* 2009;38:52–61.

41. Stripp ST, Happe T. How algae produce hydrogen—news from the photosynthetic hydrogenase. *Dalton Trans* 2009: 9960–9969.

42. Winter G, Buhrke T, Lenz O, Jones AK, Forgber M, Friedrich B. A model system for [NiFe] hydrogenase maturation studies: purification of an active site-containing hydrogenase large subunit without small subunit. *FEBS Lett* 2005;579:4292–4296.

43. Dementin S, Burlat B, Fourmond V, Leroux F, Liebgott P-P, Hamdan AA, Léger C, Rousset M, Guigliarelli B, Bertrand P. Rates of intra- and intermolecular electron transfers in hydrogenase deduced from steady-state activity measurements. *J Am Chem Soc* 2011;133:10211–10221.

44. Rousset M, Montet Y, Guigliarelli B, Forget N, Asso M, Bertrand P, Fontecilla-Camps JC, Hatchikian EC. 3Fe-4S to 4Fe-4S cluster conversion in *Desulfovibrio fructosovorans* [NiFe] hydrogenase by site-directed mutagenesis. *Proc Natl Acad Sci USA* 1998;95:11625–11630.

45. Pershad HR, Duff JLC, Heering HA, Duin EC, Albracht SPJ, Armstrong FA. Catalytic electron transport in *Chromatium vinosum* [NiFe]-hydrogenase: application of voltammetry in detecting redox-active centers and establishing that hydrogen oxidation is very fast even at potentials close to the reversible H^+/H_2 value. *Biochemistry* 1999;38:8992–8999.

46. Dementin S, Belle V, Bertrand P, Guiglliarelli B, Adryanczyk-Perrier G, De Lacey AL, Fernandez VM, Rousset M, Léger C. Changing the ligation of the distal [4Fe4S] cluster in NiFe hydrogenase impairs inter- and intramolecular electron transfers. *J Am Chem Soc* 2006;128:5209–5218.

47. Page CC, Moser CC, Chen XX, Dutton PL. Natural engineering principles of electron tunnelling in biological oxidation–reduction. *Nature* 1999;402:47–52.

48. Burgdorf T, Lenz O, Buhrke T, van der Linden E, Jones AK, Albracht SPJ, Friedrich B. [NiFe]-hydrogenases of *Ralstonia eutropha* H16: modular enzymes for oxygen-tolerant biological hydrogen oxidation. *J Mol Microb Biotechnol* 2005;10:181–196.

49. Silva PJ, de Castro B, Hagen WR. On the prosthetic groups of the [NiFe] sulfhydrogenase from *Pyrococcus furiosus*: topology, structure, and temperature-dependent redox chemistry. *J Biol Inorg Chem* 1999;4:284–291.

50. Kubas GJ. Fundamentals of H_2 binding and reactivity on transition metals underlying hydrogenase function and H_2 production and storage. *Chem Rev* 2007;107:4152–4205.

51. Stripp ST, Goldet G, Brandmayr C, Sanganas O, Vincent KA, Haumann M, Armstrong FA, Happe T. How oxygen attacks [FeFe] hydrogenases from photosynthetic organisms. *Proc Natl Acad Sci USA* 2009;106:17331–17336.

52. Jones AK, Lamle SE, Pershad HR, Vincent KA, Albracht SPJ, Armstrong FA. Enzyme electrokinetics: electrochemical studies of the anaerobic interconversions between active and inactive states of *Allochromatium vinosum* [NiFe]-hydrogenase. *J Am Chem Soc* 2003;125:8505–8514.

53. Vincent KA, Cracknell JA, Lenz O, Zebger I, Friedrich B, Armstrong FA. Electrocatalytic hydrogen oxidation by an enzyme at high carbon monoxide or oxygen levels. *Proc Natl Acad Sci USA* 2005;102:16951–16954.

54. Lamle SE, Albracht SPJ, Armstrong FA. Electrochemical potential-step investigations of the aerobic interconversions of [NiFe]-hydrogenase from *Allochromatium vinosum*: insights into the puzzling difference between unready and ready oxidized inactive states. *J Am Chem Soc* 2004;126:14899–14909.

55. Léger C, Dementin S, Bertrand P, Rousset M, Guigliarelli B. Inhibition and aerobic inactivation kinetics of *Desulfovibrio fructosovorans* [NiFe]-hydrogenase studied by protein film voltammetry. *J Am Chem Soc* 2004;126:12162–12172.

56. Bleijlevens B, van Broekhuizen FA, De Lacey AL, Roseboom W, Fernandez VM, Albracht SPJ. The activation of the [NiFe]-hydrogenase from *Allochromatium vinosum*. An infrared spectro-electrochemical study. *J Biol Inorg Chem* 2004;9:743–752.

57. Kurkin S, George SJ, Thorneley RNF, Albracht SPJ. Hydrogen-induced activation of the [NiFe]-hydrogenase from *Allochromatium vinosum* as studied by stopped-flow infrared spectroscopy. *Biochemistry* 2004;43:6820–6831.

58. Millo D, Hildebrandt P, Pandelia ME, Lubitz W, Zebger I. SEIRA spectroscopy of the electrochemical activation of an immobilized [NiFe] hydrogenase under turnover and non-turnover conditions. *Angew Chem Int Ed* 2011;50:2632–2634.

59. Lenz O, Ludwig M, Schubert T, Bürstel I, Ganskow S, Goris T, Schwarze A, Friedrich B. H_2 conversion in the presence of O_2 as performed by the membrane-bound [NiFe]-hydrogenase of *Ralstonia eutropha*. *ChemPhysChem* 2010;11:1107–1119.

60. Pandelia M-E, Fourmond V, Tron-Infossi P, Lojou E, Bertrand P, Léger C, Guidici-Orticoni M-T, Lubitz W. Membrane-bound hydrogenase I from the hyperthermophilic bacterium *Aquifex aeolicus*: enzyme activation, redox intermediates and oxygen tolerance. *J Am Chem Soc* 2010;132:6991–7004.

61. Pandelia M-E, Nitschke W, Infossi P, Giudici-Ortoni M-T, Bill E, Lubitz W. Characterization of a unique [FeS] cluster in the electron transfer chain of the oxygen tolerant [NiFe] hydrogenase from *Aquifex aeolicus*. *Proc Natl Acad Sci USA* 2011;108: 6097–6102.

62. Lukey MJ, Parkin A, Roessler MM, Murphy BJ, Harmer J, Palmer T, Sargent F, Armstrong FA. How *Escherichia coli* is equipped to oxidize hydrogen under different redox conditions. *J Biol Chem* 2010;285:3928–3938.

63. Lukey MJ, Roessler MM, Parkin A, Evans RM, Davies RA, Lenz O, Friedrich B, Sargent F, Armstrong FA. Oxygen-tolerant [NiFe]-hydrogenases: the individual and collective importance of supernumerary cysteines at the proximal Fe–S cluster. *J Am Chem Soc* 2011;133:16881–16892.

64. Fritsch J, Scheerer P, Frielingsdorf S, Kroschinsky S, Friedrich B, Lenz O, Spahn CMT. The crystal structure of an oxygen-tolerant hydrogenase uncovers a novel iron–sulphur centre. *Nature* 2011;479:249–252.

65. Goris T, Wait AF, Saggu M, Fritsch J, Heidary N, Stein M, Zebger I, Lendzian F, Armstrong FA, Friedrich B, Lenz O. A unique iron–sulfur cluster is crucial for oxygen tolerance of a [NiFe]-hydrogenase. *Nat Chem Biol* 2011;7:310–318.

66. Mertens R, Greiner L, van den Ban ECD, Haaker H, Liese A. Practical applications of hydrogenase I from *Pyrococcus furiosus* for NADPH generation and regeneration. *J Mol Catal B: Enzym* 2003;24–25:39–52.

67. Reeve HA, Lauterbach L, Ash PA, Lenz O, Vincent KA. A modular system for regeneration of NAD cofactors using graphite particles modified with hydrogenase and diaphorase moieties. *Chem Commun* 2012;48:1589–1591.

68. Greiner L, Schroder I, Mulle DH, Liese A. Utilization of adsorption effects for the continuous reduction of $NADP^+$ with molecular hydrogen by *Pyrococcus furiosus* hydrogenase. *Green Chem* 2003;5:697–700.

69. Limoges B, Marchal D, Mavre F, Saveant J-M. Electrochemistry of immobilized redox enzymes: kinetic characteristics of NADH oxidation catalysis at diaphorase monolayers affinity immobilized on electrodes. *J Am Chem Soc* 2006;128:2084–2092.

70. Germer F, Zebger I, Saggu M, Lendzian F, Schulz R, Appel J. Overexpression, isolation, and spectroscopic characterization of the bidirectional [NiFe] hydrogenase from *Synechocystis* sp. PCC 6803. *J Biol Chem* 2009;284:36462–36472.

71. Lauterbach L, Liu J, Horch M, Hummel P, Schwarze A, Haumann M, Vincent KA, Lenz O, Zebger I. The hydrogenase subcomplex of the NAD^+-reducing [NiFe] hydrogenase from *Ralstonia eutropha* — insights into catalysis and redox interconversions. *Eur J Inorg Chem* 2011;7:1067–1079.

72. Hambourger M, Gervaldo M, Svedruzic D, King PW, Gust D, Ghirardi M, Moore AL, Moore TA. [FeFe]-hydrogenase-catalyzed H_2 production in a photoelectrochemical biofuel cell. *J Am Chem Soc* 2008;130:2015–2022.

73. Woolerton TW, Vincent KA. Oxidation of dilute H_2 and H_2/O_2 mixtures by hydrogenases and Pt. *Electrochim Acta* 2009;54:5011–5017.

74. Karyakin AA, Morozov SV, Karyakina EE, Zorin NA, Perelygin VV, Cosnier S. Hydrogenase electrodes for fuel cells. *Biochem Soc Trans* 2005;33:73–75.

75. Karyakin AA, Morozov SV, Voronin OG, Zorin NA, Karyakina EE, Fateyev VN, Cosnier S. The limiting performance characteristics in bioelectrocatalysis of hydrogenase enzymes. *Angew Chem Int Ed* 2007;46:7244–7246.

76. Tsujimura S, Fujita M, Tatsumi H, Kano K, Ikeda T. Bioelectrocatalysis-based dihydrogen/dioxygen fuel cell operating at physiological pH. *Phys Chem Chem Phys* 2001;3:1331–1335.

77. Kortlüke C, Horstmann K, Schwartz E, Rohde M, Binsack R, Friedrich B. A gene complex coding for the membrane-bound hydrogenase of *Alcaligenes eutrophus* H16. *J Bacteriol* 1992;174:6277–6289.

78. Saggu M, Zebger I, Ludwig M, Lenz O, Friedrich B, Hildebrandt P, Lendzian F. Spectroscopic insights into the oxygen-tolerant membrane-associated [NiFe] hydrogenase of *Ralstonia eutropha* H16. *J Biol Chem* 2009;284:16264–16276.

79. Pohlmann A, Fricke WF, Reinecke F, Kusian B, Liesegang H, Cramm R, Eitinger T, Ewering C, Pötter M, Schwartz E, Strittmatter A, Voß I, Gottschalk G, Steinbüchel A, Friedrich B, Bowien B. Genome sequence of the bioplastic-producing "knallgas" bacterium *Ralstonia eutropha* H16. *Nat Biotechnol* 2006;24:1257–1262.

80. Schwartz E, Voigt B, Zühlke D, Pohlmann A, Lenz O, Albrecht D, Schwarze A, Kohlmann Y, Krause C, Hecker M, Friedich B. A proteomic view of the facultatively chemolithoautotrophic lifestyle of *Ralstonia eutropha* H16. *Proteomics* 2009;9: 5132–5142.

81. Albracht SPJ. Nickel hydrogenases—in search of the active site. *Biochim Biophys Acta: Bioenerg* 1994;1188:167–204.

82. Thauer RK, Kaster A-K, Seedorf H, Buckel W, Hedderich R. Methanogenic archaea: ecologically relevant differences in energy conservation. *Nat Rev Microbiol* 2008;6: 579–591.

83. Vincent KA, Cracknell JA, Clark JR, Ludwig M, Lenz O, Friedrich B, Armstrong FA. Electricity from low-level H_2 in still air—an ultimate test for an oxygen tolerant hydrogenase. *Chem Commun* 2006;48:5033–5035.

84. Wait AF, Parkin A, Morley GM, dos Santos L, Armstrong FA. Characteristics of enzyme-based hydrogen fuel cells using an oxygen-tolerant hydrogenase as the anodic catalyst. *J Phys Chem C* 2010;114:12003–12009.

85. Gao F, Viry L, Maugey M, Poulin P, Mano N. Engineering hybrid nanotube wires for high-power biofuel cells. *Nat Commun* 2010;1(2):2–7.

86. Alonso-Lomillo MA, Rüdiger O, Maroto-Valiente A, Velez M, Rodríguez-Ramos I, Muñoz FJ, Fernández VM, De Lacey AL. Hydrogenase-coated carbon nanotubes for efficient H_2 oxidation. *Nano Lett* 2007;7:1603–1608.

87. Kwan P, Schmitt D, Volosin AM, McIntosh CL, Seo D-K, Jones AK. Spectroelectrochemistry of cytochrome *c* and azurin immobilized in nanoporous antimony-doped tin oxide. *Chem Commun* 2011;47:12367–12369.

88. Baur J, Le Goff A, Dementin S, Holzinger M, Rousset M, Cosnier S. Three-dimensional carbon nanotube–polypyrrole–[NiFe] hydrogenase electrodes for the efficient electro-catalytic oxidation of H_2. *Int J Hydrogen Energy* 2011;36:12096–12101.

89. Healy AJ, Reeve HA, Parkin A, Vincent KA. Electrically conducting particle networks in polymer electrolyte as three-dimensional electrodes for hydrogenase electrocatalysis. *Electrochim Acta* 2011;56:10786–10790.

90. Gutiérrez-Sánchez C, Olea D, Marques M, Fernández VM, Pereira IAC, Vélez M, De Lacey AL. Oriented immobilization of a membrane-bound hydrogenase onto an electrode for direct electron transfer. *Langmuir* 2011;27:6449–6457.

91. Lojou E. Hydrogenases as catalysts for fuel cells: strategies for efficient immobilization at electrode interfaces. *Electrochim Acta* 2012;56:10385–10397.

92. Zipoli F, Car R, Cohen MH, Selloni A. Simulation of electrocatalytic hydrogen production by a bioinspired catalyst anchored to a pyrite electrode. *J Am Chem Soc* 2010;132:8593–8601.

93. Shastik ES, Vokhmyanina DV, Zorn NA, Voronin OG, Karyakin AA, Tsygankov AA. Demonstration of hydrogenase electrode operation in a bioreactor. *Enzyme Microb Technol* 2011;49:453–458.

94. Lee HS, Vermaas WFJ, Rittmann BE. Biological hydrogen production: prospects and challenges. *Trends Biotechnol* 2010;28:262–271.

95. Reisner E, Fontecilla-Camps JC, Armstrong FA. Catalytic electrochemistry of a NiFeSe-hydrogenase on TiO_2 and demonstration of its suitability for visible-light driven H_2 production. *Chem Commun* 2009;5:550–552.

96. Reisner E, Powell DJ, Cavazza C, Fontecilla-Camps JC, Armstrong FA. Visible light-driven H_2 production by hydrogenases attached to dye-sensitized TiO_2 nanoparticles. *J Am Chem Soc* 2009;131:18457–18466.

97. Parkin A, Goldet G, Cavazza C, Fontecilla-Camps JC, Armstrong FA. The difference a Se makes? Oxygen-tolerant hydrogen production by the [NiFeSe]-hydrogenase from *Desulfomicrobium baculatum. J Am Chem Soc* 2008;130:13410–13416.

98. Bard AJ, Faulkner LR. *Electrochemical Methods: Fundamentals and Applications.* John Wiley & Sons, Inc., New York, 2001.

99. Lenz O, Gleiche A, Strack A, Friedrich B. Requirements for heterologous production of a complex metalloenzyme: the membrane-bound [NiFe] hydrogenase. *J Bacteriol* 2005;187:6590–6595.

100. Shepard EM, Boyd ES, Broderick JB, Peters JW. Biosynthesis of complex iron–sulfur enzymes. *Curr Opin Chem Biol* 2011;15:319–327.

101. Kuchenreuther JM, George SJ, Grady-Smith CS, Cramer SP, Swartz JR. Cell-free H-cluster synthesis and [FeFe] hydrogenase activation: all five CO and CN^- ligands derive from tyrosine. *PLoS One* 2011;6:e20346.

102. Kuchenreuther JM, Grady-Smith CS, Bingham AS, George SJ, Cramer SP, Swartz JR. High-yield expression of heterologous [FeFe] hydrogenases in *Escherichia coli. PLoS One* 2010;5:e15491.

103. Tard C, Pickett CJ. Structural and functional analogues of the active sites of the Fe-, [NiFe]-, and [FeFe]-hydrogenases. *Chem Rev* 2009;109:2245–2274.

104. Tard C, Liu X, Ibraham SK, Bruschi M, De Gioia L, Davies SC, Yang X, Wang L-S, Sawers G, Pickett CJ. Synthesis of the H-cluster framework of iron-only hydrogenase. *Nature* 2005;433:610–613.

105. Morvan D, Capon J-F, Gloaguen F, Le Goff A, Marchivie M, Michaud F, Schollhammer P, Talarmin J, Yaouanc J-J. N-heterocyclic carbene ligands in nonsymmetric diiron models of hydrogenase active sites. *Organometallics* 2007;26:2042–2052.

106. Orain P-Y, Capon J-F, Kervarec N, Gloaguen F, Pétillon F, Pichon R, Schollhammer P, Talarmin J. Use of 1,10-phenanthroline in diiron dithiolate derivatives related to the [Fe–Fe] hydrogenase active site. *Dalton Trans* 2007;34:3754–3756.

107. Ezzaher S, Capon J-F, Gloaguen F, Pétillon FY, Schollhammer P, Talarmin J. Evidence for the formation of terminal hydrides by protonation of an asymmetric iron hydrogenase active site mimic. *Inorg Chem* 2007;46:3426–3428.

108. Barton BE, Rauchfuss TB. Terminal hydride in [FeFe]-hydrogenase model has lower potential for H_2 production than the isomeric bridging hydride. *Inorg Chem* 2008;47: 2261–2263.

109. Li P, Wang M, Chen L, Liu J, Zhao Z, Sun L. Structures, protonation, and electrochemical properties of diiron dithiolate complexes containing pyridyl-phosphine ligands. *Dalton Trans* 2009;11:1919–1926.

110. Liu T, Darensbourg, MY. A Mixed-Valent, Fe(II)Fe(I), Diiron Complex Reproduces the Unique Rotated State of the [FeFe] Hydrogenase Active Site. *J. Am Chem. Soc.* 2007;129:7008–7009.

111. Camara JM, Rauchfuss TB. Combining acid-base, redox and substrate binding functionalities to give a complete model for the [Fe–Fe]-hydrogenase. *Nat Chem* 2011;4(1):26–30.

112. Lai CH, Reibenspies JH, Darensbourg MY. Thiolate bridged nickel–iron complexes containing both iron(0) and iron(II) carbonyls. *Angew Chem Int Ed* 1996;35:2390–2393.

113. Zhu WF, Marr AC, Wang Q, Neese F, Spencer DJE, Blake AJ, Cooke PA, Wilson C, Schröder M. Modulation of the electronic structure and the Ni–Fe distance in hetero-bimetallic models for the active site in [NiFe] hydrogenase. *Proc Natl Acad Sci USA* 2005;102:18280–18285.

114. Verhagen JAW, Lutz M, Spek AL, Bouwman E. Synthesis and characterisation of new nickel–iron complexes with an S-4 coordination environment around the nickel centre. *Eur J Inorg Chem* 2003;20:3968–3974.

115. Li ZL, Ohki Y, Tatsumi K. Dithiolato-bridged dinuclear iron–nickel complexes $[Fe(CO)_2(CN)_2(\mu\text{-}SCH_2CH_2CH_2S)Ni(S_2CNR_2)]^-$. Modeling the active site of [NiFe] hydrogenase. *J Am Chem Soc* 2005;127:8950–8951.

116. Sellmann D, Geipel F, Lauderbach F, Heinemann FW. $(C_6H_4S_2)Ni(\mu\text{-}`S\text{-}3')Fe(CO)(PMe_3)_2$: a dinuclear [NiFe] complex modeling the $(RS)_2Ni(\mu\text{-}SR)_2Fe(CO)(L)_2$ core of [NiFe] hydrogenase centers. *Angew Chem Int Ed* 2002;41:632–634.

117. Barton BE, Rauchfuss TB. Hydride-containing models for the active site of the nickel–iron hydrogenases. *J Am Chem Soc* 2010;132:14877–14885.

118. Carroll ME, Barton BE, Gray DL, Mack AE, Rauchfuss TB. Active-site models for the nickel–iron hydrogenases: effects of ligands on reactivity and catalytic properties. *Inorg Chem* 2011;50:9554–9563.

119. Lauderbach F, Prakash R, Götz AW, Munoz M, Heinemann FW, Nickel U, Hess BA, Sellmann D. Alternative synthesis, density functional calculations and proton reactivity study of a trinuclear [NiFe] hydrogenase model compound. *Eur J Inorg Chem* 2007;21:3385–3393.

120. Sellmann D, Lauderbach F, Geipel F, Heinemann FW, Moll M. A trinuclear [NiFe] cluster exhibiting structural and functional key features of [NiFe] hydrogenases. *Angew Chem Int Ed* 2004;43:3141–3144.

121. Perra A, Davies S, Hyde JR, Wang Q, McMaster J, Schröder M. Electrocatalytic production of hydrogen by a synthetic model of [NiFe] hydrogenases. *Chem Commun* 2006;10:1103–1105.

122. Canaguier S, Vaccaro L, Artero V, Ostermann R, Pécaut J, Field MJ, Fontecave M. Cyclopentadienyl ruthenium–nickel catalysts for biomimetic hydrogen evolution: electrocatalytic properties and mechanistic DFT studies. *Chem Eur J* 2009;15:9350–9364.

123. Oudart Y, Artero V, Pecaut J, Fontecave M. Ni(xbsms)Ru(CO)$_2$Cl$_2$: a bioinspired nickel–ruthenium functional model of [NiFe] hydrogenase. *Inorg Chem* 2006;45:4334–4336.

124. Oudart Y, Artero V, Pecaut J, Lebrun C, Fontecave M. Dinuclear nickel–ruthenium complexes as functional bio-inspired models of [NiFe] hydrogenases. *Eur J Inorg Chem* 2007;18:2613–2626.

125. Oudart Y, Artero V, Norel L, Traini C, Pécaut J, Fontecave M. Synthesis, crystal structure, magnetic properties and reactivity of a Ni–Ru model of NiFe hydrogenases with a pentacoordinated triplet ($S = 1$) Ni(II) center. *J Organomet Chem* 2009;694:2866–2869.

126. Vaccaro L, Artero V, Canaguier S, Fontecave M, Field MJ. Mechanism of hydrogen evolution catalyzed by NiFe hydrogenases: insights from a Ni–Ru model compound. *Dalton Trans* 2010;39:3043–3049.

127. Reynolds MA, Rauchfuss TB, Wilson SR. Ruthenium derivatives of NiS$_2$N$_2$ complexes as analogues of bioorganometallic reaction centers. *Organometallics* 2003;22:1619–1625.

128. Ogo S, Kabe R, Uehara K, Kure B, Nishimura T, Menon SC, Harada R, Fukuzumi S, Higuchi Y, Ohhara T, Tamada T, Kuroki R. A dinuclear Ni(μ-H)Ru complex derived from H$_2$. *Science* 2007;316:585–587.

129. Kure B, Matsumoto T, Ichikawa K, Fukuzumi S, Higuchi Y, Yagi T, Ogo S. pH-dependent isotope exchange and hydrogenation catalysed by water-soluble NiRu complexes as functional models for [NiFe] hydrogenases. *Dalton Trans* 2008;35:4747–4755.

130. Ogo S. Electrons from hydrogen. *Chem Commun* 2009;23:3317–3325.

131. Curtis CJ, Miedaner A, Ciancanelli R, Ellis WW, Noll BC, DuBois MR, DuBois DL. Ni(Et$_2$PCH$_2$NMeCH$_2$PEt$_2$)$_2$$^{2+}$ as a functional model for hydrogenases. *Inorg Chem* 2003;42:216–227.

132. McNamara WR, Han Z, Alperin PJ, Brennessel WW, Holland PL, Eisenberg R. A cobalt-dithiolene complex for the photocatalytic and electrocatalytic reduction of protons. *J Am Chem Soc* 2011;133:15368–15371.

133. Stubbert B D, Peters JC, Gray HB. Rapid water reduction to H$_2$ catalyzed by a cobalt bis (iminopyridine) complex. *J Am Chem Soc* 2011;133:18070–18073.

134. Wilson AD, Newell RH, McNevin MJ, Muckerman JT, Dubois MR, Dubois DL. Hydrogen oxidation and production using nickel-based molecular catalysts with positioned proton relays. *J Am Chem Soc* 2006;128:358–366.

135. Wilson AD, Shoemaker RK, Miedaner A, Muckerman JT, DuBois DL, Dubois MR. Nature of hydrogen interactions with Ni(II) complexes containing cyclic phosphine ligands with pendant nitrogen bases. *Proc Natl Acad Sci USA* 2007;104:6951–6956.

136. Helm ML, Stewart MP, Bullock RM, DuBois MR, DuBois DL. A synthetic nickel electrocatalyst with a turnover frequency above 100,000 s^{-1} for H$_2$ production. *Science* 2011;333:863–866.

137. Jacques PA, Artero V, Pecaut J, Fontecave M. Cobalt and nickel diimine–dioxime complexes as molecular electrocatalysts for hydrogen evolution with low overvoltages. *Proc Natl Acad Sci USA* 2009;106:20627–20632.

138. Jain A, Lense S, Linehan JC, Raugei S, Cho H, DuBois DL, Shaw WJ. Incorporating peptides in the outer-coordination sphere of bioinspired electrocatalysts for hydrogen production. *Inorg Chem* 2011;50:4073–4085.

139. Jones AK, Lichtenstein BR, Dutta A, Gordon G, Dutton PL. Synthetic hydrogenases: incorporation of an iron carbonyl thiolate into a designed peptide. *J Am Chem Soc* 2007;129:14844–14845.

140. Roy S, Shinde S, Hamilton GA, Hartnett HE, Jones AK. Artificial [FeFe]-hydrogenase: on resin modification of an amino acid to anchor a hexacarbonyldiiron cluster in a peptide framework. *Eur J Inorg Chem* 2011;7:1050–1055.

141. Sano Y, Onoda A, Hayashi T. A hydrogenase model system based on the sequence of cytochrome *c*: photochemical hydrogen evolution in aqueous media. *Chem Commun* 2011;47:8229–8231.

142. Le Goff A, Artero V, Jousselme B, Tran PD, Guillet N, Métayé R, Fihri A, Palacin S, Fontecave M. From hydrogenases to noble metal-free catalytic nanomaterials for H_2 production and uptake. *Science* 2009;326:1384–1387.

143. Tran PD, Le Goff A, Heidkamp J, Jousselme B, Guillet N, Palacin S, Dau H, Fontecave M, Artero V. Noncovalent modification of carbon nanotubes with pyrene-functionalized nickel complexes: carbon monoxide tolerant catalysts for hydrogen evolution and uptake. *Angew Chem Int Ed* 2011;50:1371–1374.

144. DeLano WL. The PyMOL Molecular Graphics System. 2002 (http://www.pymol.org).

7

PROTEIN ENGINEERING FOR ENZYMATIC FUEL CELLS

ELLIOT CAMPBELL AND SCOTT BANTA

Department of Chemical Engineering, Columbia University, New York, NY, USA

7.1 ENGINEERING ENZYMES FOR CATALYSIS

Enzymes possess a number of properties that make them attractive alternatives to conventional precious metal catalysts for the construction of fuel cells. They are highly efficient, exhibiting rapid kinetics and exquisite substrate specificity. They operate under mild conditions and can be produced cheaply and on a large scale. With the advent of protein engineering, several techniques have been developed that allow the properties of enzymes to be tuned for a wide range of applications. From a kinetic standpoint, these improvements can be in the affinity of the enzyme for the desired substrate (captured by the Michaelis constant, K_M) and the turnover rate of the enzyme (k_{cat}). In a biological fuel cell (BFC), the local concentration of the substrates is generally high; thus, the affinity of the enzyme for the substrate is less important than the enzymatic turnover rate. For this reason, the primary goal of protein engineering in BFCs has often been to maximize k_{cat} with a desired substrate. However, as the field of biocatalysis continues to grow, it has become clear that many other factors significantly impact the performance of BFCs (Figure 7.1). Here, we will focus on recent advances in the field of enzymatic fuel cells and the major considerations involved in engineering enzymes for efficient bioelectrocatalysis.

The field of protein engineering relies on two main approaches for obtaining proteins with desired properties. The first, rational design, attempts to take advantage of existing knowledge of the protein (sequence and structural information, structure–function relationships, computational modeling, etc.) to identify specific amino acid

Enzymatic Fuel Cells: From Fundamentals to Applications, First Edition. Edited by Heather R. Luckarift, Plamen Atanassov, and Glenn R. Johnson.
© 2014 John Wiley & Sons, Inc. Published 2014 by John Wiley & Sons, Inc.

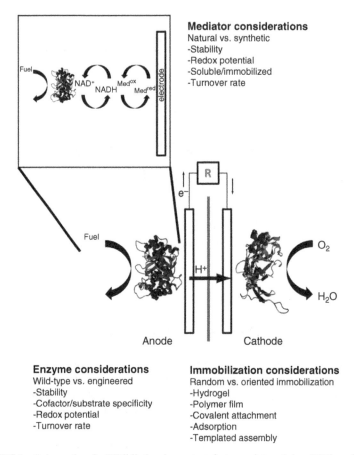

Mediator considerations
Natural vs. synthetic
-Stability
-Redox potential
-Soluble/immobilized
-Turnover rate

Enzyme considerations
Wild-type vs. engineered
-Stability
-Cofactor/substrate specificity
-Redox potential
-Turnover rate

Immobilization considerations
Random vs. oriented immobilization
-Hydrogel
-Polymer film
-Covalent attachment
-Adsorption
-Templated assembly

FIGURE 7.1   Schematic of a BFC listing important features determining BFC performance.

mutations that will elicit the desired changes. For instance, to increase the activity of an ethanol-oxidizing alcohol dehydrogenase with methanol, the size of the substrate-binding pocket could be decreased by replacing residues lining the binding pocket with larger residues [1]. Rational design has been successfully applied to several enzyme engineering problems, including increasing activity [2], altering specificity [3], and improving stability [4]. However, the complexities of the structure–function relationship can just as often lead to detrimental mutations, which are impossible to predict *a priori*. Further, the requirement for detailed structural and functional information excludes a large number of proteins.

The second approach, directed evolution, attempts to mimic natural evolution on a laboratory timescale. The basis of directed evolution is repeated rounds of mutagenesis or recombination followed by a screening or selection step (while maintaining a genotype–phenotype linkage) to identify proteins with improved properties. This process requires no structural or functional information and is widely applicable. Currently, the main drawback of directed evolution is the lack of availability of high-throughput selection schemes to identify improved mutants [5]. Although it is relatively straightforward to select for high-affinity binders to a target or even for proteins with increased stability, identifying enzymes with improved activities is more difficult. A wide array of enzymatic selection steps have been developed [5–10], but each is highly specialized for a particular reaction and none is applicable to BFCs. Instead, most of the successes in using directed evolution to improve enzyme performance have come through tedious and time-consuming screening methods, whereby each mutant is assayed individually. Such screens are low to medium throughput and therefore are only able to examine a small fraction of the possible mutants that are created in a directed evolution library. Regardless, directed evolution is still extremely useful and has been used to engineer enzymes for improved activity [11], altered specificity [12], altered optimum temperature and pH [13], and other beneficial properties [6–8,14]. Hybrid approaches using both techniques have also proved successful in a number of instances [15].

Electrons can be transferred between an enzyme and electrode directly, termed *direct electron transfer* (DET), or through a mediator, termed *mediated electron transfer* (MET). Enzymes capable of DET contain a reactive metal center or other redox center, such as flavin adenine dinucleotide (FAD), fixed in the active site. When the active site is located within a short distance of the electrode (less than ~20 Å), electrons are transferred between them through electron tunneling [16]. Only a few classes of enzymes have demonstrated DET, notably cytochrome *c*, some hydrogenases, peroxidases, oxidases, and laccases [17,18]. In BFCs, laccases and oxidases are the most commonly found enzymes capable of DET.

Laccases are used at the cathode of a BFC to catalyze the four-electron electroreduction of O_2 (which acts as the terminal electron acceptor) to water. They belong to a group of enzymes called "blue copper oxidases," due to the presence of a type 1 (T1) copper site in the enzyme. This copper site acts as the primary electron acceptor and imparts laccases with a blue color. Additional copper ions in type 2 and type 3 (T2/T3) sites form a trinuclear cluster that acts as a binding site for molecular oxygen. Here, the electrons transferred from the T1 site reduce oxygen to water. Laccases have been the subject of study for several decades, and have found use in fields ranging from wastewater treatment to the paper industry [19,20]. However, it is their DET ability and high redox potential that have led to their use in biosensors and BFCs.

Laccases are extremely efficient and operate near the thermodynamic redox potential for the dioxygen/water couple [21]. Further, studies have shown a high correlation between the potential of the T1 copper site and the catalytic rate of the enzyme [22]. In a BFC, a high redox potential is required to minimize thermodynamic losses, and a high enzymatic rate leads to high current densities. Thus, many studies have looked at ways to increase the redox potential and catalytic rates of laccases through protein engineering. Site-directed mutagenesis has been used to alter the residues surrounding the T1 copper site based on sequence alignments of laccases with known redox potentials [23], or through molecular dynamics simulations of the T1 copper site [24]. These approaches have yielded slight increases in redox potential, but the multiple direct and indirect interactions of the T1 copper site coordination sphere affecting the redox potential and catalytic rate of the enzyme have proven difficult to predict.

Recently, directed evolution techniques have been developed to address these issues. CotA-laccase from *Bacillus subtilis* has been engineered for an increased redox potential using a high-throughput screen based on the reaction of mutant enzymes with a number of high redox potential synthetic dyes [25]. Thus, mutants with increased redox potentials could be selected because of their increased decolorization rate of the dyes with higher redox potentials. A similar approach has been used to evolve other high redox potential laccases with improved expression levels and stabilities [25–28].

Glucose oxidase (GOx) catalyzes the oxidation of D-glucose to D-gluconolactone with the concomitant reduction of its bound FAD cofactor to $FADH_2$ (reduced form of flavin adenine dinucleotide). The reduced cofactor is then oxidized by oxygen (producing hydrogen peroxide) or is able to directly transfer electrons to a nearby electrode or mediator such as pyrroloquinoline quinone (PQQ). GOx has been extensively studied for use in glucose biosensors and BFCs, because of its linear amperometric response to glucose and suitability for DET [29]. To this end, novel polymers have been engineered to covalently attach the required flavin cofactor to the surface of an electrode through an intermediate mediator, PQQ. Reconstitution of apo-GOx on the immobilized FAD electrode produces active enzyme electrically "wired" from the active site FAD, to PQQ, to the electrode surface [30–32]. More recently, GOx has been engineered to have a free cysteine side chain near its active site, which has been used to bind a maleimide-modified gold nanoparticle. This also enables the enzyme to perform DET [33].

Similar to laccase, a number of methods have been developed to engineer GOx for increased activity and specificity. An enzymatic assay coupled with site-directed mutagenesis was used to alter the substrate specificity of GOx to increase the dynamic range of glucose biosensors [34], and a directed evolution screen was developed to improve GOx activity in BFCs [35].

Most enzymes are not capable of DET, however, and must instead rely on a mediator or cofactor to shuttle electrons between the enzyme active site and the electrode. This includes the multitude of NAD(P)(H) (nicotinamide adenine dinucleotide (phosphate) (reduced))-dependent oxidoreductase enzymes, such as dehydrogenases, which are commonly used in BFCs. These cofactors are further discussed in Section 7.2.3.

7.2 ENGINEERING OTHER PROPERTIES OF ENZYMES

7.2.1 Stability

To be commercially viable, BFCs must have a long shelf life and must be able to operate for a significant period of time. The useful lifetime of BFCs is often limited by the stability of the enzymes, cofactors, and mediators used in the system. Enzymes may denature or become inactivated because of temperature, pH, or exposure to various solvents or chemical species. Thermostable enzymes have been shown to be resistant to these environmental effects; thus, thermostability is a desirable property in enzymes for BFCs. One source of such enzymes is the large number of proteins that have been identified in hyperthermophiles as being homologous to more commonly used mesostable proteins [36,37]. Additionally, the previously discussed protein engineering techniques can be used to obtain modest improvements in stability. The stability of mediators and cofactors is discussed later in the chapter.

7.2.2 Size

Enzymes are relatively large compared with conventional catalysts, and the active site only occupies a small area of the overall protein. Therefore, to maximize the number of active sites per area, smaller enzymes are preferred. Unnecessary protein domains or surface attachments (i.e., glycosylation) can also inhibit mass transfer and prevent efficient electron transfer between the active site and electrode [38]. For example, when the GOx enzyme was deglycosylated, its ability to perform DET was improved [39]. Protein engineering has been used to identify the minimal functional units of enzymes and remove unnecessary domains [40]. In enzymes capable of DET, the distance between the active site and the current collector strongly affects the efficiency of electron transfer. For instance, truncating several residues from the C-terminus of a laccase was hypothesized to move the T1 copper site closer to the electrode surface, resulting in more efficient DET [41]. However, this was accompanied by a decrease in redox potential of the active site, inhibiting activity.

7.2.3 Cofactor Specificity

Many redox enzymes used in BFCs require cofactors for catalysis. For example, yeast alcohol dehydrogenase catalyzes the oxidation of ethanol to acetaldehyde, with the concomitant reduction of NAD^+ to NADH. Most commonly, cofactor specificity has been engineered to increase the activity of NADP(H)-dependent enzymes with NAD(H) or to correct a cofactor imbalance in a metabolic process. As the NAD(H) cofactor is more stable and less expensive than NADP(H), this generally improves the economics of a process [42–44].

However, these natural cofactors are not ideal for use in bioelectrocatalytic systems. In addition to being expensive, natural cofactors are only moderately stable and have a limited lifetime in enzymatic systems (Table 7.1). The relatively large size of the cofactors can lead to mass transfer limitations in some BFC architectures. Finally, the high (~1 V) overpotential required to oxidize NAD(P)H at an electrode significantly reduces the efficiency of the reaction to far below that of the theoretical limit. The overpotential can be lowered through the addition of redox mediators to the system, but they also possess relatively poor stability and have a limited lifetime.

To address these limitations, the use of alternative biomimetic cofactors has been investigated. In this way, a cofactor with the desired size, charge, redox potential, regeneration kinetics, or any other features deemed important for the

TABLE 7.1 Properties of Cofactors Commonly Used in Biological Fuel Cells

Cofactor	Cost ($\$\,g^{-1}$)[a]	Molecular Weight[b]	Standard Redox Potential (E_0', V)	Comments
NAD+/NADH	63	663.43	−0.320	Weakly bound, acts as shuttle for MET, commonly found in dehydrogenases and reductases
NADP+/ NADPH	250	743.41	−0.320	Weakly bound, acts as shuttle for MET, commonly found in dehydrogenases and reductases
FAD/FADH$_2$	365	829.51	−0.180	Strongly bound to enzyme, capable of DET, commonly found in oxidases
PQQ	45,000	330.21	+0.090	Found in some oxidases/de hydrogenases, can also act as an electron mediator

[a]Prices from Sigma-Aldrich, January 2012.
[b]Molecular weight reported for oxidized species.

application is selected. Once an improved cofactor is selected, the enzyme will likely need to be engineered to use the new cofactor efficiently. Once again, the techniques of rational design and/or directed evolution can be applied to address this problem.

There have been few reports in the literature of using non-natural cofactors for catalysis. Notably, in 2002, Lo and Fish synthesized a variety of *N*-benzylnicotinamide derivatives with the goal of improving turnover and cofactor stability in chiral synthesis applications [45]. Whereas the wild-type enzymes tested displayed poor turnover with the cofactor derivatives, a series of engineered mutants with altered cofactor specificity were able to use the novel cofactors at rates approaching those of the natural cofactors [46,47].

The first reported use of an alternative cofactor for bioelectrocatalysis used the truncated nicotinamide cofactor nicotinamide mononucleotide (NMN+) to address mass transport limitations in Nafion-immobilized BFCs [48]. The rate of cofactor diffusion has been shown to be rate limiting in a similar Nafion-immobilized architecture [49], so a smaller cofactor was hypothesized to increase performance. Again, the wild-type enzyme investigated in this study displayed poor turnover with the truncated cofactor, but a previously engineered double mutant with broadened cofactor specificity was able to use NMN+ at a much higher rate (Figure 7.2a). Although the turnover of the double mutant enzyme with the truncated cofactor was still orders of magnitude slower than the wild-type enzyme with its native cofactor, BFC performance was increased (Figure 7.2b). Thus, the increase in the rate of diffusion of NMN+ versus NAD+ (and also likely the rate of cofactor oxidation at the electrode surface) was able to overcome the significant decrease in turnover rate. This surprising result highlights the substantial role of the cofactor in MET and paves the way for further increases in BFC performance through improved cofactor properties.

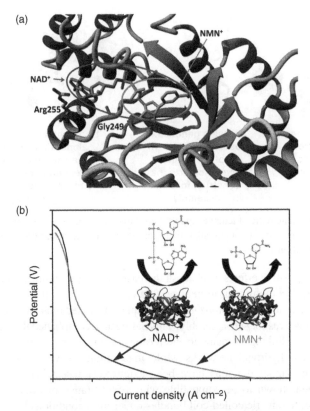

FIGURE 7.2 (a) The cofactor binding pocket of *Pyrococcus furiosus* alcohol dehydrogenase D K249G/H255R double mutant, indicating the position of bound NMN$^+$ (orange) and NAD$^+$ (purple), as well as the mutated residues. (b) These mutations led to an increase in turnover with the minimal cofactor NMN$^+$ versus the natural NAD$^+$ cofactor, and resulted in a significant increase in performance when employed in a BFC. (Reproduced with permission from Ref. [48]. Copyright 2012, Royal Society of Chemistry.) (Please see the color version of this figure in Color Plates section.)

7.3 ENZYME IMMOBILIZATION AND SELF-ASSEMBLY

An important consideration in the design of a BFC is the method for retaining enzymes in a device, in most cases in close proximity to the electrode surface. A BFC that requires constant addition of soluble enzyme would be prohibitively costly to operate, and the rate of electron transfer between the enzymes and electrode surface would likely be poor. Instead, enzymes are commonly immobilized in a BFC through adsorption [50], covalent attachment [51], confinement within a polymer matrix (such as Nafion [52], chitosan [53], or other synthetic polymers) [54], or other means [55]. The method of immobilization used can have a profound effect on enzyme activity through denaturation or inactivation of the enzyme, steric hindrance (blocking) of the active site, imposing mass transport limitations on substrate and/or cofactor diffusion, or significant modification of the local environment of the enzyme. Immobilization

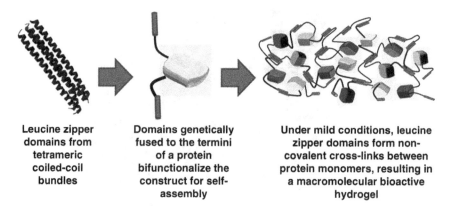

| Leucine zipper domains from tetrameric coiled-coil bundles | Domains genetically fused to the termini of a protein bifunctionalize the construct for self-assembly | Under mild conditions, leucine zipper domains form non-covalent cross-links between protein monomers, resulting in a macromolecular bioactive hydrogel |

FIGURE 7.3 Schematic of leucine zipper fusions to enable self-assembly of various proteins of interest into a robust macromolecular hydrogel. (Please see the color version of this figure in Color Plates section.)

can also have positive effects, such as an increase in enzyme stability and lifetime and better electrical communication between the enzyme and electrode.

An alternative to the immobilization techniques discussed thus far is to engineer the enzymes themselves for self-assembly. This has been accomplished through the fusion of various cross-linking domains to the termini of enzymes, enabling them to form macromolecular hydrogel networks while maintaining enzymatic activity. Alpha-helical leucine zipper domains have been extensively researched for this purpose and have been shown to be compatible with a wide range of proteins [56]. These domains form tetrameric coiled-coil bundles under mild conditions [57], allowing for the possibility of immobilizing enzymes *in situ* using changes in protein concentration or pH (Figure 7.3). Leucine zipper fusions to a thermostable alcohol dehydrogenase were shown to form a thermostable, enzymatically active hydrogel at temperatures of up to 60 °C [58]. Mixed hydrogels have also been demonstrated using a leucine zipper–laccase fusion combined with a leucine zipper–redox mediator fusion to create a conductive, redox-active hydrogel [59]. A similar domain fusion approach may also be used to "wire" enzymes capable of DET directly to electrodes. For instance, the small laccase from *Streptomyces coelicolor* (SLAC) could be fused to a carbon nanotube (CNT) binding peptide or otherwise functionalized to bind to CNTs to increase electrical communication between the enzyme active site and the electrode [60].

7.3.1 Engineering for Supermolecular Assembly

The energy density of a fuel can be maximized through the use of multienzyme cascades, wherein the product of one enzyme in the pathway is the substrate for the next. This approach has been demonstrated in BFCs capable of the complete oxidation of methanol to carbon dioxide [61], using a three-enzyme cascade consisting of alcohol dehydrogenase (oxidation of methanol to formaldehyde), aldehyde dehydrogenase (oxidation of formaldehyde to formate), and formate dehydrogenase (oxidation of formate to CO_2), and later in the complete oxidation of glucose to carbon dioxide [62].

In Nature, such multienzyme cascades, or metabolons, are often spatially oriented such that sequential enzymes in the cascade are in close proximity to each other [63]. This allows the substrate to be efficiently channeled from one enzyme to the next along the cascade, maximizing throughput. Several techniques have been developed to implement such an organization in a BFC [58,64], and these have yielded impressive results. A pyruvate/air BFC using a group of Krebs cycle enzymes isolated from mitochondria demonstrates a 49% increase in current density and 32% increase in power density when cross-linked into their native ordered conformation, versus the same enzymes free in solution [65].

7.4 ARTIFICIAL METABOLONS

7.4.1 DNA-Templated Metabolons

The design of artificial metabolons requires a general but robust method of specifically ordering enzymes in a cascade [65]. One approach could be to use enzyme fusions to zinc finger DNA binding domains to allow ordered assembly onto double-stranded DNA templates [66]. The fusion of zinc domains to SLAC (Figure 7.4) was

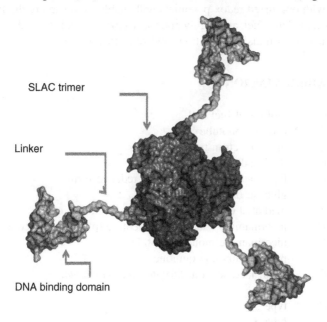

SLAC trimer

Linker

DNA binding domain

FIGURE 7.4 Schematic diagram of the SLAC–zinc finger fusion protein. The SLAC enzyme is trimeric and the C-terminus of each monomer (blue, bluish gray, purple) is fused to a DNA binding zinc finger domain (brown) via a flexible peptide linker (green). The DNA binding domain is a three zinc finger module from a mouse transcription factor Zif268 that specifically binds double-stranded DNA having the sequence 5′-GCGTGGGCG-3′. (Reproduced with permission from Ref. [66]. Copyright 2011, Royal Society of Chemistry.) (Please see the color version of this figure in Color Plates section.)

shown to specifically bind to a desired DNA sequence while maintaining catalytic activity. This approach can be extended by fusing zinc finger domains specific to different DNA sequences to the other enzymes in the cascade. Thus, the enzymes can be ordered in a metabolon-like complex by the fusion proteins binding to a template DNA strand containing the zinc finger recognition sites in the appropriate sequence. The DNA strands containing the enzymes could then be immobilized on nano-structured electrodes to create complex anode systems.

7.5 OUTLOOK

The advances in understanding the factors affecting BFC performance discussed in this chapter highlight the tremendous progress being made in the field. BFC research is an extremely interdisciplinary field, and increased collaboration between protein engineers, electrochemists, physicists, biologists, and device designers has made these advances possible. The future of this research will likely undergo a paradigm shift owing to the variety of tools available to engineer every aspect of a BFC. Instead of trying to build a device around suboptimal, wild-type enzymes, we will build the enzyme (and required cofactors or mediators) to suit the desired device. Cofactors and mediators with optimized redox potentials will enable us to capture the full thermo-dynamic potential of a fuel, and highly optimized enzymes and electrode architectures will minimize mass transfer limitations and resistance losses.

LIST OF ABBREVIATIONS

BFC	biological fuel cell
CNT	carbon nanotube
DET	direct electron transfer
FAD	flavin adenine dinucleotide
$FADH_2$	flavin adenine dinucleotide, reduced form
GOx	glucose oxidase
MET	mediated electron transfer
NAD(P)(H)	nicotinamide adenine dinucleotide (phosphate) (reduced)
NMN	nicotinamide mononucleotide
PQQ	pyrroloquinoline quinone
SLAC	small laccase from *Streptomyces coelicolor*
T1	type 1
T2	type 2
T3	type 3

REFERENCES

1. Höög JO, Eklund H, Jörnvall H. A single-residue exchange gives human recombinant beta beta alcohol dehydrogenase gamma gamma isozyme properties. *Eur J Biochem* 1992;205:519–526.

2. Nixon AE, Firestine SM, Salinas FG, Benkovic SJ. Rational design of a scytalone dehydratase-like enzyme using a structurally homologous protein scaffold. *Proc Natl Acad Sci USA* 1999;96:3568–3571.

3. Woodyer R, van der Donk WA, Zhao H. Relaxing the nicotinamide cofactor specificity of phosphite dehydrogenase by rational design. *Biochemistry* 2003;42: 11604–11614.

4. Eijsink VGH, Bjørk A, Gåseidnes S, Sirevåg R, Synstad B, van den Burg B, Vriend G. Rational engineering of enzyme stability. *J Biotechnol* 2004;113:105–120.

5. Turner NJ. Directed evolution drives the next generation of biocatalysts. *Nat Chem Biol* 2009;5:567–573.

6. Cherry JR, Fidantsef AL. Directed evolution of industrial enzymes: an update. *Curr Opin Biotechnol* 2003;14:438–443.

7. Sen S, Dasu V, Mandal B. Developments in directed evolution for improving enzyme functions. *Appl Biochem Biotechnol* 2007;143:212–223.

8. Zhao H, Chockalingam K, Chen Z. Directed evolution of enzymes and pathways for industrial biocatalysis. *Curr Opin Biotechnol* 2002;13:104–110.

9. Amstutz P, Pelletier JN, Guggisberg A, Jermutus L, Cesaro-Tadic S, Zahnd C, Pluckthun A. *In vitro* selection for catalytic activity with ribosome display. *J Am Chem Soc* 2002;124:9396–9403.

10. Takahashi F, Ebihara T, Mie M, Yanagida Y, Endo Y, Kobatake E, Aizawa M. Ribosome display for selection of active dihydrofolate reductase mutants using immobilized methotrexate on agarose beads. *FEBS Lett* 2002;514:106–110.

11. Griffiths AD, Tawfik DS. Directed evolution of an extremely fast phosphotriesterase by *in vitro* compartmentalization. *EMBO J* 2003;22:24–35.

12. Yano T, Oue S, Kagamiyama H. Directed evolution of an aspartate aminotransferase with new substrate specificities. *Proc Natl Acad Sci USA* 1998;95:5511–5515.

13. Khurana J, Singh R, Kaur J. Engineering of *Bacillus* lipase by directed evolution for enhanced thermal stability: effect of isoleucine to threonine mutation at protein surface. *Mol Biol Rep* 2010; 1–8.

14. Liang L, Zhang J, Lin Z. Altering coenzyme specificity of *Pichia stipitis* xylose reductase by the semi-rational approach CASTing. *Microb Cell Fact* 2007;6:36.

15. Chica RA, Doucet N, Pelletier JN. Semi-rational approaches to engineering enzyme activity: combining the benefits of directed evolution and rational design. *Curr Opin Biotechnol* 2005;16:378–384.

16. Kim J, Jia H, Wang P. Challenges in biocatalysis for enzyme-based biofuel cells. *Biotechnol Adv* 2006;24:296–308.

17. Shleev S, Tkac J, Christenson A, Ruzgas T, Yaropolov AI, Whittaker JW, Gorton L. Direct electron transfer between copper-containing proteins and electrodes. *Biosens Bioelectron* 2005;20:2517–2554.

18. Ghindilis AL, Atanasov P, Wilkins E. Enzyme-catalyzed direct electron transfer: fundamentals and analytical applications. *Electroanalysis* 1997;9:661–674.

19. Messerschmidt A. Blue copper oxidases. In: Sykes AG (ed.), *Advances in Inorganic Chemistry*. Academic Press, San Diego, CA, 1993, pp. 121–185.

20. Solomon EI, Sundaram UM, Machonkin TE. Multicopper oxidases and oxygenases. *Chem Rev* 1996;96:2563–2606.

21. Rodgers CJ, Blanford CF, Giddens SR, Skamnioti P, Armstrong FA, Gurr SJ. Designer laccases: a vogue for high-potential fungal enzymes? *Trends Biotechnol* 2010;28:63–72.

22. Kamitaka Y, Tsujimura S, Kataoka K, Sakurai T, Ikeda T, Kano K. Effects of axial ligand mutation of the type I copper site in bilirubin oxidase on direct electron transfer-type bioelectrocatalytic reduction of dioxygen. *J Electroanal Chem* 2007;601:119–24.

23. Xu F, Berka RM, Wahleithner JA, Nelson BA, Shuster JR, Brown SH, Palmer AE, Solomon EI. Site-directed mutations in fungal laccase: effect on redox potential, activity and pH profile. *Biochem J* 1998;334:63–70.

24. Hong G, Ivnitski DM, Johnson GR, Atanassov P, Pachter R. Design parameters for tuning the type 1 Cu multicopper oxidase redox potential: insight from a combination of first principles and empirical molecular dynamics simulations. *J Am Chem Soc* 2011;133:4802–4809.

25. Brissos V, Pereira L, Munteanu F-D, Cavaco-Paulo A, Martins LO. Expression system of CotA-laccase for directed evolution and high-throughput screenings for the oxidation of high-redox potential dyes. *Biotechnol J* 2009;4:558–563.

26. Festa G, Autore F, Fraternali F, Giardina P, Sannia G. Development of new laccases by directed evolution: functional and computational analyses. *Proteins* 2008;72:25–34.

27. Maté D, García-Burgos C, García-Ruiz E, Ballesteros AO, Camarero S, Alcalde M. Laboratory evolution of high-redox potential laccases. *Chem Biol* 2010;17:1030–1041.

28. Bulter T, Alcalde M, Sieber V, Meinhold P, Schlachtbauer C, Arnold FH. Functional expression of a fungal laccase in *Saccharomyces cerevisiae* by directed evolution. *Appl Environ Microbiol* 2003;69:987–995.

29. Wilson R, Turner APF. Glucose oxidase: an ideal enzyme. *Biosens Bioelectron* 1992;7:165–185.

30. Willner I, Heleg-Shabtai V, Blonder R, Katz E, Tao G, Bückmann AF, Heller A. Electrical wiring of glucose oxidase by reconstitution of FAD-modified monolayers assembled onto Au-electrodes. *J Am Chem Soc* 1996;118:10321–10322.

31. Zayats M, Katz E, Willner I. Electrical contacting of glucose oxidase by surface-reconstitution of the apo-protein on a relay-boronic acid-FAD cofactor monolayer. *J Am Chem Soc* 2002;124:2120–2121.

32. Katz E, Willner I, Kotlyar AB. A non-compartmentalized glucose–O_2 biofuel cell by bioengineered electrode surfaces. *J Electroanal Chem* 1999;479:64–68.

33. Holland JT, Lau C, Brozik S, Atanassov P, Banta S. Engineering of glucose oxidase for direct electron transfer via site-specific gold nanoparticle conjugation. *J Am Chem Soc* 2011;133:19262–19265.

34. Sode K, Kojima K. Improved substrate specificity and dynamic range for glucose measurement of *Escherichia coli* PQQ glucose dehydrogenase by site directed mutagenesis. *Biotechnol Lett* 1997;19:1073–1077.

35. Zhu Z, Momeu C, Zakhartsev M, Schwaneberg U. Making glucose oxidase fit for biofuel cell applications by directed protein evolution. *Biosens Bioelectron* 2006;21:2046–2051.

36. Littlechild JA, Guy J, Connelly S, Mallett L, Waddell S, Rye CA, Line K, Isupov M. Natural methods of protein stabilization: thermostable biocatalysts. *Biochem Soc Trans* 2007;35:1558–1563.

37. Unsworth LD, van der Oost J, Koutsopoulos S. Hyperthermophilic enzymes—stability, activity and implementation strategies for high temperature applications. *FEBS J* 2007;274:4044–4056.

38. Cracknell JA, Vincent KA, Armstrong FA. Enzymes as working or inspirational electro-catalysts for fuel cells and electrolysis. *Chem Rev* 2008;108:2439–2461.

39. Courjean O, Gao F, Mano N. Deglycosylation of glucose oxidase for direct and efficient glucose electrooxidation on a glassy carbon electrode. *Angew Chem* 2009;48:5897–5899.

40. Böhm I, Holzbaur IE, Hanefeld U, Cortési J, Staunton J, Leadlay PF. Engineering of a minimal modular polyketide synthase, and targeted alteration of the stereospecificity of polyketide chain extension. *Chem Biol* 1998;5:407–412.

41. Gelo-Pujic M, Kim H-H, Butlin NG, Palmore GTR. Electrochemical studies of a truncated laccase produced in *Pichia pastoris*. *Appl Environ Microbiol* 1999;65:5515–5521.

42. Woodyer R, van der Donk WA, Zhao H. Relaxing the nicotinamide cofactor specificity of phosphite dehydrogenase by rational design. *Biochemistry* 2003;42:11604–11614.

43. Banta S, Boston M, Jarnagin A, Anderson S. Mathematical modeling of *in vitro* enzymatic production of 2-keto-L-gulonic acid using NAD(H) or NADP(H) as cofactors. *Metab Eng* 2002;4:273–284.

44. Hurley JH, Chen R, Dean AM. Determinants of cofactor specificity in isocitrate dehydrogenase: structure of an engineered $NADP^+ \rightarrow NAD^+$ specificity-reversal mutant. *Biochemistry* 1996;35:5670–5678.

45. Lo HC, Fish RH. Biomimetic NAD^+ models for tandem cofactor regeneration, horse liver alcohol dehydrogenase recognition of 1,4-NADH derivatives, and chiral synthesis. *Angew Chem Int Ed* 2002;41:478–481.

46. Lutz J, Hollmann F, Ho TV, Schnyder A, Fish RH, Schmid A. Bioorganometallic chemistry: biocatalytic oxidation reactions with biomimetic $NAD^+/NADH$ co-factors and $[Cp^*Rh(bpy)H]^+$ for selective organic synthesis. *J Organomet Chem* 2004;689:4783–4790.

47. Ryan JD, Fish RH, Clark DS. Engineering cytochrome P450 enzymes for improved activity towards biomimetic 1,4-NADH cofactors. *ChemBioChem* 2008;9:2579–2582.

48. Campbell E, Meredith M, Minteer SD, Banta S. Enzymatic biofuel cells utilizing a biomimetic cofactor. *Chem Commun* 2012;48:1898–1900.

49. Moore CM, Minteer SD, Martin RS. Microchip-based ethanol/oxygen biofuel cell. *Lab Chip* 2005;5:218–225.

50. Mano N, Mao F, Heller A. A miniature biofuel cell operating in a physiological buffer. *J Am Chem Soc* 2002;124:12962–12963.

51. Fischback MB, Youn JK, Zhao X, Wang P, Park HG, Chang HN, Kim J, Ha S. Miniature biofuel cells with improved stability under continuous operation. *Electroanalysis* 2006;18:2016–2022.

52. Thomas TJ, Ponnusamy KE, Chang NM, Galmore K, Minteer SD. Effects of annealing on mixture-cast membranes of Nafion® and quaternary ammonium bromide salts. *J Membr Sci* 2003;213:55–66.

53. Klotzbach T, Watt M, Ansari Y, Minteer SD. Effects of hydrophobic modification of chitosan and Nafion on transport properties, ion-exchange capacities, and enzyme immobilization. *J Membr Sci* 2006;282:276–283.

54. Gregg BA, Heller A. Cross-linked redox gels containing glucose oxidase for amperometric biosensor applications. *Anal Chem* 1990;62:258–263.

55. Barlett PN, Cooper JM. A review of the immobilization of enzymes in electropolymerized films. *J Electroanal Chem* 1993;362:1–12.

56. Wheeldon IR, Calabrese Barton S, Banta S. Bioactive proteinaceous hydrogels from designed bifunctional building blocks. *Biomacromolecules* 2007;8:2990–2994.

57. Shen W, Lammertink RGH, Sakata JK, Kornfield JA, Tirrell DA. Assembly of an artificial protein hydrogel through leucine zipper aggregation and disulfide bond formation. *Macromolecules* 2005;38:3909–3916.

58. Wheeldon IR, Campbell E, Banta S. A chimeric fusion protein engineered with disparate functionalities. Enzymatic activity and self-assembly. *J Mol Biol* 2009;392:129–142.

59. Wheeldon IR, Gallaway JW, Calabrese Barton S, Banta S. Bioelectrocatalytic hydrogels from electron-conducting metallopolypeptides coassembled with bifunctional enzymatic building blocks. *Proc Natl Acad Sci USA* 2008;105:15275–15280.

60. Ramasamy RP, Luckarift HR, Ivnitski DM, Atanassov PB, Johnson GR. High electrocatalytic activity of tethered multicopper oxidase–carbon nanotube conjugates. *Chem Commun* 2010;46:6045–6047.

61. Palmore GTR, Bertschy H, Bergens SH, Whitesides GM. A methanol/dioxygen biofuel cell that uses NAD$^+$-dependent dehydrogenases as catalysts: application of an electroenzymatic method to regenerate nicotinamide adenine dinucleotide at low overpotentials. *J Electroanal Chem* 1998;443:155–161.

62. Arechederra RL, Minteer SD. Complete oxidation of glycerol in an enzymatic biofuel cell. *Fuel Cells* 2009;9:63–69.

63. Srere PA. Complexes of sequential metabolic enzymes. *Annu Rev Biochem* 1987;56: 89–124.

64. Moehlenbrock MJ, Toby TK, Pelster LN, Minteer SD. Metabolon catalysts: an efficient model for multi-enzyme cascades at electrode surfaces. *ChemCatChem* 2011;3:561–570.

65. Moehlenbrock MJ, Toby TK, Waheed A, Minteer SD. Metabolon catalyzed pyruvate/air biofuel cell. *J Am Chem Soc* 2010;132:6288–6289.

66. Szilvay GR, Brocato S, Ivnitski D, Li C, Iglesia PDL, Lau C, Chi E, Werner-Washburne M, Banta S, Atanassov P. Engineering of a redox protein for DNA-directed assembly. *Chem Commun* 2011;47:7464–7466.

8

PURIFICATION AND CHARACTERIZATION OF MULTICOPPER OXIDASES FOR ENZYME ELECTRODES

D. Matthew Eby

Booz Allen Hamilton, Atlanta, GA, USA; Airbase Sciences Branch, Air Force Research Laboratory, Tyndall Air Force Base, FL, USA

Glenn R. Johnson

Airbase Sciences Branch, Air Force Research Laboratory, Tyndall Air Force Base, FL, USA

8.1 INTRODUCTION

All *in vitro* investigations of protein structure, function, or catalysis benefit from a homogeneous protein preparation. The protein purification method is not the highlight of most enzymology projects, yet it is an essential step for all fields of study that involve protein analysis or application *in vitro*. With respect to enzymatic fuel cells (EFCs), enzyme preparation and characterization is crucial to advancing fundamental and applied research projects in this emerging biotechnology field. This chapter will discuss the details of multicopper oxidase (MCO) purification and characterization, the commonly used cathode catalyst in biological fuel cells (BFCs).

MCO proteins are redox enzymes that catalyze the oxidation of substrates with the concomitant reduction of molecular oxygen. The MCO redox centers typically comprise four copper atoms that coordinately transfer electrons from a substrate to a bound oxygen molecule [1]. MCOs have broad industrial applications in

Enzymatic Fuel Cells: From Fundamentals to Applications, First Edition. Edited by Heather R. Luckarift, Plamen Atanassov, and Glenn R. Johnson.

processes including pulp and paper clarification, organic synthesis reactions, and bioremediation of wastes [2–4]. The MCOs are effective oxygen reduction reaction (ORR) catalysts in EFC cathodes because of their intrinsic stability and high redox potential [5]; the MCOs most often applied in EFC are phenol oxidase (laccase), ascorbate oxidase (AOx), and bilirubin oxidase (BOx) [6]. In many cases, the methods and techniques used to construct enzyme electrodes and evaluate BFC operation have not been standardized. If enzyme preparations are not standardized, then it becomes difficult to discern whether poor reproducibility or performance is due to the experimental techniques, variations in protein purification, or substandard enzyme preparations. Purification methods that produce highly active, stable, homogenous preparations of the MCOs will allow further experimentation and analysis to be completed with confidence. The general characteristics of the target protein should be met to guide effective purification methods (e.g., obtaining the optimal raw protein source, maintaining native enzyme structure, having reliable assay for enzyme activity). In addition, metalloenzymes such as MCOs require some extra considerations during purification and specific assays for characterization. Copper center stability is of utmost concern, as the copper centers are crucial to electron transfer and are assembled from complex biological pathways and may need particular care to remain functional during protein production.

This chapter begins with an introduction of general considerations for protein purification and then focuses on purification strategies and characterization technique for MCOs. Examples of MCO purification and characterization will regularly be referenced and the chapter will pay particular attention to the characterization of metals centers, as expression, purification, and characterization of copper-dependent MCOs require some unique methods. Techniques to characterize MCO activity and behavior after immobilization onto electrode surfaces will not be covered, as these techniques are described elsewhere in the book.

8.2 GENERAL CONSIDERATIONS FOR MCO EXPRESSION AND PURIFICATION

Before heading to the laboratory bench, one must consider the basic conditions for MCO production and isolation. The following are questions to consider.

1. Can the MCO be isolated from the native source?
2. How much protein can be harvested from the organism and what is its fraction of total protein (purity of crude preparation)?
3. Is the protein retained intracellularly, in the membrane fraction, or secreted from the cell?
4. If the native organism is unsuitable for propagation, has the MCO gene been isolated and what is an appropriate recombinant system?
5. Will codon usage be an issue when expressing the recombinant gene in an alternate host?

6. How do the recombinant expression system and the host affect the mature protein structure and function? Do posttranslational modifications (e.g., glycosylation, pre- and pro-polypeptide processing) need to be addressed?

7. What contaminants need to be removed from the crude protein preparation? What is the final scale for purification?

8. What is the final intent for the enzyme (e.g., analytical characterization or practical applications)?

These questions will be specifically addressed in following sections. First, the enzyme source must be identified. If the enzyme cannot be effectively isolated from the native organism, then a recombinant expression system must be engineered that is able to produce a functional enzyme in abundant quantities. Next, the enzyme must be separated from the host proteins and other cellular components and prepared for column chromatography. Most purification protocols require multiple separation techniques to remove unwanted protein and cellular debris, specific steps for retention of redox activity, and additional conditioning to allow transferring the protein from one solution to another as necessary for subsequent steps. One must take care to minimize the number of steps, as each will cause some loss of product and may stress enzyme stability. Consequently, the goal is to reach the optimum yield and purity using a minimum number of steps and the simplest possible protocol. Finally, procedures should be completed in a rational sequence to reduce the need for buffer exchanges and other conditioning steps. After the enzyme is sufficiently purified, specific methods to store the protein and to retain activity must also be considered.

8.3 MCO PRODUCTION AND EXPRESSION SYSTEMS

The MCOs are well distributed in Nature and come from a wide variety of organisms. A simple search of the National Center for Biotechnology Information protein database reveals more than 14,000 characterized and putative "multicopper oxidases" among the annotated sequences (April 2012). The defined sources are principally bacteria and fungi; however, archaea, green plants, animals, and other eukaryotes also synthesize MCO proteins (or at least carry genes encoding putative MCOs). As such, there are many potential sources of MCOs and there will be a number of different methods to produce these enzymes either in the native organism or in engineered recombinant expression systems. The inherent advantage of isolating protein from the native source is that the protein is sure to be correctly synthesized and processed and also likely maintained in optimal conditions for stability, because the enzyme evolved to function within the organism and its niche. With fungal hosts, many of the MCOs (e.g., laccases) are secreted from the cell, which can be an advantage for protein expression and purification, but conditions and handling of the culture media must be considered. In many cases, growth, protein synthesis, and purification have not been extensively studied or optimized in the native host; hence, recombinant expression systems can be an advantage.

Here, a suitable host must be considered, in particular whether the native MCO is from a prokaryote or eukaryote. Special care must be considered for post-translational modification of the protein and that the synthesis of the complex metal centers requires accessory pathways within the organism. Accordingly, simply cloning the MCO gene and transferring it to a recombinant host may not yield an active enzyme. In these cases, complications may be overcome by using recombinant expression systems that are similar to the native organisms (e.g., expressing eukaryotic MCOs in yeast expression systems). Ultimately, the advantages and disadvantages for each must be examined and the approach tailored to the best course of action for the particular MCO of interest.

MCO enzymes for EFC technology have come from a wide variety of plant, fungal, bacterial, and even human sources. To best describe and understand these expression systems, the application of several published expression systems will be detailed. The majority of MCOs used for cathodic redox reactions in EFC studies have been laccases, AOx, and BOx; their heterologous expression will be outlined in the next section, with a brief summary of other MCOs such as ceruloplasmin (CP) and bacterial MCOs.

Laccase (benzenediol:oxygen oxidoreductase, EC 1.10.3.2) is the most commonly used MCO in EFC studies. Laccase was first discovered in the Japanese lacquer tree *Rhus vernicifera* and is widely distributed among plants; its physiological function is to catalyze oxidation of phenolic compounds to assist resin and lignin polymerization [7]. More recently, other so-called laccases have been isolated from various fungi, where the catalytic role is the degradation of lignin, breaking the polymer down into smaller hydrocarbons for metabolism. Many of the fungal laccases exhibit high redox potential, and their isolation from the growth of these microorganisms makes them attractive sources for fuel cell development. Although the growth of fungal strains is not usually amenable to enzyme purification, many fungi will secrete laccases after synthesis, and protein can be purified from the culture supernatant without need to disrupt cells or mycelia [8–13]. For example, a procedure was completed in 1998 by Grotewold et al. for the purification of a laccase from culture supernatants of *Neurospora crassa*, which allowed concentration of the extracellular enzyme from large culture volumes using only one column chromatography step [8]. Higher (or woody) plants synthesize laccases and then secrete the enzyme into vascular tissues to catalyze lignin polymerization [14]. Woody plants are not a conventional source for protein purification, but laccases have been purified to homogeneity from cultured angiosperm tissue [15], and the angiosperm tissues have been used in a few instances as a recombinant host [13].

Because of low levels of protein synthesis and undesirable preparation requirements, plant and fungal laccases are often recombinantly expressed using recombinant yeast systems. An exhaustive list and description of heterologously produced laccases can be found in Ref. [13]. *Saccharomyces cerevisiae* and *Pichia pastoris* are two common hosts for laccase expression. The yeasts have advantage over native sources because they will grow to high density in liquid cultures and still complete eukaryote-specific posttranslational modifications [13].

AOx (L-ascorbate:dioxygen oxidoreductase, EC 1.10.3.3) is another specific example of an MCO that has been applied as an electrocatalyst in EFC [6]. These enzymes have the greatest substrate specificity toward L-ascorbic acid but will use other electron donors including carbon electrodes in direct electron transfer (DET) processes [16]. Squash and cucumber have been traditional sources for AOx, but AOx has also been isolated from various other fruits, spinach, celery, and wheat [17–20]. The enzyme is generally found associated with cell wall material in the plant and has been isolated from crude juice preparations of the fruiting body (cucurbit) [20]. With respect to heterologous expression, several expression systems have been devised for AOx using transgenic tobacco. Cucumber and pumpkin AOx have also been expressed recombinantly in tobacco leaves and protoplasts, respectively, with the latter system secreted into the culture medium [21,22]. AOx from *Brassica napus* (rapeseed) was transferred to *Escherichia coli*, but recombinant gene expression and protein synthesis inhibited growth of the host bacteria. That same gene product could be effectively synthesized in transgenic tobacco plants and subsequently isolated [23].

BOx (bilirubin:oxygen oxidoreductase, EC 1.3.3.5) catalyzes the oxidation of tetrapyrroles, such as bilirubin to biliverdin in heme metabolism [4,24]. BOx is an effective ORR catalyst for EFC because it exhibits activity under neutral and slightly alkaline conditions. By comparison, laccases typically provide highest activity at acidic pH (5–6). The BOx activity near neutral pH is more compatible with typical anodic biocatalysts, which simplifies device design and improves output. BOx is typically purified from fungal strains, including *Myrothecium*, *Trachyderma*, and *Penicillium* [6,25–30]. *Myrothecium verrucaria* BOx is produced in high quantities by the native organism and isolated from the culture filtrate. In other studies, recombinant variants of the *M. verrucaria* BOx were produced using an *Aspergillus oryzae* expression system [27,28,31].

Some other MCOs have specific properties that are useful to electrocatalysis or have unusual attributes that are instructive for purification methodologies. CP (ferroxidase; iron(I):dioxygen oxidoreductase, EC 1.16.3.1), for example, oxidizes a wide variety of organic compounds. Attempts to use CP as an electrocatalyst through DET have produced mixed results [30]. Human CP can be isolated from blood serum and an interesting purification approach has been developed whereby the feedstock for large amounts of human CP originates from medical waste, either from outdated human plasma or from by-products of plasma-fractioning processes [32,33].

More recently, bacterial MCOs have gained attention, as they have many of the same catalytic characteristics as MCOs from eukaryotes, the strains can be readily cultivated, or the proteins are amenable to well-established bacterial-based, recombinant expression and purification systems. A few *Streptomyces* strains produce highly active laccases. *Streptomyces* share some physical characteristics and growth conditions with filamentous fungi (filamentous growth, spore formation, etc.), but the prokaryotic genetic basis enables recombinant expression in the well-defined host *E. coli*, although *Streptomyces coelicolor* and *Streptomyces lividans* also yield large amounts of recombinant laccases [34]. Another bacterially derived MCO used in EFC analysis and engineering is CotA, which exhibits catalysis similar to both laccase and BOx. CotA has been isolated from thermophilic *Bacillus* sp. and shown to be very stable in BFC applications [35,36]. CotA is usually not produced in significant quantities in the native organism; thus, it is often produced in *E. coli* through recombinant expression systems. The expression profile in the native *Bacillus* spp., however, lends itself to some unique approaches for purification and application. CotA is synthesized during sporulation and is found in the outer coat of the bacterial spore [37]. It is active within this matrix; hence, the spore constitutes a highly stable microparticle that can protect CotA activity in a wide range of diverse and harsh environmental conditions [38]. After expression and integration into EFC, CotA may provide highly stable and long-life cathodic enzyme reactions. CueO is another bacterial MCO implicated in periplasmic detoxification of copper as its native function; it has also been tested as a catalyst for EFC [2,39–41]. CueO was first isolated from *E. coli* and more recently from *Salmonella enterica* serovar Typhimurium [39,42,43]. Each was subsequently produced recombinantly using laboratory strains with overexpression systems. In electrocatalytic studies, CueO has been used for DET reactions on mesoporous carbon cathodes [39,41].

8.4 MCO PURIFICATION

Purification of MCOs is similar to many other proteins, and conventional techniques are applicable and often effective. Yet there are select features of MCOs that can be used advantageously and must be carefully considered to maintain structure and activity. Because most native MCOs contain a colored Cu I site (ranging from blue to green), they are easily monitored during column chromatography and fractionation, provided that the copper has been successfully integrated with the enzyme prior to purification. Copper loading and complex stability will be addressed later in this chapter.

The keys to successful and efficient protein purification are to select the most appropriate techniques, optimize their performance to suit the requirements, and combine them in a logical way to maximize yield and minimize the number of steps in the protocol. In many cases, multiple purification steps are necessary to reach the desired purity. In turn, purity plays a significant role in protein immobilization on the electrode. Contaminated preparations will foul electrode surfaces with unwanted protein and the current density and interfacial conductivity will suffer as a result. Many electrode immobilization procedures are based on generic methods for protein attachment and are not specific to the redox enzymes. Consequently, any proteins in the preparation will likely be included in attachment reactions and electrode engineering, and if contaminating proteins have greater affinity for electrode integration, these proteins will compete with MCO attachment and result in low MCO loading. As a rule, proteins should be purified to the highest extent possible so that subsequent activity and characterization methods will be reflective of the intended enzyme. Furthermore, it can be very difficult to measure the electrocatalytically active surface area for specific enzymes when a mixture of protein is immobilized on the surface. As a result, low-purity enzyme preparations will typically result in low power densities and the quantity of electrochemically active enzyme on the electrode surface will be overestimated.

The first step in the process is to develop a strategy that separates the protein from cells in a manner that will retain the structure and activity of the MCO. As mentioned in the previous section, many fungal laccases are secreted from the cells. Much of the unwanted protein can be removed by centrifugation, and filtration of the culture media separates cells and large cellular debris from bulk media. After that clarification step, specific molecular weight cutoff filters can also be used to concentrate the filtrates prior to column chromatography. In one such approach in 2006, Lyashenko et al. purified the laccase from *Cerrena maxima* by first filtering culture liquids using standard Whatman filter paper and then concentrating the bulk medium by ultrafiltration through a membrane that retained proteins larger than 15,000 Da [44]. Most of the contaminating proteins were then precipitated with ammonium sulfate at 90% saturation. Ammonium sulfate precipitation is an easy method to remove unwanted protein from the preparation, as long as the MCO is stable and will remain in solution at high concentrations of ammonium sulfate. Laccases tend to be highly stable and can withstand this relatively harsh treatment. In another example, Singh et al. collected the laccase from the γ-proteobacterium JB from an extracellular culture fraction and purified it threefold after precipitation of unwanted proteins in ammonium sulfate solutions at 60% saturation [45]. Other MCOs are purified in a similar manner. The fungus *Acremonium* sp. HI-25 secretes an AOx extracellularly [46]. The gene encoding the AOx protein was isolated from the native organism and a recombinant expression system was engineered so that the AOx would be produced in *Aspergillus nidulans* and released in the medium. After cultivation of the recombinant *A. nidulans*, the mycelia were removed from the growth medium by filtration and the AOx was precipitated in 80% ammonium sulfate. Dialysis into acetate buffer reconstituted the active enzyme for subsequent column chromatography.

For MCOs that are not secreted to extracellular environment, a method is required to disrupt cell membranes in order to release the proteins. Cell disruption can be separated into either mechanically or nonmechanically driven processes [47]. Non-mechanical systems rely on a chemical, enzymatic, or osmotic shock to destabilize cell membranes. Organic solvents can also destabilize the hydrophilic and hydrophobic properties of the phospholipid bilayer. Osmotic shock induced by high-sucrose medium will cause rapid changes in the internal pressure of cells, causing them to rupture. In other instances, ethylenediaminetetraacetic acid (EDTA) is used to destabilize the outer membrane, but this can be detrimental for purification of MCOs as EDTA is a potent metal chelator and may effectively remove copper atoms from the MCOs [18,48].

Mechanical systems include homogenizers, French press cells, ball mills for dental amalgamators and rotary blenders, and probe ultrasonicators [47]. Homogenizers and French press cells use high pressure to force cell suspensions through a small channel or orifice. Cell membrane degradation occurs because of high liquid shear through the confined volume and the sudden pressure change when exiting the high-pressure homogenizer cell. Ball mills also disrupt cells with shear force as glass or silicon beads collide with cells. This technique can be successfully used for yeast and bacterial spores, plant tissue, and large-scale disruption of fungal mycelia. Probe sonication is used at the laboratory scale and is effective in disrupting fungal and bacterial cell membranes. Ultrasonication causes cavitation, which is the rapid formation and bursting of microbubbles that disrupt cell membranes. Rotary blenders work well on cellulose-rich plant material and fungal mycelia [49–52].

One important consideration in methods of cell disruption for MCO purification is the use of additives to stabilize cell-free extracts and inhibit the activity of proteases. Once cells are lysed, proteases are released and can degrade the protein of interest. Many proteases are metalloenzymes and EDTA is commonly added to commercially available protease inhibitor cocktails. The effect is not selective, so the EDTA can also chelate copper and inhibit MCO activity. As an alternative, one may use protease inhibitor cocktails that are designed for metal-based affinity purification systems, such as SigmaFAST Protease Inhibitor Cocktail Tablets, EDTA-Free (Sigma-Aldrich, Inc.), Proteoloc Protease Inhibitor Cocktail EDTA-Free (Protein Discovery, Inc.), and Protease Inhibitor Cocktail Set V, EDTA-Free (Millipore, Inc.). These products are free of metal chelating chemicals, because metal–enzyme complex formation is the key principle of the metal affinity chromatographic separations. The preparatory conditions are not the only consideration for retaining copper in MCOs; copper atoms in the metal centers can diffuse from the enzyme for a number of reasons. If using a recombinant expression system, the recombinant host may require an optimized copper transport and accessory pathways in order to build the complex copper centers in MCOs. The Cu atoms in metal center can diffuse or not be completely loaded into folded enzymes for a number of reasons. If using a recombinant expression system, for example, the recombinant host may not have optimized Cu transport and accessory pathways needed to build the complex Cu centers in MCOs. It is critical to complete spectrophotometric and biochemical assessments of the final protein preparations to ensure that the copper cofactors are present; otherwise, specific activity will be significantly reduced.

Various chromatography methods are used to separate MCOs from unwanted proteins in protein suspensions. Column chromatography based on ion exchange, hydrophobic interaction, size exclusion, and affinity properties has been frequently used in purification strategies for MCOs. Contemporary fast protein liquid chromatography (FPLC) has become an essential laboratory tool for protein purification. FPLC systems containing multiple pumps for varying the mobile phases can combine sequential chromatography steps into a single, automated protocol. Combining chromatography techniques with different selectivity can yield very effective processes for protein purification.

Fungal laccases have been isolated more than any other MCO. They are usually isolated in dimeric or tetrameric forms; the molecular weight of the monomers is typically between 50 and 100 kDa and they have an overall negative charge on the surface. Hence, anion exchange resins are usually used as a first step in chromatographic separation from cell-free suspensions and culture supernatants. The MCOs are relatively stable enzymes and their exchange into different buffers for transition from one chromatographic separation to another does not usually inhibit activity or denature the enzyme. The stability is partly due to the high level of glycosylation, which can be up to 50% of the total weight of the protein. The extent of glycosylation may vary depending on growth conditions as well as the native or recombinant host that is used to produce the protein. The glycosylation can be detrimental to use of MCOs as electrocatalysts. The specific activity of the enzyme is inherently decreased because the polysaccharide has no catalytic role. Moreover, the modification will occupy "space" on the electrode and will reduce the electrochemically available surface area. Nonetheless, there are few reports of studies that systematically evaluated applications using unmodified protein compared with the native proteins [53]. Determining electrocatalytic activity for proteins that are deglycosylated after isolation or the recombinant proteins that are synthesized in hosts that will not go through posttranslational glycosylation may advance the application of MCOs and help purification methodologies.

Affinity-based chromatography encompasses a wide range of technical approaches. In its most basic function, affinity-based separations rely on a highly specific interaction between the subject protein and ligand. In the past, these pairings mainly comprised antigen–antibody, enzyme–substrate, or receptor–ligand interactions. This has now expanded beyond physiological interactions to include any interaction or property that can be exploited to selectively separate the protein of interest from the mix: precipitation reactions based on protein interaction with an insoluble matrix, multiphase extractions that rely on protein interactions with a particular solvent, and membrane-based purification strategies where the affinity ligand is immobilized on a filter matrix [54].

Interactions between dyes and proteins have been successfully used to isolate useful enzymes and have become a routine step in protein purification [55]. Dyes have been designed and used in both traditional chromatographic methods and new methods such as affinity precipitation, polymer aqueous two-phase separations, and expanded bed chromatography [56]. Triazinyl dyes interact with proteins containing a nicotinamide (reduced forms of nicotinamide adenine dinucleotide and nicotinamide adenine dinucleotide phosphate (NADH and NADPH)), nucleotide (adenosine triphosphate (ATP) and guanosine triphosphate (GTP)), or other natural biological ligand (folate, 3-hydroxy-3-methylglutaryl-coenzyme A (HMG-CoA)), as they mimic the heterocyclic functional groups. The triazinyl-based technique is promising for laccase purification, as many of the laccases used in the textile dye industry are active toward these substrates. Dyes have been used to track MCO activity during purification [57]. For example, dyes that are substrates can be used to detect enzyme activity in polyacrylamide gel electrophoresis (Figure 8.1).

In many cases, proteins can be modified with affinity "tags" (e.g., polyhistidine), which are fused to the protein via molecular genetic manipulations to facilitate purification by using affinity chromatography. Proteins modified with histidine residues are isolated from mixtures by using the polyhistidine motif's affinity for nickel or cobalt as the selective step. The modification is made by making genetic

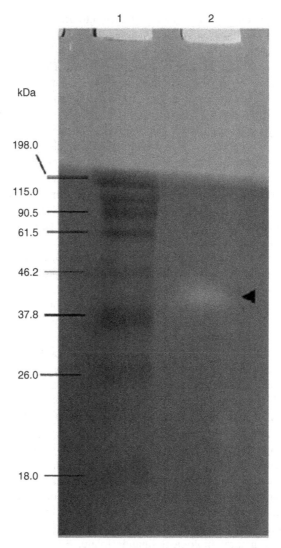

FIGURE 8.1 *Ganoderma lucidum* laccase activity after SDS-PAGE. Gel was incubated in Remazol Brilliant Blue R solution and decolorization of dye is seen surrounding the laccase band (arrow, lane 2). Molecular weight markers are in lane 1. (Adapted with permission from Ref. [57]. Copyright 2007, Elsevier.) (Please see the color version of this figure in Color Plates section.)

fusions that add five or more histidine residues at the N- or C-terminus of the protein. It is critical to test whether the modification affects catalysis. If so, the polyhistidine modification may be removed from the protein after purification using peptidases or chemical cleavage processes [58]. Expression and synthesis of the gene fusions are most commonly done using bacterial hosts, but commercial and experimental cloning vectors are also available that can be transferred to yeast strains. The eukaryotic background may provide posttranslational modification or chaperonins that may be required for synthesis of active fungal- or plant-derived MCOs.

The degree of enzyme purity will ultimately affect fuel cell performance, particularly when enzyme preparations are used to form immobilized films on electrode surfaces in DET reactions. Contaminating proteins that do not provide electron transfer effectively foul the electrode. When enzyme immobilization techniques are specific to the enzyme, then enzyme purity may not be as much as an issue, but rarely the immobilization technique is absolutely specific to the cathodic or anodic enzyme. For example, an attractive immobilization strategy is to link a particular enzyme to an electrode via its cofactor (e.g., flavin adenine dinucleotide (FAD), nicotinamide adenine dinucleotide (NAD), etc.) [59]. The cofactor is linked to the electrode material first and then the apoenzyme is allowed to "naturally" bind to the cofactor; all other proteins in the enzyme preparation that cannot bind the cofactor remain unbound and can be removed. Enzymes used in fuel cells are not so unique, and proteins in the immobilizing preparation may use the same cofactor but not the same fuel during fuel cell analysis or operation.

A 2009 study completed by Gao et al. specifically addressed the effect of enzyme purity in fuel cell systems by comparing the effects of unpurified and purified glucose oxidase (GOx) on current in a membraneless EFC [60]. A commercial preparation of GOx from *Aspergillus niger* was used as is, and after a two-step purification. The commercial preparation was measured to be ~80% pure and the remaining material was considered to be nonproteinaceous. After hydrophobic interaction and anion exchange chromatography, comparison of the specific activity between the purified and commercial forms $(337 \pm 32\,U\,mg^{-1}$ vs. $437 \pm 27\,U\,mg^{-1})$ suggests that the purified preparation is nearly pure. The catalytic current when operating (5 mM glucose, O_2-saturated electrolyte) was threefold higher when using purified enzyme than with the stock (nonpurified) enzyme as catalyst; the result strongly suggests that the contaminants in the commercial preparation significantly affect the efficiency of electron transfer in the system.

8.5 COPPER STABILITY AND SPECIFIC CONSIDERATIONS FOR MCO PRODUCTION

Metal coordination is essential to the function of electron transfer in MCOs. Accordingly, metal binding and metal complex stability during protein synthesis and subsequent purification is a critical factor [61]. A few questions outline the metal complex issues: How efficient is copper loading during protein expression? When the metal is in complex with the protein, how does its coordination affect the purification of the enzyme and its stability? How stable are the copper centers during purification, storage, and when used in fuel cell applications?

In Nature, microorganisms have developed copper transport systems to complete delivery of copper to target proteins [61]. Isolation of MCO enzymes produced in the native organism will best ensure that the holoenzyme has full complement of copper as the transport mechanisms and MCO synthesis is appropriately tuned in the evolved host. Nonetheless, the adaptation does not ensure that copper ions will be stably maintained in the protein for *in vitro* applications. Furthermore, recombinant host expression systems where copper transport mechanisms are nonexistent or not synchronized with foreign MCO synthesis can result in MCO enzymes that do not have their full complement of copper.

Supplementing culture media, crude extracts, storage, and assay buffers with copper salts can assist copper complex formation in MCO preparations. In many cases, copper can be added to bacterial and fungal culture supernatants to increase MCO activity and holoenzyme yield [62–67]. Free copper ions are toxic to micro-organisms, so copper concentrations must remain low enough during cultivation so that growth is not inhibited [61].

Various forms of MCOs containing different coordinated states of copper can affect purification methods. Furthermore, manipulation of the enzyme *in vitro* can disrupt protein structure and copper dissociates from the protein. The result is a mixture of inactive apo forms of the protein in the preparation. The amount of copper bound to the enzyme will affect the properties of the enzyme, which will change its retention and migration in column chromatography. For example, the predicted p*I* of the small laccase from *S. coelicolor* in its apo form is 6.2 versus 8.2 in the oxidized, four-copper holoenzyme [68]. Varying surface changes on the surface of the enzyme will cause the isoforms to chromatograph separately on ion exchange and hydrophobic interaction columns and, in turn, will reduce protein yield and purity.

In many instances, the copper complex can be reconstituted by simple treatments. Apoenzyme with incomplete copper coordination may be loaded by dialyzing the protein preparations against copper-containing buffers [68–70]. The process is accomplished by first dialyzing the protein against a buffer containing copper salt, and then dialyzing briefly with copper-free buffer to remove adventitiously bound copper from the enzyme. Briefly, the MCO preparations are placed in a dialysis bag or cartridge with a lower molecular weight cutoff membrane to retain MCOs and allow copper ions to diffuse from bulk solute to the protein. In some cases, the treatment can doubly serve as an additional purification step; the lower molecular weight proteins and materials will diffuse into the bulk dialysis buffer. Additionally, copper-sensitive protein may denature and precipitate from the suspension, easing removal in subsequent steps. An intermediary dialysis step against buffer containing a low concentration of EDTA can also been completed to remove excess copper [68,71]. Because copper is coordinated in MCOs in the oxidized copper(II) form, the process does not need to be completed under anaerobic conditions.

Copper salt solubility varies widely and the buffer combinations must be considered carefully in the protocols. For example, under physiological conditions, copper phosphate is minimally soluble. Phosphate is typically included in cell lysis and protein storage buffers, so addition of any additional copper salts to the protein suspensions containing phosphate must be carefully considered to limit precipitates. Furthermore, excess copper and copper precipitates can be troublesome in downstream processes and analyses. Additional purification may be necessary or the precipitate will be incorporated within the fuel cell architecture. The excess copper will confound electrochemistry assays. For example, free copper ions will react with MCO substrates, such as ascorbate to produce peroxide. The noncatalytic reaction has led to false positives and inaccurate determination of MCO activity [72].

Another special consideration for MCO purification is the tendency to form anomalous multimers. The phenomenon has been observed with AOx from plants [19,73,74] and some laccases from fungal and bacterial species (Figure 8.2) [68,75,76]. The origin of the different forms is unknown, but the inconsistent maturation also appears to occur *in vivo*; accordingly, the variants may have distinct physiological functions [77]. The inconsistent multimerization can lead to specific complications during purification. If the subunits separate during ion exchange, hydrophobic interaction, and size-exclusion chromatography, a subsequent determination must be made if components can be consolidated to reconstitute the intended MCO.

Identification and quantification of copper in MCO enzymes is essential for accurate analysis of enzyme activity and function, particularly because of the issues delineated above (incomplete complements of copper atoms and formation of different multimers). There are a few methods that are commonly used for copper

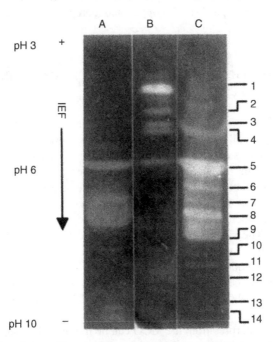

FIGURE 8.2 Ascorbate oxidase isoforms from melon cucurbits separated by native isoelectric focusing. The soluble (lane A), apoplastic (lane B), and total crude protein (lane C) fractions from cell-free extracts of cucurbits are shown. Numbers (right) identify isoenzymes. (Adapted with permission from Ref. [76]. Copyright 2003, Elsevier.)

determination and each has different sensitivities. The bicinchoninic acid (BCA) assay is a relatively easy method to determine copper content of MCO preparations [78]. The BCA assay directly quantifies copper through formation of an intense purple copper–BCA complex absorbing at 562 and 354 nm. If a homogenous MCO preparation is available, the stoichiometry of copper:protein can be subsequently calculated to ascertain whether complete "loading" has occurred. The total protein concentration is first determined by one means and then the total copper concentration in the preparation after its release from the protein via precipitation with acetone or trichloroacetic acid.

Other sensitive methods for copper determination involve elemental analysis by using atomic absorption spectrometry (AAS) and mass spectrometry (MS) to directly measure copper content in the protein. In pure protein preparations of known concentration, the molecular weight of the protein is predicted by using amino acid sequence and compared with the measured molecular weight derived from MS; by comparison, the number of copper atoms per monomer can be calculated. MS can also determine the absolute ratios of copper to protein constituents (C, N, and O) to determine specific copper loading in the MCO. Other methods, such as separation based on molecular weight and comparison with known standards (acrylamide gel or size-exclusion chromatography), are not as accurate. Additional care must be taken to ensure that adventitiously bound copper on the protein surface is eliminated since it will absorb like the type 1 (T1) copper and complicate—or even invalidate—determination.

8.6 SPECTROSCOPIC MONITORING AND CHARACTERIZATION OF COPPER CENTERS

The coordinated copper cofactors in MCOs provide an exceptional feature that is used to analyze the protein's biochemical and physical properties, determine reaction kinetics, and explore structure–function relationships. For most MCOs, the coordination chemistry of the T1 copper produces a blue-green protein in the oxidized form, due to an intense absorption band in the visible spectrum at ~600 nm [79]. The absorption arises from the electron arrangement of the coordination sphere, specifically the interaction between the copper and cysteinyl-S outer electron shells (Figure 8.3) [80]. The trigonal geometry allows for the $d_{x^2-y^2}$ ground state of copper to overlap with the $p\pi$ of the Cys-S ligand, resulting in a strong $p\pi_{(cys)}$-to-Cu^{2+} charge transfer band around 600 nm in the visible spectrum [80–82]. In the trinuclear site, the type 2 (T2) copper center does not have any obvious absorption bands in the ultraviolet (UV)/visible region, whereas the type 3 (T3) center exhibits an absorption band at 330 nm.

In its simplest interpretation, the MCO can be monitored throughout processing and purification by following the blue-green color of the enzyme. For example, as the protein elutes during column chromatography, the fractions that likely contain the target MCO will be blue-green. The characteristic color is useful in work to examine redox processes as well. Because absorbance of the T1 site is linked to

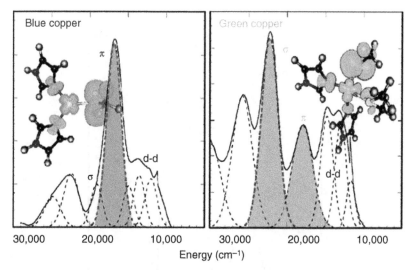

FIGURE 8.3 Representative spectroscopic and electronic features of blue and green copper centers. (Adapted with permission from Ref. [80]. Copyright 2006, American Chemical Society.) (Please see the color version of this figure in Color Plates section.)

electron transfer (i.e., the T1 site is blue when oxidized and colorless when reduced), complex analysis into the mechanism of electron transfer can be obtained. Electron transfer events from a substrate or electrode to the T1 site to the trinuclear center can be characterized spectroscopically with fine detail. A series of instruments and methods have been applied to take advantage of the electromagnetic changes to understand the MCO catalytic mechanism and structure.

Electron paramagnetic resonance (EPR) spectroscopy is used to characterize the copper cofactors in MCOs. The T1 blue copper centers have a small parallel hyperfine splitting ($A_{\parallel} < 100 \times 10^{-4} \, cm^{-1}$) in the EPR spectrum, whereas T2 copper centers normally provide a strong signal at $A_{\parallel} > 150 \times 10^{-4} \, cm^{-1}$ [4,79]. The T3 copper centers produce no EPR signal (Figure 8.4) [4,79,83]. The resolution of paramagnetic copper center is sufficient in EPR so that it can also be used to quantify copper in proteins [84]. The MCO spectral properties arise from the oxidized state of the copper centers but do not exhibit paramagnetic resonance when reduced. Thus, changes in EPR spectra and changes in the spectroscopic properties can be used to distinguish oxidized and reduced states of the enzymes [79].

Resonance Raman (RR) spectroscopy is frequently used to characterize copper site structure at atomic resolution. RR spectroscopy can probe and identify subtle differences in the structures and the quantum dynamics of MCO copper centers than other spectroscopic methods [69]. Cryogenic RR analysis is used to discern differences in Cu coordination vibrational modes between MCOs that otherwise cannot be resolved using spectral characteristics or X-ray crystallography. For example, RR spectroscopy resolves many differences between the homologous copper centers of

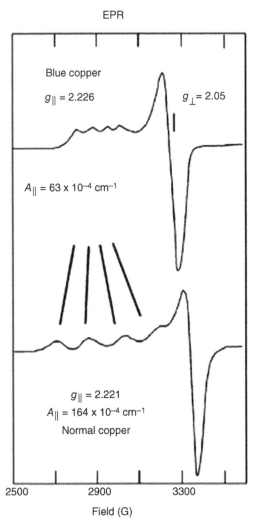

FIGURE 8.4 X-band EPR spectrum of the type 1 copper center from plastocyanin and a tetragonal Cu(II) complex representing a type 2 copper center. (Adapted with permission from Ref. [83]. Copyright 2004, American Chemical Society.)

AOx, CP, and tree and fungal laccases (Figure 8.5) [69]. Comparisons among the ground-state distribution of the $Cu-S(cys)$ stretching vibration in the T1 copper site in each of these proteins show that they differ in strength. The $Cu-S(cys)$ bond is one of the three essential coordination complexes of copper that defines the T1 site in MCOs and other copper proteins and is the major factor in electron transfer from the substrate to the trinuclear center [40,85]. Hence, RR spectroscopy can provide valuable information on how subatomic structural variations at this coordination can drive the functional properties of electron transfer at the T1 site.

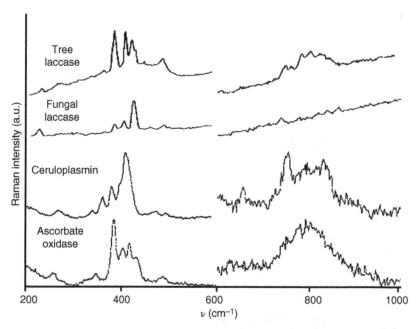

FIGURE 8.5 RR spectra of tree (12 K) and fungal laccases (77 K), human ceruloplasmin (277 K), and zucchini squash ascorbate oxidase (277 K). The overtone regions (>600 nm) are scale-expanded to show finer features. Laser wavelengths are 647.1 nm (tree laccase) and 620 nm (fungal laccase, ceruloplasmin, and zucchini ascorbate oxidase). (Adapted with permission from Ref. [69]. Copyright 1985, American Chemical Society.)

8.7 OUTLOOK

The MCOs are a technologically relevant class of enzymes due to broad range and extent of industrial uses. Their classic applications include waste stream processing for degrading phenolic compounds and decolorization of various industrial dyes. More recent advances in the ability to couple biomolecules and materials promote MCOs as electrocatalysts in fuel cells and sensors. The effects that contaminants produce at an electronic interface make protein purification a critical aspect in bioelectrocatalysis development. Many contemporary production and chromatography methods are amenable to MCO purification. Ensuring the stability of the copper cofactor complexes is the key to MCO redox activity. The structure and function of the MCOs are well examined using established spectroscopy and biochemistry methods. The abundance of available MCOs provides additional opportunity and challenge. Genomic sequence data reveal many MCOs as potential catalysts, establishing effective recombinant systems for synthesis of the gene products and then efficient purification methods will further promote development of EFCs and bioelectrochemical systems.

ACKNOWLEDGMENT

This research was prepared as an account of work sponsored by the United States Air Force Research Laboratory, Materials and Manufacturing Directorate, Airbase Technologies Division (AFRL/RXQ), but the views of authors expressed herein do not necessarily reflect those of the United States Air Force.

LIST OF ABBREVIATIONS

AAS	atomic absorption spectrometry
AOx	ascorbate oxidase
ATP	adenosine triphosphate
BCA	bicinchoninic acid
BFC	biological fuel cell
BOx	bilirubin oxidase
CotA	bacterial MCO
CP	ceruloplasmin
CueO	bacterial MCO
DET	direct electron transfer
EDTA	ethylenediaminetetraacetic acid
EFC	enzymatic fuel cell
EPR	electron paramagnetic resonance
FAD	flavin adenine dinucleotide
FPLC	fast protein liquid chromatography
GOx	glucose oxidase
GTP	guanosine triphosphate
HMG-CoA	3-hydroxy-3-methylglutaryl-coenzyme A
MCO	multicopper oxidase
MS	mass spectrometry
NAD	nicotinamide adenine dinucleotide
NADH	nicotinamide adenine dinucleotide, reduced form
NADP	nicotinamide adenine dinucleotide phosphate
NADPH	nicotinamide adenine dinucleotide phosphate, reduced form
ORR	oxygen reduction reaction
RR	resonance Raman
T1	type 1
T2	type 2
T3	type 3
UV	ultraviolet

REFERENCES

1. Stoj CS, Kosman DJ. Copper proteins: oxidases. In: *Encyclopedia of Inorganic Chemistry*. John Wiley & Sons, Ltd, Chichester, UK, 2006.

2. Sakurai T, Kataoka K. Basic and applied features of multicopper oxidases, CueO, bilirubin oxidase, and laccase. *Chem Rec* 2007;7:220–229.

3. Sirim D, Wagner F, Wang L, Schmid RD, Pleiss J. The Laccase Engineering Database: a classification and analysis system for laccases and related multicopper oxidases. *Database: J Biol Databases Curation* 2011: bar006.

4. Solomon EI, Sundaram UM, Machonkin TE. Multicopper oxidases and oxygenases. *Chem Rev* 1996;96:2563–2606.

5. Leech D, Kavanagh P, Schuhmann W. Enzymatic fuel cells: recent progress. *Electrochim Acta* 2012;84:223–234.

6. Calabrese Barton S. Enzyme catalysis in biological fuel cells. In: *Handbook of Fuel Cells*. John Wiley & Sons, Ltd, Chichester, UK, 2010.

7. Yoshida H. Chemistry of lacquer (Urushi). Part 1. *J Am Chem Soc* 1883;43:472–486.

8. Grotewold E, Taccioli GE, Aisemberg GO, Judewicz ND. A single-step purification of an extracellular fungal laccase. *World J Microbiol Biotechnol* 1988;4:357–363.

9. Nakade K, Nakagawa Y, Yano A, Sato T, Sakamoto Y. Characterization of an extracellular laccase, PbLac1, purified from *Polyporus brumalis*. *Fungal Biol* 2010;114:609–618.

10. Sahay R, Yadav RSS, Yadav KDS. Purification and characterization of extracellular laccase secreted by *Pleurotus sajor-caju* MTCC 141. *Chin J Biotechnol* 2008;24:2068–2073.

11. Wu Y-R, Luo Z-H, Kwok-Kei Chow R, Vrijmoed LLP. Purification and characterization of an extracellular laccase from the anthracene-degrading fungus *Fusarium solani* MAS2. *Bioresour Technol* 2010;101:9772–9777.

12. Zouari-Mechichi H, Mechichi T, Dhouib A, Sayadi S, Martínez AT, Martínez MJ. Laccase purification and characterization from *Trametes trogii* isolated in Tunisia: decolorization of textile dyes by the purified enzyme. *Enzyme Microb Technol* 2006;39:141–148.

13. Piscitelli A, Pezzella C, Giardina P, Faraco V, Giovanni S. Heterologous laccase production and its role in industrial applications. *Bioeng Bugs* 2010;1:254–264.

14. Dean JFD, LaFayette PR, Rugh C, Tristram AH, Hoopes JT, Eriksson K-EL, Merkle SA. Laccases associated with lignifying vascular tissues. In: Lewis NG, Sarkanen S (eds), *Lignin and Lignan Biosynthesis*, ACS Symposium Series, Vol. 697.American Chemical Society, Washington, DC, 1998, Chapter 8, pp. 96–108.

15. Ranocha P, McDougall G, Hawkins S, Sterjiades R, Borderies G, Stewart D, Cabanes-Macheteau M, Boudet A-M, Goffner D. Biochemical characterization, molecular cloning and expression of laccases—a divergent gene family—in poplar. *Eur J Biochem* 1999;259:485–495.

16. Ivnitski DM, Khripin C, Luckarift HR, Johnson GR, Atanassov P. Surface characterization and direct bioelectrocatalysis of multicopper oxidases. *Electrochim Acta* 2010;55:7385–7393.

17. Every D. Purification and characterisation of ascorbate oxidase from flour and immature wheat kernels. *J Cereal Sci* 1999;30:245–254.

18. Mosery O, Kanellis AK. Ascorbate oxidase of *Cucumis melo* L. var. *reticulatus*: purification, characterization and antibody production. *J Exp Bot* 1994;45:717–724.

19. Saari NB, Fujita S, Haraguchi K, Tono T. Purification and characterisation of basic ascorbate oxidase from satsuma mandarin (*Citrus unshiu* Marc). *J Sci Food Agric* 1994;65:153–156.

20. Kroneck P. Redox properties of blue multi-copper oxidases. In: Messerschmidt A (ed.), *Multi-Copper Oxidases*. World Scientific, Singapore, 1997, pp. 391–402.

21. Sanmartin M, Drogoudi P, Lyons T, Pateraki I, Barnes J, Kanellis A. Over-expression of ascorbate oxidase in the apoplast of transgenic tobacco results in altered ascorbate and glutathione redox states and increased sensitivity to ozone. *Planta* 2003;216:918–928.

22. Kato N, Esaka M. Expansion of transgenic tobacco protoplasts expressing pumpkin ascorbate oxidase is more rapid than that of wild-type protoplasts. *Planta* 2000;210: 1018–1022.

23. Albani D, Sardana R, Robert LS, Altosaar I, Arnison PG, Fabijanski SF. A *Brassica napus* gene family which shows sequence similarity to ascorbate oxidase is expressed in developing pollen. Molecular characterization and analysis of promoter activity in transgenic tobacco plants. *Plant J* 1992;2:331–342.

24. Christenson A, Shleev S, Mano N, Heller A, Gorton L. Redox potentials of the blue copper sites of bilirubin oxidases. *Biochim Biophys Acta: Bioenerg* 2006;1757:1634–1641.

25. Guo J, Liang XX, Mo PS, Li GX. Purification and properties of bilirubin oxidase from *Myrothecium verrucaria*. *Appl Biochem Biotechnol* 1991;31:135–143.

26. Seki Y, Takeguchi M, Okura I. Purification and properties of bilirubin oxidase from *Penicillium janthinellum*. *J Biotechnol* 1996;46:145–151.

27. Shimizu A, Kwon JH, Sasaki T, Satoh T, Sakurai N, Sakurai T, Yamaguchi S, Samejima T. *Myrothecium verrucaria* bilirubin oxidase and its mutants for potential copper ligands. *Biochemistry* 1999;38:3034–3042.

28. Shimizu A, Sasaki T, Kwon JH, Odaka A, Satoh T, Sakurai N, Sakurai T, Yamaguchi S, Samejima T. Site-directed mutagenesis of a possible type 1 copper ligand of bilirubin oxidase: a Met467Gln mutant shows stellacyanin-like properties. *J Biochem* 1999;125: 662–668.

29. Xu F, Shin W, Brown SH, Wahleithner JA, Sundaram UM, Solomon EI. A study of a series of recombinant fungal laccases and bilirubin oxidase that exhibit significant differences in redox potential, substrate specificity, and stability. *Biochim Biophys Acta* 1996;1292:303–311.

30. Shleev S, Tkac J, Christenson A, Ruzgas T, Yaropolov AI, Whittaker JW, Gorton L. Direct electron transfer between copper-containing proteins and electrodes. *Biosens Bioelectron* 2005;20:2517–2554.

31. Shimizu A, Samejima T, Hirota S, Yamaguchi S, Sakurai N, Sakurai T. Type III Cu mutants of *Myrothecium verrucaria* bilirubin oxidase. *J Biochem* 2003;133:767–772.

32. Oosthuizen MMJ, Nel L, Myburgh JA, Crookes RL. Purification of undegraded ceruloplasmin from outdated human plasma. *Anal Biochem* 1985;146:1–6.

33. Kouoh Elombo F, Radosevich M, Poulle M, Descamps J, Chtourou S, Burnouf T, Catteau JP, Bernier JL, Cotelle N. Purification of human ceruloplasmin as a by-product of C1-inhibitor. *Biol Pharm Bull* 2000;23:1406–1409.

34. Dube E, Shareck F, Hurtubise Y, Daneault C, Beauregard M. Homologous cloning, expression, and characterisation of a laccase from *Streptomyces coelicolor* and enzymatic decolourisation of an indigo dye. *Appl Microbiol Biotechnol* 2008;79:597–603.

35. Beneyton T, El Harrak A, Griffiths AD, Hellwig P, Taly V. Immobilization of CotA, an extremophilic laccase from *Bacillus subtilis*, on glassy carbon electrodes for biofuel cell applications. *Electrochem Commun* 2011;13:24–27.

36. Reiss R, Ihssen J, Thöny-Meyer L. *Bacillus pumilus* laccase: a heat stable enzyme with a wide substrate spectrum. *BMC Biotechnol* 2011;11:9–9.

37. Hullo MF, Moszer I, Danchin A, Martin-Verstraete I. CotA of *Bacillus subtilis* is a copper-dependent laccase. *J Bacteriol* 2001;183:5426–5430.

38. Cho E-A, Seo J, Lee D-W, Pan J-G. Decolorization of indigo carmine by laccase displayed on *Bacillus subtilis* spores. *Enzyme Microb Technol* 2011;49:100–104.

39. Grass G, Rensing C. CueO is a multi-copper oxidase that confers copper tolerance in *Escherichia coli*. *Biochem Biophys Res Commun* 2001;286:902–908.

40. Sakurai T, Kataoka K. Structure and function of type I copper in multicopper oxidases. *Cell Mol Life Sci* 2007;64:2642–2656.

41. Tsujimura S, Miura Y, Kano K. CueO-immobilized porous carbon electrode exhibiting improved performance of electrochemical reduction of dioxygen to water. *Electrochim Acta* 2008;53:5716–5720.

42. Achard MES, Tree JJ, Holden JA, Simpfendorfer KR, Wijburg OLC, Strugnell RA, Schembri MA, Sweet MJ, Jennings MP, McEwan AG. The multi-copper-ion oxidase CueO of *Salmonella enterica* serovar Typhimurium is required for systemic virulence. *Infect Immun* 2010;78:2312–2319.

43. Outten FW, Outten CE, Hale J, O'Halloran TV. Transcriptional activation of an *Escherichia coli* copper efflux regulon by the chromosomal MerR homologue, cueR. *J Biol Chem* 2000;275:31024–31029.

44. Lyashenko AV, Zhukhlistova NE, Gabdoulkhakov AG, Zhukova YN, Voelter W, Zaitsev VN, Bento I, Stepanova EV, Kachalova GS, Koroleva OV, Cherkashyn EA, Tishkov VI, Lamzin VS, Schirwitz K, Morgunova EY, Betzel C, Lindley PF, Mikhailov AM. Purification, crystallization and preliminary X-ray study of the fungal laccase from *Cerrena maxima*. *Acta Crystallogr F* 2006;62:954–957.

45. Singh G, Capalash N, Goel R, Sharma P. A pH-stable laccase from alkali-tolerant γ-proteobacterium JB: purification, characterization and indigo carmine degradation. *Enzyme Microb Technol* 2007;41:794–799.

46. Takeda K, Itoh H, Yoshioka I, Yamamoto M, Misaki H, Kajita S, Shirai K, Kato M, Shin T, Murao S, Tsukagoshi N. Cloning of a thermostable ascorbate oxidase gene from *Acremonium* sp. HI-25 and modification of the azide sensitivity of the enzyme by site-directed mutagenesis. *Biochim Biophys Acta: Protein Struct Mol Enzymol* 1998;1388:444–456.

47. Seetharam R, Sharma SK. *Purification and Analysis of Recombinant Proteins*. CRC Press, New York, 1991.

48. Wang C, Zhao M, Lu L, Wei X, Li T. Characteristics of spore-bound laccase from *Bacillus subtilis* WD23 and its use in dye decolorization. *Afr J Biotechnol* 2011;10:2186–2192.

49. Jung H, Xu F, Li K. Purification and characterization of laccase from wood-degrading fungus *Trichophyton rubrum* LKY-7. *Enzyme Microb Technol* 2002;30:161–168.

50. Lettera V, Piscitelli A, Leo G, Birolo L, Pezzella C, Sannia G. Identification of a new member of *Pleurotus ostreatus* laccase family from mature fruiting body. *Fungal Biol* 2010;114:724–730.

51. Perez J, Martinez J, de la Rubia T. Purification and partial characterization of a laccase from the white rot fungus *Phanerochaete flavido-alba*. *Appl Environ Microbiol* 1996;62:4263–4267.

52. Richardson A, Duncan J, McDougall GJ. Oxidase activity in lignifying xylem of a taxonomically diverse range of trees: identification of a conifer laccase. *Tree Physiol* 2000;20:1039–1047.

53. Meredith MT, Minteer SD. Biofuel cells: enhanced enzymatic bioelectrocatalysis. *Annu Rev Anal Chem* 2012;5:157–179.

54. Mondal K, Gupta MN. The affinity concept in bioseparation: evolving paradigms and expanding range of applications. *Biomol Eng* 2006;23:59–76.

55. Denizli A, Piskin E. Dye–ligand affinity systems. *J Biochem Biophys Methods* 2001;49:391–416.

56. Garg N, Galaev IY, Mattiasson B. Dye-affinity techniques for bioprocessing: recent developments. *J Mol Recognit* 1996;9:259–274.

57. Murugesan K, Nam I-H, Kim Y-M, Chang Y-S. Decolorization of reactive dyes by a thermostable laccase produced by *Ganoderma lucidum* in solid state culture. *Enzyme Microb Technol* 2007;40:1662–1672.

58. Malhotra A. Tagging for protein expression. *Methods Enzymol* 2009;463:239–258.

59. Cosnier S. Biomolecule immobilization on electrode surfaces by entrapment or attachment to electrochemically polymerized films: a review. *Biosens Bioelectron* 1999;14:443–456.

60. Gao F, Courjean O, Mano N. An improved glucose/O_2 membrane-less biofuel cell through glucose oxidase purification. *Biosens Bioelectron* 2009;25:356–361.

61. Palm-Espling ME, Niemiec MS, Wittung-Stafshede P. Role of metal in folding and stability of copper proteins *in vitro*. *Biochim Biophys Acta* 2012;1823:1594–1603.

62. Collins PJ, Dobson A. Regulation of laccase gene transcription in *Trametes versicolor*. *Appl Environ Microbiol* 1997;63:3444–3450.

63. Galhaup C, Goller S, Peterbauer CK, Strauss J, Haltrich D. Characterization of the major laccase isoenzyme from *Trametes pubescens* and regulation of its synthesis by metal ions. *Microbiology* 2002;148:2159–2169.

64. Karahanian E, Corsini G, Lobos S, Vicuna R. Structure and expression of a laccase gene from the ligninolytic basidiomycete *Ceriporiopsis subvermispora*. *Biochim Biophys Acta* 1998;1443:65–74.

65. Palmieri G, Giardina P, Bianco C, Fontanella B, Sannia G. Copper induction of laccase isoenzymes in the ligninolytic fungus *Pleurotus ostreatus*. *Appl Environ Microbiol* 2000;66:920–924.

66. Brown MA, Zhao Z, Grant Mauk A. Expression and characterization of a recombinant multi-copper oxidase: laccase IV from *Trametes versicolor*. *Inorg Chim Acta* 2002;331: 232–238.

67. Martins LO, Soares CM, Pereira MM, Teixeira M, Costa T, Jones GH, Henriques AO. Molecular and biochemical characterization of a highly stable bacterial laccase that occurs as a structural component of the *Bacillus subtilis* endospore coat. *J Biol Chem* 2002;277:18849–18859.

68. Machczynski MC, Vijgenboom E, Samyn B, Canters GW. Characterization of SLAC: a small laccase from *Streptomyces coelicolor* with unprecedented activity. *Protein Sci* 2004;13:2388–2397.

69. Blair DF, Campbell GW, Cho WK, English AM, Fry HA, Lum V, Norton KA, Schoonover JR, Chan SI. Resonance Raman studies of blue copper proteins: effects of temperature and isotopic substitutions. Structural and thermodynamic implications. *J Am Chem Soc* 1985;107:5755–5766.

70. Miyazaki K. A hyperthermophilic laccase from *Thermus thermophilus* HB27. *Extremophiles* 2005;9:415–425.

71. Greenaway FT, O'Gara CY. Copper and connective tissue: the mechanism of lysyl oxidase. In: Sorenson JRJ (ed.), *Biology of Copper Complexes*. Humana Press, Clifton, NJ, 1987.

72. Messerschmidt A. Spatial structures of ascorbate oxidase, laccase, and related proteins: implications for the catalytic mechanism. In: Messerschmidt A (ed.), *Multi-Copper Oxidases*, World Scientific, Singapore, 1997.

73. Amon A, Markakis P. Properties of ascorbate oxidase isozymes. *Phytochemistry* 1973;12:2127–2132.

74. Messerschmidt A, Rossi A, Ladenstein R, Huber R, Bolognesi M, Gatti G, Marchesini A, Petruzzelli R, Finazzi-Agró A. X-ray crystal structure of the blue oxidase ascorbate oxidase from zucchini: analysis of the polypeptide fold and a model of the copper sites and ligands. *J Mol Biol* 1989;206:513–529.

75. Thurston CF. The structure and function of fungal laccases. *Microbiology* 1994;140:19–26.

76. Al-Madhoun AS, Sanmartin M, Kanellis AK. Expression of ascorbate oxidase isoenzymes in cucurbits and during development and ripening of melon fruit. *Postharvest Biol Technol* 2003;27:137–146.

77. Marbach I, Harel E, Mayer AM. Molecular properties of extracellular *Botrytis cinerea* laccase. *Phytochemistry* 1984;23:2713–2717.

78. Brenner AJ, Harris ED. A quantitative test for copper using bicinchoninic acid. *Anal Biochem* 1995;226:80–84.

79. Solomon EI, Machonkin TE, Sundaram UM. Spectroscopy of multi-copper oxidases. In: Messerschmidt A (ed.), *Multi-Copper Oxidases*. World Scientific, Singapore, 1997, pp. 103–128.

80. Solomon EI. Spectroscopic methods in bioinorganic chemistry: blue to green to red copper sites. *Inorg Chem* 2006;45:8012–8025.

81. Clark KM, Yu Y, Marshall NM, Sieracki NA, Nilges MJ, Blackburn NJ, van der Donk WA, Lu Y. Transforming a blue copper into a red copper protein: engineering cysteine and homocysteine into the axial position of azurin using site-directed mutagenesis and expressed protein ligation. *J Am Chem Soc* 2010;132:10093–10101.

82. LaCroix LB, Shadle SE, Wang Y, Averill BA, Hedman B, Hodgson KO, Solomon EI. Electronic structure of the perturbed blue copper site in nitrite reductase: spectroscopic properties, bonding, and implications for the entatic/rack state. *J Am Chem Soc* 1996;118:7755–7768.

83. Solomon EI, Szilagyi RK, DeBeer George S, Basumallick L. Electronic structures of metal sites in proteins and models: contributions to function in blue copper proteins. *Chem Rev* 2004;104:419–458.

84. Minakata K, Suzuki O, Horio F. Quantification of copper in biological materials by use of electron spin resonance. *Clin Chem* 2001;47:1863–1865.

85. Andrew CR, Yeom H, Valentine JS, Karlsson BG, van Pouderoyen G, Canters GW, Loehr TM, Sanders-Loehr J, Bonander N. Raman spectroscopy as an indicator of Cu–S bond length in type 1 and type 2 copper cysteinate proteins. *J Am Chem Soc* 1994;116:11489–11498.

9

MEDIATED ENZYME ELECTRODES

JOSHUA W. GALLAWAY

The CUNY Energy Institute, The City College of New York, New York, NY, USA

9.1 INTRODUCTION

Enzymatic biological fuel cells (BFCs) use enzyme catalysts at one or both electrodes. The naturally evolved catalysis of enzymes allows either oxidation (at an anode) or reduction (at a cathode) to proceed at a rate high enough to be useful. By removing enzymes from their natural environment (a physiological one) and adapting them to an environment of our own design (a solid electrode surface), we benefit from their evolved catalytic functions. However, enzymes are surrounded by insulating protein layers and are therefore not specifically adapted to transmit electrons directly to a solid interface. An insulating layer of only a few tens of angstroms can be enough to reduce the likelihood of electron tunneling to essentially zero. Thus, one engineering challenge is effective electron transfer between enzyme active sites and electrodes.

Direct electron transfer (DET) is the field in which enzymes are attached to solid surfaces in such a way as to allow electron exchange between the enzyme active site and the surface. The goal of DET is to engineer a method of electron exchange around the problem of the insulating protein layer. Considering another strategy, electron exchange in the natural world of enzymes is accomplished molecule to molecule, and molecules can diffuse through insulating protein layers. Using small molecules, which *can* exchange electrons at a solid interface,

Enzymatic Fuel Cells: From Fundamentals to Applications, First Edition. Edited by Heather R. Luckarift, Plamen Atanassov, and Glenn R. Johnson.

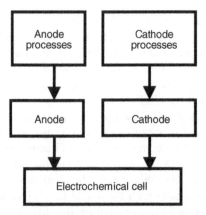

FIGURE 9.1 Concept of an electrochemical cell. The cell will have a rate-limiting process.

mediating electrons to the enzyme active site is known as mediated electron transfer (MET).

This chapter is concerned with methods of MET and their design to improve the performance of an enzyme electrode, however *performance* is defined. Figure 9.1 illustrates a way to think of an electrochemical cell, in which anode and cathode must have the same current (with opposite signs). For this reason, whichever electrode produces the lowest current will limit the cell. Each electrode involves several molecular processes, such as reaction kinetics, electric migration, and mass transport. In turn, each electrode will have a rate-limiting process. Enzyme electrodes are often limited by one of a few processes: mass transport of substrate, catalytic rate toward the substrate, or rate of electron mediation. If the rate of electron mediation is the limiting process, we would like to devise a molecular strategy to increase this rate, in hopes of approaching the enzyme maximum turnover rate.

9.2 FUNDAMENTALS

9.2.1 Electron Transfer Overpotentials

An MET system increases the rate of electron transfer between the electrode and enzyme active site by introducing a fast redox couple capable of interacting with the electrode at a rate far higher than the enzyme itself. This redox couple is the mediator, which takes the place of one of the natural substrates of the enzyme. The choice of an appropriate mediator can result in reaction rates exceeding those of a DET system by several orders of magnitude. The *cost* of an MET scheme derives from increased complexity, as the mediator species must now be considered. Also, an additional *overpotential* is introduced: the potential difference between the enzyme active site and the mediator. An overpotential is a deviation from the cell equilibrium potential. In an MET system, the anode or cathode potentials do not directly control the enzyme redox state, but they do affect the relative concentrations of the mediator redox state at the electrode surface, as characterized by the Nernst equation. This additional overpotential is generally compensated by the higher currents afforded by MET.

Figure 9.2 illustrates an MET design for a glucose/oxygen BFC [1,2]. In this device, the enzymes and osmium redox couples that serve as mediators are co-embedded in films on the electrode surfaces. Glucose oxidase (GOx) catalyzes the oxidation of glucose at the anode, and laccase catalyzes the reduction of oxygen at the cathode. For this electrochemical device, electrons (and their corresponding ions) originate with the anode fuel, glucose. Glucose is oxidized by GOx across an overpotential defined by the difference in potential between the glucose/gluconolactone redox couple and the $FAD/FADH_2$ (oxidized/reduced forms of flavin adenine dinucleotide) active site of the enzyme. This active site is oxidized by the Os^{III}/Os^{II} couple of the anode redox polymer mediator, with a similarly defined overpotential, and the osmium couple is oxidized at the anode current collector surface. Current flows from anode to cathode across the cell potential ΔE_{cell}. Electrons proceed from the cathode in an analogous pathway through a series of reductions, finally with oxygen reduction to water. At the anode, these electron transfer steps may be written by Equations 9.1a–9.1c. Here, the osmium mediator takes the place of the natural oxidizing

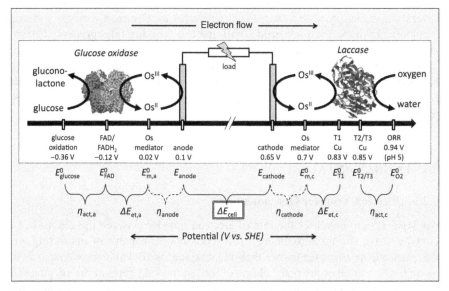

FIGURE 9.2 A mediated glucose–oxygen biological fuel cell. At both anode and cathode, the mediator is an osmium redox couple. Potential difference or overpotential for each electron transfer step is shown.

substrate of GOx, which is molecular oxygen (O_2). In Nature, GOx reduces oxygen to H_2O_2:[1]

$$GOx\text{-}FAD + glucose \rightarrow GOx\text{-}FADH_2 + gluconolactone \qquad (9.1a)$$

$$GOx\text{-}FADH_2 + 2Os^{III} \rightarrow GOx\text{-}FAD + 2Os^{II} + 2H^+ \qquad (9.1b)$$

$$Os^{II} \rightarrow Os^{III} + e^- \text{ (at electrode–film interface)} \qquad (9.1c)$$

The cathode enzyme laccase contains four copper ions, which are classified as three distinct types: a type 1 (Lac-T1), a type 2 (Lac-T2), and two type 3 (Lac-T3). The Lac-T2 and Lac-T3 pair is within 4 Å of each other and forms a trinuclear cluster (Lac-T2/T3). The Lac-T1 copper is 13 Å away from this cluster and serves as the site where electrons are accepted from the reducing substrate. Four electrons passing through the Lac-T1 site reduce all four copper ions, and the Lac-T2/T3 cluster then serves as the binding site for reduction of molecular oxygen to water. The somewhat complex reaction mechanism can be approximated as follows (Equations 9.2a–9.2d) (the charges of the copper sites are omitted for clarity) [3]:

$$Os^{III} + e^- \rightarrow Os^{II} \text{ (at current collector–film interface)} \qquad (9.2a)$$

$$Lac\text{-}T1_{ox} + Os^{II} \rightarrow Lac\text{-}T1_{red} + Os^{III} \qquad (9.2b)$$

$$3Lac\text{-}T1_{red} + Lac\text{-}T2/T3_{ox} \rightarrow 3Lac\text{-}T1_{ox} + Lac\text{-}T2/T3_{red} \qquad (9.2c)$$

$$O_2 + 4H^+ + Lac\text{-}T1_{red} + Lac\text{-}T2/T3_{red} \rightarrow 2H_2O + Lac\text{-}T1_{ox} + Lac\text{-}T2/T3_{ox} \qquad (9.2d)$$

In Nature, laccases are found in many higher plants, fungi, and microorganisms, where they catalyze the oxidation of phenolic compounds to *o*- and *p*-quinones, coupled to the four-electron reduction of atmospheric oxygen to water [4]. Here, the osmium mediator takes the place of quinines as the natural reducing substrate.

---

[1] The element osmium is often associated with toxicity, as osmium tetroxide, the volatile Os^{VIII} form, is highly acutely toxic. OsO_4 is a yellowish solid that sublimes at room temperature and can cause blindness by attacking the cornea. However, there is some indication that exposure to osmium is not a chronic hazard. Refiners, histologists, and workers continually exposed to low levels of osmium tetroxide vapors show no apparent long-term consequences, and, in fact, osmic acid is injected into the body to treat rheumatoid arthritis [5]. Osmium redox polymers possibly share none of the toxicity of the tetroxide material, especially as the osmium salts themselves used to produce redox polymers are not classified as toxic. Coordination polymers of osmium, formerly known as osmium black, were first reported by Hanker et al. in 1967 [6]. The oligomeric compound $[Os(SCH_2CHOHCH_2OH)_4]_4$ was injected into albino mice. The dark brown solution colored the blood, ears, and eyes, but cleared after 45 min and was excreted in the urine. No acute toxicity to the mice was found.

Equations 9.1 and 9.2 are added for an overall cell reaction (Equation 9.3).

$$2 \text{ glucose} + O_2 \rightarrow 2 \text{ gluconolactone} + 2H_2O \tag{9.3}$$

The theoretical equilibrium cell potential is given as (Equation 9.4)

$$\Delta E_{eq} = E^0_{O_2} - E^0_{glucose} = 1.3 \text{ V} \tag{9.4}$$

However, this potential will not be observed because neither oxidant (glucose) nor reductant (oxygen) interacts with the anode and cathode directly. Rather, this interaction is mediated by the osmium compounds. In practice, the open-circuit potential observed will be slightly higher than the difference in the mediator redox potentials (Equation 9.5):

$$\Delta E_{OC} > E^0_{m,c} - E^0_{m,a} = 0.68 \text{ V} \tag{9.5}$$

This value does not exactly equal the redox potential difference in anode and cathode mediators because of a mixed potential, as there is some DET potential at equilibrium even when mediators are present. When the cell is not at equilibrium and current flows, the electrode potentials are separated from the mediator redox potentials by an electrode polarization overpotential η_{anode} or $\eta_{cathode}$. For example, at the cathode (Equation 9.6),

$$E_{cathode} = E^0_{m,c} - \eta_{cathode} \tag{9.6}$$

The overpotential $\eta_{cathode}$ is the driving force for the reaction given in Equation 9.2a. The current, in $A\,cm^{-2}$, passed to or from the electrode, arrives through interaction with the mediator and is related to the electrode polarization through a current–overpotential equation. For the cathode (Equation 9.7),

$$i = i_0 \left[\frac{C^0_m - C_m}{C^0_m} \exp\left(\frac{\eta_{cathode}}{b}\right) - \frac{C_m}{C^0_m} \exp\left(\frac{\eta_{cathode}}{b}\right) \right] \tag{9.7}$$

This equation includes both Butler–Volmer kinetics between mediator and electrode and mediator mass transport. The exchange current density is i_0, $C^0_{m,c}$ is the mediator concentration, $C_{m,c}$ is the reduced mediator concentration, and b is a Butler–Volmer kinetic term that essentially captures the peak separation in a cyclic voltammogram of the mediator in the absence of catalysis. A set of equations analogous to Equations 9.6 and 9.7 may be written for the anode with minor modifications. At an electrode of area A, Equation 9.7 is related to the net rate of reaction (9.2a) v_{2a} by Faraday's constant F (Equation 9.8):

$$i = \frac{-Fv_{2a}}{A} \tag{9.8}$$

If a DET scheme were used, the electrode polarizations would be relative to the enzyme active site redox potentials, rather than to the mediators. The extra potential loss at each electrode, added by MET, is ΔE_{et}, which is the driving force for electron transfer between the enzyme active site and the mediator, that is, the driving force for mediation.

9.2.2 Electron Transfer Rate

Mediators are meant to carry electrons and transfer them to another species, which could be either another redox-active atom or an electrode. The rate of electron transfer at an interface is typically described by the Butler–Volmer equation, which is included in Equation 9.7. The two exponential terms describe anodic and cathodic charge transfer reactions from a species in solution to the electrode. Electron transfer between species in solution is given by Equation 9.9, where the rate k_{ET} is the product of two exponential terms [7]:

$$k_{ET} = \nu \exp\left[-\beta(d-3)\right]\exp\left[-\frac{\lambda}{4RT}\left(1+\frac{\Delta G^0}{\lambda}\right)^2\right] \qquad (9.9)$$

The first term involves d, the distance between the reduced and oxidized species in angstroms, and β, which characterizes how rapidly rate decays with increased distance in a given system. In 1988, for example, Cowan and Gray reported a β of 0.9 for magnesium–ruthenium electron transfer in myoglobins [8]. The second term in Equation 9.9 involves the free energy driving force for the electron transfer, ΔG^0, which is related to the redox potentials of the two species by $\Delta G^0 = -nF\Delta E^0$. The term λ is the energy for changes in nuclear configuration around the redox sites. The proportionality term ν has units of s^{-1}, accounting for nuclear frequency. Equation 9.9 can be used to assess the effect of distance (d) and driving force (ΔE^0, given by overpotentials η or ΔE_{et} in Figure 9.2) changes on electron transfer to and from mediators. Note that if the squared terms in the second term are comparatively small, Equation 9.9 can be reduced to a Butler–Volmer form [9].

9.2.3 Enzyme Kinetics

The choice of an effective mediator involves several criteria: The mediator must be stable in both oxidized and reduced forms, must engage in rapid electron transfer with the biocatalyst and at the electrode, and must have a redox potential that allows the electrode to be poised appropriately to avoid unwanted reactions and minimize overpotential. The enzyme undergoes a reaction with both the oxidizing and reducing substrates, one of which is a mediator. Michaelis–Menten kinetics can be assumed for each of these two reactions. For a cathode enzyme (E), in which the mediator (M) is the reducing substrate and we denote the oxidizing substrate simply as S (Equation 9.10),

$$S + E_{red} \underset{k_{-1S}}{\overset{k_{1S}}{\rightleftharpoons}} ES \xrightarrow{k_{2S}} P + E_{ox} \tag{9.10}$$

$$M_{red} + E_{ox} \underset{k_{-1m}}{\overset{k_{1m}}{\rightleftharpoons}} EM \xrightarrow{k_{2m}} M_{ox} + E_{red} \tag{9.11}$$

Equation 9.10 represents the kinetics for the reaction given in Equation 9.2d. In the same way, Equation 9.11 gives the kinetics for Equation 9.2b (equivalent versions can be written for Equations 9.1a and 9.1b by flipping the red/ox subscripts). Electrons are passed from the reduced mediator to the enzyme, and the product (P) is produced by reduction of S. The concentrations of the intermediate complexes ES and EM are assumed to be low and at steady state. Expressions for individual kinetic rates can be written for both substrate and mediator. When coupled, these result in the bi-bi ping-pong expression for total reaction rate with respect to the enzyme v_e (Equation 9.12) [10,11]:

$$v_e = \frac{k_{cat}K_m^{-1}C_eC_mC_s}{C_mK_m^{-1}(C_s + K_s) + C_s} \tag{9.12}$$

where C is concentration of enzyme (e), reduced mediator (m), or substrate (s). The enzyme turnover rate k_{cat} and the mediator and substrate Michaelis constants K_x are collections of the kinetic constants in Equations 9.10 and 9.11 [11–15]. Writing Equation 9.12 in this form is useful, as the term $C_mK_m^{-1}$ indicates the kinetic saturation of the mediation reaction in Equation 9.11.

9.3 TYPES OF MEDIATION

9.3.1 Freely Diffusing Mediator in Solution

The electron transfer between enzyme active sites and small, nonbiological redox-active molecules has a long history. Methylene blue was often used to replace oxygen as an enzyme-oxidizing substrate in early research [16]. Measurement of the electron transfer rate in these studies was often colorimetric, for example, the loss of blue color when titrating fungal laccase. Efforts to determine the redox potential of coordinated metal atoms within enzyme active sites involved titration with redox molecules of known potential, such as potassium hexacyanoferrate or molybdenum octacyanide [17,18]. Starting in 1974, McArdle et al., for example, employed chelates of group VIII elements, using tris(1,10-phenanthroline)Co(III), to study the electron transfer kinetics from ferrocytochrome c and the mononuclear blue copper proteins stellacyanin and azurin [19,20]. With the introduction of an electrode, these small molecules become mediators, provided they are redox active at the electrode as well as at the enzyme.

In 1964, Yahiro et al. demonstrated enzyme mediation to platinum electrodes [21]. They used three enzymes: GOx, D-amino acid oxidase, and alcohol dehydrogenase (ADH). If iron was introduced into the system with either GOx or D-amino acid oxidase, an electrical potential was observed. However, this was not the case with ADH, which is dependent on a nicotinamide adenine dinucleotide (NAD) cofactor (see below).

Electrocatalytic enzyme mediation has been demonstrated using quinones, viologens, 2,2-azinobis(3-ethylbenzothiazoline-6-sulfonate) (ABTS), and complexes of iron, ruthenium, cobalt, osmium, and many other compounds [22–24]. Much early work concerned the GOx anode, intended for a glucose sensor. In 1974, Schläpfer et al. tested 11 different mediators for a GOx electrode with a semipermeable membrane [25]. Ten years later, Cass et al. reported membrane-bound electrodes that operated in whole blood [26]. In 1986, Bourdillon et al. presented an analysis arguing that immobilized enzyme electrodes have higher efficiency than those with free enzyme in solution [27]. These examples demonstrate several possible enzyme/mediator configurations. Both enzyme and mediator can exist as free species in the liquid electrolyte, or one or both can be immobilized on the electrode surface. As an alternative to immobilization, enzyme and mediator can also be confined near the electrode by a semipermeable membrane.

Some enzymes require nonprotein cofactor molecules for catalysis, as with NAD mentioned above. A cofactor may be covalently bound to the enzyme, such as FAD in the case of GOx, but others are diffusing freely in solution. Figure 9.3 shows the mechanism of aldehyde dehydrogenase (ALDH), in which an aldehyde is oxidized to a carboxylic acid in conjunction with an NAD cofactor [28]. The NAD cofactor simultaneously binds in the enzyme active site and is released as the reduced form NADH. In this case, NAD acts as the oxidizing substrate of the enzyme and is regenerated back to NAD elsewhere in the system, independently of ALDH. Thus, NAD and other freely diffusing redox cofactors may be thought of as natural mediators. Pyrroloquinoline quinone (PQQ) is another frequently encountered redox cofactor.

Solution-phase BFCs have been developed by Palmore et al., with anode and cathode compartments separated by a Nafion membrane. Starting in 1998, Palmore et al. used laccase in a mediated cathode and cascades of multiple enzymes designed to fully oxidize methanol to carbon dioxide at a mediated anode [22,29]. The cathode used in Palmore et al.'s early work was ABTS, which was chosen for the value of $E_{m,c}^0$

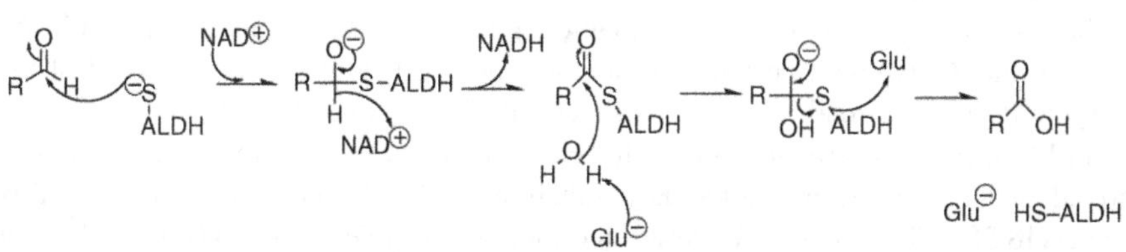

FIGURE 9.3 Mechanism for oxidation of an aldehyde to a carboxylic acid by aldehyde dehydrogenase, in conjunction with an NAD cofactor. The cofactor must bind and be reduced while the aldehyde substrate is bound. Glu is glutamate on the enzyme.

it affords. In 2005, Farneth et al. used this ABTS–laccase system to demonstrate that mediated electrodes, like those depicted in Figure 9.2, can be modeled using the intrinsic physical and chemical properties of the constituents, for example, the kinetic constants of the enzyme and the diffusion coefficient of the mediator [30]. Similarly, in 2001, the Ikeda group demonstrated ABTS mediation toward bilirubin oxidase, another oxygen-reducing enzyme [31].

9.3.2 Mediation in Cross-Linked Redox Polymers

When in contact with a fluid under convection, enzyme and mediator must be confined to the electrode surface to avoid being lost to the bulk fluid. Although semipermeable membranes are appropriate in some cases, in other cases they introduce unwanted complexity or block convection of fuel, oxidant, or analytes to the electrode surface. As a result, many other methods have been devised to attach both enzyme and mediator to the electrode surface:

1. Enzymes embedded in a cross-linked redox hydrogel film.
2. Enzymes and small mediator molecules co-immobilized in a cross-linked film such as polypyrrole or Nafion.
3. Multiple cross-linked layers in contact, that is, the first containing enzymes and a cofactor, and the second containing mediator species.
4. Enzymes attached to mediator-containing monolayers.
5. Any of the above incorporated in a high surface area matrix such as carbon paper or fabric.

In the first of these examples, a mediating redox polymer is cross-linked into a redox hydrogel, which serves to both entrap the enzyme and provide mediation to the electrode. The molecular innovations that lead to this strategy are worth considering in some detail and are presented in the following section [32].

9.3.2.1 The "Wired" Glucose Oxidase Anode   For implanted applications such as an *in vivo* glucose sensor, enzyme and mediator must be confined to a volume near the electrode, otherwise freely diffusing enzyme would be lost into the bloodstream and the device would no longer operate without catalyst. Beginning in 1987, the Adam Heller group at the University of Texas at Austin achieved a breakthrough in MET while attempting to develop an implanted glucose sensor. They began with the premise that GOx could be bound to an electrode surface if the protein shell of the enzyme itself were made sufficiently electronically conducting [33,34].

In 1988, Gray and coworkers reported electron transfer rates between redox centers bound in large macromolecules as a function of the distance between the redox centers [8,35]. Electron transfer between ruthenium complexes $[Ru(NH_3)_5Hist]^{3+}$, for example, where Hist is a histidine group in a protein, was found to increase by a factor of 10 when the complexes were 2.1 Å closer. This corresponds to a $\beta = 0.91$ in Equation 9.9.

In 1987, the Heller group reasoned that binding redox centers throughout the enzyme protein shell would effectively act as molecular "relays" and create an electron transport path such that the adsorbed enzyme could transfer electrons to an electrode surface at a useful rate. The unaltered electron transfer distance of GOx is 30 Å, and by inserting one relay at the midpoint, the transfer rate could theoretically be increased by a factor of nearly 10^7. The study found that it was possible to attach an average of 12 ferrocene carboxylate groups per GOx enzyme. The resulting GOx retained catalytic activity and specificity for glucose and produced an anodic current at an electrode surface [34,36–38]. However, this small number of relays is not sufficient to render an enzyme's entire protein shell conductive, and therefore orientation of the enzyme on the electrode surface remains a limitation. Unless the enzyme is allowed to diffuse freely, only a small portion will have the correct orientation; this need for free diffusion requires a semipermeable membrane to contain the enzyme.

As a second strategy, the Heller group sought to "wire" GOx enzymes to an electrode surface. Instead of being bound to the enzyme shell, redox complexes were bound to a polymer or macromolecule: polyvinylpyridine (PVP). Instead of ferrocene redox complexes, which are neutral in the reduced state, osmium bis-bipyridine was used, making the redox sites positively charged in both oxidized and reduced states. At physiological pH, GOx has a net negative charge, and this electrostatic affinity between the redox centers and the enzyme was found to keep them bound to one another [39]. Thus, the *redox polymer* $[Os(2,2'\text{-bipyridine})_2(PVP)Cl]^{+/2+}$ formed an electrostatic adduct with the enzyme and strongly adsorbed to the electrode surface, as shown in Figure 9.4 [40–43]. As a result, the enzyme is effectively wired to the electrode surface.

Anodic current produced by wired enzymes, such as that in Figure 9.4, was found to be a function of ionic strength, which proved unacceptable for sensor

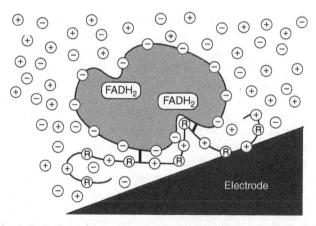

FIGURE 9.4 A FAD-dependent enzyme "wired" to an electrode surface by a positively charged redox polymer. Electrons may move by self-exchange along the polymer from R to R. The negatively charged enzyme is joined to the polymer electrostatically. (Reproduced with permission from Ref. [32]. Copyright 1990, American Chemical Society.)

applications [39]. This was due to screening of the redox center positive charges on the PVP backbone, which occurred at an electrolyte concentration above ~0.5 M. At high ionic strength, the redox polymers tended to form coils rather than linear chains, which are preferred for serving as electron transfer paths. This problem was solved by introducing covalent bonds to make a cross-linked structure, keeping the electron transport paths extended [39]. By making this cross-linked structure three-dimensional, it was found that enzyme–redox polymer films up to 1 μm thick could be formed. These films retained their catalytic activity and wiring functionality. Current densities were 10 times greater than those of two-dimensional wired enzymes, suggesting that more than 10 equivalent enzyme layers could be made active in this configuration [44]. These wired enzyme electrodes had several commercial advantages for use in glucose sensors: fabrication was simple, they were durable, they were highly reproducible, their response time to changes in glucose concentration was less than 1 s, and, due to the extremely small mass of material required, they were inexpensive. As with most enzyme electrodes, they were highly specific to glucose. The discovery of these electrodes led to the startup company TheraSense, which was acquired by Abbott Laboratories in 2004. As a result, enzyme–redox polymer electrodes are today a commercial product widely available in drug stores and pharmacies.

9.3.3 Further Redox Polymer Mediation

Redox species can be incorporated into polymers (then termed *redox polymers*) either at the monomer stage or after polymerization. These macromolecules have distinct electrochemically active sites at one or more locations on the chain. Examples of redox polymers are shown in Table 9.1. In the previous section, it was described how the redox polymer shown in panel A was used to wire enzymes to electrode surfaces [39].

The wired GOx electrode was made using redox polymers based on osmium complexes [45]. This is because, unlike the corresponding iron and ruthenium complexes, osmium is photochemically stable. Like most group VIII transition metals, osmium generally has a coordination number of six, meaning it will form six dative bonds to make an octahedral complex. This complex may be written in the form OsL_6, where L represents ligands, which donate electron pairs to the bond. Pyridine is a common ligand, donating one nitrogen lone pair. Bipyridine, which consists of two joined pyridine rings, can bond with a metal atom in two locations and is therefore "bidentate" as it bites the metal atom twice. (It is also called a *chelating ligand*, as it holds the metal atom like a claw.)

Beginning in 1949, Cassidy and coworkers presented a series of papers demonstrating redox polymers [46–49]. In these early demonstrations, the redox sites were based on hydroquinones. If the redox site is a transition metal complex, however, the redox polymer is also a *metallopolymer* [50–53]. Ruthenium and osmium metallopolymers with polyvinylimidazole backbones, for example, are shown in panels B and C of Table 9.1. By participating in oxidation–reduction reactions with other species and through self-exchange, a redox polymer in solution can conduct electrons

TABLE 9.1 Examples of Redox Polymer Structures

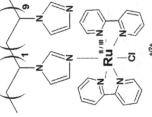

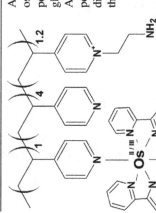

A: Redox polymer based on osmium bis-bipyridine and polyvinylpyridine used to wire glucose oxidase [39,54,55]. An earlier ferrocene redox polymer was uncharged and did not electrostatically bind to the enzyme [33,34].

B: Ruthenium metallopolymer based on ruthenium bis-bipyridine and polyvinylimidazole [52].

C: Osmium bis-bipyridine-based analog to the metallopolymer in panel B. Whereas ruthenium undergoes photochemistry, osmium does not [52].

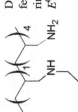

D: Redox polymer based on polyallylamine and ferrocene. Ferrous ion is 2+ and cyclopentadienyl rings are 1− each, giving a neutral molecule. $E^0 = 0.55$ V [56].

E: Vitamin K and polyallylamine redox polymer. $E^0 = -0.05$ V, with typical vitamin K being $E^0 = -0.08$ V [57].

F: Polyethylenimine polymer modified with 3-(dimethylferrocenyl) propyl redox centers, shown to be an effective glucose oxidase mediator. $E^0 = 0.41$ V [58].

G: Osmium-based redox polymer with the redox site at the end of a 13-atom chain. This molecular design is meant to increase the redox site range of motion. $E^0 = 0.01$ V [59].

H: The proposed structure of poly(methylene blue) in the oxidized form. Addition of two protons and two electrons reduces to a form with a N−N double bond. In this case, the redox site is not pendant, but part of the chain [60].

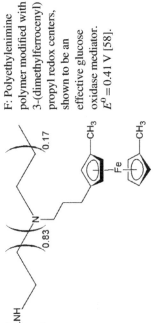

Note: Counter ions for charged species are not shown. Potentials are in V versus SHE.

158

within this ionic phase. These are distinct from conducting polymers such as polypyrrole, in which electrons are delocalized through a conjugated p-orbital system [61].

In 1955, it was demonstrated that redox polymers could be cross-linked into redox hydrogels [62]. Since the advent of the wired GOx anode described in the previous section, BFC electrodes using an osmium redox polymer hydrogel have appeared extensively in the literature [63–66]. Redox polymer mediators based on species other than osmium have also been demonstrated in recent BFC systems. Redox polymers based on ferrocene, for example, are a continuing area of research [56]. Sato et al. reported a glucose dehydrogenase BFC with a redox mediator based on vitamin K (see panel E of Table 9.1) in 2005 [57]. Six years later, Meredith et al. reported that a polyethylenimine polymer modified with 3-(dimethylferrocenyl)propyl redox centers is a durable and efficient mediator to GOx despite being a neutral molecule (see panel F of Table 9.1) [58].

Electron transport in cross-linked redox polymer films occurs by electron self-exchange between adjacent redox sites [50,67–71]. This is termed *electron hopping* and can be described under some conditions as an apparent electron diffusion D_{app}. As such, in 1992, Blauch and Saveant derived the relation described in Equation 9.13 [72]:

$$D_{app} = \frac{1}{6}k_{ex}\left(\delta^2 + 3\lambda^2\right)C_m^0 \qquad (9.13)$$

where k_{ex} is a self-exchange rate constant, δ is the electron tunneling distance, λ is the range of bounded (or "wiggle") motion by the redox site, and C_m^0 is the concentration of redox sites. In regimes where this equation holds, apparent electron diffusion is proportional to the concentration of redox sites. It should be noted that this linear relation is not always observed [50]. However, for redox polymers of similar structure, a linear relation between D_{app} and C_m^0 can hold. For a collection of osmium and polyvinylimidazole redox polymers like that shown in panel C of Table 9.1, the relation found was $D_{app} = (4.5 \times 10^{-9})C_m^0$, with C_m^0 in M and D_{app} in $cm^2 s^{-1}$ [73]. Apparent electron diffusion is typically on the order of 10^{-9} to 10^{-11} $cm^2 s^{-1}$, several orders of magnitude less than typical diffusion in solution.

To exploit the λ term in Equation 9.13, Mao et al. introduced an osmium redox polymer in 2003 in which the redox center was a pendant on the polymer backbone at the end of a 13-atom molecular tether [59]. The increased range of motion for the redox site led to an increase in D_{app} (see panel G of Table 9.1). A version with a higher redox potential, more appropriate to mediate a cathode enzyme, was also produced [74].

Three-dimensional redox hydrogels containing enzymes can be prepared on material with a large surface area such as carbon cloth or paper, thereby increasing the catalytic rate per geometric surface area [65]. Figure 9.5 demonstrates that many more effective layers of enzyme can be made active in such a configuration, as opposed to a planar hydrogel film. Enzyme electrodes with large surface areas are usually porous, allowing mass transport of substrate throughout, and have been

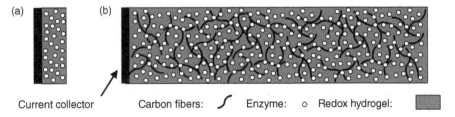

Current collector   Carbon fibers: $\int$ Enzyme: o   Redox hydrogel:

FIGURE 9.5 (a) A cross-linked redox hydrogel with immobilized enzymes, $\sim 1\,\mu m$ in thickness. (b) A redox hydrogel with immobilized enzymes cross-linked over a high surface area porous support such as carbon cloth, felt, or paper, $\sim 1\,mm$ in thickness.

modeled as plug flow reactors [75,76]. Methods to further increase surface area have included incorporating carbon nanotubes and related nanoscale architectures [77,78]. To date, the highest performing enzymatic BFC electrodes are MET systems based on osmium-wired enzymes and a high surface area support [79,80].

9.3.4 Mediation in Other Immobilized Layers

Entrapping enzymes in a cross-linked redox hydrogel is not the only method for immobilizing both enzyme and mediator at an electrode surface. For example, in 2007, the Palmore group immobilized ABTS by incorporation into a polypyrrole and laccase film, thus trapping a small mediator molecule in a cross-linked film [81]. Similarly, starting in 2001, the Calvo group created a layer-by-layer method using osmium redox polymers based on a polyallylamine backbone [82–85]. This allows fine control of enzyme–mediator layer thickness [86].

Another method of immobilizing enzyme and mediator is the attachment of a mediator-containing monolayer to an electrode surface. If this monolayer is also functionalized with an enzyme cofactor meant to bind the enzyme, a catalytic electrode results. For example, gold electrode surfaces can be covered with a PQQ monolayer. The monolayer retains the redox properties of PQQ and can be functionalized further. If FAD cofactors are then covalently bound, GOx apoenzyme can bind the FAD centers. A covalent molecular chain results and electrons may "hop" from FAD to PQQ to the electrode surface, thus mediating from the enzyme [87]. PQQ monolayers can also link to covalently bound NAD^+, for mediation to NAD^+-dependent dehydrogenase enzymes [88,89]. Functionalizing with monolayers of cytochrome c has also been demonstrated [90].

Beginning in 2003, Katz and Willner used this architecture to engineer signal-switchable BFCs, in which power output can be activated or deactivated by an external signal, such as potential, magnetic field, pH, or the presence of certain biochemical species [91–94]. For example, a GOx anode produced from a PQQ–FAD monolayer responds to a parallel 0.92 T magnetic field with a doubling of the catalytic rate [95,96].

Also starting in 2003, the Minteer group demonstrated immobilization of enzymes within a modified Nafion cation exchange film, with retention of activity for up to

90 days [97–99]. This is accomplished by introducing quaternary ammonium salts such as tetrabutylammonium bromide into the Nafion. This large cation displaces protons in the film, buffering the pH to a milder level and increasing the hydrophilic pore size. Thus, enzymes may fit within at the appropriate pH. Mediators can also be introduced, such as $Ru(bpy)_3^{2+}$, which is bound by attraction to sulfonate groups in the pores. In a hydrogel film such as this, the mediator will be present at concentrations on the order of 1 M, and render the pore channels electronically as well as ionically conducting. An ethanol–oxygen BFC with this design operated for 20 days before mediation efficiency began to diminish, demonstrating that the enzyme and mediator were well immobilized [98].

Redox films can be produced by electropolymerization of small redox molecules, which retain redox chemistry in the polymerized state. The suggested structure of a polymeric version of poly(methylene blue) is shown in panel H of Table 9.1 [60]. In 1996, Zhou et al. demonstrated that methylene green could be electropolymerized on an electrode surface, resulting in a film with a half-wave potential in the vicinity of ~0.21 V versus the standard hydrogen electrode (SHE) [100]. Furthermore, their work showed that this film could mediate to NADH, reducing the required oxidation overpotential by 400 mV. Similar polymers have been demonstrated to mediate to NAD/NADH, such as poly(neutral red) [101].

The Minteer group has shown that a layer of modified Nafion containing enzymes deposited over a layer of poly(methylene green) mediator produces a versatile immobilized MET electrode [102–104]. A modified Nafion film containing NAD and an NAD-dependent dehydrogenase produce NADH within the film, which is reoxidized to NAD at the interface with the poly(methylene green) film and provides mediation to the electrode beneath. This MET platform has been used to demonstrate several BFC anodes based on enzyme cascades designed to fully oxidize complex fuels such as ethanol. A cascade of enzymes designed to mimic the citric acid (Krebs) cycle, for example, accomplished ethanol oxidation, with the dehydrogenase enzymes incorporated in a deacidified Nafion film. NADH mediated the dehydrogenases and was regenerated by a poly(methylene green) film, which had been electrodeposited directly over the current collector [105–108]. This versatile method was also used in 2010 to revisit the development of a methanol–air BFC [109,110]. Further, a Krebs cycle metabolon cross-linked *in situ* in mitochondria and then isolated has been incorporated into a BFC in this manner [111]. A metabolon is a structural complex of sequential enzymes in a metabolic pathway. By cross-linking *in situ*, the proximity and orientation of the metabolon enzymes with respect to each other is maintained close to that found in nature. This leads to maximal flux of substrate from enzyme to enzyme within the cascade, and thus higher catalytic current when connecting this Natural catalyst to an electrode via mediation.

The examples of MET herein are intended to give an overview of the history, development, and breadth of mediated enzyme electrodes. The examples given here are representative but not exhaustive, and the reader is directed to a selection of review papers, which taken together will provide a more complete view [12,94,112,113].

9.4   ASPECTS OF MEDIATOR DESIGN I: MEDIATOR OVERPOTENTIALS

As the driving force for electron exchange reactions is a potential difference, the redox potentials of mediators is an important consideration. In this section, we present an example to demonstrate that even though an electron transfer step causes an overpotential, which is a potential penalty, introducing additional electron transfer steps can result in a higher overall cell power.

9.4.1   Considering Species Potentials in a Methanol–Oxygen BFC

Consider the methanol–oxygen fuel cell, in which methanol is oxidized to carbon dioxide and oxygen is reduced to water, for the overall reaction (Equation 9.14):

$$CH_3OH + \tfrac{3}{2}O_2 \rightarrow 2H_2O + CO_2 \tag{9.14}$$

The theoretical maximum potential of this electrochemical cell is $\Delta E_{eq} = E_r$ oxygen $- E_o$ methanol $= 1.19\,V$ (see Figure 9.6). As a power-producing device, a BFC must generate high current at high cell potential in order to maximize power. Device power P in watts is given by Equation 9.15:

$$P = \Delta E_{cell} I \tag{9.15}$$

where ΔE_{cell} and I are cell potential and current, respectively [114]. Cell potential ΔE_{cell} is determined by the difference between the cathode potential $E_{cathode}$ and anode potential E_{anode} (Equation 9.16).

$$\Delta E_{cell} = E_{cathode} - E_{anode} \tag{9.16}$$

These potentials are those at the electrode current collector surfaces. Considering this, a disadvantage to MET in comparison to DET becomes apparent. In DET, the overpotential at the anode would be relative to the redox potential of the enzyme active site. In MET, the overpotential is relative to the mediator redox potential. For an anode, we may write an equation similar to Equation 9.6 (Equation 9.17):

$$E_{anode} = E_{m,a}^0 - \eta_{anode} \tag{9.17}$$

where $E_{m,a}^0$ is the mediator redox potential.

9.4.2   The Earliest Methanol-Oxidizing BFC Anodes

Methanol-oxidizing BFC anode designs first appeared in the early 1980s, in which enzymes were used instead of noble metal catalysts. In one design, ADH was used to oxidize methanol to formic acid [115]. This enzyme was linked to the cofactor PQQ, which was mediated to the electrode by tetramethylphenylenediamine (TMPD or Wurster's blue). The redox potential of TMPD is 0.12 V versus the saturated calomel electrode (SCE). If the oxygen cathode has a potential of 0.55 V versus SCE at the relevant pH, this results in a maximum cell potential of 0.48 V. This maximum cell potential is illustrated in Figure 9.6.

A second methanol anode was demonstrated with methanol dehydrogenase, which was mediated by one of two species: phenazine methosulfate (PMS) or phenazine ethosulfate (PES) [116]. The redox potentials of these mediators are 0.08 and 0.05 V versus SCE (also shown in Figure 9.6). It can be seen that all three mediators have a large potential difference from the methanol oxidation potential. Eliminating this potential loss was an engineering problem waiting to be solved.

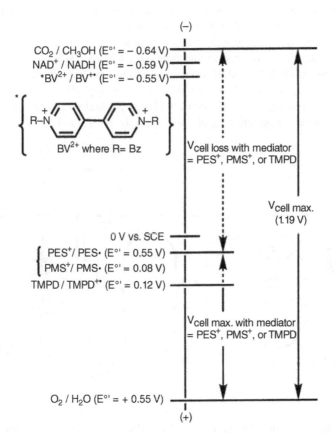

FIGURE 9.6 Redox potentials in various methanol–oxygen biological fuel cells. Potentials are at pH 7.5 and given versus SCE. (Reproduced with permission from Ref. [29]. Copyright 1998, Elsevier.)

9.4.3 A Four-Enzyme Methanol-Oxidizing Anode

In 1998, Palmore et al. reported a methanol–oxygen fuel cell that carried out the full six-electron oxidation of methanol to carbon dioxide in its anode compartment (Equation 9.18) [29]:

$$CH_3OH + H_2O \rightarrow CO_2 + 6H^+ + 6e^- \tag{9.18}$$

A multienzyme pathway accomplished the multistep oxidation of methanol as follows:

1. Methanol to formaldehyde by ADH.
2. Formaldehyde to formate by ALDH.
3. Formate to carbon dioxide by formate dehydrogenase.

The cathode current collector was platinum gauze, which directly catalyzed the reduction of oxygen. The reaction scheme of this BFC is shown in Figure 9.7. All of these dehydrogenase enzymes rely on the natural cofactor NAD^+, which binds at the enzyme active site and is reduced to NADH during substrate oxidation. For example, the mechanism of ALDH involves binding formaldehyde and NAD^+. NADH and formic acid are then produced in that order (see Figure 9.3) [28].

In this case, the NAD^+/NADH couple is operating as a natural mediator, with a redox potential of $E^0_{NAD} = -0.59$ V versus SCE, which is approximately $E^0_{NAD} = -0.32$ V versus SHE. However, the oxidation of NADH at an electrode surface is a kinetically hindered reaction, requiring a η_a of approximately 1 V [117–121]. Thus, relying on

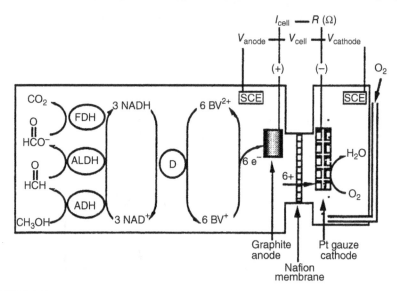

FIGURE 9.7 A methanol–oxygen biological fuel cell with mediators and enzymes freely diffusing in solution. This cell used an enzyme cascade of ADH $\rightarrow$ ALDH $\rightarrow$ FDH to completely oxidize methanol to carbon dioxide. (Reproduced with permission from Ref. [29]. Copyright 1998, Elsevier.)

TABLE 9.2   Electrolyte Components of a Methanol–Oxygen BFC (Figure 9.7)

Anode Compartment	Cathode Compartment
1 M LiCl	1 M LiCl
0.1 M Tris buffer (pH 7.5)	0.1 M Tris buffer (pH 7.5)
0.1 M CH_3OH	
5 U formate dehydrogenase	
40 U ALDH	
11 U ADH	
1 mM NAD^+	
1.7 U diaphorase	
50 mM benzyl viologen (BV^{2+})	

electrode oxidation of NADH would cause the cell potential ΔE_{cell} to fall to almost zero, rendering the cell useless as a power-producing device. Palmore et al. solved this problem by introducing a second redox couple, benzyl viologen ($BV^{2+}/BV^{•+}$). The redox potential of this mediator couple is $E_{BV}^0 = -0.55$ V versus SCE, which is very close to the potential of $NAD^+/NADH$. The overpotential between these two couples is only 40 mV and proceeds at high rates in the presence of the enzyme diaphorase (D). Unlike NADH, the reduced benzyl viologen form $BV^{•+}$ requires only a minimal anode overpotential η_a to be reoxidized to BV^{2+}. Figure 9.6 illustrates these potentials and demonstrates the benefit of adding the diaphorase/benzyl viologen electron transfer step. The anode and cathode components of this fuel cell are listed in Table 9.2 [29].

All components of the anode were diffusing freely in the electrolyte, and the anode compartment size was 75 ml with a 2 cm^2 graphite current collector. The cathode compartment was 50 ml, and the total fuel cell volume was 250 cm^3. The cell open-circuit potential E_{OCP} was 0.8 V, and produced maximum power of 670 μW cm^{-2} (normalized to anode area) at 0.49 V. This is higher than the previous BFC presented in Section 9.4.2, which produced 28 nW cm^{-2} and 19 μW cm^{-2}, respectively.

From the point of view of electron flow in the fuel cell, electrons originate with methanol, which is reduced to carbon dioxide by the three dehydrogenase enzymes in concert. All six electrons per methanol molecule are carried from these enzymes by NADH. NADH is oxidized by BV at diaphorase enzymes, and $BV^{•+}$ carries the electrons to the graphite current collector, where it is reoxidized. Although the addition of each electron transfer step introduces an additional overpotential, this design results in a higher ΔE_{cell} than designs with fewer steps.

9.5   ASPECTS OF MEDIATOR DESIGN II: SATURATED MEDIATOR KINETICS

The complexity of a mediated enzyme electrode means that careful consideration of the system parameters is required when designing and optimizing for a particular task. One might need to take different approaches depending on the purpose of the electrode—for example, electrodes in amperometric biosensors and electrochemical power sources might require quite different engineering. Also, the parameters amenable to design depend on the configuration of the enzyme electrode. In solution, changing mediator concentration may be quite easy. In a cross-linked mediator film, such a change could be difficult or impossible, depending on how the film is produced.

9.5.1   An Immobilized Laccase Cathode

What follows is a practical example demonstrating an engineering effort to improve a redox hydrogel-mediated laccase cathode meant to reduce oxygen to water in a BFC [65]. The laccase cathode is produced by combining an osmium redox polymer, laccase from *Trametes versicolor*, and a diepoxide cross-linker in the proportions 61, 32, and 7% by mass, respectively. After curing in a dry environment, a catalytic film is produced, which will reduce oxygen to water on an electrode surface [73,79].

We begin by making a few assumptions. The first is that *the anode of the BFC will not limit the device current* [64,122,123]. For this reason, we wish to raise the current output of the cathode, as it is by default the limiting factor. The second assumption is that *the laccase cathode will be in contact with an electrolyte containing dissolved oxygen* (which has a typical saturation of 1 mM). The third is that *there will be significant convection in the electrolyte*. This is so the laccase cathode will not be limited by oxygen mass transport outside the film. The fourth is that *Michaelis–Menten kinetics apply for both mediator and oxygen* reactions with laccase, meaning that Equations 9.10–9.12 should hold true. Fifth, we assume that *diffusion and partition of counter ions such as Cl^- and H^+ do not limit the current*. Sixth, we assume that *ionic strength in the laccase cathode is high enough that electric migration may be neglected during steady-state operation*.

The laccase cathode in Figure 9.2 illustrates that transport of two species to laccase enzymes in the film will determine the film reaction rate: reduced mediator species (Os^{II}) and molecular oxygen (O_2). Concentration of these species is given by C_m and C_s in Equation 9.12. Following the path of electron flow, oxidized mediator complexes are reduced at the cathode current collector surface, introducing electrons into the film. The reduced mediator is transported through the film by electron hopping, and reduces a laccase T1 copper. This occurs four times until all copper centers in the laccase are reduced [3,4]. Oxygen partitions into the film from the electrolyte and is transported into the film by diffusion. This oxygen and four protons react with a fully reduced laccase enzyme, producing two water molecules. Overall, electrons from the cathode current collector reduce oxygen to water. This reaction scheme has been given previously in Equations 9.2a–9.2d, which may be added to give the overall oxygen reduction reaction (ORR) (Equation 9.19):

$$O_2 + 4H^+ + 4e^- \rightarrow 2H_2O \qquad (9.19)$$

9.5.2   Potential of the Osmium Redox Polymer

Although the mediator has already been specified as a cross-linked osmium redox polymer, there are many molecular structures that fulfill this requirement. The first engineering issue is to select the best available choice for the application.

The driving force $\Delta E_{et,c}$ between the mediator and enzyme active site determines the rate of electron exchange between them. As described by Marcus in Equation 9.9, rates of electron transfer between species in solution are exponentially dependent on the free energy (potential) difference between the two species, which may be thought of as an overpotential [7,9,124]. If we neglect the quadratic term and assume d and β will be identical for similar redox polymers, we may collect constants and write a Tafel-like form of this equation (Equation 9.20):

$$\frac{k_{cat}}{K_m} = \left(\frac{k_{cat}}{K_m}\right)^0 \exp\left[\frac{\alpha F}{RT}\ \Delta E_{et}\right] \tag{9.20}$$

where $(k_{cat}/K_m)^0$ is a standard bimolecular rate constant and α is the transfer coefficient. If $\Delta E_{et,c}$, the redox potential difference between the species undergoing electron exchange at the cathode, is small, then the rate of exchange will be low (see Figure 9.2). Thus, a design trade-off can be seen: increasing $\Delta E_{et,c}$ increases rate, but this increase will require the potential of the cathode current collector $E_{cathode}$ to be lower, which will lower the overall device cell potential ΔE_{cell}. Changing mediator redox potential will affect $\Delta E_{et,c}$. Note that the corresponding overpotential for the oxygen substrate, which we will refer to as cathode activation overpotential $\eta_{act,c}$, will be affected by pH, as the ORR is proton dependent [124]. This effect, in conjunction with hydroxide inhibition at high pH values, explains the variation in pH activities of different laccases, which have a range of potentials [125,126]. The required pH of device operation affects the choice of laccase species that will be appropriate, as high pH requires a lower potential enzyme.

If the mediator is a transition metal complex, the redox potential of the mediator can be tuned to a desired value by changing the electron-withdrawing characteristics of the ligands bound to the central atom [127]. Making use of this effect, the influence of $\Delta E_{et,c}$ on electron transfer kinetics has been quantified for several systems freely diffusing in solution. In 1992, Zakeeruddin et al. synthesized a series of osmium- and ruthenium-based complexes covering a range of redox potentials and found that [Os(diamino-bpy)$_2$(dimethyl-bpy)]Cl$_2$ was the optimum mediator for GOx [23]. Nine years later, Nakabayashi et al. conducted a similar study for osmium mediators with different ligand constructions, covering a mediator potential range of 200–850 mV versus SHE [128].

For an osmium redox polymer, we select a form based on a polyvinylimidazole backbone (such as in panel C of Table 9.1), in which imidazole rings serve as ligands for attached osmium complexes. The nature of the other ligands can be varied in such a way as to produce a series of mediators with redox potentials spanning a wide range [127]. Such a series is shown in Figure 9.8, in which the highest potential mediator has essentially the same redox potential as the laccase enzyme.

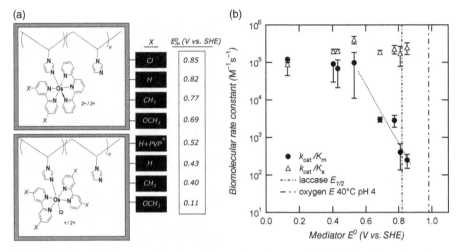

FIGURE 9.8 (a) A series of redox polymers based on polyvinylimidazole and osmium bipyridine and terpyridine. Variation of the groups marked as X allows variation of the redox potential. (b) Kinetic parameters measured for redox hydrogel-mediated laccase cathodes. The cathodes were in a planar film configuration (cf. Figure 9.5a) and reduced dissolved oxygen to water at pH 4. By assuming k_{cat} is a fixed value of $300\,s^{-1}$, mediator and oxygen Michaelis constants K_m and K_s can be calculated (see Equation 9.12). (Panel (b) reproduced with permission from Ref. [73]. Copyright 2008, American Chemical Society.)

Experimentally determined electron exchange kinetics between these mediators and laccase are shown in Figure 9.8 [73]. When $\Delta E_{et,c}$ is small, the bimolecular rate constant is on the order of $100–200\,M^{-1}\,s^{-1}$. As $\Delta E_{et,c}$ increases to near 0.4 V, the rate constant increases to $100{,}000\,M^{-1}\,s^{-1}$ and plateaus.

Fitting Equation 9.20 results in a $(k_{cat}/K_m)^0$ of $490\,M^{-1}\,s^{-1}$ and a transfer coefficient of $\alpha = 0.48$. This suggests reversible electron exchange and is comparable to mediator–enzyme systems in solution [24,63,129–131]. It should be noted that the observed maximum rate constant of $100{,}000\,M^{-1}\,s^{-1}$ is at least an order of magnitude lower than that for cases reported in solution. Thus, enzyme mobility may be low and active sites somewhat hindered in a cross-linked film [23,24,125,132,133].

Recall from Equations 9.15 and 9.16 that power is a function of cell potential ΔE_{cell}, which is the difference between the anode and cathode potentials. Good design of a power-producing laccase cathode involves increasing $E_{cathode}$ as high as possible to maximize ΔE_{cell}. However, the potential of laccase from *T. versicolor* is fixed at $\sim 0.83\,V$ (T1 copper potential; see Figure 9.2), and $E_{cathode}$ cannot increase beyond that value. Indeed, $E_{cathode}$ must be lower than the laccase potential to account for two potential drops. The first is the mediator–enzyme electron transfer driving force $\Delta E_{et,c}$. The second is the cathode polarization or overpotential $\eta_{cathode}$, which is a deviation from the osmium mediator redox potential, causing the overall reaction at the current collector to be reduction of osmium (described by Equation 9.6). We may assume that similar mediators will require similar values of $\eta_{cathode}$ to be sufficiently reduced at the cathode current collector.

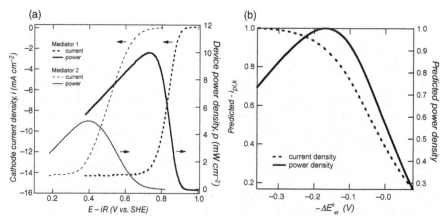

FIGURE 9.9 (a) Polarization curves showing the E–I relationship for redox hydrogel-mediated laccase cathodes on high surface area carbon paper. Data were collected on a rotating disk electrode at a rotation rate of 900 rpm at pH 4 and saturated oxygen concentration. The data are ohmic corrected. Power curves calculated from Equation 9.15 assuming an anode at 0 V. (b) Power curve maximum value as a function of $\Delta E_{et,c}$ calculated from the laccase cathode model in Equations 9.21 and 9.22, using the kinetic data in Figure 9.8 and assuming typical values for all other parameters. (Panel (b) is reproduced with permission from Ref. [73]. Copyright 2008, American Chemical Society.)

Figure 9.9a shows *polarization curves* for laccase cathodes in oxygen-saturated electrolyte. The S-shaped curves begin at zero current, and as $E_{cathode}$ is swept to lower values, negative current is produced as the electrodes reduce oxygen through an electron transport path such as that shown in Figure 9.2. At some value of $E_{cathode}$, the current reaches a maximum magnitude and is no longer a function of potential. This plateau current is determined by some limiting process or collection of processes in the cathode, as in the concept shown in Figure 9.1. The two laccase cathodes shown have similar polarizations but are shifted in potential because of different values of $E^0_{m,c}$. If power curves are calculated using Equation 9.15, and assuming a hydrogen anode at 0 V, the cathode produced from the mediator with higher $E^0_{m,c}$ has a higher power maximum.

We can construct a one-dimensional model of a redox hydrogel-mediated enzyme film by performing a mass balance around the reduced mediator species, making use of the kinetic expression in Equation 9.12, resulting in Equation 9.21.

$$D_{app} \frac{\partial^2 C_m}{\partial x^2} = \frac{k_{cat} K_m^{-1} C_e C_m C_s}{C_m K_m^{-1}(C_s + K_s) + C_s} \tag{9.21}$$

Appropriate boundary conditions for the mediator at the electrode interface ($x = 0$) and electrolyte interface ($x = \phi$) (Equation 9.22) are

$$x = 0, \quad i = -FD_{app} \frac{dc_m}{dx}; \quad x = \phi, \quad \frac{dc_m}{dx} = 0 \tag{9.22}$$

This is related to electrode potential through Equation 9.7. It is a reasonable assumption that oxygen concentration C_s and enzyme concentration C_e are constants throughout the film.

Turning our attention to the effect of $\Delta E_{et,c}$ on plateau current, the information in Figure 9.8 allows us to predict the effect that changing the mediator potential will have on reaction rate (current). As $\Delta E_{et,c}$ decreases to zero, plateau current of the laccase cathode will be limited by the electron exchange reaction from mediator to enzyme, which will become small. Using the laccase cathode model to predict plateau current using the kinetic rates in Figure 9.8 and assuming all other parameters to be equal (e.g., mediator concentration, electron diffusion coefficient, film thickness), we can determine a value of $\Delta E_{et,c}$ that will produce the maximum power. Figure 9.9b shows this maximum power plotted as a function of $\Delta E_{et,c}$, and is calculated by solving Equation 9.21 numerically. The optimum E_m^0 is predicted to be ~0.66 V versus SHE.

9.5.3 Concentration of Redox Sites in the Mediator Film

Based on the analysis of mediator potential, we select an osmium redox polymer mediator with a structure of poly(vinylimidazole$_n$[Os(terpyridine)(dimethyl-bipyridine)]$^{2+/3+}$). In Figure 9.8, this structure is presented as having a redox potential of 0.77 V versus SHE. The corresponding value of $\Delta E_{et,c}$ falls within 90% of the maximum power in Figure 9.9, indicating it as a good mediator for a laccase cathode. A laccase cathode produced using an *early* version of the redox polymer mediator was produced, which yielded 7 mA cm^{-2} on high surface area carbon paper [79]. However, it was desirable to engineer a *modified* version of this redox polymer to produce higher BFC power. One aspect of this mediator that can be engineered is the value of n, the number of imidazole groups on the polymer backbone present per attached osmium complex. An early version of this mediator was presented in the literature with a value of $n = 23$ [79]. The laccase cathode produced using this mediator had a film thickness of 540 nm and an osmium complex concentration C_m^0 of 240 mM due to the compactness of the film. The mediator–enzyme electron exchange rate analysis for this polymer shown in Figure 9.8 shows that if the enzyme has a turnover rate of 300 s^{-1}, the Michaelis constant K_m for the mediator is approximately 110 mM. Considering kinetics, we note that the mediator concentration is larger than the Michaelis constant, but only by a factor of 2.2. The mediator–enzyme reaction will approach its maximum rate V_{max} only as C_m^0 exceeds K_m by a factor of 5–10. Therefore, higher current could be realized by increasing C_m^0.

Compare this with the mediator structure having a redox potential of 0.43 V, which has a K_m of 3 mM. For a similar concentration, this mediator would be 80 times the Michaelis constant, and the mediator–enzyme reaction would be at its maximum rate. It is clear that performance of the chosen mediator would be increased by a higher value of C_m^0. This increase can be accomplished by altering the synthesis conditions of the redox polymer. Table 9.3 shows conditions of syntheses for the early and modified versions of the chosen mediator. When solvent

TABLE 9.3 Redox Polymer Synthesis Conditions and Resulting Measured Properties

Redox Polymer Synthesis Conditions				Measured Properties		
	ml	VI (mM)	Os (mM)	n	ρ (g cm^{-3})	C_m^0 (mM)	D_m (cm^2 s^{-1})
Early	46	16	3.4	23	1.1	240	1.2×10^{-9}
Modified	26	93	20	20	1.5	340	1.7×10^{-9}

volume for the refluxing reaction was decreased, and concentrations of the vinyl-imidazole function group and osmium precursor complex were increased (at a constant ratio of 4.7:1), osmium loading on the polymer backbone was enhanced from a value of $n = 23$ to $n = 20$.

This seemingly small change in the value of n in the mediator structure had a profound effect on the performance of the resulting laccase cathode (Figure 9.10). The plateau current was increased by a factor of 1.9, nearly doubling the effectiveness of the cathode. This occurs because mediation kinetics respond positively to more mediators at this value of K_m, and because apparent electron transport is more rapid. More osmium on the mediator backbone made the film denser and more osmium sites were available to perform electron transfer, both to each other and to the enzyme. Compare this with the mediator structure having a redox potential of 0.43 V, which would have benefited from greater apparent electron transport as in Equation 9.13, but already displayed fully saturated mediation kinetics.

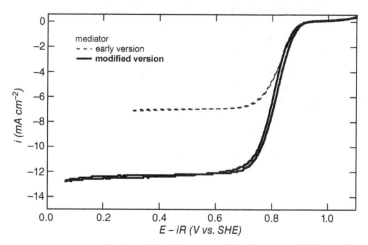

FIGURE 9.10 Polarization curves comparing performance of early and modified laccase cathodes. Engineering the redox polymer to have a higher osmium loading (Table 9.3) increased current nearly two times. Cathodes produced on high surface area carbon paper. Data were collected on a rotating disk electrode at a rotation rate of 900 rpm at pH 4 and saturated oxygen concentration. The data are ohmic corrected. (Adapted with permission from Ref. [79]. Copyright 2009, Elsevier.)

9.6 OUTLOOK

Mediated enzyme electrodes are currently found in commercially successful consumer products, and continue to be the subject of intense research. In the coming years, as humankind searches for solutions to important problems such as the efficient, inexpensive, and distributed production of energy, the high catalytic rates made possible by enzyme mediation hold much promise. The combination of enzyme cascades with mediation could make complex, targeted electrocatalysis economically useful.

LIST OF ABBREVIATIONS

ABTS	2,2-azinobis(3-ethylbenzothiazoline-6-sulfonate)
ADH	alcohol dehydrogenase
ALDH	aldehyde dehydrogenase
BFC	biological fuel cell
BV	benzyl viologen
D	diaphorase
DET	direct electron transfer
FAD	flavin adenine dinucleotide, oxidized form
$FADH_2$	flavin adenine dinucleotide, reduced form
GOx	glucose oxidase
MET	mediated electron transfer
NAD	nicotinamide adenine dinucleotide, oxidized form
NADH	nicotinamide adenine dinucleotide, reduced form
ORR	oxygen reduction reaction
PES	phenazine ethosulfate
PMS	phenazine methosulfate
PQQ	pyrroloquinoline quinone
PVP	polyvinylpyridine
SCE	saturated calomel electrode
SHE	standard hydrogen electrode
TMPD	tetramethylphenylenediamine

REFERENCES

1. Chen T, Calabrese Barton S, Binyamin G, Gao ZQ, Zhang YC, Kim HH, Heller A. A miniature biofuel cell. *J Am Chem Soc* 2001;123:8630–8631.
2. Kim HH, Mano N, Zhang XC, Heller A. A miniature membrane-less biofuel cell operating under physiological conditions at 0.5 V. *J Electrochem Soc* 2003;150: A209–A213.
3. Palmer AE, Kyu Lee S, Solomon EI. Decay of the peroxide intermediate in laccase: reductive cleavage of the O–O bond. *J Am Chem Soc* 2001;123(27):6591–6599.

4. Solomon EI, Sundaram UM, Machonkin TE. Multicopper oxidases and oxygenases. *Chem Rev* 1996;96:2563–2605.

5. Smith I, Carson B, Ferguson T. Osmium: an appraisal of environmental exposure. *Environ Health Perspect* 1974;8:201–213.

6. Hanker JS, Kasler F, Bloom MG, Copeland JS, Seligman AM. Coordination polymers of osmium—nature of osmium black. *Science* 1967;156:1737–1738.

7. Marcus RA, Sutin N. Electron transfers in chemistry and biology. *Biochim Biophys Acta* 1985;811:265–322.

8. Cowan JA, Gray HB. Long-range electron transfer in metal-substituted myoglobins. *Chem Scr* 1988;28A:21–26.

9. Calvo EJ. Interfacial kinetics and mass transport. In: Bard AJ, Stratmann M (eds), *Encyclopedia of Electrochemistry*, Vol. 2.Wiley-VCH Verlag GmbH, Weinheim, 2003, pp. 3–30.

10. Lueck JD, Fromm HJ. Analysis of exchange rates for ping-pong bi-bi mechanism and concept of substrate synergism. *FEBS Lett* 1973;32:184–186.

11. Bartlett PN, Pratt KFE. Theoretical treatment of diffusion and kinetics in amperometric immobilized enzyme electrodes. 1. Redox mediator entrapped within the film. *J Electroanal Chem* 1995;397:61–78.

12. Kano K, Ikeda T. Fundamentals and practices of mediated bioelectrocatalysis. *Anal Sci* 2000;16:1013–1021.

13. Andrieux CP, Dumasbouchiat JM, Saveant JM. Catalysis of electrochemical reactions at redox polymer electrodes—kinetic model for stationary voltammetric techniques. *J Electroanal Chem* 1982;131:1–35.

14. Limoges B, Marchal D, Mavre F, Saveant JM. Electrochemistry of immobilized redox enzymes: kinetic characteristics of NADH oxidation catalysis at diaphorase monolayers affinity immobilized on electrodes. *J Am Chem Soc* 2006;128:2084–2092.

15. Limoges B, Moiroux J, Saveant JM. Kinetic control by the substrate and the cosubstrate in electrochemically monitored redox enzymatic immobilized systems. Catalytic responses in cyclic voltammetry and steady state techniques. *J Electroanal Chem* 2002;521:8–15.

16. Theorell H. Nonproteolytic enzymes. *Annu Rev Biochem* 1940;9:663–690.

17. Fee JA, Malmström BG. Redox potential of fungal laccase. *Biochim Biophys Acta* 1968;153:299–302.

18. Reinhamm BR. Oxidation–reduction potentials of electron acceptors in laccases and stellacyanin. *Biochim Biophys Acta* 1972;275:245–259.

19. McArdle JV, Coyle CL, Gray HB, Yoneda GS, Holwerda RA. Kinetics studies of oxidation of blue copper proteins by tris(1,10-phenanthroline)cobalt(III) ions. *J Am Chem Soc* 1977;99:2483–2489.

20. McArdle JV, Gray HB, Creutz C, Sutin N. Kinetic studies of oxidation of ferrocytochrome *c* from horse heart and *Candida krusei* by tris(1,10-phenanthroline)cobalt(III). *J Am Chem Soc* 1974;96:5737–5741.

21. Yahiro AT, Lee SM, Kimble DO. Bioelectrochemistry. I. Enzyme utilizing bio-fuel cell studies. *Biochim Biophys Acta* 1964;88:375–383.

22. Palmore GTR, Kim HH. Electro-enzymatic reduction of dioxygen to water in the cathode compartment of a biofuel cell. *J Electroanal Chem* 1999;464:110–117.

23. Zakeeruddin SM, Fraser DM, Nazeeruddin MK, Gratzel M. Towards mediator design—characterization of tris-(4,4'-substituted-2,2'-bipyridine) complexes of iron(II), ruthenium(II) and osmium(II) as mediators for glucose oxidase of *Aspergillus niger* and other redox proteins. *J Electroanal Chem* 1992;337:253–283.

24. Takagi K, Kano K, Ikeda T. Mediated bioelectrocatalysis based on NAD-related enzymes with reversible characteristics. *J Electroanal Chem* 1998;445:211–219.

25. Schläpfer P, Mindt W, Racine P. Electrochemical measurement of glucose using various electron acceptors. *Clin Chim Acta* 1974;57:283–289.

26. Cass AEG, Davis G, Francis GD, Hill HAO, Aston WJ, Higgins IJ, Plotkin EV, Scott LDL, Turner APF. Ferrocene-mediated enzyme electrode for amperometric determination of glucose. *Anal Chem* 1984;56:667–671.

27. Bourdillon C, Laval JM, Thomas D. Enzymatic electrocatalysis—controlled potential electrolysis and cosubstrate regeneration with immobilized enzyme modified electrode. *J Electrochem Soc* 1986;133:706–711.

28. Liu ZJ, Sun YJ, Rose J, Chung YJ, Hsiao CD, Chang WR, Kuo I, Perozich J, Lindahl R, Hempel J, Wang BC. The first structure of an aldehyde dehydrogenase reveals novel interactions between NAD and the Rossmann fold. *Nat Struct Biol* 1997;4:317–326.

29. Palmore GTR, Bertschy H, Bergens SH, Whitesides GM. A methanol/dioxygen biofuel cell that uses NAD^+-dependent dehydrogenases as catalysts: application of an electroenzymatic method to regenerate nicotinamide adenine dinucleotide at low overpotentials. *J Electroanal Chem* 1998;443:155–161.

30. Farneth WE, Diner BA, Gierke TD, D'Amore MB. Current densities from electrocatalytic oxygen reduction in laccase/ABTS solutions. *J Electroanal Chem* 2005;581:190–196.

31. Tsujimura S, Tatsumi B, Ogawa J, Shimizu S, Kano K, Ikeda T. Bioelectrocatalytic reduction of dioxygen to water at neutral pH using bilirubin oxidase as an enzyme and 2,2'-azinobis(3-ethylbenzothiazolin-6-sulfonate) as an electron transfer mediator. *J Electroanal Chem* 2001;496:69–75.

32. Heller A. Electrical wiring of redox enzymes. *Acc Chem Res* 1990;23:128–134.

33. Degani Y, Heller A. Direct electrical communication between chemically modified enzymes and metal electrodes. 1. Electron transfer from glucose oxidase to metal electrodes via electron relays, bound covalently to the enzyme. *J Phys Chem* 1987;91:1285–1289.

34. Degani Y, Heller A. Direct electrical communication between chemically modified enzymes and metal electrodes. 2. Methods for bonding electron-transfer relays to glucose oxidase and D-amino-acid oxidase. *J Am Chem Soc* 1988;110:2615–2620.

35. Cowan JA, Upmacis RK, Beratan DN, Onuchic JN, Gray HB. Long-range electron transfer in myoglobin. *Ann NY Acad Sci* 1988;550:68–84.

36. Bartlett PN, Whitaker RG. Electrochemical immobilization of enzymes. 1. Theory. *J Electroanal Chem* 1987;224:27–35.

37. Bartlett PN, Whitaker RG. Electrochemical immobilization of enzymes. 2. Glucose oxidase immobilized in poly-*N*-methylpyrrole. *J Electroanal Chem* 1987;224:37–48.

38. Bartlett PN, Whitaker RG, Green MJ, Frew J. Covalent binding of electron relays to glucose oxidase. *J Chem Soc, Chem Commun* 1987; 1603–1604.

39. Degani Y, Heller A. Electrical communication between redox centers of glucose oxidase and electrodes via electrostatically and covalently bound redox polymers. *J Am Chem Soc* 1989;111:2357–2358.

40. Oyama N, Anson FC. Facile attachment of transition metal complexes to graphite electrodes coated with polymeric ligands—observation and control of metal–ligand coordination among reactants confined to electrode surfaces. *J Am Chem Soc* 1979; 101:739–741.

41. Oyama N, Anson FC. Polymeric ligands as anchoring groups for the attachment of metal complexes to graphite electrode surfaces. *J Am Chem Soc* 1979;101:3450–3456.

42. Pishko MV, Katakis I, Lindquist SE, Ye L, Gregg BA, Heller A. Direct electrical communication between graphite electrodes and surface adsorbed glucose oxidase redox polymer complexes. *Angew Chem Int Ed* 1990;29:82–89.

43. Heller A. Electrical connection of enzyme redox centers to electrodes. *J Phys Chem* 1992;96:3579–3587.

44. Gregg BA, Heller A. Cross-linked redox gels containing glucose oxidase for amperometric biosensor applications. *Anal Chem* 1990;62:258–263.

45. Mano N, Mao F, Heller A. On the parameters affecting the characteristics of the "wired" glucose oxidase anode. *J Electroanal Chem* 2005;574:347–357.

46. Cassidy HG. Electron exchange polymers. 1. *J Am Chem Soc* 1949;71:402–406.

47. Updegraff IH, Cassidy HG. Electron exchange polymers. 2. Vinylhydroquinone monomer and polymer. *J Am Chem Soc* 1949;71:407–410.

48. Cassidy HG, Ezrin M, Updegraff IH. Electron exchange polymers. 4. Countercurrent applications. *J Am Chem Soc* 1953;75:1615–1617.

49. Ezrin M, Updegraff IH, Cassidy HG. Electron exchange polymers. 3. Polymers and copolymers of vinylhydroquinone. *J Am Chem Soc* 1953;75:1610–1614.

50. Facci JS, Schmehl RH, Murray RW. Effect of redox site concentration on the rate of electron transport in a redox co-polymer film. *J Am Chem Soc* 1982;104:4959–4960.

51. Gao ZQ, Binyamin G, Kim HH, Calabrese Barton S, Zhang YC, Heller A. Electrodeposition of redox polymers and co-electrodeposition of enzymes by coordinative crosslinking. *Angew Chem Int Ed* 2002;41:810–813.

52. Forster RJ, Vos JG. Synthesis, characterization, and properties of a series of osmium-containing and ruthenium-containing metallopolymers. *Macromolecules* 1990;23:4372–4377.

53. Wilbert G, Wiesemann A, Zentel R. New ferrocene-containing copolyesters. *Macromol Chem Phys* 1995;196:3771–3788.

54. Gregg BA, Heller A. Redox polymer films containing enzymes. 1. A redox conducting epoxy cement—synthesis, characterization, and electrocatalytic oxidation of hydroquinone. *J Phys Chem* 1991;95:5970–5975.

55. Gregg BA, Heller A. Redox polymer films containing enzymes. 2. Glucose oxidase containing enzyme electrodes. *J Phys Chem* 1991;95:5976–5980.

56. Calvo EJ, Danilowicz C, Etchenique R. Measurement of viscoelastic changes at electrodes modified with redox hydrogels with a quartz crystal device. *J Chem Soc, Faraday Trans* 1995;91:4083–4091.

57. Sato F, Togo M, Islam MK, Matsue T, Kosuge J, Fukasaku N, Kurosawa S, Nishizawa M. Enzyme-based glucose fuel cell using vitamin K3-immobilized polymer as an electron mediator. *Electrochem Commun* 2005;7:643–647.

58. Meredith MT, Kao DY, Hickey D, Schmidtke DW, Glatzhofer DT. High current density ferrocene-modified linear poly(ethylenimine) bioanodes and their use in biofuel cells. *J Electrochem Soc* 2011;158:B166–B174.

59. Mao F, Mano N, Heller A. Long tethers binding redox centers to polymer backbones enhance electron transport in enzyme "wiring" hydrogels. *J Am Chem Soc* 2003;125: 4951–4957.

60. Liu JC, Mu SL. The electrochemical polymerization of methylene blue and properties of polymethylene blue. *Synth Met* 1999;107:159–165.

61. Foulds NC, Lowe CR. Enzyme entrapment in electrically conducting polymers— immobilization of glucose oxidase in polypyrrole and its application in amperometric glucose sensors. *J Chem Soc, Faraday Trans 1* 1986;82:1259–1264.

62. Ezrin M, Cassidy HG. Electron exchange polymers. 6. Preparation of water-soluble and water-swellable polymers. *J Am Chem Soc* 1956;78:2525–2526.

63. Barriere F, Ferry Y, Rochefort D, Leech D. Targeting redox polymers as mediators for laccase oxygen reduction in a membrane-less biofuel cell. *Electrochem Commun* 2004;6:237–241.

64. Barriere F, Kavanagh P, Leech D. A laccase–glucose oxidase biofuel cell prototype operating in a physiological buffer. *Electrochim Acta* 2006;51:5187–5192.

65. Calabrese Barton S, Kim HH, Binyamin G, Zhang YC, Heller A. Electroreduction of O_2 to water on the "wired" laccase cathode. *J Phys Chem B* 2001;105:11917–11921.

66. Calabrese Barton S, Pickard M, Vazquez-Duhalt R, Heller A. Electroreduction of O_2 to water at 0.6 V (SHE) at pH 7 on the 'wired' *Pleurotus ostreatus* laccase cathode *Biosens Bioelectron* 2002;17:1071–1074.

67. Surridge NA, Sosnoff CS, Schmehl R, Facci JS, Murray RW. Electron and counterion diffusion constants in mixed-valent polymeric osmium bipyridine films. *J Phys Chem* 1994;98:917–923.

68. Majda M. Dynamics of electron transport in polymeric assemblies of redox centers. In: Murray RW (ed.), *Molecular Design of Electrode Surfaces*. John Wiley & Sons, Inc., New York, 1992, pp. 159–206.

69. White HS, Leddy J, Bard AJ. Polymer-films on electrodes. 8. Investigation of charge-transport mechanisms in Nafion polymer modified electrodes. *J Am Chem Soc* 1982;104:4811–4817.

70. Forster RJ, Vos JG. Redox site loading, electrolyte concentration and temperature effects on charge transport and electrode kinetics of electrodes modified with osmium containing poly(4-vinylpyridine) films in sulfuric acid. *J Electroanal Chem* 1991;314: 135–152.

71. Sosnoff CS, Sullivan M, Murray RW. Electron self-exchange rates in a site-dilutable osmium bipyridine redox polymer. *J Phys Chem* 1994;98:13643–13650.

72. Blauch DN, Saveant JM. Dynamics of electron hopping in assemblies of redox centers— percolation and diffusion. *J Am Chem Soc* 1992;114:3323–3332.

73. Gallaway JW, Calabrese Barton S. Kinetics of redox polymer-mediated enzyme electrodes. *J Am Chem Soc* 2008;130:8527–8536.

74. Mano N, Soukharev V, Heller A. A laccase-wiring redox hydrogel for efficient catalysis of O_2 electroreduction. *J Phys Chem B* 2006;110:11180–11187.

75. Bonnecaze RT, Mano N, Nam B, Heller A. On the behavior of the porous rotating disk electrode. *J Electrochem Soc* 2007;154:F44–F47.

76. Calabrese Barton S. Oxygen transport in composite mediated biocathodes. *Electrochim Acta* 2005;50:2145–2153.

77. Calabrese Barton S, Sun YH, Chandra B, White S, Hone J. Mediated enzyme electrodes with combined micro- and nanoscale supports. *Electrochem Solid State Lett* 2007;10: B96–B100.

78. Zhou HJ, Zhang ZP, Yu P, Su L, Ohsaka T, Mao LQ. Noncovalent attachment of NAD$^+$ cofactor onto carbon nanotubes for preparation of integrated dehydrogenase-based electrochemical biosensors. *Langmuir* 2010;26:6028–6032.

79. Gallaway JW, Calabrese Barton SA. Effect of redox polymer synthesis on the performance of a mediated laccase oxygen cathode. *J Electroanal Chem* 2009;626:149–155.

80. Little SJ, Ralph SF, Mano N, Chen J, Wallace GG. A novel enzymatic bioelectrode system combining a redox hydrogel with a carbon NanoWeb. *Chem Commun* 2011;47:8886–8888.

81. Fei JF, Song HK, Palmore GTR. A biopolymer composite that catalyzes the reduction of oxygen to water. *Chem Mater* 2007;19:1565–1570.

82. Calvo EJ, Danilowicz CB, Wolosiuk A. Supramolecular multilayer structures of wired redox enzyme electrodes. *Phys Chem Chem Phys* 2005;7:1800–1806.

83. Etchenique R, Calvo EJ. Electrochemical quartz crystal microbalance gravimetric and viscoelastic studies of nickel hydroxide battery electrodes. *J Electrochem Soc* 2001;148: A361–A367.

84. Flexer V, Forzani ES, Calvo EJ. Structure and thickness dependence of "molecular wiring" in nanostructured enzyme multilayers. *Anal Chem* 2006;78:399–407.

85. Flexer V, Ielmini MV, Calvo EJ, Bartlett PN. Extracting kinetic parameters for homogeneous [Os(bpy)$_2$ClPyCOOH]$^+$ mediated enzyme reactions from cyclic voltammetry and simulations. *Bioelectrochemistry* 2008;74:201–209.

86. Calvo EJ, Etchenique R, Pietrasanta L, Wolosiuk A, Danilowicz C. Layer-by-layer self-assembly of glucose oxidase and Os(Bpy)$_2$ClPyCH$_2$NH-poly(allylamine) bioelectrode. *Anal Chem* 2001;73:1161–1168.

87. Willner I, Heleg-Shabtai V, Blonder R, Katz E, Tao G, Buckmann AF, Heller A. Electrical wiring of glucose oxidase by reconstitution of FAD-modified monolayers assembled onto Au-electrodes. *J Am Chem Soc* 1996;118:10321–10322.

88. Blonder R, Willner I, Buckmann AF. Reconstitution of apo-glucose oxidase on nitrospiropyran and FAD mixed monolayers on gold electrodes: photostimulation of bioelectrocatalytic features of the biocatalyst. *J Am Chem Soc* 1998;120:9335–9341.

89. Bardea A, Katz E, Buckmann AF, Willner I. NAD$^+$-dependent enzyme electrodes: electrical contact of cofactor-dependent enzymes and electrodes. *J Am Chem Soc* 1997;119:9114–9119.

90. Pardo-Yissar V, Katz E, Willner I, Kotlyar AB, Sanders C, Lill H. Biomaterial engineered electrodes for bioelectronics. *Faraday Discuss* 2000;116:119–134.

91. Katz E. Biofuel cells with switchable power output. *Electroanalysis* 2010;22:744–756.

92. Katz E, Willner I. A biofuel cell with electrochemically switchable and tunable power output. *J Am Chem Soc* 2003;125:6803–6813.

93. Tam TK, Strack G, Pita M, Katz E. Biofuel cell logically controlled by antigen–antibody recognition: towards immune-regulated bioelectronic devices. *J Am Chem Soc* 2009;131:11670–11671.

94. Willner I, Yan YM, Willner B, Tel-Vered R. Integrated enzyme-based biofuel cells—a review. *Fuel Cells* 2009;9:7–24.

95. Lioubashevski O, Katz E, Willner I. Magnetic field effects on electrochemical processes: a theoretical hydrodynamic model. *J Phys Chem B* 2004;108:5778–5784.

96. Katz E, Lioubashevski O, Willner I. Magnetic field effects on bioelectrocatalytic reactions of surface-confined enzyme systems: enhanced performance of biofuel cells. *J Am Chem Soc* 2005;127:3979–3988.

97. Akers NL, Moore CM, Minteer SD. Development of alcohol/O_2 biofuel cells using salt-extracted tetrabutylammonium bromide/Nafion® membranes to immobilize dehydrogenase enzymes. *Electrochim Acta* 2005;50:2521–2525.

98. Topcagic S, Minteer SD. Development of a membraneless ethanol/oxygen biofuel cell. *Electrochim Acta* 2006;51:2168–2172.

99. Thomas TJ, Ponnusamy KE, Chang NM, Galmore K, Minteer SD. Effects of annealing on mixture-cast membranes of Nafion® and quaternary ammonium bromide salts. *J Membr Sci* 2003;213:55–66.

100. Zhou DM, Fang HQ, Chen HY, Ju HX, Wang Y. The electrochemical polymerization of methylene green and its electrocatalysis for the oxidation of NADH. *Anal Chim Acta* 1996;329:41–48.

101. Arechederra MN, Addo PK, Minteer SD. Poly(neutral red) as a NAD^+ reduction catalyst and a NADH oxidation catalyst: towards the development of a rechargeable biobattery. *Electrochim Acta* 2011;56:1585–1590.

102. Rincón RA, Artyushkova K, Mojica M, Germain MN, Minteer SD, Atanassov P. Structure and electrochemical properties of electrocatalysts for NADH oxidation. *Electroanalysis* 2010;22:799–806.

103. Rincón RA, Artyushkova K, Atanassov P, Germain MN, Minteer SD, Lau C, Cooney MJ. Integrating poly-azine catalysts for NADH oxidation in biofuel cell anodes. *Abstr Pap Am Chem Soc* 2009; 238.

104. Moore CM, Akers NL, Hill AD, Johnson ZC, Minteer SD. Improving the environment for immobilized dehydrogenase enzymes by modifying Nafion with tetraalkylammonium bromides. *Biomacromolecules* 2004;5:1241–1247.

105. Sokic-Lazic D, Arechederra RL, Treu BL, Minteer SD. Oxidation of biofuels: fuel diversity and effectiveness of fuel oxidation through multiple enzyme cascades. *Electroanalysis* 2010;22:757–764.

106. Sokic-Lazic D, de Andrade AR, Minteer SD. Utilization of enzyme cascades for complete oxidation of lactate in an enzymatic biofuel cell. *Electrochim Acta* 2011;56:10772–10775.

107. Sokic-Lazic D, Minteer SD. Citric acid cycle biomimic on a carbon electrode. *Biosens Bioelectron* 2008;24:939–944.

108. Sokic-Lazic D, Minteer SD. Pyruvate/air enzymatic biofuel cell capable of complete oxidation. *Electrochem Solid State Lett* 2009;12:F26–F28.

109. Addo PK, Arechederra RL, Minteer SD. Evaluating enzyme cascades for methanol/air biofuel cells based on NAD^+-dependent enzymes. *Electroanalysis* 2010;22:807–812.

110. Addo PK, Arechederra RL, Minteer SD. Towards a rechargeable alcohol biobattery. *J Power Sources* 2011;196:3448–3451.

111. Moehlenbrock MJ, Toby TK, Waheed A, Minteer SD. Metabolon catalyzed pyruvate/air biofuel cell. *J Am Chem Soc* 2010;132:6288–6289.

112. Calabrese Barton S, Gallaway J, Atanassov P. Enzymatic biofuel cells for implantable and microscale devices. *Chem Rev* 2004;104:4867–4886.

113. Meredith MT, Minteer SD. Biofuel cells: enhanced enzymatic bioelectrocatalysis. *Annu Rev Anal Chem* 2012;5:157–179.

114. Bullen RA, Arnot TC, Lakeman JB, Walsh FC. Biofuel cells and their development. *Biosens Bioelectron* 2006;21:2015–2045.

115. Davis G, Hill HAO, Aston WJ, Higgins IJ, Turner APF. Bioelectrochemical fuel cell and sensor based on a quinoprotein, alcohol dehydrogenase. *Enzyme Microb Technol* 1983;5:383–388.

116. Plotkin EV, Higgins IJ, Hill HAO. Methanol dehydrogenase bioelectrochemical cell and alcohol detector. *Biotechnol Lett* 1981;3:187–192.

117. Rodkey FL. Oxidation–reduction potentials of the diphosphopyridine nucleotide system. *J Biol Chem* 1955;213:777–786.

118. Moiroux J, Elving PJ. Effects of adsorption, electrode material, and operational variables on oxidation of dihydronicotinamide adenine dinucleotide at carbon electrodes. *Anal Chem* 1978;50:1056–1062.

119. Moiroux J, Elving PJ. Adsorption phenomena in the NAD^+–NADH system at glassy carbon electrodes. *J Electroanal Chem* 1979;102:93–108.

120. Moiroux J, Elving PJ. Optimization of the analytical oxidation of dihydronicotinamide adenine dinucleotide at carbon and platinum electrodes. *Anal Chem* 1979;51:346–350.

121. Moiroux J, Elving PJ. Mechanistic aspects of the electrochemical oxidation of dihydronicotinamide adenine dinucleotide (NADH). *J Am Chem Soc* 1980;102:6533–6538.

122. Kamitaka Y, Tsujimura S, Setoyama N, Kajino T, Kano K. Fructose/dioxygen biofuel cell. *Phys Chem Chem Phys* 2007;9:1793–1801.

123. Mano N, Mao F, Heller A. A miniature membrane-less biofuel cell operating at +0.60 V under physiological conditions. *ChemBioChem* 2004;5:1703–1705.

124. Bard AJ, Faulkner LR. *Electrochemical Methods*, 2nd edition. John Wiley & Sons, Inc., New York, 2001.

125. Xu F. Oxidation of phenols, anilines, and benzenethiols by fungal laccases: correlation between activity and redox potentials as well as halide inhibition. *Biochemistry* 1996;35:7608–7614.

126. Xu F. Effects of redox potential and hydroxide inhibition on the pH activity profile of fungal laccases. *J Biol Chem* 1997;272:924–928.

127. Lever ABP. Electrochemical parametrization of metal complex redox potentials, using the ruthenium(III)/ruthenium(II) couple to generate a ligand electrochemical series. *Inorg Chem* 1990;29:1271–1285.

128. Nakabayashi Y, Omayu A, Yagi S, Nakamura K. Evaluation of osmium(II) complexes as electron transfer mediators accessible for amperometric glucose sensors. *Anal Sci* 2001;17:945–950.

129. Kulys JJ, Cenas NK. Reagent redox potential and pH effects on the enzymatic electron transfer. *J Mol Catal* 1988;47:335–341.

130. Xu F, Shin WS, Brown SH, Wahleithner JA, Sundaram UM, Solomon EI. A study of a series of recombinant fungal laccases and bilirubin oxidase that exhibit significant

differences in redox potential, substrate specificity, and stability. *Biochim Biophys Acta* 1997;1341:99.

131. Coury LA, Murray RW, Johnson JL, Rajagopalan KV. Electrochemical study of kinetics of electron transfer between synthetic electron acceptors and reduced molybdoheme protein sulfite oxidase. *J Phys Chem* 1991;95:6034–6040.

132. Brandi P, D'Annibale A, Galli C, Gentili P, Pontes ASN. In search for practical advantages from the immobilisation of an enzyme: the case of laccase. *J Mol Catal B: Enzym* 2006;41:61–69.

133. Xu F. Dioxygen reactivity of laccase—dependence of laccase source, pH, and anion inhibition. *Appl Biochem Biotechnol* 2001;95:125–133.

10

HIERARCHICAL MATERIALS ARCHITECTURES FOR ENZYMATIC FUEL CELLS

GUINEVERE STRACK

Oak Ridge Institute of Science and Engineering, Oak Ridge, TN, USA; Airbase Sciences Branch, Air Force Research Laboratory, Tyndall Air Force Base, FL, USA

GLENN R. JOHNSON

Airbase Sciences Branch, Air Force Research Laboratory, Tyndall Air Force Base, FL, USA

10.1 INTRODUCTION

Many systems found in Nature or assembled in a laboratory exhibit hierarchical structure or organization. In the assembly, all of the individual elements are connected in some manner. In a family, that connection is shared genetics; in a complex structure, physical links hold all of the elements together. In another sense, one may assume that all subelements serve the apex or purpose of the hierarchy. For example, individual cells differentiate to yield complex organisms and all the pieces of a corporation coordinate to generate profit.

Extending that metaphor of structure and function at each hierarchical level, materials possessing such architectures are oftentimes presented as a model for bionanocomposite materials. The analogy is particularly fitting in instances in which bionanomaterials are coupled with bulk material to form a component such as an electrode or transducer. All the elements (catalysts, interfacial associations, nano-materials, bulk materials) must be properly integrated to move electrons from the outermost element throughout the hierarchy. In such a construct, the performance of

Enzymatic Fuel Cells: From Fundamentals to Applications, First Edition. Edited by Heather R. Luckarift, Plamen Atanassov, and Glenn R. Johnson.
© 2014 John Wiley & Sons, Inc. Published 2014 by John Wiley & Sons, Inc.

the electrochemical system depends on the flow of electrons from the donor to the conductive interface and through materials structured to maximize the performance of the electrochemical system.

This chapter provides examples of various biocatalysts, nanomaterials, and fabrication processes that yield functional bioelectrodes for anodic or cathodic processes. Most of the descriptions of electrode materials in this chapter focus on the fabrication of electrode architectures suitable for direct electron transfer (DET) processes, with an emphasis on enzyme-based electrodes; however, examples of materials that are also suitable for microbial anodes are also included because of the parallels in development of such conductive architectures (see Sections 10.4.2 and 10.5.2).

Advances in materials science, the ability to control assembly of materials and biological catalysts, and a movement toward standard evaluation of performance are key developments in biological fuel cell (BFC) technology [1,2]. Projected uses of the concept for BFCs range from miniature implantable devices [2–8] to environmental devices that may be deployed in marine settings [9–12] or sewage treatment facilities [13–15]. Although considerable increases in performance and operational lifetime have been achieved in these areas, there are still many design and scaling challenges for BFCs to become practical technology. On all scales, complex problems must be addressed to optimize mass transport and charge transfer to sustain bioelectrocatalysis. For example, enzymatic fuel cells (EFCs) take advantage of aspects of nanomaterials for improved performance [2,4,16–19]. On the other hand, the dimensions and architecture of microbial fuel cells (MFCs) must be able to accommodate the microbial population [20–22].

The conceptual theme of hierarchically structured materials that are specifically designed to maximize anode or cathode performance can be separately considered in the following components: (1) the bulk electrode materials, (2) pores in the bulk material that accommodate fluid flow, (3) mesoporous or structural material to transition from the bulk material to the nanometric dimensions, (4) the nanostructured transducer, (5) the bio–nano interface, and (6) the biocatalyst (Figure 10.1) [23]. All of the elements must be coupled to achieve electron flow from the electron donor through the architecture to the final electron acceptor. In the anodic half-cell, fuel is oxidized and the liberated electrons are transferred to the electrode and through an external load. The cathode provides the BFC with electromotive force by catalyzing the reduction of oxygen at relatively positive potentials.

The complexity of the bio–electronic interface necessitates that material selection and electrode fabrication be understood to obtain an optimal result. In the design of an anode or a cathode, the materials should enhance stability of the immobilized biocatalyst and foster electron transfer (ET) between the biocatalyst and the conductive material. As a result, various materials and methodologies have been described in the literature to coordinate redox biocatalysts and electrodes [24,25]. In addition, several factors should be considered when selecting materials, including the electrochemically accessible surface area (EASA), mechanical stability, and conductivity. The electrode architecture can be designed to enhance mass transport of fuel to the biocatalyst, and, in some instances, to include mediators to shuttle electrons between

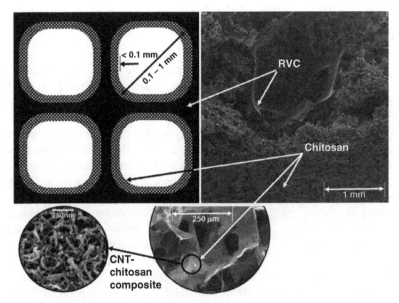

FIGURE 10.1 Schematic of hierarchically structured carbon architecture with macroscale porosity, mesoscale component responsible for interconnectivity, and nanomaterials to provide a large surface area in contact with the biocatalyst. (Reproduced with permission from Ref. [23]. Copyright 2012, Elsevier.)

catalyst and electrode [23]. In other words, porous architectures incorporated with nanometric structures will accommodate fluid flow and provide increased surface area and an interface suitable for direct bioelectrocatalysis.

The performance of BFCs hinges on several factors. One in particular—operational charge transfer—is tied to the organization and coupling of components within the bioelectrode architecture. Charge transfer between redox enzymes and electrode interfaces occurs via one of two mechanisms: DET, in which electrons tunnel or hop between a biocatalyst and a conductive surface, or mediated electron transfer (MET), in which a redox-active molecule acts as an electron shuttle between a catalyst and an electrode interface. The mediator may diffuse through the electrolyte or be immobilized within the electrode architecture. Although MET offers some advantages over DET, such as high current density, DET eliminates drawbacks associated with mediators, such as limited life span and lower thermodynamic potentials [26]. Furthermore, the removal of dissolved chemical mediators simplifies system design, which broadens the range of applications. Therefore, DET-centered bioelectrocatalysis is of great interest when designing BFC electrodes. Although the examples and discussion here are focused primarily on DET processes, the architectures and principles are also applicable to MET, in particular when mediators are immobilized within the electrode structure. In other words, characteristics such as EASA, material porosity, conductivity, and biocompatibility will contribute to the effectiveness of electron transport, regardless of the mode.

10.2   CARBON NANOMATERIALS AND THE CONSTRUCTION OF THE BIO–NANO INTERFACE

Carbon nanostructures often serve as components in hierarchically structured electrode materials. Conductive carbon, for example, is densely packed, sp^2-bonded, and classified based on its dimensions. A major asset of nanostructured carbon is its high relative surface area, which facilitates an increase of both MET and DET.

When considering enzymes as electrocatalysts, it is critical to recognize that the redox process is typically centered at cofactors within the folded protein structure. Accordingly, interfacial ET between the catalytic center of the enzyme and the electrode must overcome the electron transfer distance and the "insulating" nature of the protein. Nanostructured electrode architectures may overcome the challenges by penetrating the protein structure and decreasing the distance between the redox center and the conductive support. According to the 1985 theory of Marcus and Sutin, a major hurdle of electron tunneling is the distance between the redox center of the protein and the conductive surface; therefore, nanostructured architectures with appropriate dimensions could enhance the ET step [27]. Yet even if the distance between the redox centers and electrode is minimized, other aspects of the system influence performance, such as the reorganizational energy of the protein and the potential difference between the redox centers and electrode.

Several types of conductive nanomaterials have been explored in the construction of BFCs, and a large emphasis has been placed on carbon materials with nanometric dimensions. Specifically, carbon black nanomaterials (CBNs), graphene, and carbon nanotubes (CNTs) have been used to construct bio–nano interfaces with the intent to optimize ET processes. In this section, these three nanostructured carbon materials will be briefly discussed, followed by two archetypal carbon assemblies that approach hierarchical structures: CNT-decorated porous carbon architectures and buckypaper (BP).

10.2.1   Carbon Black Nanomaterials

CBNs are spherical, carbon-based nanomaterials that possess high surface area, conductivity, and porosity that allows for the physisorption and immobilization of biomolecules. Examples of biocatalysts immobilized on CBN are hemoglobin [28] and cuprous oxidase (CuO) [29]. In this context, the redox center of the physisorbed hemoglobin underwent reversible redox processes by contacting the CBN, but direct bioelectrocatalysis was not demonstrated. In the case of physisorbed CuO, DET between the CBN and the enzyme was revealed by the production of cathodic current when oxygen was available as an electron acceptor. The biocatalytic reduction of oxygen in solution is limited by the low concentration of dissolved oxygen and a modest diffusion coefficient; however, proper application of CBN can mitigate the effects of low oxygen concentration.

CBNs are also suitable for construction of gas diffusion electrodes (GDEs). The modification of CBN with emulsions of Teflon or other highly hydrophobic polymers, such as polyvinylidene difluoride, results in a composite material with a "triphase" interface [29]. Enhanced hydrophobicity of the composite material changes the surface energy to increase the interaction between the carbon composite and the gaseous bulk phase. Properly constructed GDEs exhibit higher current density when exposed to the gaseous phase compared with the aqueous phase. To design a biocatalytic GDE, the hydrophobic–hydrophilic properties of the electrode must accommodate the biocatalyst, which, although immobilized on the carbon surface, needs the aqueous phase to maintain an active conformation. Therefore, a careful balance between electrolyte, carbon, and hydrophobic surface must be achieved to construct a triphasic interface [29–31].

10.2.2 Carbon Nanotubes

Because of electrode surface area enhancement and their high aspect ratio, CNTs support the production of high current density when used to construct enzyme cathodes and anodes. CNTs are constructed as multiwalled (MWCNT) or single-walled (SWCNT) configurations with various chiral vectors. Engineering the bio–nano interface involves careful selection and application of CNTs with specific properties that enhance DET between the conductive carbon architecture and the biocatalyst. Electrodes composed of (or decorated with) CNTs have a larger surface area for enzyme immobilization than conventional carbon surfaces: the enlarged electrode surface area increases the current density.

Current density is typically expressed as electrical current per unit area. The unit area used to calculate the electron flux, however, depends on the conductor and can be a cross-sectional area or simple geometric dimensions. Current density for biocatalytic electrodes is typically expressed as amperes per square meter; however, the surface area of the electrode is based on a geometric area and does not account for the roughness of the electrode. For example, some seemingly "flat" surfaces, such as glassy carbon, have a degree of roughness. Electrode roughness can be determined using imaging techniques (such as atomic force microscopy; see Chapter 14). The total surface area, however, may not be representative of the EASA, which is defined as the conductive surface area accessible to the surrounding fluid and dissolved ions involved in electrical double-layer formation. Variation of EASA is highly dependent on surface wetting and is a source of error when comparing the performance of biocatalytic electrodes, especially electrodes with high porosity and surface area. Therefore, determining the EASA for each electrode will reduce the error. Reported current densities normalized to geometric area do not represent EASA as well as factors such as mass, volume, and material composition. As a result, comparisons of electrode performance based solely on current density defined by geometric area may lead to the erroneous conclusion that the performance of one electrode is superior to another.

In addition to increased surface area, an important attribute of CNTs is that the high aspect ratio potentially allows direct access to the redox center of a biocatalyst [24,32]. Depending on the type of CNT, the carbon architecture can be tailored to specific attributes of the redox enzyme. For example, MWCNTs typically have larger dimensions than SWCNTs (Figure 10.2). Given that enzymes are generally 5–10 nm in diameter, a large CNT (>10 nm; typically MWCNT) is effectively

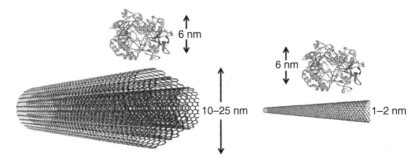

FIGURE 10.2 Relative dimensions of GOx molecule (6 nm) compared with typical MWCNT and SWCNT. The MWCNT has a diameter of 10–25 nm, whereas that of the SWCNT is 1–2 nm.

"flat" from the perspective of the protein. On the other hand, SWCNTs can be fabricated with diameters as small as 1–2 nm; it is plausible that the small dimension will enable superior ET rates due to penetration of the SWCNT close to the buried redox center. The interactions between SWCNTs and two multicopper oxidases (MCOs), laccase from *Trametes versicolor* and small laccase from *Streptomyces coelicolor* (SLAC), were modeled in 2012 by Trohalaki et al. using molecular dynamics (MD) simulations. These results revealed that CNTs with optimal diameters minimize the distance between the type 1 (T1) center and the CNT and increase the interaction energy between the protein and the CNT [32]. Additional simulations suggest that interaction energy between the CNT and the MCO also depends on the physical properties of the protein itself, for example, the hydrophobicity of the cleft near the T1 site [32,33]. Calculated interaction energies between a SWCNT surface and SLAC are greater than energies between a SWCNT and laccase because of SLAC's higher hydrophobicity.

Another factor influencing electrode performance is the process of enzyme attachment, that is, how the enzyme is stabilized and linked with the CNT surface. The tethering process is dictated (to an extent) by the properties of the CNT, such as the hydrophobicity or hydrophilicity of the CNT surface. Aspect ratio is also a factor—for example, short CNTs have a large fraction of open ends amenable to modification, and provide a relatively high percentage of oxidized surface area [34]. Proteins may physically adsorb onto the CNT surface and remain relatively stable under certain conditions [35]; however, covalent attachment can result in increased current density and enhance electrode functionality [36]. A widely used tethering method is carbodiimide chemistry. First, the CNTs are oxidized to generate surface carboxyl groups, which are available to form covalent attachments with primary amines (typically lysine) [24]. The carbodiimide tethering technique is commonly used because lysine residues are available on the surface of virtually all soluble proteins. It should be noted, however, that the location of lysine residues does not necessarily facilitate orientation of the enzyme redox center (with respect to the CNT surface) to promote DET. In fact, MD simulations predict that SWCNTs are predisposed to adsorb onto the hydrophobic cleft near the laccase T1 site (and two other available hydrophobic regions) [32]. A drawback of the carbodiimide method is that the chemical oxidation process decreases material conductivity by creating surface defects and decreasing availability of π-electrons. As an alternative method, heterobifunctional molecules—1-pyrenebutyric acid N-hydroxysuccinimidyl ester (PBSE), for example—will interact strongly with CNTs and react with the protein catalyst. Such linkers can be used with unmodified CNTs and preserve high conductivity of the material [36,37]. The pyrene moiety interacts with the CNTs via π–π stacking, whereas the succinimidyl ester reacts with lysine residues on the surface of the enzyme resulting in covalent attachment. The interaction of the enzyme with the CNT surface is primarily dictated by physical interactions; however, the location of surface lysine residues determines the number of covalent attachment points and their presence decreases the probability of desorption and rearrangements. Therefore, the PBSE tethering method can enhance biocatalytic activity by stabilizing the enzyme on the surface and decreasing the distance from the T1 site to the CNT surface, although this is dependent on the CNT's chiral indices as well as number and positions of covalent attachment points.

10.2.3 Graphene

Graphene is a single, flat (two-dimensional) sheet of graphite one atom thick. MWCNTs comprise multiple rolled layers of graphene sheets. The graphene sheets can be arranged in the concentric nested structure, in which a SWCNT is within another SWCNT, or in the parchment structure, in which a single graphene sheet is rolled around itself. The *IUPAC Compendium of Chemical Terminology* asserts that the word *graphene* should not be used to describe layers of graphite or carbon sheets: graphene is inherently a 2D structure, whereas multiple layers of graphene comprise a 3D structure [38]. Care should be taken when working with carbon sheets to ensure that a single sheet of graphite is indeed incorporated into the selected system.

Although graphene is being explored as an engineering material, its utility for bioelectrodes remains to be determined. Given that graphene possesses physical properties similar to CNTs, the coupling of biocatalysts to graphene may result in high catalytic current density [39–42]. For example, in 2010 Liu et al. incorporated graphene into glucose oxidase (GOx) sol–gel 3D matrices, which rendered the composite material conductive and supported a twofold increase in power output when compared with composite materials containing CNTs [39]. The enhanced conductivity of the composite material indicated that several junctions of electrical contact between graphene flakes were formed; therefore, even if a fraction of the conductive carbon retained its single-layer character, the interaction between several flakes would qualify the material as graphite, not graphene. In similar examples, DET was demonstrated with GOx–graphene–chitosan electrodes [42], and with graphene–cytochrome c composites [40]. In both cases, DET was evidenced by the production of anodic current in the presence of the appropriate substrate.

One approach used to disperse graphene is suspending graphene oxide flakes with an amphiphilic material. The graphene oxide flakes are then reduced, resulting in suspended graphene [43]. This can be accomplished by chemically reducing graphene oxide [41], or, interestingly, by exploiting bacterial respiration [43]. In 2011,

Wang et al. showed that certain *Shewanella* sp. including *Shewanella oneidensis* MR-1 can use graphene oxide as a terminal electron acceptor, ultimately yielding conductive graphene [43].

10.2.4   CNT-Decorated Porous Carbon Architectures

Porous carbon materials such as carbon paper and carbon fabric (CF) are highly conductive, 3D, interconnected, macroporous structures. Their high porosity accommodates fluid flow while providing surface for the deposition of CNTs. The associated process is often described as *decoration of the bulk material*. Vertically aligned CNTs grown on CF, for example, support construction of hierarchically structured electrodes with high specific capacitance [44]. This high capacitance derives from the EASA. Functionally, the decoration of porous carbon with CNTs increases the surface area available for biocatalyst immobilization while providing transport of dissolved electron donors and acceptors throughout the electrode body. The deposition of CNTs onto porous conductive carbon surfaces can be accomplished through drop casting [37] or chemical vapor deposition (CVD) [24,25,45]. Drop casting is a simple process in which prefabricated CNTs are suspended in a volatile solvent and applied to a conductive surface. The solvent evaporates leaving behind a mesh of CNTs held to the surface by physical adsorption. CVD directly deposits CNTs onto a surface using metal catalyst nanoparticles, high temperature, a processing gas, and a carbon precursor gas. This process results in a "forest" of CNTs covering the surface of the carbon paper (Figure 10.3). Direct bioelectrocatalysis was successfully demonstrated on carbon paper that had been decorated with CNT using both drop casting and CVD [24,25,37,45]. In these systems, the presence of CNTs "wired" to the carbon paper surface was apparent by increased capacitive current during cyclic voltammetry. Moreover, catalytic current (for both anodic and cathodic processes) was increased by the presence of the CNTs compared with the bulk electrodes.

10.2.5   Buckypaper

BP is a thin film of interlocking CNTs held together by van der Waals attraction and other non-covalent interactions. BP electrodes are porous and do not have fillers or binders that may influence conductivity and surface accessibility. Figure 10.4a, for example, depicts the mechanical integrity and flexibility of a BP electrode and Figure 10.4b and c display the architecture of pressed carbon mats composed of MWCNT bundles (with diameters ranging from 10 to 200 nm) that combine to form a 3D porous network. This matrix of nanometric material can provide an exceptional base for bioelectrode construction in that BP is relatively simple to handle and can be cut like conventional paper to desired dimensions. BP is highly conductive, which eliminates the need to incorporate current collectors such as metallic mesh into the material. Another advantage is that BP is relatively easy to fabricate, and, as such, allows different approaches to obtain a matrix with specific physical and electrochemical properties [35]. Accordingly, small-scale preparation of a specific BP can be

FIGURE 10.3 (a) Toray carbon paper with (b) CVD-deposited CNTs. (Reproduced with permission from Ref. [25]. Copyright 2006, Elsevier.)

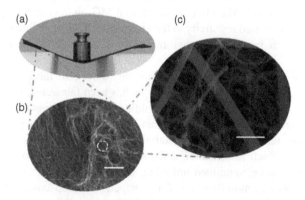

FIGURE 10.4 Imaging of buckypaper electrode material. (a) Photograph showing mechanical integrity and flexibility of buckypaper. SEM images of buckypaper: (b) scale bar 1 μm; (c) scale bar 0.1 μm. (SEM image in (c) is a contribution from the group of Professor Kenneth Sandhage at the Georgia Institute of Technology.)

made for in-house experimental work. One approach begins by suspending MWCNTs with nonionic surfactants. The suspension process can be enhanced through mechanical stirring and ultrasonication. After the CNT agglomerates are removed by centrifugation, the suspended MWCNTs are deposited onto a nylon membrane and vacuum filtered. The surfactant is removed with the solvent and then the BP is dried in a vacuum oven.

To demonstrate the previously described BP as a good candidate for direct bioelectrocatalysis, in 2011 Hussein et al. immobilized MCOs on a BP electrode via physisorption [35]. MCOs are often applied as oxygen reduction catalysts in BFCs [5,40,46–52]. By using BP as an electrode material, DET from the conductive surface to the T1 redox site is achieved and eliminates the need for mediators, thus simplifying design [53,54]. Compared with MCO-coated CNT aggregates, BP cathodes fabricated in this manner exhibited superior performance when normalized to the average mass of the CNT and BP electrodes, respectively [55].

For anodic electrocatalyst development, BP was applied as a high surface area conductive scaffold for mediator immobilization. Poly(methylene green) (poly(MG))-coated BP was fabricated to catalyze the oxidation of NADH (reduced form of nicotinamide adenine dinucleotide) at relatively low potentials [56]. The biocatalytic oxidation of fuel results in the transfer of electrons to nicotinamide adenine dinucleotide (NAD^+) [57]. The methylene green (MG) monomer was incorporated into the BP during the synthesis process. In this case, the SWCNTs were suspended in isopropyl alcohol (IPA) to produce a dispersion of bundles and then filtered through polytetrafluoroethylene (PTFE) filter paper and vacuum dried at 80 °C. The MG was added to the SWCNT/IPA solution to incorporate the MG monomer into the BP. Following fabrication of the MG-impregnated BP, additional MG was electrochemically polymerized onto the BP, which resulted in the formation of poly(MG) particles. Performance of the poly(MG)–BP electrode fabricated by first incorporating the MG into the paper was better than that of BP with MG polymerized after fabrication. Hence, NADH oxidation is enhanced by the additional MG incorporated into the BP. The effect of additional MG on the catalytic performance compensated for increased resistivity of the electrode. This simple fabrication process can be modified to incorporate other materials into the CNT mesh for specific applications.

BP is amenable to surface modification not only during the fabrication process, as demonstrated with MG, but also after the paper has been fabricated. Oxidation of the BP can destroy the physical interactions holding the CNTs together; therefore, the application of heterobifunctional linkers can be advantageous. As previously mentioned, the PBSE tether enhances biocatalytic activity by stabilizing the enzyme on the CNT surface. Such stabilization can prevent the enzyme from desorbing and possibly mitigate surface-confined unfolding and aggregation. This has been demonstrated on CNTs immobilized on Toray paper, via drop casting [37], and on prefabricated BP [12,36]. The presence of PBSE on the BP surface also enhances surface wetting. The wetting suggests that in addition to anchoring the protein to the CNT, the amphiphilic PBSE tether allows diffusion of electrolyte into the porous body. Long-term stability of the laccase-modified PBSE–BP electrode was tested in a flow-through cell configuration [36]. The bioelectrode was subjected to a series of N_2- and O_2-saturated buffer inputs at a flow rate of $5\,ml\,min^{-1}$ in a stacked-cell configuration while a potential of 0.4 V versus Ag/AgCl was applied. The amperometric response to the fluctuating O_2 input was recorded for at least 1000 cycles during a trial: response fidelity lasted for a period of 20 days. Given the rapid transition between the current values produced by the fluctuating input feed, the input sequence of the buffer was programmed to produce a binary code. The resultant amperometric response was directly readable by a conventional barcode scanner and further treatment of the data was not required. It should be noted that although several unconventional barcodes have been developed [58–62], the robust bioelectronic interface BP allows for continuous flow of chemical information. This result demonstrates that PBSE-modified BP electrodes are good materials to incorporate into products requiring extended and dynamic operating conditions, including environmental settings such as seawater [12].

One inherent advantage of BP is that it can be specifically tailored for the desired application. For example, the type of CNT can be selected (e.g., MWCNT, SWCNT, carboxylated CNT) and the fabrication process optimized to produce the desired physical properties. For instance, surface wettability, thickness, and porosity can be adjusted to obtain maximum enzyme coverage and/or facilitate diffusion of substrate and/or mediator to the surface. Conductivity can also be enhanced by fabricating CNTs to be individually suspended or in bundles. The shape and size of BPs can also be readily controlled for facile engineering of electrodes and their integration with fuel cells.

10.3 BIOTEMPLATING: THE ASSEMBLY OF NANOSTRUCTURED BIOLOGICAL–INORGANIC MATERIALS

Several approaches can be taken to fabricate ordered structures with controlled geometries on the nanoscale. Two general classifications are given to nano-fabrication approaches: top-down and bottom-up (as typically defined for development of patterned surfaces in semiconductor technology). A commonly used top-down approach involves the transfer of computer-generated patterns onto a surface. The pattern is then etched into the substrate using various methods. Another top-down approach is scanning probe lithography, which deposits patterns on a surface using a nanometric probe [63]. In contrast, bottom-up approaches generally start with a templating substrate that initiates nucleation and/or guides the self-assembly of colloidal structures [64].

An interesting frontier of bottom-up synthesis is "biotemplating" [65]. In general, biotemplating employs a biological structure to guide the assembly of a material that replicates the morphological characteristics and/or functionality of a biological structure. A diverse range of nanostructured materials can be fabricated from complex biological templates. For instance, various biological structures (diatoms, bacteria, viruses, butterfly wings) have been used as templates to fabricate a positive or negative copy. For example, in 2006, Huang et al. used butterfly wings as a template for the deposition of Al_2O_3 and produced nanostructured architectures that retained the optical properties of the original biological material and were applied for the development of photonic structures [66].

Guiding the assembly of nanomaterials using biological templates is another process in which, natural biological systems are used to nucleate the formation of specific structures. The biomolecules promote assembly of materials through unique physical or chemical attributes that guide structure formation via specific morphological interactions (influenced by attributes such as size, shape, charge, and the location of functional groups). In essence, self-assembled architectures can be derived from a wide variety of biomolecules, such as lipids, proteins, and DNA.

10.3.1 Protein-Mediated 3D Biotemplating

Various aspects of protein chemistry and structure can be exploited to act as template molecules. For example, structural proteins can be used as scaffolds to form nanostructures including wires, particle arrays, and spirals [65]. In another approach, the self-assembly of protein monomers was controlled to produce tunable silver nanoparticles of a desired structure [67]. Structural proteins such as actin can also be used for biotemplating and, in fact, self-assembly of bacterial S-layer proteins can result in crystalline arrangements of great complexity and exactness. Biotemplating is applicable not only in 2D or surface-confined systems, but 3D matrixes and structures can also be formed using bio-based suspensions and scaffolds. This process can be used to fabricate porous hierarchical structures with nanometric dimensions while incorporating the desired material into the matrix. For example, nanoparticles and colloidal systems can be formed with the aid of proteins or peptides. In this case, the biomaterial can be selected or modified to obtain a surface (with the appropriate charge or functional group) suitable for the precipitation of inorganic materials. Given that proteins and peptides are amphiphilic, they can also serve as stabilizers in phase-separated systems. Moreover, additional nanometric materials and biomolecules can be incorporated or entrapped in such matrixes.

Protein- and peptide-mediated biomineralization processes are commonly used for organizing inorganic nanomaterials and the immobilization of enzymes [68]. The most common example is the *in vitro* process based on principles that marine diatoms use to promote the condensation of silica from silicic acid [69]. Cationic peptides catalyze the rapid condensation of silica nanospheres (~8 nm diameter), which subsequently agglomerate to form larger framboid spheres (~500 nm diameter) (Figure 10.5) [70]. The peptides also act as a scaffold to organize the material due to the peptide's ionic and amphiphilic properties. The rapid condensation reaction can allow additional nanometric materials and biomolecules to be fortuitously incorporated or entrapped in the matrix as it forms. An early example demonstrated entrapment of an active hydrolase (butyrylcholinesterase) [69]. The benign reaction conditions allowed a high proportion of enzyme activity to be retained compared with conventional silicification (sol–gel) processes. The encapsulated enzyme sustained activity compared with free enzyme and was more resistant to heat inactivation. The process has subsequently been applied to immobilize enzymes for various

FIGURE 10.5 SEM image of biosilica-immobilized butyrylcholinesterase. (Reproduced with permission from Ref. [69]. Copyright 2004, Nature Publishing Group.)

applications, including sensors, BFCs, packed reactor beds, and microfluidic systems for biocatalytic synthesis [68].

The peptide-driven silicification method was demonstrated for construction of bioelectrocatalytic hierarchical architectures [71]. The first example of the approach entrapped GOx on different carbon interfaces to develop anodic catalysts that may be used in EFCs [71]. The architecture was assembled by first adsorbing lysozyme to a screen-printed carbon electrode. Subsequently, CNTs and GOx were suspended in buffered silicification reagents and then applied to the lysozyme-modified electrode surface. Silica rapidly condensed and coated the carbon surface, trapping the GOx and CNT on the interface. The lysozyme enhanced surface wetting of the carbon electrode, acted as a binder for GOx–CNT, and provided the cationic scaffold/ template necessary for rapid silica formation. Electrodes prepared in the same manner without CNT exhibited a comparatively low capacitance current during cyclic voltammetry, demonstrating that the CNTs were in electrical contact with each other and the screen-printed carbon electrode [71]. In addition, electrodes prepared with CNT exhibited quasi-reversible peaks in potential ranges associated with flavin adenine dinucleotide (FAD), the catalytic center of GOx. In the presence of glucose, a cathodic wave associated with H_2O_2 reduction indicated that the immobilized GOx retained activity. Retention of GOx activity was also confirmed by assays that revealed a minimal decrease in activity of the electrodes fabricated with silica over 30 days, whereas the activity of electrodes fabricated without silica rapidly decayed with time. The decrease is likely due to a combination of protein denaturation and enzyme loss when not confined by the silica matrix.

In another example, long-term retention of biocatalytic activity was demonstrated via the physisorption of papain to MWCNTs—functionalized with carboxyl and amine groups—followed by biomimetic silicification [72]. Another benefit of the silicification process was the retention of activity over a broad pH range (3–12), whereas free papain and noncoated bioconjugates clearly exhibited activity only in relatively narrow pH ranges. This result is tentatively attributed to a "confinement effect," in which the silica matrix stabilizes the native conformation of the protein and prevents pH-induced unfolding and deactivation.

The approach of peptide-mediated biomineralization and entrapment addresses several priorities in the construction of bioelectronic interfaces, such as limiting the dissociation of enzymes from the conductive interface [73], improving the long-term stability of the redox enzyme activity [40,72], and providing a robust and versatile method for associating redox catalysts and conductive materials within the electronic system [74,75]. Although silica-based architectures dominate among the reported examples, other metal oxide sol–gels such as protein–titania composites can be used for fabrication and incorporation of biocatalyst and MWCNTs into porous electrodes [76,77].

10.4 FABRICATION OF HIERARCHICALLY ORDERED 3D MATERIALS FOR ENZYME AND MICROBIAL ELECTRODES

The previous sections focused on methods to establish connectivity between the redox center of the protein and the conductive interface. Further design aspects must also be considered to improve BFC output. For example, two critical aspects of BFC electrodes are the conductivity of the system and its 3D structure. With respect to the conductivity of the BFC, one can consider both the ionic conductivity between the electrodes and the electrical conductivity of the electrode architecture. The material discussed here will focus on the conductivity of the electrode itself, as opposed to the ionic conductivity between the electrodes.

The proper design and organization of hierarchically structured electrodes optimize the bioelectronic performance and enhance mass transport of fuel to the biocatalysts. By using nanostructured materials at the interface, the relative surface area for catalyst loading is increased compared with bulk materials and the distance between catalyst and the conductive surface may be decreased to enhance DET. Nanostructured modification or functionalization of 2D surfaces may enhance power output; however, the essentially planar surface can limit scalability.

One solution may be to stack several functionalized planar electrodes. The stacked electrode layers will retain high conductivity—with high enzyme loading and efficient DET—but the power output will be limited by mass transport of electron acceptors/donors to the bioelectronic surface. As an alternative, porous 3D electrode architectures can allow effective mass transport of electrochemical species while the topology of the electronic interface maximizes surface area, which promotes efficient catalyst loading and ET [78]. Porous electrode structures have been described as having three levels of scale: macro, meso, and nano (Figure 10.6) [23]. Macroscale pores allow for fluid flow, whereas mesoscale structures are responsible for connectivity and transition of material properties from bulk to nanoscale. The nanomaterials relevant for BFCs can be described as discrete, prefabricated, conductive structures

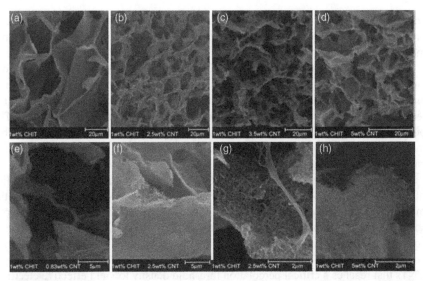

FIGURE 10.6   SEM images of chitosan–MWCNT scaffolds of different chitosan/CNT ratios: (a–d) macroporous structures at lower magnification; (e–g) high magnification of scaffold walls presenting CNT distribution. Panels (a) and (e) show chitosan only; panels (b) and (f), (c) and (g), and (d) and (h) depict chitosan structures with 2.5, 3.5, and 5.0 wt% CNT, respectively. (Reproduced with permission from Ref. [79]. Copyright 2008, American Chemical Society.)

with dimensions less than 100 nm that are applied to the surface of a macroscale architecture. The dimensions noted here are conceptual guidelines. The exact dimensions of the architecture can vary and other factors, such as the material, fabrication process, and operational parameters, may influence design. Ultimately the design should enhance catalytic performance, mass transport, and surface wetting, and all material properties—including pore size—can be considered in electrode construction.

The application of porous, conductive scaffolds as components of anode architectures can enhance power production by three processes: (1) transporting fuel from the bulk solution to the surface-confined biocatalysts, (2) increasing the amount of 3D surface area suitable for enzyme immobilization or biofilm formation, and (3) enhancing conductivity and, in turn, ET between the biocatalyst and the electrode. When considering electrode design, the support should be biocompatible and highly conductive at the same time. Additionally, the scaffold should have enough porosity to enhance mass transport of fuel.

10.4.1   Chitosan–CNT Conductive Porous Scaffolds

Chitosan is a polyionic biopolymer that has been used as a core scaffold for constructing porous electrode architectures. Its 3D structure is biocompatible and has reactive functional groups available for covalent attachment of materials or biomolecules. In addition, the scaffolds can be tailored to maximize enzyme loading, control porosity, and enhance surface wetting [79]. The material architecture and properties promote high enzyme loading, and chitosan can be incorporated into mediated and nonmediated systems and integrated with biosensors and BFCs.

One method used to modify the structure of chitosan to yield a porous scaffold architecture is thermally induced phase separation, which precipitates the polymer and ice crystals [79]. The ice crystals are subsequently removed under vacuum, which leaves a macroporous structure suitable as a scaffold for electrode construction. The porous structure is amenable to designs for MET and DET processes: soluble mediators can freely diffuse within the macroporous matrix and the open structure can accommodate additional conductive nanomaterials.

CNTs are effectively incorporated with chitosan by exploiting various properties of the biopolymer [79]. Chitosan, after protonation of amino groups to increase the material's amphiphilicity, enhances dispersion of unmodified CNTs in aqueous solvents. The physical properties of chitosan compensate CNT surface energy in aqueous solutions by suspending them in the biopolymer matrix. For electrode architectures, the process should be optimized to create a conductive network of CNTs that reaches a critical percolation threshold density to yield effective electrical contact throughout the matrix, even in the presence of a nonconductive chitosan layer and macroscopic pores. Therefore, the weight percentage of CNT in the composite electrode will influence several critical material properties, including conductivity, surface area, pore size, and porosity.

Scanning electron microscopy (SEM) images of the composite chitosan–CNT material verify that the majority of the CNTs are fixed within the chitosan precipitate (Figure 10.6) [79]. The phase-separated chitosan has macropores and the relative surface area (roughness) of the composite material increases upon addition of CNTs: higher weight percentages of CNTs result in rougher surfaces. In addition, mean pore size and resistivity decrease with increased CNT weight percentages. Although pure chitosan scaffolds are compressible, scaffolds with high weight percentages of CNTs are relatively brittle. Given that the proportion of CNTs influences several composite material properties—conductivity, porosity, and physical robustness—experimentation is needed to optimize composition for electrode fabrication.

10.4.2 Polymer/Carbon Architectures Fabricated Using Solid Templates

Another method to fabricate 3D porous electrodes is the use of a removable template or "porogen" to yield ordered voids within the material architecture. In this method, conductive nanostructured materials are combined with a macro-sized particulate porogen and a matrix is formed by compressing the mixture within a solid mold. The particles are then "glued" together by percolating a polymer through the matrix. After polymer deposition, the template is dissolved, leaving behind conductive nano-structured material embedded in a porous matrix (Figure 10.7).

One specific demonstration of porogen templating by Luckarift et al. in 2012 was the incorporation of a biocompatible polymer—poly(3-hydroxybutyrate-*co*-3-hydroxyvalerate) (PHBV)—into a conductive scaffold [21]. PHBV is naturally produced by bacteria and was selected for its biocompatible properties, deemed

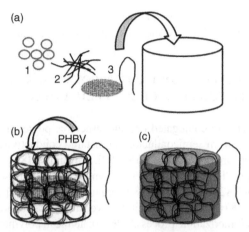

FIGURE 10.7 Fabrication of polymer/carbon architectures for bioelectrodes. (a) (1) Sucrose template is mixed with (2) carbon nanofibers and placed in a (3) Teflon mold along with a Ni mesh current collector attached to a Ti wire. (b) The mixture is coated with liquid PHBV polymer. (c) After the polymer solidifies, the composite is submerged in water to dissolve the sucrose template leaving behind a conductive porous architecture [21].

useful for bioelectrode fabrication. The polymer is stable under typical operating conditions, such as neutral pH and ambient temperatures; furthermore, PHBV provides a hydrophilic local environment that allows good retention of protein or bacterial cells as electrocatalysts. The approach was effectively demonstrated using sucrose as porogen and CF as conductive interface for an electrode ultimately used as an MFC anode. The size of the porogen particles affected the material characteristics: small porogen particles yielded a densely packed scaffold that limited polymer penetration. Larger particles yielded an open, porous structure, but the matrix was relatively brittle because of larger voids in the final architecture. These findings are consistent with chitosan–CNT architectures: pore size influences many physical and performance properties of the material. CF incorporated within the porous structure contributed a key aspect to the hierarchical order. The CF matrix provides high relative surface area for catalyst immobilization and a conductive interface for ET. CF deposition around the pores allowed effective fuel transport to the catalyst and provided conductivity. The 3D porous structure can be controlled by the pore-forming agent, which can be carefully tuned to fulfill the requirements of fuel cell design. Moreover, the mold used to form the electrode can be tailored to achieve the desired shape and size of the electrode.

When comparing the templating process with chitosan composite fabrication, phase separation parameters can also be optimized to control pore size. In both examples, the nonconductive scaffold/binder was composed of a biopolymer to enhance biofilm formation for MFCs. In summary, a carbon-based hierarchical structure illustrates three aspects of material design and fabrication process: a nanometric and/or compatible interface for the immobilization of biomaterials, high conductivity within the conductive architecture, and sufficient porosity to enhance mass transport of fuel to the conductive surface.

10.5 INCORPORATING CONDUCTIVE POLYMERS INTO BIOELECTRODES FOR FUEL CELL APPLICATIONS

Conductive polymers are conjugated organic structures possessing characteristics of conductivity ranging from semiconductors to metallic conductivity. Commonly used conductive polymers are polypyrrole (PPy) [80], polyaniline [81], and poly(3,4-ethyl-enedioxythiophene) (PEDOT) [40]. These polymers are used in several applications, including biosensors [82–84], actuators [80], organic solar cells [85], organic LEDs [86], and supercapacitors [87]. One reason behind the versatility of conductive polymers is the electrochemical polymerization process. Electrochemical polymerization effectively deposits a layer of polymer conforming to the topography of the polarized conductive surface. The approach provides conductive polymer layers of controllable thickness ranging from nano- to macroscale. Typically, this process is performed by dissolving the monomer along with a doping salt and then applying an oxidative potential to the conductive surface. For example, the electropolymerization of PPy takes place via the Diaz mechanism [88]. This mechanism is initiated by oxidation of the monomer by the polarized electrode to generate a radical cation. Oxidation of the monomer is faster than its diffusion back into the bulk phase; thus, two oxidized monomers will dimerize. This can be verified by showing that polymer film growth is linear with respect to t (as opposed to $t^{1/2}$, which indicates mass transport limitations). After the loss of two protons, the dimer can be readily oxidized by the electrode and coupled with another oxidized monomer. Thus, polymerization propagates and the polarized material is coated through the repeated sequence of oxidation, coupling, and deprotonation.

The process of electrochemical deposition of a polymer is affected by the type of solvent, composition and concentration of doping agent, temperature, and applied current density. The π-electron system is formed during the oxidative polymerization process and the negatively charged ion is incorporated into the polymer matrix as a counter ion to compensate the positive charges on the polymer backbone: the positive charges are charge carriers for electrical conduction. The resulting conductivity involves charge transport along the polymer chains along with intrachain electron hopping. Film thickness and morphology also play a role in shaping the conductivity and physical properties of the conductive polymer. Thus, their adjustment allows progressive improvement to the electrochemical deposition conditions and, in turn, the tuning of composite materials for the desired application.

10.5.1 Conductive Polymer-Facilitated DET Between Laccase and a Conductive Surface

Biomolecules are immobilized on conductive polymers through several mechanisms, including covalent attachment, affinity interactions, and physical entrapment [82,83]. In most examples of enzyme–conductive polymer systems, DET is not accomplished;

thus, the systems include a mediator to shuffle electrons between the conductive surface and redox catalyst [82]. An advantage of using conductive polymers as a component in bioelectrodes is that the *in situ* electrochemical polymerization can form conductive polymer "glue" around the adsorbed enzymes. The sequence can increase enzyme loading, enhance composite material conductivity, and prevent biocatalyst loss.

One method used to increase enzyme loading *and* facilitate DET between redox enzymes and a glassy carbon electrode is the application of electrochemically synthesized conducting polymer bilayers. Using this method in 2011, Wang et al. immobilized laccase in a dual-layer architecture of PEDOT films, with NO_3^- as the counter ion in the first layer (PEDOT-NO_3) and polystyrene sulfonate (PSS) anions (PEDOT-PSS) in the second layer (Figure 10.8) [46]. First, the PEDOT-NO_3 layer was deposited on the glassy carbon through electrochemical deposition. Next, laccase was cast on the PEDOT-NO_3 layer, and, finally, the capping PSS layer was deposited through electro-polymerization to prevent enzyme leaching and facilitate DET between the conducting polymer and the T1 copper center of the laccase. The multilayered construct was electrocatalytically active: when oxygen was present in the half-cell, cathodic waves were evident at approximately 0.55 V (vs. Ag/AgCl), an indication of bioelectrocatalytic oxygen reduction. Other bilayer carbon paper film combinations (PEDOT-PSS/PEDOT-PSS and PEDOT-NO_3/PEDOT-NO_3) do not exhibit the biocatalytic redox achieved with the PEDOT-NO_3/PEDOT-PSS bilayer combination.

The thickness and density of the PEDOT composite were varied to evaluate whether the layer thicknesses/morphologies influence the biocatalytic oxygen reduction [46]. SEM analysis of the conductive polymer layers indicated that the first layer

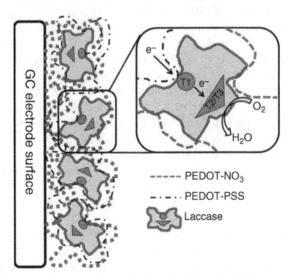

FIGURE 10.8 Schematic ET pathway in the dual-layer architecture enzyme electrode PEDOT-NO_3/PEDOT-PSS with laccase. The PEDOT-NO_3 is electrochemically deposited on the glassy carbon surface followed by the physisorption of laccase. PEDOT-PSS is then polymerized as a capping layer to retain laccase and facilitate direct bioelectrocatalysis. (Reproduced with permission from Ref. [46]. Copyright 2011, American Chemical Society.)

accommodates the casting of the enzyme and has a compact morphology with nanometric pores. The capping layer has a dense appearance, which helps to trap the enzyme in the polymer matrix. Varying thickness (through deposition charge) of both the PEDOT-NO$_3$ and the PEDOT-PSS layers affects electrocatalytic results, although cathodic current was observed for all tested thicknesses. In addition, the measured cathodic current was directly proportional to the amount of laccase immobilized in the architecture. Therefore, the accommodation layer thickness, the capping layer thickness, and the amount of deposited enzyme should be examined and adjusted to obtain optimal current density for a selected application.

Given that DET is not observed in the case of PEDOT-NO$_3$/PEDOT-NO$_3$, the PEDOT-PSS layer appears to play an important role in allowing the redox catalyst to associate with the electrode. It is suggested that the physical interaction of PSS with the T1 binding pocket is facilitated during electropolymerization by incorporation of the PSS counter ion. The interaction between the styrene molecules and the hydrophobic substrate-binding pocket of laccase causes the PEDOT chains to grow in close proximity to the laccase T1 copper site. The explanation is corroborated in work described by Vaz-Dominguez et al. in 2008, in which conductive surfaces that were functionalized with hydrophobic molecules resembling laccase substrates show enhanced DET [89].

10.5.2   Materials Design for MFC

MFCs use bacterial respiration to generate anodic current by using the anode as the electron acceptor. Commonly, MFC reactors are designed so that bacteria will form a biofilm on the electrode surface to promote current generation. Microbial electrocatalysis is subject to similar challenges of connectivity, EASA, and mass transport as EFCs. To overcome these limitations, hierarchically ordered electrode architectures can be fabricated to allow appropriate biofilm formation and enable mass transport of the electron donor. Additional complexity is introduced by requiring that the appropriate, active, bacterial community be established and maintained in the MFC anode to yield current. As an alternative, methods have been developed that engineer "artificial" biofilms using cells cultivated with standardized laboratory techniques and then trapping the cells on the surface in silica sol–gels [21,22].

In 2011, Yu et al. developed another approach to improve MFC performance aimed to surround active cells with a conductive architecture described as a conductive artificial biofilm (CAB) using a mixture of graphite particles (<20 μm in diameter) and the bacterium, *Shewanella oneidensis* [20]. *S. oneidensis* has been applied in several MFCs as a model anodic catalyst [90–93]. The bacteria–graphite particle mixture was deposited onto a carbon cloth and dried. PPy was electrochemically deposited on the cloth, effectively forming a conductive sheath around the *S. oneidensis*-covered graphite [20]. Using this technique to form the CAB, the film thickness can be controlled to gain optimal benefits, such as providing a thick film with good access to electrical contact while maintaining a porous structure that facilitates mass transport. The confined architecture may enhance both MET and DET mechanisms that bacteria use for extracellular ET [94]. After lactate is metabolized by *S. oneidensis*, excreted electrons/mediators can interact with the graphite surface or the PPy, thus transferring electrons to the conductive matrix. Biofilms fabricated with PPy-coated *S. oneidensis* on carbon cloth without the graphite particles did not produce anodic current. Also, the biofilms without PPy encapsulation exhibited lower power output than architectures that included graphite. Therefore, PPy plays a vital role in stabilizing the integrity of the CAB and enhancing conductivity.

10.6 OUTLOOK

The development and design of bioelectrochemical hierarchical architectures has advanced enormously since the earliest reports of the EFC concept [95]. Advances in a wide range of scientific and technical areas have contributed to improvements in this field. Deeper understanding of redox enzyme structure allows predictable interaction between enzyme and electrode interface and a basis to rationally engineer the interactions. New combinations and methods for modification of materials have provided greater redox enzyme loading onto electrode surfaces. Likewise, robust attachment and immobilization processes provide linkage between enzyme and electrode to help sustain bioelectrocatalysis. Building from that molecular and nanomaterial foundation, we can assemble well-designed architectures that will increase performance and improve suitability for electrochemical applications. The combined steps and materials that allow appropriate physical scale, materials cost, and operational lifetime must evolve to advance EFCs from experimental devices to practical technology applications. The means and methods to control electrode structure and performance are critical aspects in the development.

ACKNOWLEDGMENT

This research was supported in part by an appointment to the Postgraduate Research Participation Program at the Air Force Research Laboratory administered by the Oak Ridge Institute for Science and Education (ORISE) through an interagency agreement between the U.S. Department of Energy and the Air Force Research Laboratory, Materials and Manufacturing Directorate, Airbase Technologies Division (AFRL/RXQ). This research was prepared as an account of work sponsored by AFRL/RXQ, but the views of authors expressed herein do not necessarily reflect those of the United States Air Force.

LIST OF ABBREVIATIONS

BFC	biological fuel cell
BP	buckypaper
CAB	conductive artificial biofilm
CBN	carbon black nanomaterial
CF	carbon fabric
CNT	carbon nanotube
CuO	cuprous oxidase
CVD	chemical vapor deposition

DET	direct electron transfer
EASA	electrochemically accessible surface area
EFC	enzymatic fuel cell
ET	electron transfer
FAD	flavin adenine dinucleotide
GDE	gas diffusion electrode
GOx	glucose oxidase
IPA	isopropyl alcohol
MCO	multicopper oxidase
MD	molecular dynamics
MET	mediated electron transfer
MFC	microbial fuel cell
MG	methylene green
MWCNT	multiwalled carbon nanotube
NAD$^+$	nicotinamide adenine dinucleotide
NADH	nicotinamide adenine dinucleotide, reduced form
PBSE	1-pyrenepyrenebutyric acid N-hydroxysuccinimidyl ester
PEDOT	poly(3,4-ethylenedioxythiophene)
PHBV	poly(3-hydroxybutyrate-co-3-hydroxyvalerate)
poly(MG)	poly(methylene green)
PPy	polypyrrole
PSS	polystyrene sulfonate
PTFE	polytetrafluoroethylene
SEM	scanning electron microscopy
SLAC	small laccase from *Streptomyces coelicolor*
SWCNT	single-walled carbon nanotube
T1	type 1

REFERENCES

1. Bullenn RA, Arnot TC, Lakeman JB, Walsh FC. Biofuel cells and their development. *Biosens Bioelectron* 2006;21:2015–2045.

2. Vaddiraju S, Tomazos I, Burgess DJ, Jain FC, Papadimitrakopoulos F. Emerging synergy between nanotechnology and implantable biosensors: a review. *Biosens Bioelectron* 2010;25:1553–1565.

3. Mano N, Mao F, Heller A. A miniature biofuel cell operating in a physiological buffer. *J Am Chem Soc* 2002;124:12962–12963.

4. Calabrese Barton S, Gallaway J, Atanasov P. Enzymatic biofuel cells for implantable and microscale devices. *Chem Rev* 2004;104:4867–4886.

5. Gallaway J, Wheeldon I, Rincón R, Atanassov P, Banta S, Calabrese Barton S. Oxygen-reducing enzyme cathodes produced from SLAC, a small laccase from *Streptomyces coelicolor*. *Biosens Bioelectron* 2008;23:1229–35.

6. Heller A. Miniature biofuel cells. *Phys Chem Chem Phys* 2004;6:209–216.

7. Heller A. Potentially implantable miniature batteries. *Anal Bioanal Chem* 2006;385:469–473.

8. Cinquin P, Gondran C, Giroud F, Mazabrard S, Pellissier A, Boucher F, Alcaraz J-P, Gorgy K, Lenouvel F, Mathé S, Porcu P, Cosnier S. A glucose biofuel cell implanted in rats. *PLoS One* 2010;5(5):e10476.

9. Reimers CE, Girguis PI, Stecher HA, Tender LM, Ryckelynck N, Whaling P. Microbial fuel cell energy from an ocean cold seep. *Geobiology* 2006;4(2):123–136.

10. Erable B, Vandecandelaere I, Faimali M, Delia M-L, Etcheverry L, Vandamme P, Bergel A. Marine aerobic biofilm as biocathode catalyst. *Bioelectrochemistry* 2010;78: 51–56.

11. Gong Y, Radachowksy SE, Wolf M, Nielsen ME, Girguis PR, Reimers CE. Benthic microbial fuel cell as direct power source for an acoustic modem and seawater oxygen/temperature sensor system. *Environ Sci Technol* 2011;45:5047–5053.

12. Strack G, Luckarift HR, Nichols R, Cozart K, Katz E, Johnson GR. Bioelectrocatalytic generation of directly readable code: harnessing cathodic current for long-term information relay. *Chem Commun* 2011;47(27):7662–7664.

13. Logan B. Generating electricity from wastewater treatment. *Water Environ Res* 2005;77 (3):211.

14. Logan BE. Simultaneous wastewater treatment and biological electricity generation. *Water Sci Technol* 2005;52(1–2):31–37.

15. Logan BE. Exoelectrogenic bacteria that power microbial fuel cells. *Nat Rev Microbiol* 2009;7:375–381.

16. Ghindilis AL, Atanasov P, Wilkins E. Enzyme catalyzed direct electron transfer: fundamentals and analytical applications. *Electroanalysis* 1997;9:661–674.

17. Minteer SD, Liaw BY, Cooney MJ. Enzyme-based biofuel cells. *Curr Opin Biotechnol* 2007;18:228–234.

18. Cooney MJ, Svoboda V, Lau C, Martin G, Minteer SD. Enzyme catalysed biofuel cells. *Energy Environ Sci* 2008;1:320–337.

19. Willner I, Yan YM, Willner B, Tel-Vered R. Integrated enzyme-based biofuel cells—a review. *Fuel Cells* 2009;9(1):7–24.

20. Yu YY, Chen HL, Yong YC, Kim DH, Song H. Conductive artificial biofilm dramatically enhances bioelectricity production in *Shewanella*-inoculated microbial fuel cells. *Chem Commun* 2011;47(48):12825–12827.

21. Luckarift HR, Sizemore SR, Farrington KE, Roy J, Lau C, Atanassov PB, Johnson GR. Facile fabrication of scalable, hierarchically structured polymer/carbon architectures for bioelectrodes. *ACS Appl Mater Interfaces* 2012;4(4):2082–2087.

22. Luckarift HR, Sizemore SR, Roy J, Lau C, Gupta G, Atanassov P, Johnson GR. Standardized microbial fuel cell anodes of silica-immobilized *Shewanella oneidensis*. *Chem Commun* 2010;46(33):6048–6050.

23. Minteer SD, Atanassov P, Luckarift HR, Johnson GR. New materials for BFCs. *Mater Today* 2012;15:166–173.

24. Ivnitski D, Atanassov P, Apblett C. Direct bioelectrocatalysis of PQQ-dependent glucose dehydrogenase. *Electroanalysis* 2007;19(15):1562–1568.

25. Ivnitski D, Branch B, Atanassov P, Apblett C. Glucose oxidase anode for biofuel cell based on direct electron transfer. *Electrochem Commun* 2006;8:1204–1210.

26. Meredith MT, Minteer SD. Biofuel cells: enhanced enzymatic bioelectrocatalysis. *Annu Rev Anal Chem* 2012;5:157–179.

27. Marcus RA, Sutin N. Electron transfers in chemistry and biology. *Biochim Biophys Acta* 1985;811:265–322.

28. Ma G-X, Lu T-H, Xia Y-Y. Direct electrochemistry and bioelectrocatalysis of hemoglobin immobilized on carbon black. *Bioelectrochemistry* 2007;71(2):180–185.

29. Kontani R, Tsujimura S, Kano K. Air diffusion biocathode with CueO as electrocatalyst adsorbed on carbon particle-modified electrodes. *Bioelectrochemistry* 2009;76(1–2):10–13.

30. Bidault F, Brett DJL, Middleton PH, Brandon NP. Review of gas diffusion cathodes for alkaline fuel cells. *J Power Sources* 2009;187(1):39–48.

31. Shleev S, Shumakovich G, Morozova O, Yaropolov A. Stable 'floating' air diffusion biocathode based on direct electron transfer reactions between carbon particles and high redox potential laccase. *Fuel Cells* 2010;10(4):726–733.

32. Trohalaki S, Pachter R, Luckarift HR, Johnson GR. Immobilization of the laccases from *Trametes versicolor* and *Streptomyces coelicolor* on single-wall carbon nanotube electrodes: a molecular dynamics study. *Fuel Cells* 2012;12(4):656–664.

33. Hong G, Ivnitski DM, Johnson GR, Atanassov P, Pachter R. Design parameters for tuning the type 1 Cu multicopper oxidase redox potential: insight from a combination of first principles and empirical molecular dynamics simulations. *J Am Chem Soc* 2011;133(13):4802–4809.

34. Meredith MT, Minson M, Hickey D, Artyushkova K, Glatzhofer D, Minteer SD. Anthracene-modified multi-walled carbon nanotubes as direct electron transfer scaffolds for enzymatic oxygen reduction. *ACS Catal* 2011;1:1683–1690.

35. Hussein L, Urban G, Kruger M. Fabrication and characterization of buckypaper-based nanostructured electrodes as a novel material for biofuel cell applications. *Phys Chem Chem Phys* 2011;13(13):5831–5839.

36. Strack G, Luckarift HR, Nichols R, Cozart K, Katz E, Johnson GR. Bioelectrocatalytic generation of directly readable code: harnessing cathodic current for long-term information relay. *Chem Commun* 2011;47(27):7662–7664.

37. Ramasamy RP, Luckarift HR, Ivnitski DM, Atanassov PB, Johnson GR. High electrocatalytic activity of tethered multicopper oxidase–carbon nanotube conjugates. *Chem Commun* 2010;46(33):6045–6047.

38. McNaught AD, Wilkinson A. *IUPAC Compendium of Chemical Terminology*, 2nd edition. Blackwell Scientific Publications, Oxford, 1997.

39. Liu C, Alwarappan S, Chen Z, Kong X, Li CZ. Membraneless enzymatic biofuel cells based on graphene nanosheets. *Biosens Bioelectron* 2010;25(7):1829–1833.

40. Wu JF, Xu MQ, Zhao GC. Graphene-based modified electrode for the direct electron transfer of cytochrome *c* and biosensing. *Electrochem Commun* 2010;12(1):175–177.

41. Wu P, Shao Q, Hu Y, Jin J, Yin Y, Zhang H, Cai C. Direct electrochemistry of glucose oxidase assembled on graphene and application to glucose detection. *Electrochim Acta* 2010;55(28):8606–8614.

42. Kang X, Wang J, Wu H, Aksay IA, Liu J, Lin Y. Glucose oxidase–graphene–chitosan modified electrode for direct electrochemistry and glucose sensing. *Biosens Bioelectron* 2009;25(4):901–905.

43. Wang G, Qian F, Saltikov CW, Jiao Y, Li Y. Microbial reduction of graphene oxide by *Shewanella. Nano Res* 2011;4(6):563–570.

44. Lv P, Zhang P, Li F, Li Y, Feng Y, Feng W. Vertically aligned carbon nanotubes grown on carbon fabric with high rate capability for super-capacitors. *Synth Met* 2012;162:1090–1096.

45. Miyake T, Yoshino S, Yamada T, Hata K, Nishizawa M. Self-regulating enzyme–nanotube ensemble films and their application as flexible electrodes for biofuel cells. *J Am Chem Soc* 2011;133:5129–5134.

46. Wang X, Latonen R-M, Sjöberg-Eerola P, Eriksson J-E, Bobacka J, Boer H Bergelin M. Direct electron transfer of *Trametes hirsuta* laccase in a dual-layer architecture of poly(3,4-ethylenedioxythiophene) films. *J Phys Chem C* 2011;115:5919–5929.

47. Calabrese Barton S, Kim H-H, Binyamin G, Zhang Y, Heller A. The "wired" laccase cathode: high current density electroreduction of O_2 to water at +0.7 V (NHE) at pH 5. *J Am Chem Soc* 2001;123:5802–5803.

48. Gupta G, Lau C, Rajendran V, Colon F, Branch B, Ivnitski D, Atanassov P. Direct electron transfer catalyzed by bilirubin oxidase for air breathing gas-diffusion electrodes. *Electrochem Commun* 2011;13:247–249.

49. Murata K, Nakamura N, Ohno H. Direct electron transfer reaction of ascorbate oxidase immobilized by a self-assembled monolayer and polymer membrane combined system. *Electroanalysis* 2007;19:530–534.

50. Ivnitski D, Khripin C, Luckarift HR, Johnson GR, Atanassov P. Surface characterization and direct bioelectrocatalysis of multicopper oxidases. *Electrochim Acta* 2010;55:7385–7393.

51. Quintanar L, Stoj C, Taylor AB, hart PJ, Kosman DJ, Solomon EI. Shall we dance? How a multicopper oxidase chooses its electron transfer partner. *Acc Chem Res* 2007;40(6):445–452.

52. Shleev S, Pita M, Yaropolov AI, Ruzgas T, Gorton L. Direct heterogeneous electron transfer reactions of *Trametes hirsuta* laccase at bare and thiol-modified gold electrodes. *Electroanalysis* 2006;18:1901–1908.

53. Shleev S, Christenson A, Serezhenkov V, Burbaev D, Yaropolov A, Gorton L, Ruzgas T. Electrochemical redox transformations of T1 and T2 copper sites in native *Trametes hirsuta* laccase at gold electrode. *Biochem J* 2005;385:745–754.

54. Shleev S, Tkac J, Christenson A, Ruzgas T, Yaropolov AI, Whittaker JW, Gorton L. Direct electron transfer between copper-containing proteins and electrodes. *Biosens Bioelectron* 2005;20:2517–2554.

55. Hussein L, Rubenwolf S, von Stetten F, Urban G, Zengerle R, Krueger M, Kerzenmacher S. A highly efficient buckypaper-based electrode material for mediatorless laccase-catalyzed dioxygen reduction. *Biosens Bioelectron* 2011;26(10):4133–4138.

56. Svoboda V, Cooney M, Liaw BY, Minteer S, Piles E, Lehnert D, Calabrese Barton S, Rincón R, Atanassov P. Standardized characterization of electrocatalytic electrodes. *Electroanalysis* 2008;20(10):1099–1109.

57. Narvaez Villarrubia CW, Rincón RA, Radhakrishnan VK, Davis V, Atanassov P. Methylene green electrodeposited on SWNTs-based "bucky" papers for NADH and L-malate oxidation. *ACS Appl Mater Interfaces* 2011;3(7):2402–2409.

58. Appleyard DC, Chapin SC, Doyle PS. Multiplexed protein quantification with barcoded hydrogel microparticles. *Anal Chem* 2010;83(1):193–199.

59. Demirok UK, Burdick J, Wang J. Orthogonal multi-readout identification of alloy nanowire barcodes. *J Am Chem Soc* 2009;131(1):22–23.

60. Gunnarsson A, Sjovall P, Hook F. Liposome-based chemical barcodes for single molecule DNA detection using imaging mass spectrometry. *Nano Lett* 2010;10(2):732–737.

61. Kim J, Seo K, Wang J. Multiplexed electrochemical protein coding based on quantum dot (QD)-bioconjugates for a clinical barcode system. *Conf Proc IEEE Eng Med Biol Soc* 2004;1:137–140.

62. Thaxton CS, Elghanian R, Thomas AD, Stoeva SI, Lee JS, Smith ND, Schaeffer AJ, Klocker H, Horninger W, Bartsch G, Mirkin CA. Nanoparticle-based bio-barcode assay redefines "undetectable" PSA and biochemical recurrence after radical prostatectomy. *Proc Natl Acad Sci USA* 2009;106(44):18437–18442.

63. Sierra-Sastre Y, Choi S, Picraux ST, Batt CA. Vertical growth of Ge nanowires from biotemplated Au nanoparticle catalysts. *J Am Chem Soc* 2008;130(32):10488–10489.

64. Stuart MA, Huck WT, Genzer J, Muller M, Ober C, Stamm M, Sukhorukov GB, Szleifer I, Tsukruk VV, Urban M, Winnik F, Zauscher S, Luzinov I, Minko S. Emerging applications of stimuli-responsive polymer materials. *Nat Mater* 2010;9(2):101–113.

65. Sotiropoulou S, Sierra-Sastre Y, Mark SS, Carl BA. Biotemplated nanostructured materials. *Chem Mater* 2008;20:821–834.

66. Huang J, Wang X, Wang ZL. Controlled replication of butterfly wings for achieving tunable photonic properties. *Nano Lett* 2006;6(10):2325–2331.

67. Guo J, Wang X, Liao X, Zhanga W, Shi S. Skin collagen fiber-biotemplated synthesis of size-tunable silver nanoparticle-embedded hierarchical intertextures with lightweight and highly efficient microwave absorption properties. *J Phys Chem B* 2012;116:8188–8195.

68. Bentacor L, Luckarift HR. Bio-inspired enzyme encapsulation for bioelectrocatalysis. *Trends Biotechnol* 2008;26:566–572.

69. Luckarift HR, Spain JC, Naik RR, Stone MO. Enzyme immobilization in a biomimetic silica support. *Nat Biotechnol* 2004;22(2):211–213.

70. Cardoso M, Luckarift H, Urban V, O'Neill H, Johnson G. Biomimetic composites: protein localization in silica nanospheres derived via biomimetic mineralization. *Adv Funct Mater* 2010;20:2963–3182.

71. Ivnitski D, Artyushkova K, Rincón RA, Atanassov P, Luckarift HR, Johnson GR. Entrapment of enzymes and carbon nanotubes in biologically synthesized silica: glucose oxidase-catalyzed direct electron transfer. *Small* 2008;4(3):357–364.

72. Wang Q, Zhou L, Jiang Y, Gao J. Improved stability of the carbon nanotubes–enzyme bioconjugates by biomimetic silicification. *Enzyme Microb Technol* 2011;49:11–16.

73. Pchelintsev NA, Neville F, Millner PA. Biomimetic silication of surfaces and its application to preventing leaching of electrostatically immobilized enzymes. *Sens Actuators B: Chem* 2008;135:21–26.

74. Vamvakaki V, Hatzimarinaki M, Chaniotakis N. Biomimetically synthesized silica–carbon nanofiber architectures for the development of highly stable electrochemical biosensor systems. *Anal Chem* 2008;80(15):5970–5975.

75. Komathi S, Gopalan AI, Lee KP. Covalently linked silica–multiwall carbon nanotube-polyaniline network: an electroactive matrix for ultrasensitive biosensor. *Biosens Bioelectron* 2009;25(4):944–947.

76. Si P, Ding S, Yuan J, Lou XW, Kim DH. Hierarchically structured one-dimensional TiO_2 for protein immobilization, direct electrochemistry, and mediator-free glucose sensing. *ACS Nano* 2011;5(9):7617–7626.

77. Jiang Y, Yang D, Zhang L, Sun Q, Sun X, Li J, Jiang Z. Preparation of protamine–titania microcapsules through synergy between layer-by-layer assembly and biomimetic mineralization. *Adv Funct Mater* 2009;19:150–156.

78. Rincón RA, Lau C, Luckarift HR, Garcia KE, Adkins E, Johnson GR, Atanassov P. Enzymatic fuel cells: integrating flow-through anode and air-breathing cathode into a membrane-less biofuel cell design. *Biosens Bioelectron* 2011;27(1):132–136.

79. Lau C, Cooney MJ, Atanassov P. Conductive macroporous composite chitosan–carbon nanotube scaffolds. *Langmuir* 2008;24(13):7004–7010.

80. Strack G, Bocharova V, Arugula MA, Pita M, Halámek J, Katz E. Artificial muscle reversibly controlled by enzyme reactions. *J Phys Chem Lett* 2010;1:839–843.

81. Bidez PR, Li S, MacDiarmid AG, Venancio EC, Wei Y, Lelkes PI. Polyaniline, an electroactive polymer, supports adhesion and proliferation of cardiac myoblasts. *J Biomater Sci Polym Ed* 2006;17(1–2):199–212.

82. Cosnier S. Biosensors based on electropolymerized films: new trends. *Anal Bioanal Chem* 2003;377(3):507–520.

83. Cosnier S and Holzinger M. Electrosynthesized polymers for biosensing. *Chem Soc Rev* 2011;40:2146–56.

84. Cosnier S, Stoytcheva M, Senillou A, Perrot H, Furriel RP, Leone FA. A biotinylated conducting polypyrrole for the spatially controlled construction of an amperometric biosensor. *Anal Chem* 1999;71(17):3692–3697.

85. Peet J, Kim JY, Coates NE, Ma WL, Moses D, Heeger AJ, Bazan GC. Efficiency enhancement in low-bandgap polymer solar cells by processing with alkane dithiols. *Nat Mater* 2007;6(7):497–500.

86. Sariciftci NS, Smilowitz L, Heeger AJ, Wudl F. Photoinduced electron transfer from a conducting polymer to buckminsterfullerene. *Science* 1992;258(5087):1474–1476.

87. Marchioni F, Yang J, Walker W, Wudl F. A low band gap conjugated polymer for supercapacitor devices. *J Phys Chem B* 2006;110(44):22202–22206.

88. Sadki S, Schottland P, Brodie N, Sabouraud G. The mechanisms of pyrrole electropolymerization. *Chem Soc Rev* 2000;29:283–293.

89. Vaz-Dominguez C, Campuzano S, Rudiger O, Pita M, Gorbacheva M, Shleev S, Fernandez VM, De Lacey AL. Laccase electrode for direct electrocatalytic reduction of O_2 to H_2O with high-operational stability and resistance to chloride inhibition. *Biosens Bioelectron* 2008;24(4):531–537.

90. Biffinger JC, Byrd JN, Dudley BL, Ringeisen BR. Oxygen exposure promotes fuel diversity for *Shewanella oneidensis* microbial fuel cells. *Biosens Bioelectron* 2008; 23(6):820–826.

91. Biffinger JC, Pietron J, Bretschger O, Nadeau LJ, Johnson GR, Williams CC, Nealson KH, Ringeisen BR. The influence of acidity on microbial fuel cells containing *Shewanella oneidensis*. *Biosens Bioelectron* 2008;24(4):906–911.

92. Biffinger JC, Pietron J, Ray R, Little B, Ringeisen BR. A biofilm enhanced miniature microbial fuel cell using *Shewanella oneidensis* DSP10 and oxygen reduction cathodes. *Biosens Bioelectron* 2007;22(8):1672–1679.

93. Biffinger JC, Ray R, Little BJ, Fitzgerald LA, Ribbens M, Finkel SE, Ringeisen BR. Simultaneous analysis of physiological and electrical output changes in an operating microbial fuel cell with *Shewanella oneidensis*. *Biotechnol Bioeng* 2009;103(3):524–531.

94. Gorby YA, Yanina S, McLean JS, Rosso KM, Moyles D, Dohnalkova A, Beveridge TJ, Chang IS, Kim BH, Kim KS, Culley DE, Reed SB, Romine MF, Saffarini DA, Hill EA, Shi L, Elias DA, Kennedy DW, Pinchuk G, Watanabe K, Ishii S, Logan B, Nealson KH, Fredrickson JK. Electrically conductive bacterial nanowires produced by *Shewanella oneidensis* strain MR-1 and other microorganisms. *Proc Natl Acad Sci USA* 2006;103(30):11358–11363.

95. Yahiro AT, Lee SM, Kimble DO. Enzyme utilizing biofuel cell studies. *Biochim Biophys Acta* 1964;88:375–383.

11

ENZYME IMMOBILIZATION FOR BIOLOGICAL FUEL CELL APPLICATIONS

LORENA BETANCOR

Laboratorio de Biotecnología, Facultad de Ingeniería, Universidad ORT Uruguay, Montevideo, Uruguay

HEATHER R. LUCKARIFT

Universal Technology Corporation, Dayton, OH, USA; Airbase Sciences Branch, Air Force Research Laboratory, Tyndall Air Force Base, FL, USA

11.1 INTRODUCTION

Enzymes are, by nature, soluble, which typically limits direct integration of proteins during device fabrication. Numerous methods of enzyme immobilization, therefore, have been explored in order to anchor proteins in a scaffold, or at a surface, in a manner that retains the native three-dimensional structure and hence the native catalytic activity [1]. Many of the bottlenecks that limit power density in enzymatic fuel cells (EFCs) are related to enzyme stability, electron transfer rate, and enzyme loading, all issues that can potentially be addressed by efficient enzyme immobilization. Efficient bioelectrocatalysis relies on controllable methods to immobilize biomolecules in a manner that optimizes orientation and interaction between protein and a transducer. Enzymes typically exhibit poor stability over continuous use. In biological fuel cells (BFCs), this leads to low power density over extended operational time. Retention of enzyme integrity (enzyme stabilization) is thus essential to maintain effective electron transfer [2]. Techniques for enzyme

Enzymatic Fuel Cells: From Fundamentals to Applications, First Edition. Edited by Heather R. Luckarift, Plamen Atanassov, and Glenn R. Johnson.

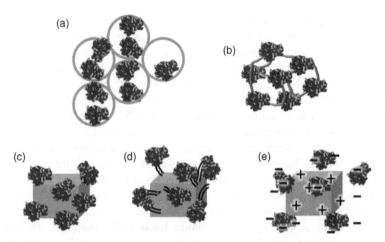

FIGURE 11.1 Typical methods for enzyme immobilization include (a) entrapment, (b) cross-linking, (c) physical adsorption, (d) covalent binding, and (e) ionic interactions.

immobilization and stabilization include ionic and hydrophobic adsorption (see Section 11.2), aggregation and entrapment (see Section 11.3), and covalent binding (see Section 11.4) (Figure 11.1). The requirements for different enzymes do vary and choice of suitable immobilization techniques is usually dictated by protein character-istics, rational design, and an element of trial and error. Significant screening and optimization of immobilization techniques is often required, and some loss in activity may be sacrificed in favor of increased stability or operational longevity [3,4]. There are examples in which the use of covalent binding has proven to be detrimental to enzyme activity and, in fact, has reduced the stability of the biocatalyst [5]. Immobilization through ionic and hydrophobic interactions provides a weaker association than covalent attachment but can confer additional stability against denaturation in nonphysiological environments and provide a protected micro-environment that enhances the catalytic properties of the biocatalyst [3].

In any approach to immobilizing enzymes, the goal is to achieve a high specific activity without compromising stability. In bioelectrocatalysis, the process is further complicated by a need to maintain close association between catalyst and electrode. In addition, the electrode architecture should exhibit conductivity and stability under continuous flow and should provide optimal mass transfer kinetics for fuel. Herein, numerous approaches to immobilizing enzymes for bioelectrocatalysis are discussed with the goal of providing guidance toward rational design.

11.2   IMMOBILIZATION BY PHYSICAL METHODS

11.2.1   Adsorption

Physical adsorption (primarily by electrostatic binding) is an attractive method for enzyme immobilization due to its simplicity and is often investigated in EFCs as a first principle. Adsorption involves the physical association of the enzyme to a support that often does not require complex surface derivatization protocols. Adsorption is dictated by physical characteristics such as electrostatic, hydrophobic, or hydrophilic interactions such that extended stability is often maintained. Physical adsorption of enzymes is optimal when the porosity of the support matrix can be controlled; mesoporous materials, for example, provide good immobilization matrices when the pore size is tailored to a similar size as the enzyme. Charge also has an effect; if the charge of the protein is opposite to that of the porous matrix, attractive forces will contribute to protein stability [6]. The charge of a scaffold matrix can be controlled, however, by changes in buffer pH or chemical modification with functional groups, such as amino or carboxyl moieties [7]. Electrocatalytic activity is retained by physical adsorption, but power density is often low because of poor protein loading. Recent reports of EFCs, however, show that fabrication of both anodes and cathodes can be achieved by simple physical adsorption. Labus et al. reported the adsorption of laccase onto polyamide membranes with no loss of activity over 10 subsequent batch experiments. The work also demonstrated that specific functionalization of the support for covalent attachment of the enzyme did not significantly improve the total activity or stability of the enzyme after immobilization [8]. Similarly, fructose dehydrogenase and laccase showed enhanced stability by physical adsorption to carbon black [9,10].

An economical advantage of physical adsorption is that the process is reversible, allowing the reuse of an electrode after the enzyme has been inactivated [11–13]. The reversibility of protein adsorption was shown in 2011 by Kihara et al., using vertically aligned single-walled carbon nanotubes (SWCNTs) as a support for hydrogenase in an electrochemical hydrogen production system [13]. The SWCNT forest was able to hydrophobically adsorb hydrogenase (in high milligram quantities) for catalytic hydrogen production; 95% of the catalyst, however, could be subsequently removed by the addition of detergent (Triton X-100) to leave a bare device that could be reused. This is a useful consideration if the electrode scaffold is a costly commodity.

Physical adsorption can provide additional stability against denaturation [14], by creating a protected microenvironment that can actually lead to changes in the catalytic properties of the enzyme [13,15–17]. Improvement in the enzymatic properties on adsorption may not be strictly related to the enzyme–support interaction but may also arise from improvements in mass transfer of substrate. For phenolic compounds, for example, the diffusion coefficient obtained by adsorption of *Trametes versicolor* laccase on multiwalled carbon nanotubes (MWCNTs) is one order of magnitude higher than that for covalent immobilization of the enzyme to derivatized MWCNTs [18]. The observation is attributed to enhanced permeability of substrate in the presence of negatively charged groups on the MWCNT support structure.

One of the possible disadvantages of adsorption as a strategy for enzyme immobilization is loss of activity due to leaching of enzyme molecules over time [19]. When adsorption occurs through specific electrostatic interactions, some control of the process can be provided by optimizing the ionic strength and pH of the enzymatic environment to prevent enzymatic desorption [20,21]. In 2012, Nicolau et al. demonstrated pH dependence of immobilization for alcohol

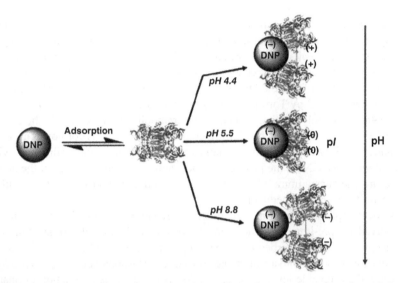

FIGURE 11.2 Schematic representation of the effect of pH on the adsorption of alcohol dehydrogenase on diamond nanoparticles. (Reproduced with permission from Ref. [21]. Copyright 2012, Elsevier.)

dehydrogenase (ADH) on diamond nanoparticles when fabricating a bioelectrochemical cell for oxidation of ethanol [21]. By choosing a pH near the isoelectric point (pI) of the enzyme, the protein loading of ADH was doubled compared with when the immobilization pH was outside the range of the pI: the observation was attributed to a reduction in repulsion between immobilized enzyme molecules (Figure 11.2).

A further advantage of immobilization by adsorption is the relatively mild conditions for immobilization that facilitate an almost universal approach for a wide range of enzymes. An excellent example of this was reported in 2011 by Sokic-Lazic et al., who demonstrated the adsorption of an enzymatic cascade of eight enzymes onto a modified Nafion® membrane [22]. The enzyme cascade mimics a metabolic pathway and catalyzes the complete oxidation of lactate, and as such provides a bioanode that was subsequently demonstrated in an EFC that produced current densities approaching $0.15\,\mathrm{mA\,cm^{-2}}$ from the catalysis of lactate and air [22]. All eight enzymes in the cascade exhibit catalytic activity on adsorption and in concert provide maximum current and power densities that are nine times higher than those for BFCs bearing only a single lactate-oxidizing anode.

11.3 ENTRAPMENT AS A PRE- AND POST-IMMOBILIZATION STRATEGY

In the field of biosensors, the benefits of diverse pre- and post-immobilization treatments on the activity and stability of enzymes are well known [23–27]. The strategies primarily include chemical modification or entrapment of enzymes within a variety of polymers. The premise is intended to prolong the half-life of the enzymes, prevent enzymatic leakage, and/or avoid activity losses that may arise from negative interactions between the enzyme molecules and the support. These advantages usually compensate for the additional steps required in preparation. This approach was reported in 2012 by Tortolini et al. by embedding laccase from *T. versicolor* in a polyazetidine prepolymer (physicochemical immobilization), which is then associated by deposition on MWCNT [18]. Compared with methods for covalent (carbodiimide and *N*-hydroxysuccinimide chemistry) or electrostatic immobilization (within Nafion) with MWCNTs, the physicochemical approach better preserved the activity of the enzyme during immobilization, and the resulting composite exhibited a higher permeability for redox mediators.

A novel application for ionic liquids in the preparation of functional materials has also served as a pre-immobilization strategy for a robust laccase electrode. In this case, the enzyme was first adsorbed to an ionic liquid-functionalized cellulose acetate that was then incorporated into a carbon paste electrode. The approach served to produce a device with acceptable levels of accuracy for methyldopa determination in pharmaceutical samples. In a similar manner, microencapsulation of laccase from *T. versicolor* in polyethylenimine (PEI) was demonstrated as a prior treatment for creating a coating on a paper support. Although first attempts showed a deleterious effect of PEI on laccase because of negative conformational changes that reduced the activity of the encapsulated enzyme [28], optimized microencapsulation conditions resulted in superior stability compared with free enzyme [27].

11.3.1   Stabilization via Encapsulation

Alternatively, methods of enzyme encapsulation can provide a means to stabilize proteins in a "protective" environment by either trapping the protein, wiring the protein to a polymer backbone, or specifically depositing enzymes within micellar pockets [29–32]. Enzymes immobilized within the pores of hydrophobically modified micellar polymers such as Nafion and chitosan, for example, have been shown to effectively stabilize enzymes at electrode surfaces and promote operation lifetimes of more than 2 years [29].

Redox catalysts can be stabilized by encapsulation during silica sol–gel formation [33–35], in which the conductivity of the silica matrix is achieved by co-immobilization of a conductive material, such as carbon nanotubes (CNTs). The cationic protein lysozyme catalyzes and templates the formation of silica directly onto a conductive carbon paper electrode. Inclusion of CNT and glucose oxidase (GOx) into the reaction mixture results in a catalytic composite that becomes encapsulated as the silica forms [36].

11.3.2   Redox Hydrogels

Redox hydrogels have provided a significant contribution to development of BFCs, particularly with application to implantable devices with implications for *in vivo* diabetes management [37,38]. Redox hydrogels are typically based on osmium or ruthenium complexes into which enzymes can be co-immobilized by covalently binding to the hydrogel [39]. Concerns of long-term stability are addressed by anchoring the hydrogel to electrodes using surface-specific functional groups, such as carboxylates or amines. The operational lifetime of these systems is now in the realm of several weeks [40,41]. Osmium-based redox hydrogels have been used for both anodes and cathodes and, in fact, their primary advantage is the ability to tune the polymer structure to a specific redox potential depending on its chemical functionality [42–47]. In 2004, Heller and coworkers, for example, reported high redox potential polymers for cathodic reactions and low redox potential polymers for anodic reactions [48,49]. By applying such polymer matrices to fabricate a full BFC, a membrane-free design with power density $>0.1 \, \text{mW cm}^{-2}$ has been reported [45,46].

11.4   ENZYME IMMOBILIZATION VIA CHEMICAL METHODS

11.4.1   Covalent Immobilization

The mechanisms of electron transfer vary depending on specific enzyme association with an electrode, which in turn is dictated by the properties and functionality of the electrode architecture. Immobilization of laccase to screen-printed electrodes, for example, varies significantly when the electrode architecture is carbon, planar gold, or carbon modified with gold nanoparticles (Figure 11.3). The variation in bioelectro-catalytic activity for different electrode materials is attributed to immobilization chemistry, which in turn is dictated by the electrode architecture. Glutaraldehyde provides effective immobilization on carbon electrodes. In contrast, dithiobis(succi-nimidyl) propionate (DSP) binds to protein via acylation of primary amines and to gold via disulfide groups, which together provide a covalent link on gold-modified surfaces. Laccase immobilized to an electrode modified with carbon and gold nanoparticles by using DSP as the cross-linker produced 100-fold higher current densities than for planar gold electrodes [50]. Incorporation of gold nanoparticles into the carbon-based electrode increases the available surface area and allows preferential binding of laccase to gold. The electrocatalytic current at the electrode begins at a potential of approximately 0.6 V (vs. Ag/AgCl) and provides direct evidence of efficient direct electron transfer (DET) in response to substrate (i.e., molecular oxygen).

11.4.2   Molecular Tethering

Direct interaction between enzyme and electrode is a particular challenge for many redox enzymes as the redox cofactor is buried deep within the protein structure. The cofactor of GOx, for example, is buried deep within the protein structure. One interesting solution to this limitation is to anchor the cofactor flavin adenine dinucleotide (FAD) directly to the electrode surface. When the apoenzyme (enzyme without cofactor) is subsequently added, the protein reforms around the anchored FAD and tethers the enzyme in close

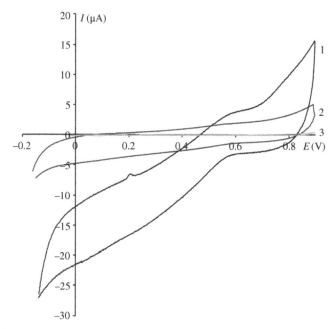

FIGURE 11.3   Oxygen reduction activity of laccase following immobilization on (1) screen-printed carbon electrode modified with gold nanoparticles, (2) screen-printed carbon electrode, and (3) screen-printed planar gold electrode. (Modified with permission from Ref. [50]. Copyright 2012, Wiley-VCH Verlag GmbH.)

proximity to the electrode. The cofactor can be anchored to various electrode architectures, such as gold nanoparticles or CNTs that can serve as an electron bridge (Figure 11.4) [51–53]. FAD, for example, can be linked to SWCNTs and used to position the apoenzyme of GOx [54].

Similarly, in 2006, Ivnitski et al. demonstrated the anchoring of GOx to CNTs and observed DET between the active site of the enzyme and MWCNTs that were grown directly on a Toray® carbon paper electrode [55]. This technique works for other cofactors such as pyrroloquinoline quinone (PQQ), which when covalently bound to a gold electrode and subsequently modified with nicotinamide adenine dinucleotide (NAD⁺) can be used to reconstitute enzymes such as PQQ-dependent lactate dehydrogenase [56,57].

The introduction of CNT as a conductive material has provided a means to develop new conductive architectures, and a range of buckypaper (CNT paper), buckygel (CNT gels), and carbon nanofiber electrodes have since been reported that demonstrate a significant enhancement in electron transfer characteristics for both anodic and cathodic catalysts [58–61]. The advantage of CNT as an electrode material is the high electrical conductivity, mechanical stability, and, to some extent, adaptability in respect to the types of materials that can be modified with CNTs. Cathodic oxygen reduction catalysts can be immobilized to buckypaper by simple physical

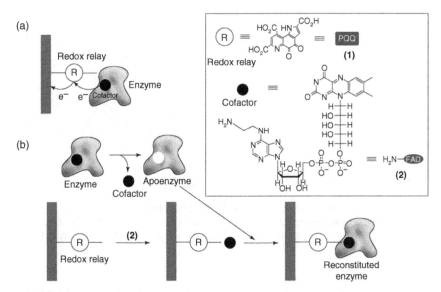

FIGURE 11.4 Electrical wiring of redox enzymes. (a) Optimal configuration for the electrical contacting of a redox enzyme with the electrode. (b) Reconstitution of an apoenzyme on a relay–cofactor monolayer for the alignment and electrical wiring of a redox enzyme. The structures of the redox relay molecule PQQ and cofactor amino-FAD are shown in the inset. (Reproduced with permission from Ref. [53]. Copyright 2006, Elsevier.)

adsorption [62], but preferential electrocatalytic activity is observed when enzyme immobilization is achieved via a bifunctional cross-linking agent (1-pyrenebutyric acid N-hydroxysuccinimidyl ester (PBSE)) that interacts directly with CNT via π–π stacking [60]. In comparison to a nonconjugated cross-linker such as DSP, which binds nonspecifically to two primary amine groups, the tethered cross-linker PBSE and a structural homolog DDPSE (4,4′-[(8,16-dihydro-8,16-dioxodibenzo[a, j]per-ylene-2,10-diyl)dioxy]dibutyric acid di(N-succinimidyl ester)) (Figure 11.5) have a defined aromatic functionality for specific π–π stacking interactions with CNT. The two additional protein binding "spacer arms" of DDPSE are proposed to provide more defined enzyme binding by drawing the protein closer to the CNT surface, 3 Å for DDPSE versus 4 Å for PBSE.

11.4.3 Self-Assembly

Some redox enzymes will fortuitously catalyze the reduction of metal salts to form discrete metal structures, such as gold nanoparticles. GOx, for example, will catalyze the reduction of gold(III) chloride with size-controllable formation of gold particles, into which the protein becomes entrained as the metal structure forms. The resulting GOx/gold composite retains the catalytic activity of the protein, and DET of the FAD cofactor is observed during cyclic voltammetry (−0.44 V vs. Ag/AgCl). In addition, a catalytic current is observed in response to physiological levels of glucose [64].

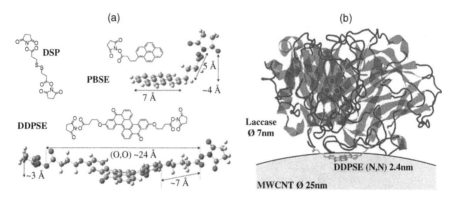

FIGURE 11.5 (a) Structure of the cross-linker DSP, PBSE, and DDPSE. (b) A size comparison of a 25 nm diameter MWCNT, 7 nm diameter laccase, and 26 Å N−N distanced DDPSE. (Reproduced with permission from Ref. [63]. Copyright 2011, Wiley-VCH Verlag GmbH.)

Cross-linked enzyme aggregates (CLEAs) rely on chemical cross-linking to form a self-assembled enzyme matrix that can be fabricated into, or onto, an electrode material. CLEAs find application in EFCs, particularly anodic catalysts [65]. GOx, for example, can be cross-linked in the presence of ammonium sulfate onto the surface of various nanomaterials. The aggregation of the enzyme results in a substantial increase in volumetric protein loading and enhanced stability (>200 days). The use of GOx as CLEA was used as the anode of a biological fuel cell and demonstrated reproducible power density and stability up to 50 °C.

11.5 ORIENTATION MATTERS

Covalent immobilization strategies typically provide superior electrocatalytic characteristics, but the tethering can sometimes hinder protein conformation [66]. In addition, the functional groups on the enzyme that are used for tethering should not be essential to catalysis, or enzyme inactivation losses will occur. The ability to tailor covalent chemistry to specific regions of a protein, however, allows some control over protein orientation. A recent tendency in protein immobilization is the specific and rational orientation of protein molecules on a selected surface [67–69]. This strategy aims to create uniformly oriented protein immobilized in a manner that may optimize catalytic efficiency [70], enhance stability [68], or increase electron transfer efficiency [71,72].

Oriented immobilization may be achieved by introducing reactive groups or covalent tethers to the protein surface that can then interact with a compatible functionalized surface [70,73,74]. Alternatively, by careful design of the immobilization conditions, it is possible to exploit the inherent physicochemical properties of an enzyme [68,72]. The exposed amine groups of lysine residues, for example, react readily with active ester groups, such as N-hydroxysuccinimide esters,

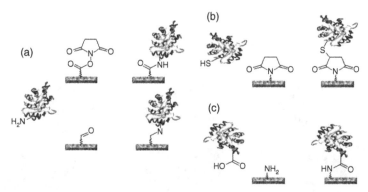

FIGURE 11.6 Covalent enzyme immobilization via reaction of (a) lysine residues with NHS esters or aldehydes, (b) cysteine residues with maleimide groups, and (c) carboxyl residue of glutamic acid or aspartic acid with amine-functionalized materials via NHS-mediated ester chemistry.

and form stable amide bonds. In most proteins, however, lysines are numerous on the surface and therefore attachments can occur at multiple reaction sites, which then confer multiple protein orientations and increased heterogeneity. Functional groups such as aldehydes, maleimides, carbodiimides, and thiols all bind preferentially to different regions of a protein structure or specific amino acids (Figure 11.6) [75]. Such covalent binding will orient the protein depending on the specific covalent binding chemistry and the availability, number, and location of functional groups [76].

For laccase, orientation can be guided by different attachment strategies that aim specifically to attach the enzyme in a particular position [77,78]. As an example, in 2011, Pita et al. developed and optimized a strategy for oriented covalent immobilization of *Trametes hirsuta* laccase on gold electrodes [72]. After optimizing the immobilization for DET via the type 1 (T1) copper site, they were able to measure current density values up to 40 μA cm^{-2} for the electrocatalytic reduction of O_2 in the absence of redox mediators.

Similarly, the orientation of laccase on gold was demonstrated by recombinant expression of laccase with a six-histidine "His-tag" linker that then specifically orients the protein with a thiol-modified monolayer, formed on the surface of gold [79]. The oriented immobilization of the enzyme preserves catalytic activity, and the catalytic reduction of dioxygen was observed in the presence of a redox mediator.

A further example of the influence of orientation in immobilization was provided by Lojou in 2011 [71], who demonstrated the specific orientation of hydrogenase on CNTs that were functionalized with amino groups (Figure 11.7). The immobilization method affected not only the efficiency of the enzymatic oxidation of H_2 but also the stability of the enzyme. The study also investigated the effect of catalytic currents for H_2 oxidation as a function of the thiol derivative used for the construction of self-assembled monolayers (SAMs) on gold electrodes. SAMs can be prepared by chemisorption of thiol functional groups, typically to a gold surface. The resulting functional groups and their associated charge are then amenable to protein

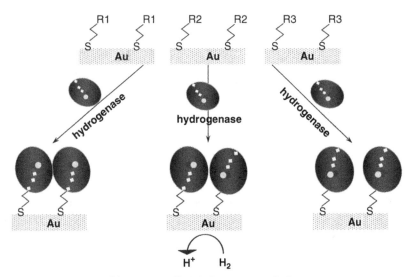

Direct or mediated electron transfer?

FIGURE 11.7   Modification of a gold electrode with functional group-specific SAM orients hydrogenase and controls electron transfer. Hydrogenase is depicted (in black and white) with the [NiFe] active site as two spheres and the FeS clusters as squares. (Reproduced with permission from Ref. [71]. Copyright 2011, Elsevier.)

immobilization. At positively charged interfaces (prepared from a cysteamine SAM), maximum efficiency for H_2 oxidation was reported via DET, due to preferential orientation of the protein's distal FeS cluster facing the electrode. On the contrary, at negatively charged interfaces (prepared from a 3-mercaptopropionate SAM), the catalytic cluster is opposite to the electrode, and as a result, catalytic currents were low. The orientation of the immobilized hydrogenase protein also inherently influenced the mechanism of electron transfer, particularly in respect to direct or mediated electron transfer [71].

11.6   OUTLOOK

The field of enzyme immobilization, although mature, still falls short of a universal protocol for every enzyme. For BFC applications, this is particularly true. In the last 5 years, however, the "toolbox" of usable methodologies has certainly expanded and a number of reliable technologies now exist for BFC fabrication. Encapsulation of biocatalysts in conductive polymer, for example, would benefit from attempts to optimize mass transport of fuel. Stability of enzyme electrodes in nonphysiological environments will also become a crux of EFC research, as advancements in fuel cell fabrication result in prototypes that can undergo realistic operational "field" testing. Only then can the true operational stability and the effects of temperature, salinity, pressure, pH, humidity, and so on be determined. Enzyme immobilization as a factor of scale-up will also become important as fuel cell development turns to three-dimensional hierarchical architectures that will ultimately be required to achieve useful power densities. Further investigation in this field, therefore, may still yield protocols to enhance enzyme immobilization in EFCs.

ACKNOWLEDGMENT

This research was prepared as an account of work sponsored by the United States Air Force Research Laboratory, Materials and Manufacturing Directorate, Airbase Technologies Division (AFRL/RXQ), but the views of authors expressed herein do not necessarily reflect those of the United States Air Force.

LIST OF ABBREVIATIONS

ADH	alcohol dehydrogenase
BFC	biological fuel cell
CLEA	cross-linked enzyme aggregate
CNT	carbon nanotube
DDPSE	4,4′-[(8,16-dihydro-8,16-dioxodibenzo[a, j]perylene-2,10-diyl) dioxy]dibutyric acid di(N-succinimidyl ester)
DET	direct electron transfer
DSP	dithiobis(succinimidyl) propionate
EFC	enzymatic fuel cell
FAD	flavin adenine dinucleotide
GOx	glucose oxidase
MWCNT	multiwalled carbon nanotube
NAD$^+$	nicotinamide adenine dinucleotide
PBSE	1-pyrenebutyric acid N-hydroxysuccinimidyl ester
PEI	polyethylenimine
pI	isoelectric point
PQQ	pyrroloquinoline quinone
SAM	self-assembled monolayer
SWCNT	single-walled carbon nanotube
T1	type 1

REFERENCES

1. Moehlenbrock MJ, Minteer SD. Introduction to the field of enzyme immobilization and stabilization. In: Minteer SD (ed.), *Enzyme Stabilization and Immobilization*. Humana Press, New York, 2011, pp. 1–7.
2. Osman MH, Shah AA, Walsh FC. Recent progress and continuing challenges in bio-fuel cells. Part 1. Enzymatic cells. *Biosens Bioelectron* 2011;26:3087–3102.

3. Mateo C, Palomo JM, Fernandez-Lorente G, Guisan JM, Fernandez-Lafuente R. Improvement of enzyme activity, stability and selectivity via immobilization techniques. *Enzyme Microb Technol* 2007;40:1451–1463.

4. Sheldon R. Enzyme immobilization: the quest for optimum performance. *Adv Synth Catal* 2007;349:1289–1307.

5. Brena BM, Irazoqui G, Giacomini C, Batista-Viera F. Effect of increasing co-solvent concentration on the stability of soluble and immobilized β-galactosidase. *J Mol Catal B: Enzym* 2003;21:25–29.

6. Kim J, Jia H, Wang P. Challenges in biocatalysis for enzyme-based biofuel cells. *Biotechnol Adv* 2006;24:296–308.

7. Lei C, Shin Y, Liu J, Ackerman EJ. Entrapping enzyme in a functionalized nanoporous support. *J Am Chem Soc* 2002;124:11242–11243.

8. Labus K, Gancarz I, Bryjak J. Immobilization of laccase and tyrosinase on untreated and plasma-treated cellulosic and polyamide membranes. *Mater Sci Eng C* 2012;32:228–235.

9. Kamitaka Y, Tsujimura S, Setoyama N, Najino T, Kano K. Fructose/dioxygen biofuel cell based on direct electron transfer-type bioelectrocatalysis. *Phys Chem Chem Phys* 2007;9:1793–1801.

10. Rincón RA, Lau C, Luckarift HR, Garcia KE, Adkins E, Johnson GR, Atanassov P. Enzymatic fuel cells: integrating flow-through anode and air-breathing cathode into a membrane-less biofuel cell design. *Biosens Bioelectron* 2011;27:132–136.

11. Pessela BCC, Fuentes M, Mateo C, Munilla R, Carrascosa AV, Fernandez-Lafuente R, Guisan JM. Purification and very strong reversible immobilization of large proteins on anionic exchangers by controlling the support and the immobilization conditions. *Enzyme Microb Technol* 2006;39:909–915.

12. Pessela BCC, Munilla R, Betancor L, Fuentes M, Carrascosa AV, Vian A, Fernandez-Lafuente R, Guisán JM. Ion exchange using poorly activated supports, an easy way for purification of large proteins. *J Chromatogr A* 2004;1034:155–159.

13. Kihara T, Liu X-Y, Nakamura C, Park K-M, Han S-W, Qian D-J, Kawasaki K, Zorin NA, Yasuda S, Hata K, Wakayama T, Miyake J. Direct electron transfer to hydrogenase for catalytic hydrogen production using a single-walled carbon nanotube forest. *Int J Hydrogen Energy* 2011;36:7523–7529.

14. Mateo C, Abian O, Fernandez-Lafuente R, Guisan JM. Reversible enzyme immobilization via a very strong and nondistorting ionic adsorption on support–polyethylenimine composites. *Biotechnol Bioeng* 2000;68:98–105.

15. Lee JY, Shin HY, Kang SW, Park C, Kim SW. Improvement of electrical properties via glucose oxidase-immobilization by actively turning over glucose for an enzyme-based biofuel cell modified with DNA-wrapped single walled nanotubes. *Biosens Bioelectron* 2011;26:2685–2688.

16. Petkar M, Lali A, Caimi P, Daminati M. Immobilization of lipases for non-aqueous synthesis. *J Mol Catal B: Enzym* 2006;39:83–90.

17. Shin HJ, Shin KM, Lee JW, Kwon CH, Lee S-H, Kim SI, Jeon J-H, Kim SJ. Electrocatalytic characteristics of electrodes based on ferritin/carbon nanotube composites for biofuel cells. *Sens Actuators B: Chem* 2011;160:384–388.

18. Tortolini C, Rea S, Carota E, Cannistraro S, Mazzei F. Influence of the immobilization procedures on the electroanalytical performances of *Trametes versicolor* laccase based bioelectrode. *Microchem J* 2012;100:8–13.

19. Rawal R, Chawla S, Pundir CS. Polyphenol biosensor based on laccase immobilized onto silver nanoparticles/multiwalled carbon nanotube/polyaniline gold electrode. *Anal Biochem* 2011;419:196–204.

20. Huajun Q, Caixia X, Xirong H, Yi D, Yinbo Q, Peiji G. Immobilization of laccase on nanoporous gold: comparative studies on the immobilization strategies and the particle size effects. *J Phys Chem C* 2009;113:2521–2525.

21. Nicolau E, Mendez J, Fonseca JJ, Griebenow K, Cabrera CR. Bioelectrochemistry of non-covalent immobilized alcohol dehydrogenase on oxidized diamond nanoparticles. *Bioelectrochemistry* 2012;85:1–6.

22. Sokic-Lazic D, De Andrade AR, Minteer SD. Utilization of enzyme cascades for complete oxidation of lactate in an enzymatic biofuel cell. *Electrochim Acta* 2011;56:10772–10775.

23. Betancor L, López-Gallego F, Hidalgo A, Alonso-Morales N, Fuentes M, Fernández-Lafuente R, Guisán JM. Prevention of interfacial inactivation of enzymes by coating the enzyme surface with dextran-aldehyde. *J Biotechnol* 2004;110:201–207.

24. Betancor L, López-Gallego F, Hidalgo A, Fuentes M, Podrasky O, Kuncova G, Guisán JM, Fernández-Lafuente R. Advantages of the pre-immobilization of enzymes on porous supports for their entrapment in sol–gels. *Biomacromolecules* 2005;6:1027–1030.

25. Bolivar JM, Rocha-Martin J, Mateo C, Cava F, Berenguer J, Fernandez-Lafuente R, Guisan JM. Coating of soluble and immobilized enzymes with ionic polymers: full stabilization of the quaternary structure of multimeric enzymes. *Biomacromolecules* 2009;10:742–747.

26. Deng C, Li M, Xie Q, Liu M, Yang Q, Xiang C, Yao S. Construction as well as EQCM and SECM characterizations of a novel Nafion®/glucose oxidase–glutaraldehyde/poly(thionine)/Au enzyme electrode for glucose sensing. *Sens Actuators B: Chem* 2007;122: 148–157.

27. Guerrero MP, Bertrand F, Rochefort D. Activity, stability and inhibition of a bioactive paper prepared by large-scale coating of laccase microcapsules. *Chem Eng Sci* 2011;66:5313–5320.

28. Zhang Y, Rochefort D. Activity, conformation and thermal stability of laccase and glucose oxidase in poly(ethyleneimine) microcapsules for immobilization in paper. *Process Biochem* 2011;46:993–1000.

29. Besic S, Minteer SD. Micellar polymer encapsulation of enzymes. In: *Methods in Molecular Biology*, Vol. 679. Humana Press, Clifton, NJ, 2011, pp. 113–131.

30. Kim H, Lee I, Kwon Y, Kim BC, Ha S, Lee H-J, Kim J. Immobilization of glucose oxidase into polyaniline nanofiber matrix for biofuel cell applications. *Biosens Bioelectron* 2011;26:3908–3913.

31. Tan Y, Deng W, Ge B, Xie Q, Huang J, Yao S. Biofuel cell and phenolic biosensor based on acid-resistant laccase–glutaraldehyde functionalized chitosan–multiwalled carbon nanotubes nanocomposite film. *Biosens Bioelectron* 2009;24:2225–2231.

32. Cosnier S. Biomolecule immobilization on electrode surfaces by entrapment or attachment to electrochemically polymerized films: a review. *Biosens Bioelectron* 1999;14:443–456.

33. Lim J, Cirigliano N, Wang J, Dunn B. Direct electron transfer in nanostructured sol–gel electrodes containing bilirubin oxidase. *Phys Chem Chem Phys* 2007;9:1809–1814.

34. Sarma AK, Vatsyayan P, Goswami P, Minteer SD. Recent advances in material science for developing enzyme electrodes. *Biosens Bioelectron* 2009;24:2313–2322.

35. Wang J. Sol–gel materials for electrochemical biosensors. *Anal Chim Acta* 1999;399: 21–27.
36. Ivnitski D, Artyushkova K, Rincón RA, Atanassov P, Luckarift HR, Johnson GR. Entrapment of enzymes and carbon nanotubes in biologically synthesized silica: glucose oxidase-catalyzed direct electron transfer. *Small* 2008;4:357–364.
37. Heller A. Potentially implantable miniature batteries. *Anal Bioanal Chem* 2006;385: 469–473.
38. Heller A, Feldman B. Electrochemistry in diabetes management. *Acc Chem Res* 2011;43:963–973.
39. Heller A. Electron-conducting redox hydrogels: design, characteristics and synthesis. *Curr Opin Chem Biol* 2006;10:664–674.
40. Boland S, Foster K, Leech D. A stability comparison of redox-active layers produced by chemical coupling of an osmium redox complex to pre-functionalized gold and carbon electrodes. *Electrochim Acta* 2009;54:1986–1991.
41. Boland S, Jenkins P, Kavanagh P, Leech D. Biocatalytic fuel cells: a comparison of surface pre-treatments for anchoring biocatalytic redox films on electrode surfaces. *J Electroanal Chem* 2009;626:111–115.
42. Calvo EJ, Danilowicz CB, Wolosiuk A. Supramolecular multilayer structures of wired redox enzyme electrodes. *Phys Chem Chem Phys* 2005;7:1800–1806.
43. Gallaway JW, Calabrese Barton SA. Kinetics of redox polymer-mediated enzyme electrodes. *J Am Chem Soc* 2008;130:8527–8536.
44. Guschin DA, Castillo J, Dimcheva N, Schuhmann W. Redox electrodeposition polymers: adaptation of the redox potential of polymer-bound Os complexes for bioanalytical applications. *Anal Bioanal Chem* 2010;398:1661–1673.
45. Rengaraj S, Kavanagh P, Leech D. A comparison of redox polymer and enzyme co-immobilization on carbon electrodes to provide membrane-less glucose/O₂ enzymatic fuel cells with improved power output and stability. *Biosens Bioelectron* 2011;30: 294–299.
46. Rengaraj S, Mani V, Kavanagh P, Rusling J, Leech D. A membrane-less enzymatic fuel cell with layer-by-layer assembly of redox polymer and enzyme over graphite electrodes. *Chem Commun* 2011;47:11861–11863.
47. Scodeller P, Carballo R, Szamocki R, Levin L, Forchiassin F, Calvo EJ. Layer-by-layer self-assembled osmium polymer-mediated laccase oxygen cathodes for biofuel cells: the role of hydrogen peroxide. *J Am Chem Soc* 132:11132–11140.
48. Soukharev V, Mano N, Heller A. A four-electron O₂-electroreduction biocatalyst superior to platinum and a biofuel cell operating at 0.88 V. *J Am Chem Soc* 2004;126:8368–8369.
49. Heller A. Miniature biofuel cells. *Phys Chem Chem Phys* 2004;6:209–216.
50. Luckarift HR, Ivnitski DM, Lau C, Khripin C, Atanassov P, Johnson GR. Gold-decorated carbon composite electrodes for enzymatic oxygen reduction. *Electroanalysis* 2012;24:931–937.
51. Willner B, Katz E, Willner I. Electrical contacting of redox proteins by nanotechnological means. *Curr Opin Biotechnol* 2006;17:589–596.
52. Xiao Y, Patolsky F, Katz E, Hainfeld JF, Willner I. "Plugging into enzymes": nanowiring of redox enzymes by a gold nanoparticle. *Science* 2003;299:1877–1881.

53. Willner B, Katz E, Willner I. Electrical contacting of redox proteins by nanotechnological means. *Curr Opin Biotechnol* 2006;17:589–596.

54. Patolsky F, Weizmann Y, Willner I. Long-range electrical contacting of redox enzyme by SWCNT connectors. *Angew Chem* 2004;43:2113–2117.

55. Ivnitski D, Branch B, Atanassov P, Apblett C. Glucose oxidase anode for biofuel cell based on direct electron transfer. *Electrochem Commun* 2006;8:1204–1210.

56. Katz E, Heleg-Shabtai V, Bardea A, Willner I, Rau HK, Haehnel W. Fully integrated biocatalytic electrodes based on bioaffinity interactions. *Biosens Bioelectron* 1998;13: 741–756.

57. Bardea A, Katz E, Buckmann AF, Willner I. NAD$^+$-dependent enzyme electrodes: electrical contact of cofactor-dependent enzymes and electrodes. *J Am Chem Soc* 1997;119:9114–9119.

58. Hussein L, Feng YL, Alonso-Vante G, Urban G, Kruger M. Functionalized-carbon nanotube supported electrocatalysts and buckypaper-based biocathodes for glucose fuel cell applications. *Electrochim Acta* 2011;56:7659–7665.

59. Kim BC, Zhao X, Ahn H-K, Kim JH, Lee H-J, Kim KW, Nair S, Hsiao E, Jia H, Oh M-K, Sang BI, Wang P, Kim J. Highly stable enzyme precipitate coatings and their electrochemical applications. *Biosens Bioelectron* 2011;26:1980–1986.

60. Ramasamy RP, Luckarift HR, Ivnitski DM, Atanassov PB, Johnson GR. High electrocatalytic activity of tethered multicopper oxidase–carbon nanotube conjugates. *Chem Commun* 2010;46:6045–6047.

61. Yu P, Zhou H, Cheng H, Qian Q, Mao L. Rational design and one-step formation of multifunctional gel transducer for simple fabrication of integrated electrochemical biosensors. *Anal Chem* 2011;83:5715–5720.

62. Hussein L, Rubenwolf S, von Stetten F, Urban G, Zengerle R, Krueger M, Kerzenmacher S. A highly efficient buckypaper-based electrode material for mediatorless laccase-catalyzed dioxygen reduction. *Biosens Bioelectron* 2011;26:4133–4138.

63. Lau C, Adkins ER, Ramasamy RP, Luckarift HR, Johnson GR, Atanassov P. Design of carbon nanotube-based gas diffusion cathode for O_2 reduction by multicopper oxidases. *Adv Energy Mater* 2011;2:162–168.

64. Luckarift HR, Ivnitski D, Rincón RA, Atanassov P, Johnson GR. Glucose oxidase catalyzed self-assembly of bioelectroactive gold nanostructures. *Electroanalysis* 2010;22:784–792.

65. Kim BC, Zhao X, Ahn H-K, Kim JH, Lee H-J, Kim KW, Nair S, Hsiao E, Jia H, Oh M-K, Sang BI, Kim B-S, Kim SH, Kwon Y, Gu MB, Wang P, Kim J. Highly stable enzyme precipitate coatings and their electrochemical applications. *Biosens Bioelectron* 2011;26:1980–1986.

66. Noll T, Noll G. Strategies for "wiring" redox-active proteins to electrodes and applications in biosensors, biofuel cells and nanotechnology. *Chem Soc Rev* 2011;40:3564–3576.

67. Godoy CA, Rivas BDL, Grazú V, Montes T, Guisàn JM, López-Gallego F. Glyoxyl-disulfide agarose: a tailor-made support for site-directed rigidification of proteins. *Biomacromolecules* 2011;12:1800–1809.

68. Gutarra MLE, Mateo C, Freire DMG, Torres FAG, Castro AM, Guisan JM, Palomo JM. Oriented irreversible immobilization of a glycosylated *Candida antarctica* B lipase on heterofunctional organoborane–aldehyde support. *Catal Sci Technol* 2011;1:260–266.

69. Puertas S, Moros M, Fernandez-Pacheco R, Ibarra MR, Grazu V, De La Fuente JM. Designing novel nano-immunoassays: antibody orientation versus sensitivity. *J Phys D: Appl Phys* 2010; 43.

70. Wiesbauer J, Bolivar JM, Mueller M, Schiller M, Nidetzky B. Oriented immobilization of enzymes made fit for applied biocatalysis: non-covalent attachment to anionic supports using Zbasic2 module. *ChemCatChem* 2011;3:1299–1303.

71. Lojou E. Hydrogenases as catalysts for fuel cells: strategies for efficient immobilization at electrode interfaces. *Electrochim Acta* 2011;56:10385–10397.

72. Pita M, Gutierrez-Sanchez C, Olea D, Velez M, Garcia-Diego C, Shleev S, Fernandez VM, Lacey ALD. High redox potential cathode based on laccase covalently attached to gold electrode. *J Phys Chem C* 2011;115:13420–13428.

73. Johnson DL, Martin LL. Controlling protein orientation at interfaces using histidine tags: an alternative to Ni/NTA. *J Am Chem Soc* 2005;127:2018–2019.

74. Ley C, Holtmann D, Mangold K-M, Schrader J. Immobilization of histidine-tagged proteins on electrodes. *Colloids Surf B: Biointerfaces* 2011;88:539–551.

75. Wong LS, Khan F, Micklefield J. Selective covalent protein immobilization: strategies and applications. *Chem Rev* 2009;109:4025–4053.

76. Rao SV, Anderson KW, Bachas LG. Oriented immobilization of proteins. *Mikrochim Acta* 1998;128:127–143.

77. Skorupska K, Lewerenz HJ, Smith JR, Kulesza PJ, Mernagh D, Campbell SA. Macromolecule–semiconductor interfaces: from enzyme immobilization to photoelectrocatalytical applications. *J Electroanal Chem* 2011;662:169–183.

78. Rodríguez-Argüelles MC, Villalonga R, Serra C, Cao R, Sanromán MA, Longo MA. A copper(II) thiosemicarbazone complex built on gold for the immobilization of lipase and laccase. *J Colloid Interface Sci* 2010;348:96–100.

79. Balland V, Hureau C, Cusano AM, Liu Y, Tron T, Limoges B. Oriented immobilization of a fully active monolayer of histidine-tagged recombinant laccase on modified gold electrodes. *Chemistry* 2008;14:7186–7192.

12

INTERROGATING IMMOBILIZED ENZYMES IN HIERARCHICAL STRUCTURES

MICHAEL J. COONEY

Hawai'i Natural Energy Institute, School of Ocean and Earth Science and Technology, Honolulu, HI, USA

HEATHER R. LUCKARIFT

Universal Technology Corporation, Dayton, OH, USA; Airbase Sciences Branch, Air Force Research Laboratory, Tyndall Air Force Base, FL, USA

12.1 INTRODUCTION

Many of the significant advances in improving the performance of enzymatic fuel cells can be attributed to the introduction of rationally designed electrode materials, particularly the incorporation of nanoscale materials in macroscale architectures in order to improve electron transfer (ET) from the biocatalyst to the electrode [1–9]. For enzymatic fuel cell electrodes, protein interaction and orientation at the nanoscale is imperative, as the enzyme must be positioned in such a way that its redox center can transfer electrons to the transducer surface [10–15]. Because the aspect ratio of a nanomaterial approaches the molecular scale, redox proteins can establish a close and direct association with the material and effectively decrease the electron tunneling distance. Carbon nanotubes (CNTs), for example, have dimensions that are uniquely amenable to close physical association with proteins and can facilitate direct electronic interactions with redox catalysts [14]. The incorporation of CNTs into

Enzymatic Fuel Cells: From Fundamentals to Applications, First Edition. Edited by Heather R. Luckarift, Plamen Atanassov, and Glenn R. Johnson.
© 2014 John Wiley & Sons, Inc. Published 2014 by John Wiley & Sons, Inc.

electrode architecture can significantly improve the conductivity of the matrix and the effective surface area [1,3,4,8,16,17].

Redox catalysts, for example, can be stabilized by encapsulation in silica sol–gels, but the conductivity of silica is inherently low. Co-immobilization of a conductive material, such as CNTs, can therefore provide electrical connectivity between the enzyme catalysts and the electrode, while providing additional surface area for enzyme immobilization. The biocatalyst stability typically improves as a result of immobilization, and the porous silica matrix will facilitate improved mass transfer kinetics [16,18].

Often, the principle of including conductive nanoparticles is to impart conductivity to the bulk three-dimensional matrix. The resulting hierarchical structure then imparts an increased active surface area that typically correlates to an increased electrochemical conversion rate. Since the volumetric catalytic activity of proteins is typically low, it is important to design high surface area materials that allow for high enzyme loading and retain a high surface area to volume (or mass) ratio. An idealized electrode material will exhibit an architecture that provides mechanical integrity but retains macroscale porosity to allow fluid flow.

The most widely used materials are carbon felts, carbon papers, and conductive gels and foams that provide bulk integrity, particularly for use in flow-through fuel cell designs. These macroporous materials, however, provide little internal surface area for the immobilization of biocatalysts. In this respect, the integration of nanomaterials adds an additional component of microporous and nanoscale architecture. A multiscale bioelectrocatalytic composite, for example, was developed in 2011 by Bon Saint Come et al. for oxidation of D-sorbitol by incorporating D-sorbitol dehydrogenase and diaphorase into an electrodeposited paint modified with gold nanoparticles [2]. A further example of such integration is the direct grafting of nanomaterials onto open-pore structured substrates, such as the growth of CNT on carbon paper, or the incorporation of preformed CNT into polymers such as chitosan [17,19]. The resulting composite structures allow substantial enzyme loading and promote the desired nanomaterial–biocatalyst interaction.

Recent research has thus been directed toward building hierarchically structured materials that incorporate three levels of scale, that is, macroscale porosity to allow for convective flow and fuel delivery, mesoscale architecture designed to integrate materials properties, and nanomaterials, such as CNTs or gold nanoparticles, to aid protein interactions (and flow of gaseous reactants) with microporous components. The mesoporous aspect of such a composite matrix is usually responsible for interconnectivity and thus ensures the electrical conductivity of the matrix. In 2008, Lau et al. formed a conductive composite of CNTs in chitosan onto reticulated vitreous carbon that serves as a conductive macroporous scaffold. The chitosan polymer matrix develops its own porosity through a freeze-drying process and, when optimized for CNT content, demonstrates substantial conductivity and provides enhanced surface area for immobilization of biocatalyst (Figure 12.1) [17]. Such hierarchically structured electrodes facilitate the fabrication of scalable enzyme anodes when modified with immobilized biocatalysts, such as oxidases or dehydrogenases [20–23].

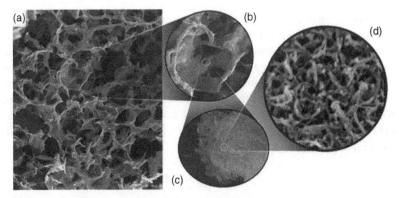

FIGURE 12.1 Schematic showing the levels of hierarchical porosity created in an electrode architecture consisting of (a) reticulated vitreous carbon coated with (b, c) a chitosan composite containing (d) a matrix of multiwalled CNTs. (Reproduced with permission from Ref. [17]. Copyright 2008, American Chemical Society.)

12.2 ESTIMATING THE BOUND ACTIVE (REDOX) ENZYME

With the introduction of hierarchical electrode architecture, it becomes necessary to develop unique tools that allow rational design based on the loading and interaction of proteins within architecture. When evaluating immobilized enzymes in hierarchical structures, it is important to determine the extent of electrobiocatalysis and whether (i) mass transport, (ii) enzyme binding, or (iii) charge transfer efficiency are responsible for limitations in electrode performance. Mass transport can be characterized by integrating experimental data with models that predict enzyme kinetics (determined as electrons released to the electrode and measured electrochemically) and by incorporating theoretical parameters that account for diffusion [24]. Enzyme binding, as measured individually, is generally determined by estimating the amount of enzyme remaining in solution after the immobilization process. Quantifying enzyme concentration is usually accomplished through (i) spectrophotometric absorbance measurements, (ii) chemical dyes that target specific proteins, or (iii) digestion methods that estimate protein concentration based on measurement of total nitrogen.

None of these techniques, however, are particularly accurate per se as they ultimately rely on a statistical difference between two measurements, that is, the quantity of enzyme in solution before, and after, the immobilization process. Generally, the amount of immobilized enzyme is small relative to the amount of enzyme in bulk solution and thus the mathematical subtraction has significant statistical error. There is also a problem that even an accurate measurement of immobilized enzyme may not reflect the percentage of enzyme that remains active once immobilized, and more importantly for electrocatalysis, the percentage of immobilized enzyme that is positioned preferentially to transfer electrons to the electrode. To address this issue, a more definitive measurement is one that calculates the charge transfer efficiency of bound enzyme.

Methods to calculate such charge transfer efficiency of bound enzymes have recently been reported [25]. One first calculates a theoretical maximum in current density based on enzyme loading, determined from spectrophotometric measurements (assuming perfect charge transfer efficiency), and then compares the value with the actual measured current density, obtained from direct current (DC) polarization experiments. For example, the direct absorption of *Pyrococcus furiosus* hydrogenase I onto carbon-based electrodes was used as a model redox enzyme system, and Faraday's law was used to calculate the maximum theoretical current density as follows (Equation 12.1):

$$I = nF \frac{dH/dt}{A} \tag{12.1}$$

where I is the current density (mA cm^{-2}), dH/dt is the substrate (i.e., H for hydrogen) with an oxidation rate (mmol s^{-1}), A is the electrode geometric surface area (cm^2), F is the Faraday constant (96,485 C mol^{-1}), and n is the moles of electrons released per mole of hydrogen oxidized. The hydrogen oxidation rate per unit electrode surface area $(dH/dt)/A$ was then determined by calculating the activity per unit electrode surface area (U cm^{-2}) based on the definition of hydrogenase enzyme activity (1 unit of enzyme activity is equal to 1 μmole of hydrogen oxidized per minute). The activity (U) can be determined directly as the measured activity per unit volume of enzyme solution in a spectrophotometric assay (Equation 12.2):

$$U = \frac{\Delta C}{\Delta t} \frac{1}{2} = \frac{\Delta A}{\Delta t} \frac{1}{\varepsilon l} \tag{12.2}$$

where $\Delta A/\Delta t$ is the slope of the measured absorbance change (e.g., ΔAbs_{580} of methyl viologen to detect hydrogenase activity) [25], and ε and l are the extinction coefficient and path length, respectively. The charge transfer efficiency of bound enzyme is then taken as the ratio of the actual measured current density to the theoretical maximum, expressed as a percentage.

In 2005, Johnston et al. used the equation for U (Equation 12.2) to calculate a loading density of 0.034 U cm^{-2}, which correlated to a maximum current density of approximately 110 μA cm^{-2} (using the equation for I; see Equation 12.1) [25]. Comparing the theoretical value with the measured current density, determined using DC polarization (~30 μA cm^{-2}), indicates a charge transfer efficiency of 28% for the electrode. This value is in line with electrodes previously demonstrated with glucose oxidase (GOx) in a polymeric base, for which a charge transfer efficiency of 20% is reported [26]. Using the calculations described, the authors were able to confirm that the *P. furiosus* hydrogenase was suitable for mediatorless bioelectrocatalysis, a conclusion that can be difficult to establish by other means. The study also revealed that the loading density of the enzyme limited the current density, thus indicating that the immobilization efficiency caused by physical adsorption was limiting the overall electrode performance. This type of information is critical in the specific design of enzyme electrodes and helps focus the direction of future designs.

Techniques such as dynamic potentiometry, DC polarization, and electrochemical impedance spectroscopy can all provide valuable insight into the performance of redox enzyme electrodes but can provide information only on enzymes that are electrochemically active, that is, actively exchanging charge with the electrode support. Conversely, spectrophotometric enzyme assays allow quantification of total bound enzyme, whether active or not. This is a critical distinction, because in the absence of charge transfer mediators, a redox enzyme must form a close physical attachment to the electrode for direct bioelectrocatalysis to occur; as discussed above, not all enzymes will be suitably orientated for effective ET. Hence, it is useful to combine measurements of enzyme capable of charge transfer (obtained via DC polarization) with determination of total enzyme loading (obtained via spectrophotometric enzyme assay) to calculate the efficiency of charge transfer between enzyme and support. Charge transfer efficiency, for example, has been calculated for GOx electrodes by directly measuring the enzyme cometabolite, H_2O_2 (via platinum rotating disk electrode) [26]. Certain enzymes, however, such as hydrogenases do not form a suitable cometabolite to make such a direct method possible, and in these cases the cumulative method discussed herein becomes imperative.

12.2.1 Modeling the Performance of Immobilized Redox Enzymes in Flow-Through Mode to Estimate the Concentration of Substrate at the Enzyme Surface

The performance of enzyme electrodes—particularly those with three-dimensional porous architecture—is affected by both enzyme kinetics and mass transfer to, and through, the electrode architecture. Understanding these mechanisms requires an approach that models the reactant concentration at the enzyme surface based on bulk concentrations, which is measured *in situ* by online spectrophotometric detection or *ex situ* by liquid chromatography. In this approach, the mass transfer characteristics of the electrodes are estimated by comparison of experimental data with model predictions of system performance in the presence of continuous flow. Thereafter, a combined mass transfer parameter can be calculated and enzyme kinetics can be evaluated. The primary advantages of this approach are to rapidly screen immobilization protocols and to provide some insight into mass transport of reactants, metabolic by-products, and electrons. Specifically, it is possible to estimate the actual V_{max} of the electrode, and hence quantify the amount of active enzyme available for bioelectrocatalysis.

In 2006, Johnston et al. provided a derivation of the Monod equation that relates measurable parameters of flow rate (F), concentration of NADH (reduced form of nicotinamide adenine dinucleotide) in bulk solution above the immobilization matrix $(NADH_b)$, and feed concentration of NAD^+ (NAD_{res}^+), by correlating the molecular diffusion coefficient for NAD^+ (M_{NAD^+}), with the maximum velocity $(V_{max}^{NAD^+})$ and

Monod constant ($K_M^{NAD^+}$) (Equation 12.3) [24]:

$$F \cdot NADH_b = \frac{V_{max}^{NAD^+} \cdot \left(NAD_{res}^+ - NADH_b - \dfrac{F \cdot NADH_b}{M_{NAD^+}} \right)}{K_M^{NAD^+} + \left(NAD_{res}^+ - NADH_b^+ - \dfrac{F \cdot NADH_b}{M_{NAD^+}} \right)} \tag{12.3}$$

This equation represents a useful model for interpreting data from flow experiments on immobilized enzyme electrodes (in this case, alcohol dehydrogenase) in which an electron mediator (i.e., NAD^+) is limiting and the substrate (i.e., ethanol) is present in excess. In this scenario, bulk NAD^+ is pumped to the electrode and its reduced counterpart (NADH) is measured spectrophotometrically in the outflow. Specifically, half of the equation is measured experimentally and the remainder is plotted in the form of a Lineweaver–Burk plot, wherein a value for the combined mass transfer parameter M_{NAD^+} is assumed. As the equation has three unknowns, $V_{max}^{NAD^+}$ ($\mu mol\, min^{-1}$), $K_M^{NAD^+}$ (mM), and M_{NAD^+} ($cm^3\, min^{-1}$), a value is selected for M_{NAD^+} and the data then plotted in Lineweaver–Burk form to solve for $V_{max}^{NAD^+}$ and $K_M^{NAD^+}$, as per their intercepts [24]. Specific values are arbitrarily dictated by the choice of M_{NAD^+} and are therefore potentially biased. If, however, the assumption is made that the value for $K_M^{NAD^+}$ will not change from the K_M measured for the enzyme in its aqueous and nonimmobilized state, then a value of M_{NAD^+} can be selected to yield a value for $K_M^{NAD^+}$ that is equivalent to the value determined experimentally in free solution.

With these assumptions in hand, interpretation of real assay data involves plotting a model-derived value for concentration of NAD^+ at the enzyme surface (NAD_e^+). The value for the M_{NAD^+} can be fitted to allow the Lineweaver–Burk plot to intercept the x-axis at a value that yields the value of K_M as determined in solution. The value for V_{max} is then read as the intercept at the y-axis (Figure 12.2). This approach permits derivation of a V_{max} for the electrode that is independent of the effects of mass transfer. If one further assumes that the immobilization process does not affect the turnover rate of the immobilized enzyme (relative to its activity in solution), then this value of V_{max} (which represents the total activity of all bound enzyme) can also be used to estimate the amount of immobilized enzyme. This model can be particularly useful when fabricating electrodes using immobilization techniques that entrap a fraction of enzyme from bulk solutions, such as direct physical absorption or co-immobilization within gels.

Although the model provides a mechanism to measure the total activity of enzyme immobilized on an electrode, it also allows estimation of the combined mass transfer parameter (M_{NAD^+}) for the diffusing substrate (in this case, NAD^+). This value, however, is limited somewhat by the fact that the term is cumulative for several variables that may individually contribute to overall electrode mass transfer: combined contributions of the molecular diffusion coefficient, effective surface area, diffusion length, pore structure, and mass transfer through the film. Unfortunately, this approach will remain limited unless additional modeling is incorporated that can specifically differentiate these parameters. If more parameters are to be added to the

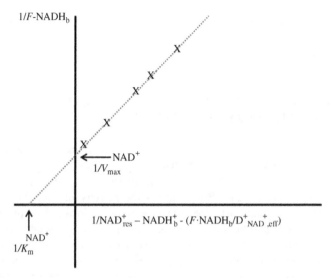

FIGURE 12.2 Graphical distribution of the Lineweaver–Burk solution to Equation 12.3.

model, however, the number of variables that can be measured becomes limited (so as to not leave the equation unsolvable).

Nevertheless, the approach has been successfully demonstrated to interpret assay data for two representative electrode types fabricated with different immobilization methods: (i) direct physical adsorption to a carbon felt electrode, and (ii) enzyme immobilized within polypyrrole electrochemically deposited on a stainless steel mesh. In each case, the data were used to estimate both V_{max} and the cumulative mass transfer parameter M (Table 12.1).

The V_{max} and effective diffusion coefficient (D'_{eff}) were both found to be higher for enzyme immobilization via direct adsorption than for immobilization with poly-pyrrole, indicating that the codeposition of enzymes in the pyrrole monomer created an architecture that hindered enzyme loading or mass transport of substrate. Further

TABLE 12.1 **Model Kinetic Parameters as a Function of Flow Rate for Immobilized Enzyme Electrodes [24]**

Flow rate, F (ml min^{-1})	K_M (mM)	V_{max} (μmol min^{-1})	D'_{eff} (cm^3 min^{-1})
	Enzyme directly absorbed to carbon felt		
1.6	0.0501	0.0636	0.91
1.6	0.0500	0.0539	0.97
2.4	0.0490	0.0684	0.68
	Enzyme immobilized in electrochemically deposited polypyrrole		
0.27	0.057	0.0025	0.26

D'_{eff}: effective diffusion coefficient.

corroboration of this hypothesis was provided by high-resolution images of the polypyrrole pore structure that showed a film structure too dense to permit adequate diffusion of substrates to the internal pores [24].

12.3 PROBING THE DISTRIBUTION OF IMMOBILIZED ENZYME WITHIN HIERARCHICAL STRUCTURES

The approaches discussed above provide methods to characterize the activity and charge transfer efficiency of an immobilized enzyme, either through the coupling of electrochemical and spectrophotometric measurements or through model-based analysis of experimental data. Both address the activity of immobilized enzyme but give little insight into why performance may be lower than the theoretical maximum. This is in part because the techniques do not directly address the spatial distribution of enzyme or the chemical microenvironments within which proteins are immobilized. In terms of rational design, it would therefore be useful to understand how electrocatalytic performances (as measured by charge transfer efficiency or enzyme kinetics) may be optimized in the context of the immobilization matrix used. The methodologies to achieve this goal currently rely primarily on the use of fluorescent probes, either freely distributed throughout the electrode matrix or tethered to enzymes entrained within the matrix. This allows a visualization of spatial distribution, aggregation state, and chemical microenvironment of the immobilized enzyme, which combined can provide a powerful tool to guide modification of the immobilization matrix.

Fluorescence microscopy can be used to track the distribution of enzyme within an immobilization matrix, particularly if the electrode architecture is a three-dimensional macroporous scaffold. Specifically, fluorophores can be tethered to a protein (e.g., Alexa Fluor 546) or included as fluorescent dyes (e.g., Nile red), added to the immobilization material, and excited at the appropriate wavelength for visualization after the immobilization procedure. Malate dehydrogenase (MDH), for example, was encapsulated in 2009 by Martin et al. in chitosan polymer, either in its native state or following hydrophobic modification by formation of an amide with butyric acid (butyl-modified) or α-linoleic acid (ALA) [27].

Hydrophobically modified chitosan is expected to be enriched in amphiphilic micelles introduced through the hydrophobic modification of the polymer prior to its molding into scaffolds (due to the modification of the native deacetylated chitosan with hydrophobic side groups along a hydrophilic chitosan backbone). The amphiphilic micelles encapsulate and retain immobilized enzyme within a unique chemical microenvironment [28], after being processed into three-dimensional scaffolds by thermally induced phase separation (TIPS) [29,30]. Using fluorescent dyes added to chitosan, as well as tagged enzymes, one can visualize the degree of amphiphilicity and distribution of tagged enzyme throughout the polymer scaffold, and quantitative data can be derived via fluorescence confocal laser scanning microscopy.

Fluorescence can also provide unique information on the chemical microenvironment surrounding the encapsulated enzyme. For example, Martin et al.

confirmed the distribution of amphiphilic micelles in 2009 through the use of a polar probe, Nile red, whose emission spectrum varies depending on the polarity of the immediate chemical microenvironment [22]. Nile red was added to the starter solutions of modified chitosan before processing by TIPS to fabricate highly porous, three-dimensional scaffolds [22,29,30]. The dried chitosan scaffolds were then imaged by fluorescence microscopy using an excitation wavelength specific for the Nile red fluorophores. The total emitted light and total emitted light below 620 nm were measured per unit pixel such that each pixel in a confocal image (512×512) represented a single data point.

The results (exemplified in Figure 12.3) are observed as the distribution of emission intensity <620 nm to total emission intensity. Emissions below 620 nm generally represent Nile red emission in hydrophobic chemical microenvironments. The total emission, by comparison, represents the total sum of Nile red emission over all wavelengths, that is, all chemical microenvironments ranging from highly polar to highly nonpolar. Consequently, the increase in the average intensity ratio from 0.22 to 0.26–0.28 for the modified chitosan polymers (relative to native deacylated chitosan) quantitatively suggests an increase in the hydrophobic nature of the chemical microenvironments provided by the amphiphilic micelles (Figure 12.3). The authors concluded that an increase in relative hydrophobic contribution to the amphiphilic nature of the micelles was likely the reason for the modified polymer's reported

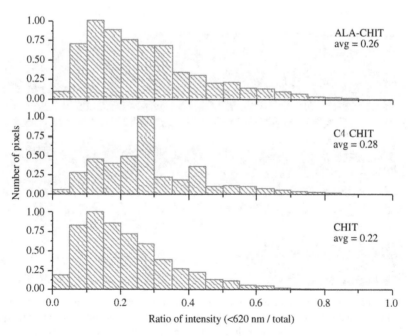

FIGURE 12.3 Distribution ratio of the emission intensity below 620 nm to the total emission intensity (each data point represents a single pixel in a 512×512 confocal image). (Reproduced with permission from Ref. [27]. Copyright 2009, American Chemical Society.)

capacity for greater enzyme stabilization [29,31]. The amphiphilic micelles are proposed to immobilize the enzymes by encapsulation, such that enzymes become entrained within the polymer during fabrication (TIPS/freeze drying). In other words, the nature of the immobilization polymer and the process by which the immobilization matrix was fabricated would have some influence on the final distribution of enzyme.

MDH (fluorescent labeled; Alexa Fluor 546) was thus prepared using a range of polymer formulations and the chitosan scaffolds were fabricated using TIPS (Figure 12.4) [29,32]. Figure 12.4a presents the spatial distribution of enzyme immobilized within a scaffold fabricated from native chitosan. The distributions of chitosan (green) and fluorescent-tagged enzyme (purple) indicate that the enzymes are predominantly located at the top of the scaffold, which is attributed to absorption of the enzyme at the surface of the precipitate. In contrast, the distribution of MDH in butyl-modified chitosan scaffolds shows a more homogenous distribution throughout the scaffold (Figure 12.4b). The same spatial distribution of enzyme is evident in the ALA-modified chitosan scaffold (Figure 12.4c) and appears to be significantly more distributed throughout the matrix than unmodified chitosan. The visual results confirm the hypothesis that the amphiphilic micelles created by the fatty acid ALA side chains would be likely to encapsulate and retain MDH and agree with observations that such fatty acids have been shown to promote incorporation and

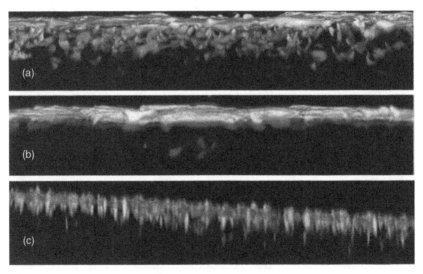

FIGURE 12.4   (a) Enzyme distribution in native chitosan scaffolds stained with fluorescein (green) combined with fluorescent (Alexa Fluor 546)-stained malate dehydrogenase (purple). (b) Enzyme distribution in butyl-modified chitosan scaffolds (same staining). (c) Enzyme distribution in ALA-modified chitosan scaffolds (same staining). (Reproduced with permission from Ref. [27]. Copyright 2009, American Chemical Society.) (Please see the color version of this figure in Color Plates section.)

stabilization to MDH in solution-based micelles [33]. Combined, these research tools permit a more accurate correlation of the effectiveness of enzyme immobilization to measured electrochemical performance of biocatalytic electrodes fabricated in multiple dimensions.

12.4 PROBING THE IMMEDIATE CHEMICAL MICROENVIRONMENTS OF ENZYMES IN HIERARCHICAL STRUCTURES

An important factor related to the activity of an immobilized enzyme is the chemical environment immediately surrounding the protein. In the previous section, we discussed the infusion of a free probe into an immobilization matrix to assay the relative hydrophobicity of the immobilization matrix. In general, it is useful to confirm that modification of the polymer prior to its formation into an immobilization matrix introduced no essential change in chemistry (i.e., did not make the matrix relatively more hydrophobic or hydrophilic). What the technique does not do, however, is definitively probe the chemical microenvironment immediately surrounding the enzyme.

Fluorescence is an excellent tool to probe chemical microenvironments. For example, in 2010, Martin et al. used a specific shift in fluorescence emission spectra of polar sensitive fluorophores, for example, acrylodan, to characterize the polarity immediately surrounding immobilized enzymes [34]. Cytoplasmic malate dehydrogenase (cMDH) and mitochondrial malate dehydrogenase (mMDH), for example, were tagged with fluorescent probes and then entrapped within macroporous three-dimensional chitosan scaffolds. A blue shift in the acrylodan emission maximum was taken to indicate a polar shift in the chemical microenvironment directly surrounding enzymes immobilized within the scaffold (Figure 12.5). The emission

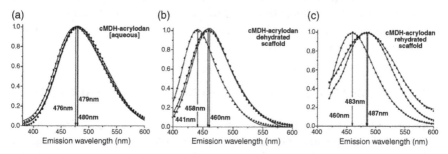

FIGURE 12.5   (a) Acrylodan-labeled cMDH in aqueous native and modified chitosan. cMDH in native and modified chitosan scaffolds: (b) dehydrated and (c) rehydrated in buffer. In all panels, modified polymers are represented by the following key: (■) chitosan; (●) ALA-modified chitosan; (▲) butyl-modified chitosan. All samples were excited at $\lambda_{ex} = 360$ nm. (Reproduced with permission from Ref. [34]. Copyright 2010, Royal Society of Chemistry.)

peak for acrylodan-tagged cMDH does not appear to change between aqueous solution and immobilization in native or modified chitosan polymers (Figure 12.5a). The results indicate that the chitosan polymers (in solution form) provide chemical microenvironments of the same polarity as the solvent, that is, neither more nor less hydrophobic. The data also suggest that the chemical environment surrounding the tagged enzyme is dominated by water and not polymer. In contrast, when the tagged enzymes were immobilized within the same chitosan polymers, the emission maximum not only shifts to a lower wavelength, but also shows significant variation across differently functionalized polymers (Figure 12.5b). Although the observed overall blue shift in the dried polymer could be attributed partially to the dehydration process (during freeze drying), removal of water molecules surrounding the immobilized enzyme would be expected to lead to a reduction in polarity. In reality, this could be viewed as a false positive, as the real working environment of the immobilized enzyme would be an aqueous environment when the immobilization matrix is resuspended in solution. Accordingly, rehydration of the scaffolds in buffer solution was investigated (Figure 12.5c). Although the rehydrated films revealed a trend toward more polar chemical microenvironments, the same trend in blue shift was retained across all forms of the polymer. The authors used this observation to conclude that the modified polymer did impact the chemical microenvironment of the enzyme when in the immobilized state, thus corroborating two separate theories: (1) modified chitosan polymers can provide altered chemical microenvironments [22,29,30,35], or (2) alteration of chemical microenvironments is most pronounced after the polymer is precipitated through freezing or drying processes into its final structure [22].

Although these measurements provide useful information individually, their full utility is achieved only if they can be correlated to performance. In the examples discussed above, MDH was immobilized within native, butyl-modified, and ALA-modified chitosan in the presence of poly(methylene green) and tested for performance in half-cell mode by amperometry. The electrochemical results confirm that the butyl-modified chitosan yields a significant (10-fold) increase in current density, compared with a modest (4-fold) increase for the ALA-modified polymer compared with unmodified native chitosan (Figure 12.6). These results provide further evidence that the activity of the immobilized is consistent with the chemical microenvironment, particularly in respect to polarity.

12.5 ENZYME AGGREGATION IN A HIERARCHICAL STRUCTURE

In the examples cited above, electrochemical performance of immobilized enzyme was correlated to polarity of the chemical microenvironment surrounding the immobilized enzyme. Whereas the amphiphilic nature of the chemical micro-environment was confirmed and correlated to increased activity, the mechanism for the increased activity was not elucidated [21,22]. Specifically, the question remains whether the increased activity was a product of favorable enzyme aggregation occurring as a result of forces provided by the modified polymer during the

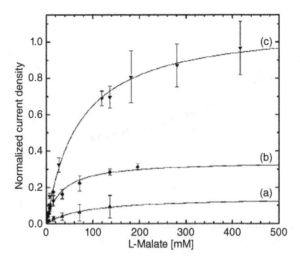

FIGURE 12.6 Electrochemical activity represented as normalized current density for immobilized MDH in different chitosan polymers: (a) native chitosan; (b) ALA-modified chitosan; (c) butyl-modified chitosan. (Reproduced with permission from Ref. [34]. Copyright 2010, Royal Society of Chemistry.)

immobilization process, or due strictly to the presence of a more amphiphilic chemical microenvironment surrounding enzymes in a nonaggregated structure.

To address this question, Förster resonance energy transfer (FRET) can be used to measure the degree of enzyme aggregation both prior to and after the immobilization process [36]. FRET is a mechanism that describes the energy transfer between two fluorophores; specifically, when the donor fluorophore and the acceptor fluorophore are within a minimum radius, R_0, the donor fluorophore, initially in its electronic excited state, may transfer energy to the acceptor fluorophore through nonradiative dipole–dipole coupling [37,38]. The measured intensity, as a function of emission wavelength, can be used to calculate the FRET efficiency, a term that can subsequently be used to estimate the distance between the two fluorophores. In particular, if those fluorophores are attached to separate enzymes, this measurement can be used to estimate the distance between proteins and hence estimate whether or not the proteins have aggregated.

Martin et al. for example, explored the general aggregation state of protein in chitosan and butyl-modified chitosan, both in aqueous solution and when immobilized in the scaffold (Figure 12.7) [36]. A constant radius of separation between enzymes with respect to time suggests a fixed state of aggregation. Reduction in the average distance of separation for proteins immobilized in chitosan ($r = 45.9 \pm 0.1$ nm) and butyl-modified chitosan ($r = 41.8 \pm 0.3$ nm), respectively, verifies protein aggregation during immobilization. The data indicate that the immobilization process induced aggregation, and suggest that this effect may be controlled by specific hydrophobic modification of the chitosan polymer. This observation may guide methods to induce protein complex formation in an immobilized state, a technique

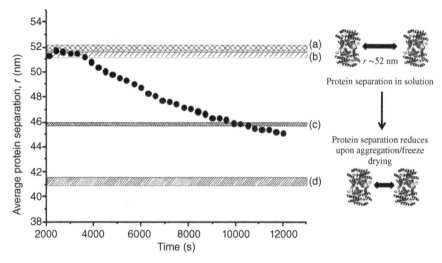

FIGURE 12.7 Aggregation of malate dehydrogenase following chitosan immobilization. Dotted line represents glutaraldehyde-induced aggregation (control: calculated radial separation for a solution of 50% Alexa Fluor 555-tagged cMDH and 50% Alexa Fluor 647-tagged cMDH in 600 mM glutaraldehyde, 50 mM Tris buffer, pH 7.4). Solid bars represent the average radial separation for an identical enzyme mixture dissolved in (a) aqueous chitosan, (b) aqueous butyl-modified chitosan, (c) freeze-dried chitosan, and (d) freeze-dried butyl-modified chitosan scaffolds. (Modified with permission from Ref. [36]. Copyright 2010, Royal Society of Chemistry).

that may find particular utility in combining sequential enzymes for oxidation of complex substrates.

12.6 OUTLOOK

By using a combination of well-defined analytical techniques, such as spectro-photometric enzyme assays, mass transport modeling, and electrochemical measure-ments, a complete characterization of the "true" electrode performance can be obtained. Specifically, determining the immobilization efficiency, loading density, and charge transfer efficiency of bound enzyme helps identify bottlenecks in electrode performance. Recent progress in the fabrication of three-dimensional hierarchical architectures has been rapid, but to date it has not been mirrored by an in-depth understanding of how specific materials characteristics contribute to incremental improvements in bioelectrocatalysis and electrode performance. Techniques to characterize three-dimensional architectures on the detailed level described herein are at the forefront of rationale design for bioelectrochemical interfaces.

Specific visualization of the spatial distribution of biocatalysts within a matrix and the corresponding catalytic activity and electrochemical kinetics permits an accurate correlation of the effectiveness of specific immobilization techniques with the choice of electrode scaffold. As materials engineering continues to provide unique materials that may contribute to advances in electrode design, understanding is critical to fully harness bioelectrocatalysis in realistic biological fuel cell devices. The techniques described could be readily extended to multienzyme systems in which individual enzymes modified with specific fluorophores can be visualized by careful choice of emission wavelengths in confocal fluorescence microscopy. Although the examples discussed are primarily focused on polymer matrices, the methodologies could readily be extended to inorganic structures (such as sol–gels), aerogels, or alternative porous substrates.

ACKNOWLEDGMENT

This research was prepared as an account of work sponsored by the United States Air Force Research Laboratory, Materials and Manufacturing Directorate, Airbase Technologies Division (AFRL/RXQ), but the views of authors expressed herein do not necessarily reflect those of the United States Air Force.

LIST OF ABBREVIATIONS

ALA	α-linoleic acid
cMDH	cytoplasmic malate dehydrogenase
CNT	carbon nanotube
DC	direct current
ET	electron transfer
FRET	Förster resonance energy transfer
GOx	glucose oxidase
MDH	malate dehydrogenase
mMDH	mitochondrial malate dehydrogenase
NAD	nicotinamide adenine dinucleotide
NADH	nicotinamide adenine dinucleotide, reduced form
TIPS	thermally induced phase separation

REFERENCES

1. Baravik I, Tel-Vered R, Ovits O, Willner I. Electrical contacting of redox enzymes by means of oligoaniline-cross-linked enzyme/CNT composites. *Langmuir* 2009;25: 13978–13983.

2. Bon Saint Come Y, Lalo H, Wang Z, Etienne M, Gajdzik J, Kohring G-W, Walcarius A, Hempelmann R, Kuhn A. Multiscale-tailored bioelectrode surfaces for optimized catalytic conversion efficiency. *Langmuir* 2011;27:12737–12744.

3. Hussein L, Feng YL, Alnoso-Vante G, Urban G, Kruger M. Functionalized-CNT supported electrocatalysts and buckypaper-based biocathodes for glucose fuel cell applications. *Electrochim Acta* 2011;56(22):7659–7665.

 4. Hussein L, Rubenwolf S, von Stetten F, Urban G, Zengerle R, Krueger M, Kerzenmacher S. A highly efficient buckypaper-based electrode material for mediatorless laccase-catalyzed dioxygen reduction. *Biosens Bioelectron* 2011;26:4133–4138.

 5. Kim BC, Zhao X, Ahn H-K, Kim JH, Lee H-J, Kim KW, Nair S, Hsiao E, Jia H, Oh M-K, Sang BI, Wang P, Kim J. Highly stable enzyme precipitate coatings and their electrochemical applications. *Biosens Bioelectron* 2011;26:1980–1986.

 6. Kim J, Jia H, Wang P. Challenges in biocatalysis for enzyme-based biofuel cells. *Biotechnol Adv* 2006;24:296–308.

 7. Kuwahara T, Ohta H, Kondo M, Shimomura M. Immobilization of glucose oxidase on carbon paper electrodes modified with conducting polymer and its application to a glucose fuel cell. *Bioelectrochemistry* 2008;74:66–72.

 8. Nazaruk E, Sadowska K, Biernat JF, Rogalski J, Ginalska G, Bilewicz R. Enzymatic electrodes nanostructured with functionalized CNTs for biofuel cell applications. *Anal Bioanal Chem* 2010;398:1651–1660.

 9. Sarma AK, Vatsyayan P, Goswami P, Minteer SD. Recent advances in material science for developing enzyme electrodes. *Biosens Bioelectron* 2009;24:2313–2322.

10. Calabrese Barton S, Gallaway J, Atanassov P. Enzymatic biofuel cells for implantable and microscale devices. *Chem Rev* 2004;104:4867–4886.

11. Mano N, Soukharev V, Heller A. A laccase-wiring redox hydrogel for efficient catalysis of O_2 electroreduction. *J Phys Chem B* 2006;110:11180–11187.

12. Noll T, Noll G. Strategies for 'wiring' redox-active proteins to electrodes and applications in biosensors, biofuel cells and nanotechnology. *Chem Soc Rev* 2011;40: 3564–3576.

13. Rubenwolf S, Strohmeier O, Kloke A, Kerzenmacher S, Zengerle R, von Stetten F. Carbon electrodes for direct electron transfer type laccase cathodes investigated by current density–cathode potential behavior. *Biosens Bioelectron* 2010;26: 841–845.

14. Willner B, Katz E, Willner I. Electrical contacting of redox proteins by nanotechnological means. *Curr Opin Biotechnol* 2006;17:589–596.

15. Xiao Y, Patolsky F, Katz E, Hainfeld JF, Willner I. "Plugging into enzymes": nanowiring of redox enzymes by a gold nanoparticle. *Science* 2003;299:1877–1881.

16. Ivnitski D, Artyushkova K, Rincón RA, Atanassov P, Luckarift HR, Johnson GR. Entrapment of enzymes and CNTs in biologically synthesized silica: glucose oxidase-catalyzed direct electron transfer. *Small* 2008;4:357–364.

17. Lau C, Cooney MJ, Atanassov P. Conductive macroporous composite chitosan–CNT scaffolds. *Langmuir* 2008;24:7004–7010.

18. Lim J, Cirigliano N, Wang J, Dunn B. Direct electron transfer in nanostructured sol–gel electrodes containing bilirubin oxidase. *Phys Chem Chem Phys* 2007;9:1809–1814.

19. Ivnitski D, Branch B, Atanassov P, Apblett C. Glucose oxidase anode for biofuel cell based on direct electron transfer. *Electrochem Commun* 2006;8:1204–1210.

20. Cooney MJ, Svoboda V, Lau C, Martin G, Minteer SD. Enzyme catalyzed biofuel cells. *Energy Environ Sci* 2008;1:320–337.

21. Lau C, Martin G, Minteer SD, Cooney MJ. Development of a chitosan scaffold electrode for fuel cell applications. *Electroanalysis* 2009;22:793–798.

22. Martin G, Ross JA, Minteer SD, Jameson DM, Cooney MJ. Fluorescence characterization of chemical microenvironments in hydrophobically modified chitosan. *Carbohydr Polym* 2009;77:695–702.

23. Tan Y, Deng W, Ge B, Xie Q, Huang J, Yao S. Biofuel cell and phenolic biosensor based on acid-resistant laccase–glutaraldehyde functionalized chitosan–multiwalled CNTs nanocomposite film. *Biosens Bioelectron* 2009;24:2225–2231.

24. Johnston W, Maynard N, Liaw B, Cooney MJ. *In situ* measurement of activity and mass transfer effects in enzyme immobilized electrodes. *Enzyme Microb Technol* 2006;39:131–140.

25. Johnston W, Cooney MJ, Liaw BY, Sapra R, Adams MWW. Design and characterization of redox enzyme electrodes: new perspectives on established techniques with application to an extremophilic hydrogenase. *Enzyme Microb Technol* 2005;36:540–549.

26. Gregg BA, Heller A. Cross-linked redox gels containing glucose oxidase for amperometric biosensor applications. *Anal Chem* 1990;62:258–263.

27. Martin G, Minteer SD, Cooney MJ. Spatial distribution of malate dehydrogenase in chitosan scaffolds. *ACS Appl Mater Interfaces* 2009;1:367–372.

28. Klotzbach T, Watt M, Ansari Y, Minteer SD. Improving the microenvironment for enzyme immobilization at electrodes by hydrophobically modifying chitosan and Nafion polymers. *J Membr Sci* 2008;311:81–88.

29. Cooney MJ, Lau L, Windmeisser M, Yann Liaw B, Klotzbachb T, Minteer SD. Design of chitosan gel pore structure: towards enzyme catalyzed flow-through electrodes. *J Mater Chem* 2008;18:667–674.

30. Cooney MJ, Liaw BY, Johnston W, Svoboda V, Quinlan F, Maynard N. *Design of macropore structure for enzyme fuel cells operation*. 230th ACS National Meeting, Washington, DC, 2005.

31. Bru R, Sanchez-Ferrer A, Garcia-Carmona F. Kinetic models in reverse micelles. *Biochem J* 1995;310:721–739.

32. Cooney MJ, Petermann J, Carolin L, Minteer SD. Characterization and evaluation of hydrophobically modified chitosan scaffolds: towards design of enzyme immobilized flow-through electrodes. *Carbohydr Polym* 2009;75:428–435.

33. Callahan JW, Kosicki GW. The effect of lipid micelles on mitochondrial malate dehydrogenase. *Can J Biochem* 1967;45:839–851.

34. Martin GL, Lau C, Minteer SD, Cooney MJ. Fluorescence analysis of chemical microenvironments and their impact upon performance of immobilized enzyme. *Analyst* 2010;135:1131–1137.

35. Klotzbach T, Watt M, Ansari Y, Minteer SD. Effects of hydrophobic modification of chitosan and Nafion on transport properties, ion-exchange capacities, and enzyme immobilization. *J Membr Sci* 2006;282:276–283.

36. Martin G, Minteer SD, Cooney MJ. Fluorescence characterization of immobilization induced enzyme aggregation. *Chem Commun* 2010;46:1–3.

37. Förster T. Zwischenmolekulare Energiewanderung und Fluoreszenz. *Ann Phys* 1948;437:55–75.

38. Lakowicz JR. *Principles of Fluorescence Spectroscopy*. Springer Science + Business Media, Inc., New York, 2004.

13

IMAGING AND CHARACTERIZATION OF THE BIO–NANO INTERFACE

KAREN E. FARRINGTON AND HEATHER R. LUCKARIFT

Universal Technology Corporation, Dayton, OH, USA; Airbase Sciences Branch, Air Force Research Laboratory, Tyndall Air Force Base, FL, USA

D. MATTHEW EBY

Booz Allen Hamilton, Atlanta, GA, USA; Airbase Sciences Branch, Air Force Research Laboratory, Tyndall Air Force Base, FL, USA

KATERYNA ARTYUSHKOVA

Department of Chemical and Nuclear Engineering and Center for Emerging Energy Technologies, University of New Mexico, Albuquerque, NM, USA

13.1 INTRODUCTION

Characterizing the interface between immobilized biomolecules and a transducer is critical for garnering insight into spatial distribution and ordered orientation of biocatalysts. In particular, the use of immobilization techniques to incorporate enzymes into hierarchical composite materials (see Chapter 10) and layered architectures requires detailed information about the composition, physical characteristics, and chemical structure of the resulting electrode [1].

Herein, we will discuss a number of tools that may be employed to characterize the bio–nano interface. Many of these techniques have been modified from other areas of research; they are often not optimized directly for biological systems. An understanding of how to modify such techniques to collect useful information from biological interfaces is a critical aspect of understanding bioelectrocatalytic activity

Enzymatic Fuel Cells: From Fundamentals to Applications, First Edition. Edited by Heather R. Luckarift, Plamen Atanassov, and Glenn R. Johnson.
© 2014 John Wiley & Sons, Inc. Published 2014 by John Wiley & Sons, Inc.

and guiding future electrode development. The techniques will be addressed from the viewpoint of resolution, from visualization of macro-architectures using techniques such as scanning electron microscopy (SEM), to nanoscale and atomic-level resolution using atomic force microscopy (AFM), to elemental and chemical information that can be derived from techniques such as X-ray photoelectron spectroscopy (XPS).

13.2 IMAGING THE BIO–NANO INTERFACE

13.2.1 Scanning Electron Microscopy

SEM allows for direct observation of the physical appearance and surface characteristics of samples that are typically not visible with the naked eye. SEM is widely used for surface characterization of inorganic substrates including metals, minerals, and ceramics. SEM also finds unique utility in studying biological surfaces and can be used to visualize tissue surfaces, intracellular surfaces, isolated organelles, or isolated molecular structures [2]. Successful SEM, however, depends on the ability of a specimen to emit secondary electrons when it is excited by a powerful electron beam that scans the surface of a sample in a raster pattern [2,3]. X-rays are produced when electrons interact with atoms within a sample and are subsequently detected (Figure 13.1). Many scanning electron microscopes are additionally equipped with energy-dispersive X-ray (EDX) spectroscopy, which enables elemental analysis of the samples.

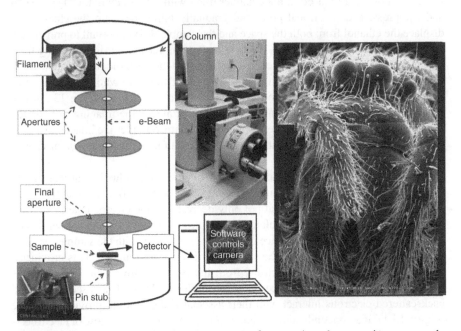

FIGURE 13.1 Schematic showing the aspects of a scanning electron microscope and a representative image of a black widow spider at 100× magnification. (Unpublished images taken by author KEF.)

During SEM analysis, samples are exposed to very dry conditions under vacuum. Dry inorganic materials are therefore readily analyzed without pretreatment. Biomolecules, however, must first be processed through a series of stabilization treatments that minimize rapid drying effects and maximize preservation of the original form and structure [4]. Many of the techniques for preservation of biological samples have been designed for plant or microbial cells, yet these techniques provide an extremely versatile toolbox for biological sample preparation. For whole cells, the initial step is fixation, typically with a chemical such as glutaraldehyde, which kills the cells while preserving their structural integrity. Subsequently, biological samples are treated with increasing concentrations of a solvent such as ethanol, to drive out residual water. The gradual transition in solvent exchange is imperative to preserve structural integrity. A common dehydration procedure involves immersing samples for 10–30 min each, depending on the sample density, in a stepwise series of solutions of ethanol in water (50, 70, 80, 90, and 100% ethanol). The effect of insufficient solvent exchange is exemplified in Figure 13.2a with *Shewanella* cells from a microbial fuel cell. These images of bacterial cells are used to highlight sample preparation methods for SEM. In addition, readers of this book on enzymatic fuel cells will undoubtedly note some parallel with the study of microbial fuel cells and the utility of some of the analytical methods described herein.

Critical point drying (CPD) is a complementary technique used to thoroughly dry samples intended for SEM analysis without leaving artifacts from the drying process. In CPD, samples are placed into a chamber filled with 100% ethanol. CO_2 is heated and compressed to its critical point and gradually introduced into the chamber to displace the ethanol from both the space and the sample. It is important to prevent the CO_2 from crossing the liquid–gas phase boundary, as surface tension resulting from the phase transition can cause severe damage to delicate biological structures. It should be noted that CPD can cause mild structural damage to whole bacterial cells, but this damage is far less severe than when cells are air dried (Figure 13.2b) [5]. Environmental SEM is a comparable technique that allows the imaging of hydrated specimens, thereby omitting the dehydration procedure. Samples are imaged in a low-pressure gaseous environment, but there are inherent limitations to the field of view at low magnifications.

For SEM observation, specimens must be electrically conductive, at least at the surface. Nonconductive surfaces must be treated with a conductive material (typically a thin layer of gold or carbon) that not only decreases static electric charge but also increases signal resolution, particularly for samples with a low atomic number. Inside the microscope's column, all surfaces that come into contact with the samples, such as the aluminum sample stubs and the stub holder itself, are grounded. This provides a path for the majority of the electrons from the beam to flow. After the electrons interact with the sample's surface, some are reflected back. These secondary and backscattered electrons interact with their associated detectors to provide an image (Figure 13.1). If a nonconductive sample is not sputter coated, the flow of electrons is interrupted. This leads to a buildup of electrons on the sample, commonly referred to as *charging*. This buildup of electrons causes a divergence of the electron beam, resulting in a shadowy, washed-out image (Figure 13.2c). Proper sputter coating of

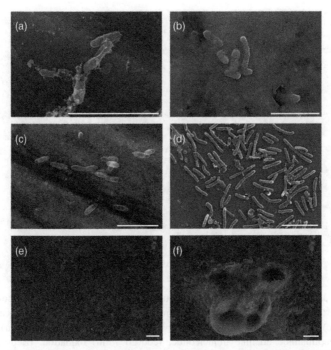

FIGURE 13.2   Bacterial cells (*Shewanella oneidensis*) on aluminum exposed to various sample preparation methods prior to SEM: (a) sample submerged directly into 100% ethanol rather than incremental increases in ethanol concentration; (b) sample air dried at room temperature instead of dehydrated with ethanol; (c) sample observed without gold coating; (d) correctly treated sample: glutaraldehyde fixation, ethanol dehydration, CPD preservation, and sputter coating with gold before observation. (e) A carbon–polymer composite anode and the same showing damage from extended exposure to the electron beam (~15 s) during SEM (f). Scale bars = 5 μm. *Note*: All images were prepared from microbial fuel cell anodes fixed in glutaraldehyde, dehydrated via ethanol exchange, and dried with a Tousimis Autosamdri™-815 critical point dryer. Samples were mounted with carbon tape onto aluminum pin stubs, sputter coated with gold (using a Denton Desk V sputter coater), and observed with the secondary electron detector of a Hitachi S2600-N microscope (accelerating voltages of 20–25 kV). (Unpublished images taken by author KEF.)

biological samples allows the bacterial cells to show up clearly and with good contrast (Figure 13.2d) against the background. Sputter coating also reduces specimen damage caused by the electron beam by dispersing surface charging (Figure 13.2e and f, undamaged vs. damaged) [3].

SEM provides a three-dimensional depth of field that is useful for understanding structural topography at a remarkable range of scale (from micro to nano). One of the primary benefits of SEM in biological fuel cell (BFC) research is the ability to investigate electrode architectures. A range of electrode materials have been studied and SEM provides crucial information on structure, porosity, and scale (Figure 13.3). Toray paper, for example, has been extensively used in electrode fabrication because

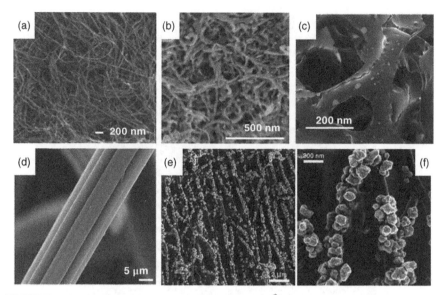

FIGURE 13.3 (a) Carbon nanotube paper (buckypaper).[*] (b) Composite of carbon nanotubes in chitosan. (Reproduced with permission from Ref. [6]. Copyright 2008, American Chemical Society.) (c) Reticulated vitreous carbon modified with poly(methylene green) for *in situ* NADH (nicotinamide adenine dinucleotide, reduced form) oxidation. (Reproduced with permission from Ref. [7]. Copyright 2011, Elsevier.) (d) Carbon fibers of Toray paper electrodes.[*] (e, f) Carbonized fibers prepared from polyacrylonitrile after silver plating, at high and low magnification. (Reproduced with permission from Ref. [8]. Copyright 2010, Elsevier.) ([*]Unpublished images taken by author KEF.)

it is highly conductive and scalable. The carbon fibers of Toray paper electrodes are pressed together during manufacturing, yielding a fragile sheet of conductive paper. Through SEM, these carbon fibers appear as long, grooved cylinders with radii of approximately 10 μm (Figure 13.3). Similarly, visualizing carbon nanotubes (CNTs) with SEM reveals scale and structural characteristics (Figure 13.3a).

Although enzymes are not directly visible by SEM, changes in surface architectures following enzyme immobilization can indeed be observed. For example, in 2011, Kim et al. immobilized glucose oxidase (GOx) to polyaniline nanofibers by cross-linking and reported a notable change in morphology due to increased enzyme loading on the surface (Figure 13.4) [9]. In addition, a visible change in porosity was observed as the enzyme aggregates filled the electrode voids. Similarly, Toray paper modified with multiwalled carbon nanotubes (MWCNTs) shows a striking change in structural morphology following immobilization in polyethylenimine (PEI) (Figure 13.4) [10].

13.2.1.1 Backscattered Electrons

Backscattered electrons are electrons that are reflected (by elastic and inelastic scattering) from the sample at an intensity that is dependent on the atomic number of the specimen. As such, backscattered electron

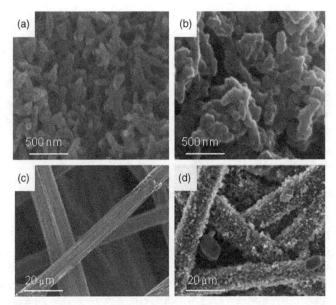

FIGURE 13.4 Polyaniline nanofibers (a) before and (b) after precipitation/adsorption of cross-linked glucose oxidase. (Reproduced with permission from Ref. [9]. Copyright 2011, Elsevier.) Toray carbon paper with CVD-deposited MWCNT, modified with (c) glucose oxidase and (d) polyethylenimine. (Reproduced with permission from Ref. [10]. Copyright 2006, Elsevier.)

imaging can provide information about the presence and distribution of specific elements in a sample and can be used to observe contrast between areas of varying chemical compositions (Figure 13.5). In the development of BFC electrodes, for example, SEM imaging using backscattered electrons can highlight the distribution of gold nanoparticles in a carbon–gold composite electrode (Figure 13.5).

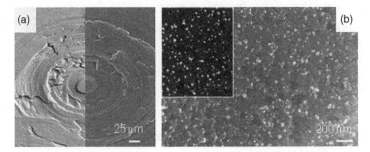

FIGURE 13.5 (a) Aluminum stub viewed by a secondary electron detector (left panel) and a backscatter electron detector (right panel).* (b) Carbon–gold composite electrode viewed by a secondary electron detector and a backscatter electron detector (inset). (Reproduced with permission from Ref. [11]. Copyright 2012, Wiley-VCH Verlag GmbH.) (*All unattributed images taken by author KEF.)

FIGURE 13.6   (a, b) *S. oneidensis* cells on Toray paper electrodes. Three-dimensional images prepared with Anaglyph Maker, ver. 1.08. Scale bars = 5 µm. (Unpublished images taken by author KEF.) 3D red cyan glasses are recommended to view this image correctly. (Please see the color version of this figure in Color Plates section.)

13.2.1.2   Three-Dimensional Imaging   The ability to use SEM to produce a three-dimensional image is a unique aspect of microscopy that adds aesthetic appeal and an ease of understanding to hierarchical electrode architectures. The three-dimensional (anaglyph) pictures included herein are of anodes taken from a BFC. The images were prepared with Anaglyph Maker [12], which overlays two separate images of the same specimen. Because the second image is taken at a slightly different tilt angle to the first, the illusion of depth is created (Figure 13.6) [13].

13.2.2   Transmission Electron Microscopy

As for SEM, similar sample information can be garnered from transmission electron microscopy (TEM), such as sample composition, topography, and morphology. In contrast to SEM, however, TEM builds an image by way of differential contrast. In TEM, the electron beam passes through the sample and forms an image based on electrons that are stopped or deflected by dense atoms in the sample. This image is magnified by the objective lens of the microscope, thus providing information about the structure of the sample. TEM is capable of imaging at significantly higher resolution than SEM, even down to the single-atom level. TEM operates under high vacuum to avoid scattering of electrons by air molecules and arcing of the high-voltage source. However, sample preparation for TEM is time consuming and analysis is often destructive to biological samples. The sample under analysis has to be thin, as the electrons are transmitted through the sample. Despite these restrictions, TEM finds some utility in BFC analysis for understanding electrode architecture and enzyme loading [14,15].

13.3   CHARACTERIZING THE BIO–NANO INTERFACE

13.3.1   X-Ray Photoelectron Spectroscopy

XPS is a powerful technique allowing estimation of the elemental and chemical composition of the upper 10 nm of a surface, and it is an effective tool to quantify the amount of protein immobilized or adsorbed during enzyme immobilization [16–21].

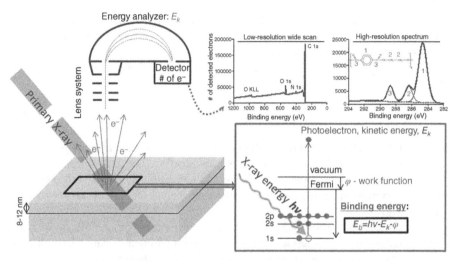

FIGURE 13.7   Basic principles of XPS.

XPS finds utility in the analysis of electrocatalysts for fuel cells [22], and an extensive review on the application of XPS to biomaterials, particularly nanobiomaterials, was published by Baer and Engelhard in 2010 [1].

In XPS, a sample is irradiated with beam of X-rays of primary energy $h\nu$ (Figure 13.7). This causes ejection of photoelectrons, which are directed to an electron energy analyzer through a series of lenses and whose kinetic energy is measured in the analyzer to produce a corresponding spectrum. Photoelectrons that are able to escape from the material must pass into the ultrahigh vacuum and reach the detector prior to losing their energy due to collisions with other electrons and atoms. As such, detectable photoelectrons typically originate from within the top 8–12 nm of the sample material.

The basic balance of energy equation relates binding energy (E_b) of an electron (which is specific to the element and its chemical environment) to kinetic energy (E_k) and primary photon energy ($h\nu$), taking into account a work function of the instrument (φ) (Equation 13.1):

$$E_b = h\nu - E_k - \varphi \tag{13.1}$$

An XPS spectrum plots the number of electrons detected (in counts per second, typically) against the binding energy of the electron in electron volts. Each element produces a characteristic spectrum at distinctive binding energies due to the configuration of electrons within the atoms, such as 1s, 2s, 2p, and so on, assuming that the primary energy of the X-ray photon is large enough to eject them. The number of electrons in each peak is related to the abundance of the element. A shift in electron binding energy occurs due to different chemical states. For carbon, for example, aromatic C, C—O, and ester (C(=O)O) bonds in polyethylene terephthalate (PET) will result in asymmetric high-resolution C 1s peaks that can be peak fitted to separate individual contributions from each chemical state. The ability to produce such distinctive chemical state information makes XPS a unique and invaluable tool for analysis of biological materials.

The sampling depth in XPS is determined by the following equation (Equation 13.2):

$$d = 3\lambda \sin \theta \qquad (13.2)$$

where λ is the inelastic mean free path of the photoelectron and θ is the angle between the sample surface and the detector acceptance direction. The term λ depends on the kinetic energy of the electron, which is specific to a particular orbital and shell [23]. For C 1s electrons, sampling depths are ~9.3 nm, whereas for N 1s the value increases to ~11.9 nm.

Angle-resolved XPS (ARXPS) is capable of giving depth profile information without removing the top surface layers [24]. By comparing the relative intensities of peaks at the same kinetic energy over a number of different takeoff angles, it is possible to calculate layer thickness. Alternately, comparing relative intensities at low and high takeoff angles indicates whether a species is enriched or depleted in the surface region. This is useful for the analysis of thin films on surfaces for which it is possible to determine the molecular orientation to the surface. In ARXPS, changing the angle of the sample with respect to the direction probed by the detector will alter the effective sampling depth (Figure 13.8) [24]. At $\theta = 90°$, for example, the sample surface is perpendicular to the line of acceptance of the analyzer, and d is the maximum sampling depth of 3λ (~9.3 nm for C 1s). As θ is reduced, the sampling depth decreases, to ~6.6 nm at 45° (see Figure 13.8).

13.3.1.1 Specific Considerations for Analysis of Enzymes Using XPS To fully take advantage of the quantitative chemical information contained within XPS spectra, several aspects of analyzing enzyme-based nanocomposites have to be considered. First, the length of time required for sample preparation and transfer

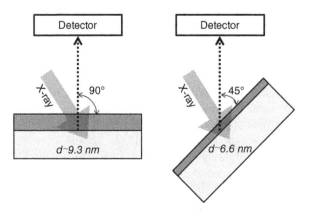

FIGURE 13.8 Schematic showing principle of angle-resolved XPS.

to vacuum must be minimized due to the relatively short lifetimes of enzymes. Additionally, high-resolution spectra may contain overlapped peaks, as different chemical species may have similar binding energy shifts; for example, $C-C^*=O$ and $N-C^*=O$ exhibit almost the same shift—2.9 and 3.1 eV, respectively—in C 1s spectra. The situation is even more complicated when studying multicomponent bioorganic or bioinorganic composites or interfaces. In these systems, it is difficult to distinguish multiple enzymes from a carbon-based support (e.g., electrode material) and an enzyme, and to determine specific interactions, such as hydrogen bonding between constituents of a composite. In practice, the quality of information achievable from overlapped peaks depends on the analyst's skill and experience. Analyzing standards such as PET gives an assessment of full-width half-maxima (FWHM) that can be used in curve-fitting high-resolution XPS spectra [25].

Furthermore, no extensive reference library exists for enzymatic materials. It is critical, therefore, to design an experiment such that individual constituents of the biocomposite, such as dry enzymes, solvent-cast enzymes, substrates, and polymers, are all analyzed to provide appropriate reference spectra. The respective widths and positions of peaks in the reference spectra can then be applied as constraints in curve fitting the spectra of the composite samples (enzyme encapsulated in matrix, layer-by-layer systems, etc.). Identification of species must be then cross-correlated and confirmed by spectra of all the elements that take part in the suspected type of chemical bond formation—for example, peaks due to $C-N=O$ must be confirmed by C 1s, O 1s, and N 1s signals.

13.3.1.2 Instrumentation and Experimental Details for XPS of Biomolecules A monochromatic X-ray source is usually employed for analysis of enzyme-containing samples, as it has much narrower natural FWHM that provides better spectral resolution. Moreover, damage introduced by a monochromatic X-ray source is typically much lower. If the sample is insulating in nature, charge composition must be performed. Quantification is done using sensitivity factors provided by the manufacturer. All the spectra must be charge calibrated to the aliphatic carbon at 284.7–285 eV. Curve fitting is carried out using a 70% Gaussian/30% Lorentzian line shape and FWHM of 0.9 ± 0.2 eV for all spectra acquired by monochromatic Al Kα sources.

13.3.1.3 Elemental Quantification for Fingerprinting Enzymes Table 13.1 shows elemental quantification for a range of biomolecules (three enzymes and one peptide) analyzed as either a powder or a soluble cast film. The use of O/C ratios

TABLE 13.1 Elemental Composition and Ratios for a Selection of Biomolecules

Biomolecule	C 1s (%)	O 1s (%)	N 1s (%)	O/C	N/C
Glucose oxidase	68.1	23.1	8.8	0.34	0.13
Lysozyme	65.2	19.1	15.7	0.29	0.24
Bilirubin oxidase	63.3	30.9	5.9	0.49	0.09
KSL peptide	68.8	16.6	14.7	0.24	0.21

for identification of proteins in a composite mixture is not as reliable as oxygen alone, because the specific signature may be attributed not only to the enzyme itself but also to contamination from possible inorganic and organic components of the composites. Nitrogen, however, is a characteristic element of enzymes and is rarely present in other constituents of the composites; therefore, the N/C ratio may be a more appropriate fingerprinting characteristic of biomolecules.

For example, in 2008, Ivnitski et al. used the N/C ratio to confirm entrapment of GOx in a silica–CNT composite obtained through lysozyme-catalyzed synthesis of silica on a Toray carbon paper electrode [26]. For samples containing both GOx and lysozyme, the N/C ratio of 0.11 was very close to that of GOx alone, indicating that the majority of enzyme detected by XPS is GOx. By comparison, the ratio in a GOx-free control was much larger (0.24), as would be expected for lysozyme alone.

13.3.1.4 High-Resolution Analysis for Fingerprinting Enzymes   The power of XPS for enzyme characterization is in the chemical speciation information that is available from high-resolution spectra. In the same system discussed above [26], C 1s spectra for GOx, lysozyme, and the composite were curve fitted (Figure 13.9). The C 1s spectrum of lysozyme (Figure 13.9a) has three major peaks due to C^*-C^* and C^*-H carbon (284.8 eV), and higher binding energy peaks at 285.8 and 288.2 eV that are attributed to $-C^*-N$ and $-C^*O-NH-$ or $-C^*OO-$ carbons, respectively. The smaller peak at 284.2 eV is thought to be due to contamination during preparation.

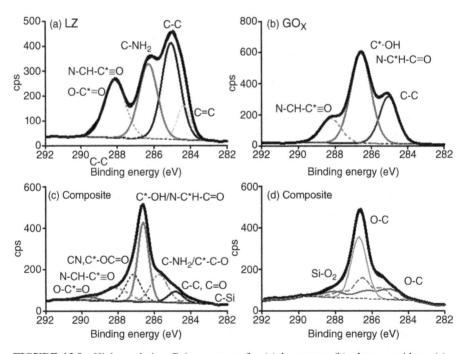

FIGURE 13.9 High-resolution C 1s spectrum for (a) lysozyme, (b) glucose oxidase, (c) composite curve fitted conventionally, and (d) composite curve fitted by model photopeaks.

The C 1s spectrum of GOx (Figure 13.9b) has three main peaks corresponding to the following bonds: aliphatic C^*-C^* (284.8 eV) that may have originated from surface contamination, oxydrilic C^*-OH, amidic $N-C^*H-CO$ (286.6 eV), and amidic $N-CH-C^*-O$ (288.3 eV). The high-resolution C 1s spectrum of the silica composite (Figure 13.9c) (containing lysozyme, GOx, and CNT) contains features of both pure enzyme samples—a dominant peak at 286.6 eV, which corresponds to GOx, and a peak at 285.8 eV, which corresponds to the primary peak in the spectrum of lysozyme. Alternatively, the spectrum of the composite can be curve fitted using experimentally obtained model photopeaks from the GOx and lysozyme, with additional peaks required for a complete curve fit (Figure 13.9d). This type of curve fit shows that the composite consists of 47% GOx, 42% lysozyme, and 11% associated with an unidentified peak at 286.7 eV that is likely due to interaction between encapsulated enzymes and the matrix. High-resolution XPS thus confirms the co-immobilization of GOx with CNT inside the silica matrix.

13.3.1.5 *Probing Molecular Interactions*
High-resolution spectral analysis of a short antimicrobial peptide (KSL) [peptide sequence KKVVFKVKFK] and KSL-catalyzed silica nanoparticles (KSL-Si) was demonstrated as a technique to determine interactions between KSL and silica. XPS is sensitive to intermolecular interactions such as hydrogen bonding and Lewis acid–base interactions and provides information describing coordination between the peptide and silica [27,28]. When the spectra of KSL and KSL-Si are compared (Figure 13.10), peak shifts in the KSL-Si spectra

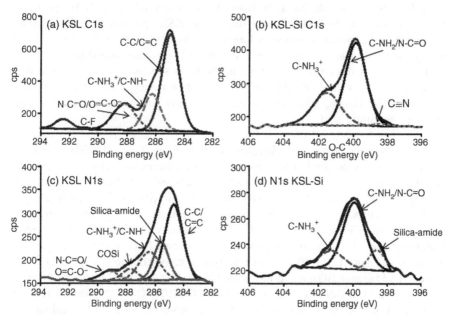

FIGURE 13.10 C 1s spectrum for (a) KSL and (c) KSL-Si; N 1s spectrum for (b) KSL and (d) KSL-Si.

TABLE 13.2 The Relative Percentage of Each Bond Type of the Purified Peptide Preparation (KSL) and Resultant Biosilica Composite (KSL-Si) from Curve Fitting of the C 1s and N 1s

Bond type	C–C/C=C	C=N–	C–NH₂/C–NH₃⁺	C–O–Si	N–C=O/O=C–O⁻	C–F	CN–/C=N–	C–NH–/C–NH₂	C–NH₃⁺
BE (eV)	284.8	285.5	286.3	287.5	288.4	292.4	398.7	400	401.5
KSL	55.1		21.5		18.8	4.6	2.5	70.6	26.9
KSL-Si	52.3	16.9	16.5	4.1	10.2		22	52.1	25.9

confirm interactions between the silica and peptide amides in the composite. The peak at 287.5 eV (Figure 13.10c) is too low to be amides (288.4 eV) and is most likely residual unreacted tetramethyl orthosilicate precursor (i.e., C–O–Si). A broader 285 eV peak in the KSL-Si spectrum suggests multiple peaks are photoemitting between the C–C/C=C and C–NH₃⁺/C–NH– peaks (284.8 and 286.3 eV, respectively). Curve-fitting analysis reveals an additional peak at 285.5 eV in the KSL spectrum, which, when taken with the decrease in amide binding energy, suggests that amides in the KSL-Si samples have been converted into bonding groups of lower energy (Table 13.2).

A difference in amide bond chemistry is also recorded in the N 1s spectra of KSL-Si (Figure 13.10d). In comparison to the KSL spectrum (Figure 13.10b), amplitude of the spectrum corresponding to peptide amides (400.0 eV) decreases by nearly 20% and a lower energy peak at 398.7 eV increases proportionally. This peak, as well as the peak at 285.5 eV in the C 1s spectrum, can be assigned to imine (C=N–), as the C=N double bond is of lower energy than a C=O double bond [29–31]. Hence, some of the peptide amides are configured in the enol tautomer as imidic acid.

13.3.1.6 Probing Physical Architecture of Thin Films Using ARXPS

Analysis by ARXPS provides nondestructive depth profile information for a wide range of samples including substrate–dye interactions, Langmuir–Blodgett films, and self-assembled monolayers [32,33]. In BFC research, ARXPS can be used to analyze multilayer architectures such as the assembly (by means of alternate electrostatic adsorption) of bilirubin oxidase (BOx) with positively charged PEI [26]. Typical high-resolution C 1s spectra of pure PEI, BOx, and layer-by-layer samples are presented in Figure 13.11. PEI has a single type of carbon due to C–NH at 285.5 eV (Figure 13.11a) and BOx alone has a unique peak at 288.5 eV in the C 1s high-resolution spectrum due to the presence of COOH/N–C=O (Figure 13.11b). A unique peak at 287.5 eV is observed for PEI on a carbon electrode; this is attributed to the interaction between amino groups of PEI and aldehyde groups at the carbon surface. Thus, each of the components of the layer-by-layer samples has a unique

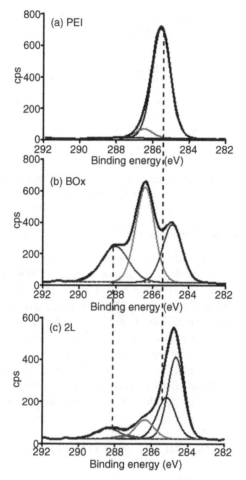

FIGURE 13.11 C 1s spectrum for (a) PEI, (b) BOx, and (c) a layer-by-layer composite of the two (2L).

signature within C 1s XPS spectra, which can be used to distinguish between individual layers on the electrode surface. Unique peaks of BOx at 288.5 eV and of carbon at 284.6 eV were used in a substrate overlayer model to estimate the relative thickness of the first layer of BOx on the carbon electrode surface. Because the peak at 287.5 eV is due to an interaction component, its uniqueness may be used to determine relative values of PEI thickness on BOx (Table 13.3).

The thicknesses of BOx on the carbon electrode and of PEI on BOx increase from a one- to a two-layered sample. However, a three-layered sample does not show further increase in either BOx or PEI thickness, as neighboring BOx and PEI layers partially interpenetrate. The reason for this is the BOx and PEI layers are intermixed, and the architectures for one-layer and two-layer samples are ordered (with some degree of intermixing) and stable due to the strong electrostatic attraction between the

**TABLE 13.3 Overall Relative Thickness of BOx and PEI Layers (1–3)
Determined from Overlayer Model**

Layer No.	BOx Thickness on Carbon (nm)	PEI Thickness on BOx (nm) (BE = 287.5 eV)
1	0.13	0.26
2	0.30	0.74
3	0.21	0.35

successive polyion and protein layers, whereas for samples with extra layers, a large
degree of intermixing and loss of order is observed.

13.3.2 Surface Plasmon Resonance

Surface plasmon resonance (SPR) is an optical technique that measures changes in
refractive index due to interactions at a sensor surface. In SPR, light passes through
the sensor (typically glass coated with a thin film of gold) and is partly reflected and
partly refracted at a specific angle. The SPR signal is sensitive to changes in mass at
the surface and the associated changes in refractive index. These changes are used to
detect and monitor the association and dissociation of analytes and ligands at the
surface. Accumulation of mass at the sensor causes an increase in the refractive index,
which is observed as a signal. As such, immobilization of enzymes at electrode
surfaces can be analyzed and characterized. SPR, for example, was used to demonstrate the stable binding of laccase to various conducting surfaces [34,35].

13.4 INTERROGATING THE BIO–NANO INTERFACE

13.4.1 Atomic Force Microscopy

AFM has established itself over recent years as an elegant and powerful tool for
characterization of electrochemically active enzymes immobilized onto a diverse
range of conductive materials for BFC applications. AFM is used to monitor the
attachment of enzymes onto conductive surfaces and to characterize the macromolecular assembly and electrochemical behavior. AFM has enabled users to
characterize the specific molecular interactions of immobilized enzymes on flat
conductive surfaces such as gold, graphene, and highly ordered pyrolytic graphite
(HOPG) and also to visualize enzymes attached to more complex surfaces, such as
enzymes decorated on CNT or complex multilayer films made from proteins and
conductive polymers. Along with imaging, electrochemical AFM techniques can be
employed to help unravel the complex structural and electrical interactions that occur
at the bio–nano interface, because the AFM probe and sample can act as a two-
electrode electrochemical cell. AFM imaging is paired with simultaneous electrochemical analyses at the molecular level to investigate voltammetry and amperometry

within a single enzyme, achieving unequaled precision and resolution. The following section provides an overview of the basic principles of AFM, highlights various AFM applications with respect to electrochemistry, and provides some pertinent examples of these techniques.

13.4.1.1 Basic Principles of AFM AFM, also known as scanning force microscopy (SFM), is a subset of scanning probe microscopy (SPM), which is a suite of microscopy techniques that involve scanning a molecularly sharp probe across the surface of a sample and monitoring the interaction between the probe and sample to image or characterize some property of the sample surface [36]. The first SPM instrument, the scanning tunneling microscope (STM), was developed in 1982 by Binnig, Rohrer, Gerber, and Weibel at IBM in Zurich, Switzerland, for which Binnig and Rohrer won the Nobel Prize in Physics in 1986. Soon afterward, the AFM was developed through collaboration between IBM and Stanford University. The basic principle of AFM is shown in Figure 13.12. AFM operates by scanning a tip attached to the end of a cantilever across the surface of a sample while monitoring the change in cantilever deflection or oscillation with a photodiode detector. Scanning, usually completed in a raster pattern, is completed either by moving the tip across the surface or by moving the sample itself while the tip is stationary. Movement at such a small scale is accomplished using piezoelectric materials.

As the tip scans the sample surface, it encounters sample features of different heights, resulting in either deflection of the tip (contact mode) or a change in frequency of an oscillating cantilever (tapping mode). Changes in the deflection or oscillation caused by sample topography interacting with the tip are commonly measured by reflecting a laser off the highly reflective tip surface. Reflected laser light is then collected by a photodiode detector, which monitors the change in movement of

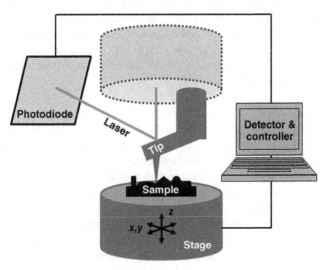

FIGURE 13.12 Basic AFM components. Scanner can be attached either to sample or to cantilever and tip (dashed).

the tip. A feedback loop is used to maintain a constant distance or oscillation (determined by a "set point") between the tip and sample surface. As topography of the sample changes during scanning, the feedback loop will position the piezo-electric scanner (controlling the tip or stage) vertically until the set point is achieved. The vertical position of the scanner is stored by a computer while the sample is scanned in the x, y direction to form a topographic image of the sample surface.

Each AFM scanning technique is built on the basic operation described above. The specific interaction between the tip and sample surface can lead to characterization of different sample properties. Using functionalizing tips with relevant molecules or holding the tip at different distances above the sample can provide information on not only the physical properties but also the electronic and chemical aspects of the sample. Force spectroscopy is an additional AFM technique that does not involve scanning across the sample surface but instead causes the tip to repeatedly connect with a specific point on the sample surface to measure the interaction or bonding strength between tip and sample. Force–distance curves are then generated, which can measure at a resolution of piconewtons due to nanoscale interactions between tip and sample, specific atomic bonding, and physical properties of a sample at a single point. Force spectroscopy is an established technique in nanotechnology and has been used extensively to characterize enzyme adhesion and interaction with conductive materials. Recent advances in AFM technology have merged force spectroscopy with scanning due to the integration of high-performance computers and faster feedback loop servos. Several competent reviews are available on this technique and will not be discussed in detail here [37–40].

13.4.1.2 AFM Techniques   In many cases, AFM techniques for enzyme-based electrodes focus on generating images at near-atomic resolution. As enzyme electrodes are created, whether through simple adsorption of enzymes on conductive surfaces or through more complex layering and assembly of proteins with conductive polymers or other nanomaterials, AFM imaging can afford a relatively quick, qualitative look at enzyme loading and morphology. There are few tools that can provide such quantitative measurements and garner electrochemical and biochemical surface data simultaneously with imaging.

Imaging Under Contact and Tapping Modes   Imaging of a sample surface at the nanoscale is the most typical use of AFM. As described in the previous section, there are two basic types of scanning the surface for imaging: contact mode and tapping mode. In contact mode, the tip is in contact with an adsorbed fluid layer on the sample surface and the degree of contact can change depending on changes in sample height, set point selection, and the sensitivity of the feedback loop response time. In contrast, tapping mode entails scanning the sample while oscillating the tip cantilever near its resonance frequency at amplitudes of up to 100 nm and monitoring changes in the oscillation frequency due to changes in tip–sample distance. Here, the tip also contacts the adsorbed water layer on the sample, but only at the trough of the oscillation wave. Each technique has its advantages and disadvantages, but tapping mode offers a particular benefit for imaging proteinaceous material (such as enzymes on an electrode surface). Such a technique was used to image laccase immobilized to gold [41]. Observing a protein layer in tapping mode lessens the chance of the tip becoming contaminated with protein. Once the tip picks up biological material from the sample surface, imaging ceases to occur and the tip must be cleaned or replaced.

Tapping mode provides the added technique of phase imaging, which provides information about variations of adhesion and viscoelasticity on the sample surface [42]. Differences in phase occur because energy dissipation from the tip will be different when it is interacting with softer or more rigid surfaces; soft "sticky" materials, for example, usually damp the tip vibration. The delay results in a phase shift in the cantilever oscillation, which is then measured and visualized as a phase image. Phase imaging can be particularly useful in discriminating between protein and more rigid electrode materials, such as metals and glassy carbon [43,44].

An additional method that is beginning to gain attention is noncontact AFM (NC-AFM) [45]. In tapping mode NC-AFM, the cantilever is oscillating above the cantilever's resonance frequency with an amplitude of only a few nanometers. In particular, the tip is maintained at a distance above the sample water layer so that tip oscillations are perturbed through van der Waals and other long-range forces. Decreases in tip oscillations due to these types of tip–sample interactions are monitored and ultimately provide surface properties beyond a topographic image of the sample surface. Imaging in NC-AFM mode, particularly in high-vacuum environments, results in tip interactions due to electrostatic forces from the sample rather than from an adsorbed water layer. This technique has some utility in producing atomic resolution for characterizing enzymes adsorbed to flat surfaces [46]. Nevertheless, enzymes do not function in a vacuum, so this imaging technique, although providing some of the highest-resolution images ever recorded [47,48], cannot be correlated to enzyme function.

The true advantage of NC-AFM for enzyme fuel cell electrodes is the measurement of local work function changes on conducting surfaces and the surface charge distribution on insulating and dielectric surfaces [45]. Electrochemically active proteins and other biomolecules usually have contrasting surface charge and electrostatic properties to the conductive surfaces they are immobilized to. Specific NC-AFM techniques, such as Kelvin probe force microscopy (KPFM) and electrostatic force microscopy (EFM), are able to efficiently map surface charge and electrostatic interactions on hybrid bioinorganic surfaces; these are discussed further below.

In recent years, technology has come to the commercial market that allows imaging of soft biological samples with higher resolution and less of the nominal friction and shear forces that cause deformation of soft surfaces. Enhanced-precision z scanners, high-speed control electronics, and fast servo feedback performance have all provided better tip control and more sensitive monitoring of the energy dissipated between tip and sample during each oscillation in both tapping and NC-AFM. Better control algorithms optimize tip interaction at each tap, reduce forces imposed on the sample surface during tapping mode, and provide no contact whatsoever in true noncontact imaging. This has allowed much faster tapping modes with better phase imaging and the performance of simultaneous force measurements to map adhesion and modulus properties of sample surfaces.

KPFM and EFM Both KPFM and EFM are subsets of NC-AFM. Each can be used to study the structural and electronic properties of functional biological and inorganic interfaces. KPFM combines the main principle of a Kelvin probe with the scanning ability of NC-AFM to generate surface potential and electrostatic interaction maps [49,50]. Combined with height imaging, the technique provides both topography and contact potential measurement between probe and sample. KPFM is applicable in measuring the electrostatic interactions of biological surfaces, such as biological ion channels, electroactive enzymes, protein folding and assembly, and cellular membranes [51].

EFM is a related technique in which the electrical field of the sample surface is detected directly. The technique is dependent upon the tip charge, because changes in cantilever oscillation are dependent upon the specific electrostatic interaction between tip and sample. In both techniques, a specific voltage can be applied to the tip to impose specific electrical interactions with the sample for certain applications.

AFM in Fluid AFM operation in fluid presents a considerable advantage over electron microscopy in imaging biological materials, because imaging can be completed without fixation and critical point drying and, in many cases, AFM imaging in fluid can be performed on active enzymes and living cells. Contact and tapping modes operate effectively in fluid, where either an O-ring or the surface tension of water between sample and cantilever holder comprises the fluid AFM cell. An optically clear cantilever holder allows the laser to transmit through the fluid and reflect off of the tip and into the detector. Enzymes and polymers can be immobilized to electrode surfaces *in situ* and monitored by AFM imaging techniques in real time. The fluid cell can also be transformed into an electrochemical cell, in which the sample surface operates as the working electrode and a counter and a reference electrode are supplied in the liquid. A specific example demonstrating simultaneous AFM imaging and electrochemical reactions in a fluid cell is presented in the next section.

Electrochemical Atomic Force Microscopy Electrochemical atomic force microscopy (ECAFM) is a subset of electrochemical scanning probe microscopy (ECSPM), also known as scanning electrochemical microscopy (SECM) (see Chapter 14) [52,53]. In its basic form, the AFM tip acts as an ultramicroelectrode. Amperometric and potentiometric responses can be measured while the tip is scanned over the sample, resulting in the simultaneous topographical and electro-chemical measurements of the sample surface. A range of information can be collected from this technique, depending on certain variables such as changes in the tip–sample distance, whether the tip is in direct contact with the sample surface, or if a voltage bias is placed on the tip or sample. Capacitance, resistivity, conductivity, and electrostatic properties can all be mapped across the sample surface. ECAFM is able to map the potential distribution across a sample surface, particularly measuring the electric double layer at a two-phase interface (such as solid–liquid).

In liquid ECAFM applications, the AFM tip and sample surface comprise the working and counter electrodes of an electrochemical cell. Additionally, a reference electrode can be incorporated into the liquid cell allowing the instrument to effectively perform as a three-electrode electrochemical cell. The technique has expanded into several applications where electrosynthesis and electrodeposition reactions can take place simultaneously with imaging [54]. Electrical connections can be made between the tip and enzymes on a surface to measure electron transfer and conductivity [44,52,55]. In addition, the tip can be functionalized with electroactive biomolecules to perform redox reactions and measure the amperometric and voltammetric response *in situ* [56].

Scanning Capacitance Microscopy   Scanning capacitance microscopy (SCM) entails contact mode operation with the tip and a sample that forms a metal–insulator–semiconductor capacitor to generate simultaneous height images and capacitance mapping [57]. An alternating current (AC) bias voltage is imposed between the tip and sample and, as the scanning image is formed, changes in the tip–sample capacitance are recorded through a sensitive high-frequency resonance circuit. SCM is usually used in the semiconductor industry for profiling doped semiconductor materials and performing failure analysis but has also been used to probe dielectric properties of biomolecules [58].

Tunneling Atomic Force Microscopy   Tunneling atomic force microscopy (TUNA) operates under the same principles as STM but is suited for highly resistive samples. TUNA is able to measure very sensitive fA currents between tip and sample. Scanning is completed in contact mode and a bias voltage is applied between a conductive tip and sample. The current passing though the tip and sample is measured by an amplifier in the fA-to-pA range. TUNA is commonly applied to complex electrode materials and dielectric films, such as silicon dioxide materials. Expanding the range of currents used in TUNA, scanning spreading resistance microscopy (SSRM) measures the resistivity of a wider range of materials from highly resistive to conducting. Similar in operation to TUNA, SSRM detects currents ranging from pA to mA, from which the resistivity of the material can be calculated using Ohm's law. The technique is known as spreading resistance, because resistance measurements include the resistance that spreads radially from the point of contact, including the bulk resistance along the entire connection between the tip and through the sample to the conductive platform. Both TUNA and SSRM are particularly effective in mapping the current of CNT films and materials regularly used in enzyme fuel cell electrodes.

13.4.1.3   *Examples of AFM Analysis and Applications*

Imaging Enzyme Fuel Cell Materials   The most common application of AFM is imaging the surface of electrode materials. Dispersion of enzyme immobilized to electrode surfaces can be qualitatively determined and, in some cases, attachment of enzyme and additives (e.g., CNT) can be quantified. Different loadings of MWCNT and GOx on indium tin oxide (ITO) electrodes, for example, were compared to determine morphological differences between enzyme and CNT on ITO surfaces [59]. The uniformity of MWCNT coatings on ITO electrodes was analyzed, and the effect of addition of MWCNT to GOx during enzyme immobilization was assayed. SEM was used to corroborate results from AFM analysis, although AFM has an advantage over SEM in that prolonged exposure to the electron beam in SEM will damage enzyme films (Figure 13.2f). The AFM and SEM images confirmed an increase in deposition of GOx to ITO surfaces in the presence of MWCNT.

Graphene oxide and graphene sheets are often used as an electrode material in electroanalysis and electrocatalysis [60]. Graphene is also an ideal substrate for the study of enzyme immobilization for electrode fabrication in enzyme fuel cells [61]. As it is essentially atomically flat, the surface is excellent for AFM imaging of arrangement and ordering of enzymes on its surface. In 2010, for example, Zhang et al. immobilized such model enzymes as horseradish peroxidase (HRP) and lysozyme onto graphene and characterized each in an AFM liquid cell [61]. Phenol and hydrogen peroxide were then used as substrates to monitor the catalytic activity of the immobilized HRP. The study found that interactions between the enzyme and functional groups at the graphene surface affected the catalytic performance of HRP, and also that the specific activity of the enzyme was not influenced by the amount of enzyme immobilized to the graphene surface. AFM images and enzyme activity results concluded that catalytic performance can be improved if immobilization methods address the full retention and conformation of the immobilized enzyme [61].

When building multiple layers of electrochemically active molecules onto fuel cell electrodes, AFM imaging can be used to confirm layer formation and characterize morphological features. For example, Noh et al. used AFM imaging in 2009 to confirm the electrochemical polymerization of iron-containing polymers based on polyterthiophene analogs onto HOPG electrodes [62]. Two different polymerization techniques were investigated on electrodes, and the surface morphology of each conductive polymer film was characterized by AFM. The AFM images showed formation of particles of varying size, depending on polymer formulation and deposition (such as the effect of consecutive polymerizations) (Figure 13.13). These findings could not be characterized by monitoring successive layers by other techniques, such as quartz crystal microbalance (QCM), or by measuring surface charge or spectroscopic characteristics. Enzymes (GOx and HRP) were then covalently linked to the conducting polymer-coated electrodes, thus comprising the anode and cathode, respectively, of a complete enzyme fuel cell.

AFM imaging of enzyme-loaded electrode materials can also be completed on rough and structurally complex surfaces. AFM characterization, including topography and roughness, was completed on hydrogenase-coated MWCNTs [63,64]. In 2011, Lojou confirmed the presence of a single layer of hydrogenase enzyme on MWCNT films using tapping mode in water. Height and roughness differences were attributed to a layer of hydrogenase 5 nm thick on the nanotube surface, which

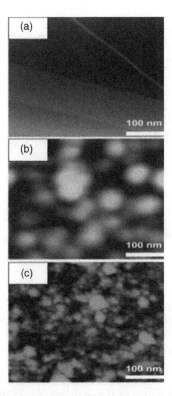

FIGURE 13.13 AFM images of (a) bare HOPG, (b) first polyterthiophene monomer, and (c) second Fe-containing polyterthiophene monomer synthesized on HOPG surfaces. (Adapted with permission from Ref. [62]. Copyright 2010, Elsevier.)

matches the three-dimensional structure of the hydrogenase used in the study (Figure 13.14) [64,65]. The authors conceded the difficulties of determining real distances between protein molecules and overall protein density on the nanotubes due to limitations of a large tip radius, and they calculated that at least one protein is present for every 20 nm of nanotube length.

Using AFM to Quantitate Enzyme Loading on Electrode Surfaces In 2010, Ivnitski et al. completed AFM imaging and force measurements on a multicopper oxidase commonly used in enzyme fuel cells [66]. Here, tapping mode AFM in liquid was employed to image the laccase from *Trametes versicolor* on screen-printed carbon electrodes. AFM imaging of the bioelectrode surface demonstrated that the laccase was attached to the electrode surface as a monolayer. Phase contrast imaging was able to resolve the enzyme from the carbon electrode surface. Height images revealed ordered structures that had an average particle spacing of 8 ± 1 nm, the calculated size for the *T. versicolor* laccase (Figure 13.15) [67,68]. The protein appeared tightly packed and ordered into rows, which permitted the rough calculation of enzyme surface coverage ($\sim$50%) through calculation of the area of a single enzyme on the

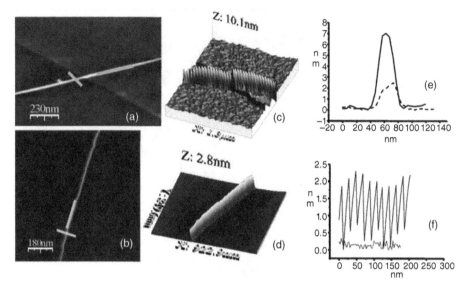

FIGURE 13.14 AFM images, height, and roughness characterization of (a, c) hydrogenase-coated MWCNTs versus (b, d) control MWCNT images. Hydrogenase-coated MWCNT (solid line) and control MWCNT (dashed line) height and roughness measurements are shown in panels (e) and (f), respectively. (Reprinted with permission from Ref. [63]. Copyright 2007, American Chemical Society.)

electrode and the theoretical area of the enzyme [66]. In the same study, AFM force curve measurements were completed to estimate protein density and protein film thickness. Force curves were obtained for an uncoated electrode and an electrode coated with laccase. Analysis on the bare electrode exhibited a typical force curve for a rigid surface with no resistance between the tip and surface interaction. Force curves of the enzyme-coated electrode demonstrated a detachment force of 0.5 nN and a gradual compression of approximately 8 nm before the tip reached the underlying rigid electrode surface, confirming a soft layer of protein on the surface of the

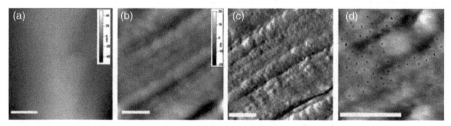

FIGURE 13.15 Tapping mode AFM images in fluid of uncoated, screen-printed carbon electrode and laccase immobilized on same. Images shown are (a) uncoated electrode height, (b) coated electrode height, (c) amplitude, and (d) enlarged section marked with black dots to designate single enzyme molecules. (Reprinted with permission from Ref. [66]. Copyright 2010, Elsevier.)

electrode. These data, the observed diameter of enzymes in height images, and the radius of the tip can be used to calculate the diameter of the enzyme. Imaging biomolecules that are smaller than the tip radius can be challenging but are estimated through the following equation (Equation 13.3):

$$D = d + 2\sqrt{r^2 - (r - h)^2} \qquad (13.3)$$

The actual diameter of an object (d) smaller than the tip radius has an imaged diameter (D) when the radius of the tip (r) and the height (h) are considered. Hence, the apparent enzyme diameter of 26 nm, calculated to be an actual radius of 4.5 nm for the enzyme, is in close agreement with an enzyme monolayer thickness of 8 nm calculated from force measurements and the reported diameter of the *T. versicolor* laccase [67]. From values of the observed electric current and electrode surface area (measured by direct electrochemistry and capacitance analysis, respectively) and the enzyme density calculated from the average area per enzyme molecule derived from AFM height images, the current per enzyme molecule was calculated to be 5×10^{-11} μA.

Electrochemical and Conductive AFM Techniques for Enzyme Fuel Cell Electrodes Conductive AFM techniques have not been extensively used for direct characterization of enzyme fuel cell electrodes, but a few studies have focused on electrical conductivity measurements of enzymes on conductive surfaces. The conductivities of self-assembled monolayer films of iron-loaded ferritin (holoferritin) and ferritin with iron removed (apoferritin), for example, were measured on gold surfaces (Figure 13.16) [55]. The analysis was completed in two different experiments. Gold-coated AFM tips were used in tapping mode to locate protein molecules deposited on the gold surface with very low coverage. Current–voltage measurements were then completed on a single molecule using the gold-coated tip. In a second experiment, a gold nanoparticle was used as the AFM tip and current–voltage

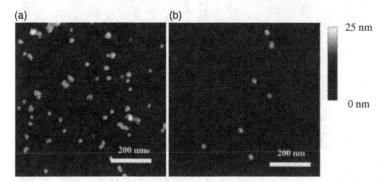

FIGURE 13.16 Tapping mode AFM images of (a) apoferritin molecules and (b) holoferritin molecules immobilized on gold surfaces. (Adapted with permission from Ref. [55]. Copyright 2005, American Chemical Society.)

measurements were completed on an area containing thousands of particles. After averaging, holoferritin was determined to be 5–15 times more conductive than apoferritin, and it was presumed that electron transfer between the AFM tip and gold substrate was completed via the ferrihydrite core.

The application of ECAFM was used in characterizing HOPG surfaces with electrosynthesized polypyrrole (PPy) and immobilized HRP [54]. In the study, PPy films were generated by cyclic voltammetry to achieve structured PPy nanoparticles on the surface of the HOPG electrode (Figure 13.17). HRP was then electroimmobilized onto the PPy nanoparticle film. The advantage of ECAFM is that electrosynthesis of PPy and ECAFM imaging were completed *in situ* using a silicon nitride tip in contact mode for typical AFM imaging within a standard electrochemical cell with a platinum counter electrode, a copper reference electrode, and HOPG as the working electrode. ECAFM was used to characterize the morphology of the thin-layer

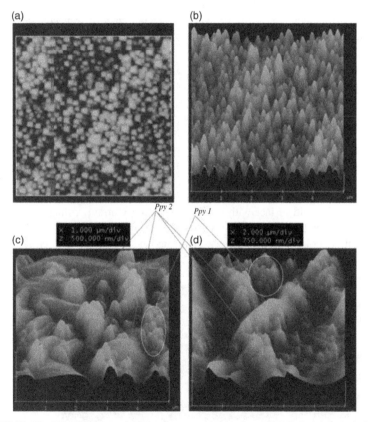

FIGURE 13.17 ECAFM image of PPy electrosynthesized on HOPG showing (a) height and (b) sectional analysis. Sectional analysis of electroimmobilization of HRP onto PPy-coated HOPG (c) after 60 s and (d) after 350 s. Sectional analysis shows two different types of PPy particles: agglomeration structures (PPy1) and individual nanoparticles (PPy2). (Adapted with permission from Ref. [54]. Copyright 2009, Elsevier.)

PPy film. The film is highly ordered and individual PPy nanoparticles are imaged in high resolution. Roughness and surface area parameters were calculated from the resulting images (not shown). Using similar conditions, HRP was immobilized onto the PPy film and imaged in contact mode. PPy nanoparticles were still apparent under a laminar structure that had coated much of the PPy film, which was attributed to crystalline HRP. The ECAFM technique described here is an elegant tool to analyze synthesis of thin conducting films *in situ*. Normally, visualization of such materials following synthesis requires removal from the electrochemical cell and additional sample preparation before analysis using more conventional SEM, TEM, or AFM.

13.5 OUTLOOK

The techniques described herein demonstrate the breadth of invaluable information that can be derived by analyzing the morphology and electrochemical properties of enzymes, redox-active biomolecules, and other polymers on conductive materials. Many of these techniques offer rapid and relatively noninvasive methods to monitor bioelectrode construction and assembly. By combining complementary analysis, we are able to characterize nearly any type of electrode surface, including biological and inert components, and understand the interfacial interactions between the two. The ability to conduct analysis in a liquid environment (such as an AFM fluid cell) provides information that can be correlated to electrocatalytic activity. In-depth analysis at the range of scale described herein will increasingly become a complementary and indispensable tool to elucidate and understand surface electrochemistry and bioelectrochemical interface chemistry.

ACKNOWLEDGMENT

This research was prepared as an account of work sponsored by the United States Air Force Research Laboratory, Materials and Manufacturing Directorate, Airbase Technologies Division (AFRL/RXQ), but the views of authors expressed herein do not necessarily reflect those of the United States Air Force.

LIST OF ABBREVIATIONS

AC	alternating current
AFM	atomic force microscopy
ARXPS	angle-resolved XPS
BFC	biological fuel cell
BOx	bilirubin oxidase
CNT	carbon nanotube
CPD	critical point drying
ECAFM	electrochemical AFM

ECSPM	electrochemical SPM
EDX	energy dispersive X-ray spectroscopy
EFM	electrostatic force microscopy
FWHM	full-width half-maximum
GOx	glucose oxidase
HOPG	highly ordered pyrolytic graphite
HRP	horseradish peroxidase
ITO	indium tin oxide
KPFM	Kelvin probe force microscopy
MWCNT	multiwalled carbon nanotube
NADH	nicotinamide adenine dinucleotide, reduced form
NC-AFM	noncontact AFM
PEI	polyethylenimine
PET	polyethylene terephthalate
PPy	polypyrrole
QCM	quartz crystal microbalance
SCM	scanning capacitance microscopy
SECM	scanning electrochemical microscopy
SEM	scanning electron microscopy
SFM	scanning force microscopy
SPM	scanning probe microscopy
SPR	surface plasmon resonance
SSRM	scanning spreading resistance microscopy
STM	scanning tunneling microscopy
TEM	transmission electron microscopy
TUNA	tunneling AFM
XPS	X-ray photoelectron spectroscopy

REFERENCES

1. Baer DR, Engelhard MH. XPS analysis of nanostructured materials and biological surfaces. *J Electron Spectrosc* 2010;178:415–432.

2. Goldberg MW. Immunolabeling for scanning electron microscopy (SEM) and field emission SEM. *Methods Cell Biol* 2008;88:109–130.

3. Brown DA, Beveridge TJ, Keevil CW, Sherriff BL. Evaluation of microscopic techniques to observe iron precipitation in a natural microbial form. *FEMS Microbiol Ecol* 1998;26:297–310.

4. Pathan AK, Bond J, Gaskin RE. Sample preparation for SEM of plant surfaces. *Mater Today* 1999;12:32–43.

5. Echlin P. *Handbook of Sample Preparation for Scanning Electron Microscopy and X-Ray Microanalysis.* Springer, New York, 2009.

6. Lau C, Cooney MJ, Atanassov P. Conductive macroporous composite chitosan–carbon nanotube scaffolds. *Langmuir* 2008;24:7004–7010.

7. Rincón RA, Lau C, Garcia KE, Atanassov P. Flow-through 3D biofuel cell anode for NAD$^+$-dependent enzymes. *Electrochim Acta* 2011;56:2503–2509.

8. Prilutsky S, Schechner P, Bubis E, Makarov V, Zussman E, Cohen Y. Anodes for glucose fuel cells based on carbonized nanofibers with embedded carbon nanotubes. *Electrochim Acta* 2010;55:3694–3702.

9. Kim H, Lee I, Kwon Y, Kim BC, Ha S, Lee J-H, Kim J. Immobilization of glucose oxidase into polyaniline nanofiber matrix for biofuel cell applications. *Biosens Bioelectron* 2011;26:3908–3913.

10. Ivnitski D, Branch B, Atanassov P, Apblett C. Glucose oxidase anode for biofuel cell based on direct electron transfer. *Electrochem Commun* 2006;8:1204–1210.

11. Luckarift HR, Ivnitski D, Lau C, Khripin C, Atanassov P, Johnson GR. Gold-decorated carbon composite electrodes for enzymatic oxygen reduction. *Electroanalysis* 2012;24: 931–937.

12. Sekitani T. Anaglyph Maker, ver. 1.08. Ed. 2001–2004.

13. Goldstein J, Newbury D, Joy D, Lyman C, Echlin P, Lifshin E, Sawyer L, Michael J. *Scanning Electron Microscopy and X-Ray Microanalysis*, 3rd edition. Kluwer Academic/ Plenum Publishers, New York, 2003.

14. Kihara T, Liu X-Y, Nakamura C, Park K-M, Han S-W, Qian D-J, Kawasaki K, Zorin NA, Yasuda S, Hata K, Wakayama T, Miyake J. Direct electron transfer to hydrogenase for catalytic hydrogen production using a single-walled carbon nanotube forest. *Int J Hydrogen Energy* 2011;36:7523–7529.

15. Jensen UB, Vagin M, Koroleva O, Sutherland DS, Besenbacher F, Ferapontova EE. Activation of laccase bioelectrocatalysis of O_2 reduction to H_2O by carbon nanoparticles. *J Electroanal Chem* 2012;667:11–18.

16. Blomberg E, Claesson PM, Froberg JC. Surfaces coated with protein layers: a surface force and ESCA study. *Biomaterials* 1998;19:371–386.

17. Curulli A, Cusma A, Kaciulis S, Padeletti G, Pandolfi L, Valentini F, Viticoli M. Immobilization of GOD and HRP enzymes on nanostructured substrates. *Surf Interface Anal* 2006;38:478–481.

18. Hoven VP, Tangpasuthadol V, Angkitpaiboon Y, Vallapa N, Kiatkamjornwong S. Surface-charged chitosan: preparation and protein adsorption. *Carbohydr Polym* 2007;68:44–53.

19. Longo L, Vasapollo G, Guascito MR, Malitesta C. New insights from X-ray photoelectron spectroscopy into the chemistry of covalent enzyme immobilization, with glutamate dehydrogenase (GDH) on silicon dioxide as an example. *Anal Bioanal Chem* 2006;385:146–152.

20. Tangpasuthadol V, Pongchaisirikul N, Hoven VP. Surface modification of chitosan films. Effects of hydrophobicity on protein adsorption. *Carbohydr Res* 2003;338:937–942.

21. Wang J, Carlisle JA. Covalent immobilization of glucose oxidase on conducting ultra-nanocrystalline diamond thin films. *Diamond Relat Mater* 2006;15:279–284.

22. Corcoran CJ, Tavassol H, Rigsby MA, Bagus PS, Wieckowski A. Application of XPS to study electrocatalysts for fuel cells. *J Power Sources* 2010;195:7856–7879.

23. Penn DR. Quantitative chemical analysis by ESCA. *J Electron Spectrosc* 1976;9:29–40.

24. Cumpson P. Angle-resolved X-ray photoelectron spectroscopy. In: Grant JT, Briggs D (eds), *Surface Analysis by Auger and X-Ray Photoelectron Spectroscopy*. IM Publications and Surface Spectra Limited, West Sussex, 2003.

25. Crist BV. *Handbook of Monochromatic XPS Spectra: Semiconductors.* John Wiley & Sons, Inc., New York, 2000.

26. Ivnitski D, Artyushkova K, Atanassov P. Surface characterization and direct electrochemistry of redox copper centers of bilirubin oxidase from fungi *Myrothecium verrucaria. Bioelectrochemistry* 2008;74:101–110.

27. Kerber SJ, Bruckner JJ, Wozniak K. The nature of hydrogen in X-ray photoelectron spectroscopy: general patterns from hydroxides to hydrogen bonding. *J Vac Sci Technol A* 1996;14:1314–1320.

28. Sarno DM, Matienzo LJ, Jones WE Jr. X-ray photoelectron spectroscopy as a probe of intermolecular interactions in porphyrin polymer thin films. *Inorg Chem* 2001;40:6308–6315.

29. Furukawa M, Yamada T, Katano S, Kawai M, Ogasawara H, Nilsson A. Geometrical characterization of adenine and guanine on Cu(110) by NEXAFS, XPS, and DFT calculation. *Surf Sci* 2007;601:5433–5440.

30. Jackson DC, Duncan DA, Unterberger W, Lerotholi TJ, Lorenzo DK, Bradley MK, Woodruff DP. Structure of cytosine on Cu(110): a scanned-energy mode photoelectron diffraction study. *J Phys Chem C* 2010;114:15454–15463.

31. Johnson PS, Cook PL, Liu XS, Yang WL, Bai YQ, Abbott NL, Himpsel FJ. Universal mechanism for breaking amide bonds by ionizing radiation. *J Chem Phys* 2011;135 (044702):1–10.

32. Schneider T, Artyushkova K, Fulghum JE, Broadwater L, Smith A, Lavrentovich OD. Oriented monolayers prepared from lyotropic chromonic liquid crystal. *Langmuir* 2005;21:2300–2307.

33. Artyushkova K, Fulghum JE. Angle resolved imaging of polymer blend systems: from images to a 3D volume of material morphology. *J Electron Spectrosc* 2005;149:51–60.

34. Mazur M, Krysinski P, Michota-Kaminska A, Bukowska J, Rogalski J, Blanchard GJ. Immobilization of laccase on gold, silver and indium tin oxide by zirconium-phosphonate-carboxylate (ZPC) coordination chemistry. *Bioelectrochemistry* 2007;71:15–22.

35. Michota-Kaminska A, Wrzosek B, Bukowska J. Resonance Raman evidence of immobilization of laccase on self-assembled monolayers of thiols on Ag and Au surfaces. *Appl Spectrosc* 2006;60:752–757.

36. Kasas S, Thomson NH, Smith BL, Hansma PK, Miklossy J, Hansma HG. Biological applications of the AFM: from single molecules to organs. *Int J Imaging Syst Technol* 1997;8:151–161.

37. Coleman AW, Lazar AN, Rousseau CF, Cecillon S, Shahgaldian P. Characterization of nanoscale systems in biology using scanning probe microscopy techniques. In: *Nanotechnologies for the Life Sciences.* Wiley-VCH Verlag GmbH, Weinheim, 2007.

38. Hölscher H, Schirmeisen A, Fuchs H. *Nanoscale Imaging and Force Analysis with Atomic Force Microscopy.* Wiley-VCH Verlag GmbH, Weinheim, 2010.

39. Puchner EM, Gaub HE. Force and function: probing proteins with AFM-based force spectroscopy. *Curr Opin Struct Biol* 2009;19:605–614.

40. Seitz M. Force spectroscopy. In: Niemeyer C, Mirkin C (eds), *Nanobiotechnology: Concepts, Applications and Perspectives.* Wiley-VCH Verlag GmbH, Weinheim, 2005, pp. 404–428.

41. Gonzalez Arzola K, Gimeno Y, Arevalo MC, Falcon MA, Hernandez Creus A. Electrochemical and AFM characterization on gold and carbon electrodes of a high redox potential laccase from *Fusarium proliferatum. Bioelectrochemistry* 2010;79:17–24.

42. García R. Phase imaging atomic force microscopy. In: García R (ed.), *Amplitude Modulation Atomic Force Microscopy*. Wiley-VCH Verlag GmbH, Weinheim, 2010, pp. 91–101.

43. Eby DM, Luckarift HR, Johnson GR. Hybrid antimicrobial enzyme and silver nanoparticle coatings for medical instruments. *ACS Appl Mater Interfaces* 2009;1:1553–1560.

44. Parra A, Casero E, Vázquez L, Jin J, Pariente F, Lorenzo E. Microscopic and voltammetric characterization of bioanalytical platforms based on lactate oxidase. *Langmuir* 2006;22:5443–5450.

45. Barth C, Foster AS, Henry CR, Shluger AL. Recent trends in surface characterization and chemistry with high-resolution scanning force methods. *Adv Mater* 2011;23:477–501.

46. Otsuka I, Yaoita M, Nagashima S, Higano M. Molecular dimensions of dried glucose oxidase on an Au(111) surface studied by dynamic mode scanning force microscopy. *Electrochim Acta* 2005;50:4861–4867.

47. Gross L, Mohn F, Moll N, Liljeroth P, Meyer G. The chemical structure of a molecule resolved by atomic force microscopy. *Science* 2009;325:1110–1114.

48. Morita S. Atom world based on nano-forces: 25 years of atomic force microscopy. *J Electron Microsc* 2011;60:S199–S211.

49. Nonnenmacher M, O'Boyle MP, Wickramasinghe HK. Kelvin probe force microscopy. *Appl Phys Lett* 1991;58:2921–2923.

50. Palermo V, Palma M, Samorì P. Electronic characterization of organic thin films by Kelvin probe force microscopy. *Adv Mater* 2006;18:145–164.

51. Rodriguez BJ, Kalinin SV. KPFM and PFM of biological systems. In: Sadewasser S, Glatzel T (eds), *Kelvin Probe Force Microscopy*. Springer, Berlin, 2012, pp. 243–287.

52. Higgins M, Wallace GG, Gelmi A, McGovern ST. Electrochemical AFM. *Imaging Microsc* 2009;11:40–43.

53. Macpherson JV, Unwin PR. Combined scanning electrochemical-atomic force microscopy. *Anal Chem* 2000;72:276–285.

54. ElKaoutit M, Naggar AH, Naranjo-Rodriguez I, Dominguez M, de Cisneros JLH-H. Electrochemical AFM investigation of horseradish peroxidase enzyme electro-immobilization with polypyrrole conducting polymer. *Synth Met* 2009;159:541–545.

55. Xu D, Watt GD, Harb JN, Davis RC. Electrical conductivity of ferritin proteins by conductive AFM. *Nano Lett* 2005;5:571–577.

56. Anne A, Cambril E, Chovin A, Demaille C, Goyer C. Electrochemical atomic force microscopy using a tip-attached redox mediator for topographic and functional imaging of nanosystems. *ACS Nano* 2009;3:2927–2940.

57. Kalinin SV, Shao R, Bonnell DA. Local phenomena in oxides by advanced scanning probe microscopy. *J Am Ceram Soc* 2005;88:1077–1098.

58. Mikamo-Satoh E, Yamada F, Takagi A, Matsumoto T, Kawai T. Electrostatic force microscopy: imaging DNA and protein polarizations one by one. *Nanotechnology* 2009;20:145102.

59. Ying L, Shen-Ming C, Ramiah S. Membraneless enzymatic biofuel cells based on multi-walled carbon nanotubes. *Int J Electrochem Sci* 2011;6:3776–3788.

60. Shao Y, Wang J, Wu H, Liu J, Aksay IA, Lin Y. Graphene based electrochemical sensors and biosensors: a review. *Electroanalysis* 2010;22:1027–1036.

61. Zhang J, Zhang F, Yang H, Huang X, Liu H, Zhang J, Guo S. Graphene oxide as a matrix for enzyme immobilization. *Langmuir* 2010;26:6083–6085.

62. Noh H-B, Won M-S, Hwang J, Kwon N-H, Shin SC, Shim Y-B. Conjugated polymers and an iron complex as electrocatalytic materials for an enzyme-based biofuel cell. *Biosens Bioelectron* 2009;25:1735–1741.

63. Alonso-Lomillo MA, Rüdiger O, Maroto-Valiente A, Velez M, Rodríguez-Ramos I, Muñoz FJ, Fernández VcM, De Lacey AL. Hydrogenase-coated carbon nanotubes for efficient H$_2$ oxidation. *Nano Lett* 2007;7:1603–1608.

64. Lojou E. Hydrogenases as catalysts for fuel cells: strategies for efficient immobilization at electrode interfaces. *Electrochim Acta* 2011;56:10385–10397.

65. Volbeda A, Charon M-H, Piras C, Hatchikian EC, Frey M, Fontecilla-Camps JC. Crystal structure of the nickel–iron hydrogenase from *Desulfovibrio gigas*. *Nature* 1995;373:580–587.

66. Ivnitski DM, Khripin C, Luckarift HR, Johnson GR, Atanassov P. Surface characterization and direct bioelectrocatalysis of multicopper oxidases. *Electrochim Acta* 2010;55:7385–7393.

67. Piontek K, Antorini M, Choinowski T. Crystal structure of a laccase from the fungus *Trametes versicolor* at 1.90-Å resolution containing a full complement of coppers. *J Biol Chem* 2002;277:37663–37669.

68. Solomon EI, Sundaram UM, Machonkin TE. Multicopper oxidases and oxygenases. *Chem Rev* 1996;96:2563–2606.

14

SCANNING ELECTROCHEMICAL MICROSCOPY FOR BIOLOGICAL FUEL CELL CHARACTERIZATION

RAMARAJA P. RAMASAMY

Nano Electrochemistry Laboratory, College of Engineering, University of Georgia, Athens, GA, USA

14.1 INTRODUCTION

Scanning electrochemical microscopy (SECM) is a variant of the scanning probe microcopy (SPM) that has evolved to specifically study electrochemical interfaces. SECM possesses the ability to map the chemical, biochemical, and electrochemical reactivity on surfaces with remarkable spatial resolution. SECM can perform the functions of a typical SPM, such as substrate surface patterning and topography imaging, while also enabling quantitative measurements on micrometer and submicrometer scales. It is a potent tool to study the morphology of microstructured surfaces and the processes occurring at solid–liquid, liquid–liquid, and liquid–vapor interfaces. SECM is also capable of probing the kinetics of solution-phase reactions [1], adsorption phenomena [2], and monitoring and quantifying heterogeneous electron transfer kinetics on conductive surfaces [3,4]. SECM is being used in an increasing number of research applications ranging from pitting corrosion and combinatorial screening of metal catalysts on one end to fingerprint imaging and bovine embryo evaluation on the other [5,6]. The technique was originally developed and pioneered by Bard et al. in 1989 [7]. An SECM setup can be custom designed with enough flexibility that allows one to operate the instrument in both aerobic and anaerobic environments. SECM integrates a number of previously disconnected analytical

Enzymatic Fuel Cells: From Fundamentals to Applications, First Edition. Edited by Heather R. Luckarift, Plamen Atanassov, and Glenn R. Johnson.
© 2014 John Wiley & Sons, Inc. Published 2014 by John Wiley & Sons, Inc.

techniques into one powerful characterization tool, the advancement of which can be attributed largely to the development of ultramicroelectrode (UME) technology. Before the development of comprehensive SPM techniques, *ex situ* spectroscopic and electrochemical measurements were the methods of choice available to obtain structural information about an electrochemical interface.

During the early years of SECM technology development, much of the theory beyond SECM was developed and expanded by Bard et al. [1,4,7–11]. SECM is capable of providing quantitative information about the topography and electrochemical characteristics of surfaces down to submicroscopic scales. In contrast to the conventional high-resolution microscopies, such as scanning electron microscopy or fluorescence microscopy, SECM can be used to investigate biological samples with no rigorous pretreatment. The sample can be submerged in the solution or embedded inside a thin film on a surface. In addition, SECM typically does not suffer from artifacts such as background fluorescence, because it is not optically based. Moreover, the system can be analytically or numerically modeled, allowing the subsequent capture of information relating to kinetics of the reaction under observation fairly easily. SECM has been applied to a diverse array of bioanalytical problems because of its nondestructive nature of the imaging process. Recently, there is a strong trend toward the development of miniature analytical devices based on enzymatic electrodes, and therefore, there is wide scope for the use of SECM for enzyme bioelectrochemical applications such as biosensors and biological fuel cells (BFCs). For an enzymatic BFC application, SECM can be used in the surface patterning (fabrication) of electrodes, characterization and optimization of bionanocomposite-based electrodes, and membrane separators. Through the determination of enzymatic activity, SECM may be used to characterize a variety of enzyme bioelectrocatalysts for fuel cell anode and cathode applications. This chapter comprehensively addresses the use of SECM for BFC characterization.

14.2 THEORY AND OPERATION

A modern SECM system consists of a piezoelectric positioning system that moves a microelectrode probe over the sample, as shown in Figure 14.1. The probe (also referred to as the *tip*) acts as an amperometric electrode and is connected to a potentiostat. A reference and a counter electrode complete the electrochemical setup. The sample (also referred to as the *substrate*) can be connected to the potentiostat as the second working electrode. The theory behind the operation of SECM has been developed for various homogenous and heterogeneous processes as well as for different tip–substrate geometries. In general, the SECM responses can be generated by numerically solving partial differential equations. In some cases, analytical approximations allow for easier generation of theoretical dependences and analysis of experimental data. The analytical approximations of SECM responses have been derived using fundamental transport and kinetics equations and have been comprehensively reviewed in 2007 by Sun et al. [12]. Theoretical descriptions over the years include but are not limited to (i) relationship between faradaic current measured at the

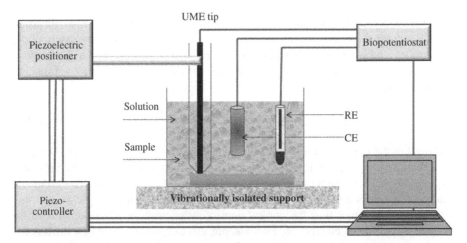

FIGURE 14.1 Schematic of the components of SECM setup. A piezoelectric positioner controls the UME movement. UME and sample are connected as working electrodes to a bipotentiostat. A computer controls both the bipotentiostat and the piezocontroller.

tip and the substrate charge transfer properties, (ii) the relationship between faradaic current detected at the tip and the tip–substrate distance, and (iii) the relationship between the mass transport and homogenous reaction kinetics in the inner tip–substrate gap [13].

SECM has become an important tool for the characterization of enzyme-modified electrodes and is being used in biosensor, biochip, and BFC applications. SECM is a chemical microscopy technique that provides information about not only the topography of the sample surface but also the reactivity of the sample (substrate) surface. When the probe is brought to the vicinity of the substrate, the electrochemical response of the tip is recorded as a function of (i) the lateral position (x, y) of the tip on the sample surface for imaging, (ii) the tip–substrate distance in approach curve measurements, or (iii) the time at a fixed-tip position in chronoamperometry measurements. For enzymatic fuel cell (EFC) applications, the SECM can be applied to the study of the kinetics of bioelectrochemical reactions on enzyme-modified electrodes in order to evaluate the local catalytic activity, to obtain a chemical image of the topography of the electrode, and to study the transport across a fuel cell membrane separator. Because the technique probes processes on a microscopic domain, the probe must be small enough to detect the changes in the microscopic domain. Therefore, UMEs are used as probes in SECM measurements.

14.3 ULTRAMICROELECTRODES

Small electrodes have been in use since the 1980s in the field of electrochemistry [14]. An electrode with at least one of its dimensions below 50 μm will exhibit noticeably different properties than macroscopic electrodes made of the same material; such electrodes are termed *ultramicroelectrodes*. UMEs form hemispherical diffusion fields around them that allows efficient mass transfer compared to macroscopic electrodes and a rapid establishment of steady-state diffusion. For a disk-shaped electrode, this steady-state diffusion-limited current ($i_{T,\infty}$) is given by Equation 14.1 [12,14]:

$$i_{T,\infty} = G(zFDC^*r_{ume})$$ (14.1)

where F is the Faraday constant, z is the number of electrons per molecule, D and C^* are the diffusivity and bulk concentration of the reacting species, respectively, and r_{ume} is the radius of the active electrode area of the disk-shaped probe. The geometric factor G assumes a value of 4 for disk-shaped electrodes. The $i_{T,\infty}$ yields a current density that is larger than the current density at macroscopic electrodes under the conditions of forced convection. For example, electrolysis of 10 μM of a small redox molecule with a diffusivity of 10^{-5} cm^2 s^{-1} at a standard 2–25 μM diameter disk probe results in an $i_{T,\infty}$ of 10 pA [15]. UMEs also offer important advantages for electro-analytical applications including greatly diminished ohmic potential drop in solution and double-layer charging currents, the ability to reach steady state in milliseconds, and the ability to perform experiments in microscopic domains.

In SECM, a UME is the most commonly used probe that acts as a primary electrode. Although the UME is disk shaped in most cases, it is referred to as the tip in electrochemical microscopy. UME parameters such as tip radius, diameter, and geometry of the insulating sheath surrounding the tip strongly influence the measurements. This is in addition to the influence from sample parameters such as sample size and topography. Their influences can be empirically quantified as explained in detail in 2007 by Wittstock et al. [16]. SECM analysis can be of two types [15]. The first type is the investigation involving no lateral movement of the UME tip or the sample. This is commonly known as the approach curve method or the probe approach method of analysis. The second type is the analysis involving the lateral movement of the tip over the sample that provides information about the sample topography (or morphology) in the form of current, potential, impedance, or capacitance. The various operation modes of SECM that are relevant to BFC application are discussed below.

14.3.1 Approach Curve Method of Analysis

A plot of the UME tip current versus the tip-to-surface (sample) distance is called the *approach curve*. The shape of the curve can be used to determine the surface reactivity of the sample. The approach curve method of analysis in SECM goes hand in hand with two-dimensional mapping analysis. It is a reliable method to operate the SECM safely by maintaining a controllable distance between the UME tip and the substrate, without which tip breakage is more likely. Approach curve analyses do not necessarily fall into the category of microscopy or topography imaging analysis. Nevertheless, the experiments using approach curve have fundamental importance for methodological development, because the understanding of any scanning probe experiment requires the knowledge of underlying signal–distance relationship.

Moreover, quantitative data can be extracted much more easily from the approach curves than from the 2D images. From the perspective of applications, such measurements are important because they enable new experiments concerning interfacial processes that are otherwise not accessible. In the approach curve analysis, the normalized tip current is plotted as a function of the tip–substrate distance at a particular coordinate (x, y) on the surface. As the tip moves toward the sample, the gap between the UME and the sample decreases; this also decreases the diffusion of species from the bulk into the gap (microenvironment), or vice versa. Further movement of the tip toward the sample will result in an exponential increase in the tip current, as diffusion of species in or out of the microenvironment is prevented by the close positioning of the tip to the sample. Ultimately, the problem of vertical positioning for 2D or 3D imaging can be circumvented because the tip current is recorded at all relevant distances. Approach curve analysis is the most suitable method for investigation of layered nanostructures on electrodes. For example, the electrochemical properties of electrodes modified with multilayered enzyme nano-composites have been investigated with ferrocene–methanol as a redox mediator using SECM approach curve method [17]. When a nonconductive substrate such as a plain glass slide is used as a sample (Figure 14.2), the tip current decreases as the UME probe approaches the sample. This is due to the hindrance of the mediator diffusion from the bulk into the tip–substrate gap (microenvironment), resulting in a negative feedback-like condition (which is discussed in detail later in this chapter). On the other hand, when a conductive surface is used as a sample, the mediator is catalytically reduced by the sample and produces a positive feedback response as the

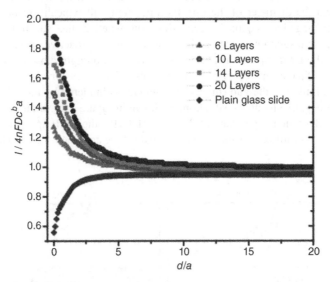

FIGURE 14.2   Approach curves measured on electrode modified with different number of SWNT–LYZ/SWNT–DNA composite layers at identical mediator concentrations. (Reproduced with permission from Ref. [17]. Copyright 2009, Elsevier.)

tip approaches the surface. The quantitative response from a conductive sample in the approach method of analysis is strongly dependent on the surface properties of the conductive sample. Accordingly, when a conducting biopolymer with high lateral conductivity such as a nanocomposite of lysozyme–carbon nanotube (LYZ–CNT) or deoxyribonucleic acid–carbon nanotube (DNA–CNT) is used, the normalized tip current as measured by the approach curve analysis varied with the number of layers of the composite [17]. As shown in Figure 14.2, a clear increase of the steady-state tip current is observed as the tip approached the film. The regeneration of the electro-active mediator at the composite-modified electrode surface is driven by the lateral conductivity within the LYZ–CNT or DNA–CNT composite films. The approach curves in Figure 14.2 also show a strong dependency of the tip current on the number of active layers of nanocomposites present on the sample.

14.4   MODES OF SECM OPERATION

Feedback (FB) mode is one of the first developed modes of operation that involves the reacting species directly in the SECM measurements. The term "feedback" does not refer to the electronic feedback of the instrument to maintain a sample–UME distance. In SECM, feedback refers to the coupling of redox reactions at the UME and the sample.

14.4.1   Negative Feedback Mode

When an UME tip probes the surface at close distances from the surface (typically smaller than the diameter of the tip), the operation will typically result in either a positive or negative feedback, depending on whether the substrate surface is conducting or insulating, respectively. Consider an electroactive species "O" being reduced at the UME tip to form "R". As the UME is brought close to an insulating substrate, the tip current decreases monotonically, since diffusion of the redox species O from the bulk into the gap (microenvironment) is hindered by the substrate. This results in a negative feedback situation, as shown in Figure 14.3. If the substrate is an insulator and is much larger than the diameter of the tip, the UME tip current decreases continuously toward zero as the tip–substrate distance approaches zero. The resulting

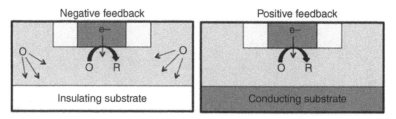

FIGURE 14.3   Schematic of negative and positive feedback situation in SECM.

approach curves will be similar to the ones obtained for the glass slide in Figure 14.2 [17]. Accordingly, a negative FB mode can be used to determine the tip–substrate distance and the substrate topography based on the values of the tip current at different (x, y) locations.

14.4.2 Positive Feedback Mode

When the substrate is a conductor that can regenerate the mediator by oxidizing R to O, the operation will result in a positive feedback situation. As shown in Figure 14.3, although the diffusion of redox-active species from the bulk to the tip is hindered at shorter tip–substrate distances, efficient redox cycling based on planar mediator diffusion in the gap amplifies the tip current as the gap becomes narrower. Under this condition, the tip responses will be similar to the ones obtained for the DNA–CNT- or LYZ–CNT-modified electrodes shown in Figure 14.2 [17].

14.4.3 Generation–Collection Mode

The generation–collection (GC) mode works in a solution that does not initially contain any substrate that can be electrochemically converted at the potential of the UME tip. Whereas the approach curves and the feedback experiments need amperometric electrodes, the GC experiments can be performed with potentiometric UMEs and biosensor probes [14,18,19], which makes GC a distinct feature of SECM for many bioelectroanalytical applications. Being passive sensors, the potentiometric probes are advantageous over amperometric UMEs as they disturb the substrate diffusion layer much less. If an electroactive compound is formed at the substrate, this compound can be detected at the potentiometric UME, if it is located close to the active region shown in Figure 14.4a. Specifically, this mode is called the substrate generation–tip collection (SG/TC) mode. On the other hand, using the UME as a local generator of the electroactive species and the sample as a collector will result in a tip generation–substrate collection (TG/SC) mode, which was demonstrated as early as 1991 by Lee et al. and is schematically shown in Figure 14.4b [20].

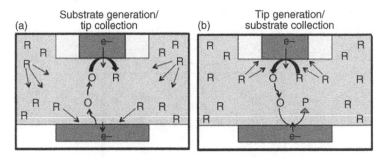

FIGURE 14.4   Schematic of GC mode of operation in SECM: (a) SG/TC mode and (b) TG/SC mode.

Seemingly a simple concept, the GC mode of operation offers some difficulty in the quantitative interpretation of the results [14]. First, if the active regions of the sample are large, steady-state condition will not be established quickly. In this case, the local concentrations will depend not only on the position of UME tip but also on the time that has passed since the onset of the reaction at the substrate surface. Second, the moving probe disturbs the macroscopic diffusion layer of the substrate through convection by hindering the diffusion of reagent to the sample region underneath the UME and/or by overlap of the diffusion layers from the sample and the UME. Therefore, when compared with FB mode, the lateral resolution is always poorer in GC mode experiments, and it is difficult to detect small active regions in the vicinity of large active areas [21]. On the positive side, GC mode offers much higher sensitivity than FB mode because the flux of reagents coming from the sample is measured essentially without a background signal. This makes GC mode particularly appropriate for the investigation of immobilized enzymes for EFC applications.

14.4.4 Induced Transfer Mode

Molecular transport across a membrane or thin porous film separating two immiscible liquid phases can be studied by induced transfer (IT) mode [15,22–26]. An ideal example of such a system is a biological cell membrane, which is a two-phase system where the intercellular and intracellular spaces are separated by a thin bilayer lipid membrane. However, the method has also been used to study liquid–liquid and liquid–vapor interfaces. Accordingly, IT mode has some potential applications in EFCs for studying the membranes that separate the anode and cathode chamber. As schematically explained in Figure 14.5, in IT mode, an UME probe is brought to the interface between two phases, where a redox-active molecule is partitioned at the equilibrium. The tip current can be enhanced even without mediator regeneration based on the redox reaction at the interface, because the bottom phase serves as a reservoir of the redox species (Figure 14.5). When the redox molecule is electrolyzed at the tip to be depleted locally, the tip-induced concentration gradient drives the molecular flux across the interface enhancing the tip current. The rate of the interfacial transfer of the redox species can be obtained from an approach curve [24]. Moreover, the diffusion coefficient and concentration of the transferred molecule at the opposite side of the interface can be determined from chronoamperometric measurements of

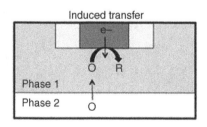

FIGURE 14.5 Schematic of IT mode of operation in SECM.

the transient and steady-state tip currents at a known tip–substrate distance without contact from the tip [15].

14.5 SECM FOR BFC ANODES

SECM has become an important tool for the characterization of enzyme-modified electrodes for BFC applications. Investigations including kinetic study of bioelectrochemical reactions, the activity distribution on the surface, and the electrochemical image of enzymes on the electrode surfaces can be precisely studied using both FB and GC modes. An advantage of SECM for enzyme electrode characterization is that the measurements are independent of the type of transducer component. Hence, the method of enzyme immobilization does not affect the measurements. The immobilized enzyme can be imaged by enzyme-mediated FB or GC mode. FB mode may offer high spatial resolution but with low sensitivity; the converse is true for substrate generation-tip collection (SG/TC) mode. Therefore, the choice of the operation mode is entirely dependent on the user, type of sample, and experimental conditions. The analytical treatment of SECM for both FB and GC modes has been comprehensively reviewed in 1992 by Zhou et al. [27].

14.5.1 Enzyme-Mediated Feedback Imaging

In FB mode, the UME tip is held at a potential where electrolysis of a soluble redox mediator proceeds under diffusion-controlled conditions and the surface does not communicate with the tip. FB mode requires that the tip be positioned at a distance d as close as possible to the surface. This requires that the probe approach currents be adjusted accordingly during the initial setup of the system. When the UME tip is positioned at a quasi-infinite distance (∞) from the surface (where $d >10$ times the tip radii), the FB current is defined by Equation 14.2 [9]:

$$i_{T, \infty} = 4zFD_m C_m r_T \qquad (14.2)$$

where r_T is the radius of the UME tip, z is the number of electrons, F is the Faraday constant ($F = 96,485 \, \text{C mol}^{-1}$), and D_m and C_m are the diffusivity and bulk concentration of the reduced form of soluble redox mediator, respectively. Currents less than $i_{T, \infty}$ represent a negative FB communication and currents greater than $i_{T, \infty}$ represent a positive FB communication between the tip and the surface. The enzyme-modified electrode surface can be characterized using SECM using the mediators in positive FB mode. There is a significant difference from the electrochemical FB described earlier and the enzymatic FB mode imaging described here. The difference primarily lies in the kinetics of enzymatic and electrochemical redox reactions that recycle the mediator back to the UME for FB detection. Any SECM feedback imaging requires that the flux of regenerated mediators from surface to UME be high enough to compete with the diffusion flux of mediators from the bulk solution (i.e., from outside the SECM microenvironment). In electrochemical FB mode on an active sample, any change in bulk mediator concentration will equally affect the rate of mediator regeneration at the sample. Therefore, electrochemical FB mode allows us to choose a fairly wide range of UME tip sizes and mediator concentrations to work with. This is not true with enzyme-mediated FB mode, as the rate of mediator regeneration by the enzymes depends on the quantity or the surface loading of the enzymes. Hence, an increase in bulk mediator concentration enhances the mass transfer of mediators from bulk to sample but does not necessarily increase the flux of regenerated mediators from the sample to UME.

Because of its low sensitivity compared with the GC mode, FB mode requires a highly active enzyme. This method of imaging involves the replacement of the enzyme cofactor or electron acceptor by a mediator. To be able to determine the enzymatic activity using the enzyme-mediated positive FB mode in SECM, the system must satisfy the following condition (Equation 14.3) [28]:

$$k_{cat}\Gamma_{enz} \geq 10^{-3}\left(\frac{D_m C_m}{r_T}\right) \tag{14.3}$$

where k_{cat} is the enzyme turnover number and Γ_{enz} is the enzyme surface concentration or enzyme loading (moles per unit area). For a given enzyme, a high Γ_{enz} is very important for a successful positive FB detection [29]. Therefore, enzymes immobilized on porous nanomaterials such as hydrogels, carbon nanotubes, and nanoparticles will be ideal surfaces for positive FB detection because of their high Γ_{enz}. Moreover, enzyme-mediated positive FB imaging requires that none of the mediator solution components interfere with the enzyme activity or undergo redox reactions at the working electrode [30]. The turnover number (k_{cat}) for first-order enzymatic kinetics can be obtained from the limiting expression of the Michaelis–Menten equation for small substrate concentrations (Equation 14.4) [29]:

$$k_{cat} = \frac{K_M k_f}{\Gamma_{enz}} \tag{14.4}$$

where k_f is the first-order apparent heterogeneous rate constant (in cm s^{-1}) and K_M is the apparent Michaelis constant, which can be estimated by a hyperbolic nonlinear fitting of the tip current for finite substrate kinetics (i_s^k) versus bulk mediator concentration (C_m^*) curve at different UME tip–substrate surface distances (d). If we define normalized parameters for tip–substrate distance (L) and the tip current for finite substrate kinetics (I_s^k) as (Equations 14.5 and 14.6)

$$L = \frac{d}{r_T} \tag{14.5}$$

$$I_s^k(L) = \frac{i_s^k(L)}{i_{T,\infty}} \tag{14.6}$$

TABLE 14.1   Common EFC Anode Enzymes Studied Using Enzyme-Mediated FB Imaging Mode in SECM

Enzyme	Substrate/Type of UME Reaction	Mediator	References
GOx	Glucose/oxidation	FcCOOH	[28]
		FcCH$_2$OH	[28]
		K$_4$[Fe(CN)$_6$]	[28]
		Hydroquinone	[28,35]
		[Os(fpy)(bpy)$_2$Cl]Cl	[36]
		Dimethylaminoferrocene	[30]
Glucose dehydrogenase	Glucose/oxidation	FcCOOH	[37]
		FcCH$_2$OH	[37,38]
		p-Aminophenol	[37]
Alcohol dehydrogenase	Ethanol/oxidation	Fe(CN)$_6{}^{3-/4-}$	[39]
		(with Os redox relay)	
Lactate oxidase	Lactate/oxidation	FcCH$_2$OH	[40]
Cytochrome c	NADH/oxidation	Fe(CN)$_6{}^{3-/4-}$	[41]

then the normalized tip current for first-order finite substrate kinetics for $(0.1 < L < 1.5)$ can be given by the following empirically derived analytical approximation (Equation 14.7) [4,29,31]:

$$I_s^k(L) = \frac{0.784}{L\left[1 + (D_m/k_f d)\right]} + \frac{0.68 + 0.332\,e^{(-1.067/L)}}{1 + \left[((11D_m/k_f d) + 7.3)/(110 - 40L)\right]} \tag{14.7}$$

Table 14.1 lists the anode enzymes studied using SECM in FB mode and reported in the literature. Although the enzyme-mediated FB has primarily been used for imaging enzymes that catalyze oxidation reactions (anode enzymes for EFCs), there are examples of using this mode for studying reduction reactions using enzymes such as nitrate reductases [32,33] and horseradish peroxidase (HRP) [34].

14.5.1.1 Imaging Glucose Oxidase Activity Using FB Mode   Glucose oxidase (GOx) is the most widely studied oxidoreductase enzyme for EFC anodes. The enzyme oxidizes glucose to gluconolactone and delivers the electrons to the native electron acceptor, oxygen. In 2001, Wittstock et al. used enzymatic FB mode imaging extensively to study GOx activity [33]. In this work, the enzyme-modified electrode was used as the substrate with platinum as the UME (Figure 14.6). An oxygen-free buffer solution of desired pH (containing glucose and a ferrocene derivative) was used as the electrolyte. The ferrocene derivatives serve as alternate electron acceptors for GOx and are oxidized at the UME and regenerated at the enzyme-modified surface. Therefore, at locations directly above the enzyme, the flux of reduced ferrocene is high, resulting in increased UME tip currents [33,42]. While the ferrocene-based compounds such as FcCOOH, FcCH$_2$OH, and dimethylaminomethyl ferrocene

(a)

(b)

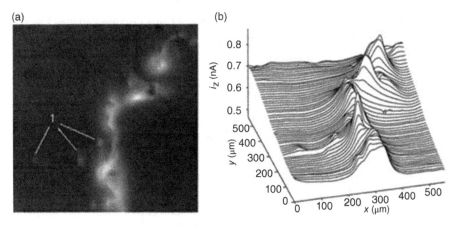

FIGURE 14.6 GOx-modified beads arranged and images using FB mode. (Reprinted with permission from Ref. [33]. Copyright 2001, Wiley-VCH Verlag GmbH.)

hydroquinone were used in 2000 by Wijayawardhana et al. [33,42], other mediators such as p-aminophenol and ferricyanide were tried 4 years later by Zhao and Wittstock for imaging GOx activity using FB mode [29].

SECM mapping of enzyme activity requires meticulous experimental setup preparation. This includes quantitative analysis of the enzyme concentration, substrate specificity, and limiting redox kinetics. High reductant specificity and sensitivity are critical for SECM imaging in FB mode. To obtain a significant positive feedback to the tip, the enzyme–mediator reaction had to be sufficiently fast. One significant drawback of imaging enzymatic activity in FB mode is the difficulty in eliminating the background contribution, especially in cases where the enzyme is immobilized on a gold surface, which results in low signal-to-noise ratio and hence low sensitivity. It is important that the signal contribution from mediator regeneration due to electrochemical reaction on the gold surface under the enzyme layer must be decoupled from the faradaic current arising because of redox reactions of the electroactive species generated by the immobilized enzyme [43,44].

14.5.2 Generation–Collection Mode Imaging

The SG/TC mode is another operation mode of SECM that has been used for imaging enzyme activity on modified electrode surfaces. Unlike FB mode, SG/TC mode does not involve mediators and the activity mapping is achieved by a direct reduction or oxidation of the enzymatic product. Therefore, the biocatalytic activity of the enzyme toward its substrate depends on neither the presence nor the location of the UME tip. In this mode, the UME tip moves across the specimen surface to map the local concentration of the species generated or consumed by the enzyme. The faradaic current at the UME tip will give a direct measure of the local enzyme activity underneath the UME. Because of the negligible background contribution, high signal sensitivity can be achieved in SG/TC mode. However, this mode lacks high spatial

TABLE 14.2   Common EFC Anode Enzymes Studied Using GC Imaging Mode in SECM

Enzyme	Substrate/Type of UME Reaction	Mediator	References
GOx	Glucose/oxidation	H_2O_2	[45–52]
		Hydroquinone	[53]
Glucose dehydrogenase	Glucose/oxidation	$Fe(CN)_6^{4-}$	[37]
Galactosidase	PAPG/oxidation	p-Aminophenol	[37]
Horseradish peroxidase	H_2O_2/oxidation	$FcCH_2OH$	[49,54–56]

resolution because the distribution of the products of biocatalytic reactions may diffuse away from the sample surface into the surrounding medium. The spatial resolution may also be affected by disturbance of diffusion layer around the enzyme region due to UME tip movement across the surrounding medium. The quality of the image will also be affected if the tip hinders the substrate diffusion from the bulk to enzyme surface, resulting in reduced tip currents. In such situations, the SG/TC mode would resemble that of a negative FB operation on an insulating surface. Also, a threshold minimum specific activity of the enzyme is required for successful imaging. To determine the enzymatic activity using the SG/TC mode in SECM, the system must satisfy the following condition (Equation 14.8):

$$k_{cat}\Gamma_{enz} \geq \frac{DC'}{r_{enz}} \qquad (14.8)$$

where r_{enz} is the radius of enzyme-modified area on the sample surface, C' is the detection limit for the species detected at UME, and D is the diffusion coefficient of that species (Table 14.2).

14.5.2.1 Imaging GOx Using SG/TC Mode

Imaging GOx using SG/TC mode is fairly straightforward. GOx at the sample surface converts glucose and dioxygen into gluconolactone and hydrogen peroxide, respectively. Hydrogen peroxide is electrochemically oxidized to water at the tip, resulting in an oxidation current. The tip is usually held at high positive potentials (>0.75 V vs. Ag/AgCl) to facilitate the oxidation of H_2O_2 [45]. A clear difference in the tip current can be noticed in the presence and absence of glucose that can be processed to obtain an electrochemical image of the surface activity (Figure 14.7).

14.6   SECM FOR BFC CATHODES

The reduction of oxygen to water is the reaction of choice for a fuel cell cathode. This reaction is one of the most widely studied reactions in the field of electrocatalysis.

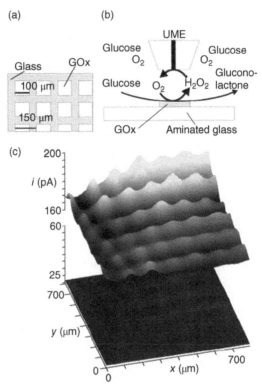

FIGURE 14.7 SECM imaging of GOx activity on patterned monolayers of glass using GC mode. (a) A mixture of GOx and carbodiimide was applied to the stamp and brought into contact with an aminated glass surface on the 100 µm × 100 µm squares; contact time, 30 min. (b) Schematic of the SECM GC experiment. (c) Top image, air-saturated 50 mM glucose, phosphate buffer, pH 7; bottom image, air-saturated phosphate buffer without glucose. (Reprinted with permission from Ref. [49]. Copyright 2002, American Chemical Society.)

14.6.1 Tip Generation–Substrate Collection Mode

Until a few years ago, the TG/SC mode of SECM operation was the most common way to image an enzyme that catalyzes oxygen reduction. TG/SC mode is well suited for imaging activity of surfaces with morphological features because it is relatively insensitive to changes in the tip–substrate distance [14]. The main difference between this mode and classical FB mode is that the feedback diffusion process is not required for TG/SC mode, which enables a direct measurement of activity in acidic solutions. This mode is the converse of SG/TC mode used for the anode catalysts. TG/SC mode has been applied to the study of the kinetics of oxygen reduction reaction (ORR) [14], evaluation of catalytically active nonprecious metal alloy compositions [57,58], optimization of Cu(II) biomimetics [59], thermodynamics-based design of catalysts [60], and analysis of wired enzyme architectures [61].

14.6.1.1 Imaging ORR by TG/SC Mode The principle of imaging ORR by TG/
SC mode is the same as any GC mode. Here, the oxygen reduction current is measured
at the sample (i_S) as a function of the UME tip position [60]. The substrate potential is
fixed at a value where O_2 should be reduced to water. In the absence of O_2, the
substrate (i_S) is negligible. On the other hand, when a UME is brought to the surface, a
constant oxidation current (i_T) is applied to the tip to oxidize water to oxygen at the tip,
and a constant flux of O_2 is generated at the tip, which diffuses to the substrate. The tip
current must be sufficiently small to prevent oxygen saturation in the solution present
in the microenvironment and/or bubble formation [60]. When O_2 reaches the substrate
surface, it is reduced at a rate depending on its substrate potential and electrocatalytic
activity. The main advantage of TG/SC mode is that the observed response does not
require the feedback process. This property allows one to study the reactions that are
inaccessible to the feedback approach and to extend the concentration range beyond
the low values required to reach diffusion control at the UME tip (Figure 14.8) [60].
However, the major drawback of this mode for studying ORR can be attributed to
the high background current, which decreases the sensitivity of i_S on large sample
surfaces. Therefore, image resolution using TG/SC mode is limited, since a decrease in
the size of the UME tip would cause the signal to be invariably concealed by the
significantly high background current contributed by the large sample. Accordingly,
TG/SC mode is considered inadequate for high-resolution imaging of catalytic activity.

The other option in GC is SG/TC mode, which has been used successfully for
imaging activity of anode catalysts such as GOx. However, in this mode, the analyte
(O_2) concentration is significantly altered, and large enzyme sample surface and small
UME tips are needed. Moreover, the application of potential pulse to the sample in
TG/SC mode is limited by capacitive charging currents, resulting in a poorly defined
UME potential. To overcome these drawbacks in GC mode for imaging cathode
catalyst activity, a new method called redox competition (RC) mode was developed in

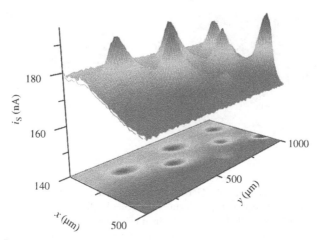

FIGURE 14.8 ORR activity imaging by TG/SC mode of an array of Pt spots supported on
glassy carbon. (Reprinted with permission from Ref. [60]. Copyright 2003, American Chemical
Society.) (Please see the color version of this figure in Color Plates section.)

2006 by Schuhmann and coworkers [62] based on the foundation of a 2003 theoretical simulation by Wittstock and coworkers [63].

14.6.1.2 Imaging Laccase by SG/TC Mode

14.6.1.2 Imaging Laccase by SG/TC Mode Encapsulation of enzyme and a conductive composite material within a sol–gel matrix is one of the well-known methods for enzyme immobilization on electrode surface for bioelectrochemical applications. Laccase, a multicopper polyphenol oxidase, has been immobilized in sol–gel-processed hydrophilic silicate films for biocathode applications earlier [64,65]. These electrodes exhibit both mediated and mediatorless bioelectrocatalysis of oxygen reduction. The stable immobilization of mediator and uniform protein distribution throughout the silicate film play an important role in the performance of biocathodes. A 2008 study by Nogala et al. showed that SECM laccase tends to aggregate within a silicate film, reducing the electroactivity of the composite [66]. The SECM images of laccase–silicate composite obtained using SG/TC mode and RC mode using 2,2'-azinobis(3-ethyl-benzothiazoline-6-sulfonic acid) (ABTS) as a mediator are shown in Figure 14.9.

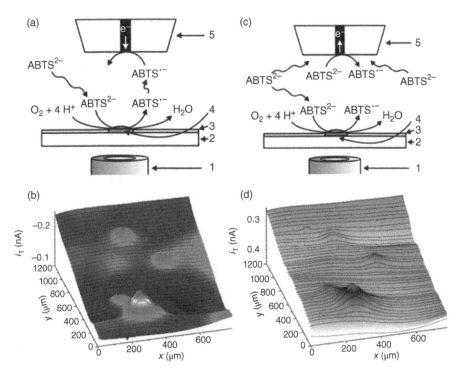

FIGURE 14.9 (a) Schematic of SG/TC mode SECM imaging of laccase activity (not to scale): (1) microscope objective; (2) glass slide; (3) laccase–silicate film; (4) laccase aggregate; (5) UME. (b) SG/TC mode SECM image obtained using 1 mM ABTS^{2-} as mediator in 0.1 M phosphate buffer (pH 4.8) at $E_T = -0.1$ V. (c) Schematic of RC mode SECM imaging of laccase activity (not to scale). (d) RC mode SECM image: $E_T = +0.4$ V and other conditions as in (b). (Reprinted with permission from Ref. [66]. Copyright 2008, Elsevier.) (Please see the color version of this figure in Color Plates section.)

14.6.2 Redox Competition Mode

In the transient RC mode, both the UME tip and the enzyme-modified surface sample compete for the same analyte present in the microenvironment between the tip and the sample. Here, both tip and sample are held at the same potential, enabling both of them to electrochemically convert the analyte. When tip and sample are not along the same vertical axis, and are sufficiently separated laterally, the currents at the tip and sample are determined by the bulk analyte concentration and kinetics at the respective electrodes. However, as the tip moves closer to the sample and/or when it is directly above the sample, both electrodes (tip and sample) compete for the limited quantity of the analyte present between them. This results in a reduced tip current, as the analyte concentration available at the tip is reduced by its reaction at the sample. The decrease in the tip current is imaged as enzymatic redox activity in RC mode of SECM [63]. RC mode does not impose any limitation on the quality of the SECM activity image. Therefore, there are no restrictions to both the sample size and the minimum size of the UME tip.

14.6.2.1 Imaging ORR by RC Mode

To image the ORR for BFC cathode application, RC mode offers the best imaging possibilities. Here, both the UME tip and the enzyme-modified electrode sample must be polarized to sufficiently negative overpotentials for ORR using a bipotentiostat, but care must be taken to avoid any hydrogen evolution in the system. For example, in their 2006 study, Eckhard et al. applied a potential of -0.6 V versus Ag/AgCl (overpotential $= -1.2$ V at pH 7) [62]. In O_2-saturated electrolyte, with the UME tip positioned far away from sample surface, the O_2 reduction current quickly reaches steady state as it depends not only on the O_2 partial pressure in the bulk electrolyte. When the tip moves over an inactive sample site, a negative feedback response may be obtained, due to the blocking of O_2 diffusion into the microenvironment between the tip and enzyme sample. Therefore, the tip–substrate distance must be smaller than the diameter of the tip itself, preferably about 50–75% of the tip diameter. While scanning the UME tip over the (x, y) plane of the sample surface at a constant z-height, the residual O_2 reduction will be constant as long as the SECM tip is located above the inactive sample site. If the tip enters the region of an active enzyme spot where there is a diffusion zone of an active O_2-consuming spot on the surface, the local O_2 concentration within the gap will be decreased, causing a diminished current measured at the SECM tip. This way, the active sites for O_2 reduction on the sample surface can be visualized, as tip current decreases as a function of the UME position on the (x, y) plane. The larger the value of the ORR current measured at the tip, the lower the ability of the enzyme-modified surface to reduce O_2. On the contrary, a relative decrease in UME tip current as it moves above the sample indicates a high-activity catalytic sample. The SECM image of a bilirubin oxidase–osmium polymer-mediated complex film on a sample imaged using the RC mode is shown in Figure 14.10 [67]. Imaging the overlapping diffusion zones in RC mode allows the SECM tip current to relay information on the catalytic activity of the sample spot that is perpendicularly located below the tip.

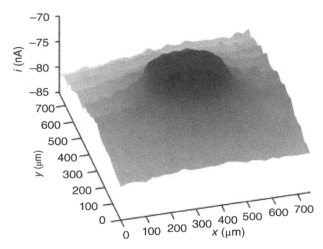

FIGURE 14.10 RC SECM image of a bilirubin oxidase/Os polymer spot on a glassy carbon chip in 20 mM phosphate citrate buffer (pH 6). The O_2 reduction current at the SECM tip (25 μm diameter Pt disk electrode) is plotted as a function of the (x, y) position. (Reprinted with permission from Ref. [67]. Copyright 2007, Elsevier.)

RC mode, however, is not limitation-free. If the sample is highly active and consumes all the locally available O_2 in the gap between tip and the surface, the tip current will become zero because of the hindrance of the diffusion of O_2 from the bulk into the gap. This will lead to incorrect imaging of the catalyst activity. Therefore, in RC mode, it is very important to control the consumption of O_2 on both the UME tip and the sample surface. This can be achieved by applying a predetermined potential pulse profile at the UME tip to avoid complete O_2 depletion [62]. In a short pulse to the water oxidation potential prior to applying the O_2 reduction potential, the analyte is produced within the gap between tip and sample. Depending on its activity, the sample is kept continuously at a potential sufficiently low to invoke O_2 reduction (e.g. −0.65 V vs. Ag/AgCl) but not low enough to deplete the O_2 completely. During the tip movement, and for a predefined conditioning time, a reasonable base potential is applied to the UME tip at which no O_2 reduction takes place. This way, the convection caused by the tip movement, local pH variations due to ORR, and the O_2 production pulse are allowed to decay before the next detection cycle. The application of this pulse method of RC-based SECM imaging for oxygen reduction using multicopper oxidases has been reviewed in detail elsewhere [62].

14.7 CATALYST SCREENING USING SECM

Fast screening of materials for their catalytic activity has gained popularity in the area of polymer electrolyte fuel cells. The method involves preparing a combinatorial library of various candidate catalysts of different composition and a quick evaluation of a key electrochemical parameter such as open-circuit potential or high-frequency impedance. This enables a quick screening of an array of materials to identify potential candidate catalytic materials for relevant applications. The timescale of the experiments can be reduced several fold when compared with traditional experimental evaluation of each material. Combinatorial libraries have been very popular in the field of biotechnology and medicine but have been used only in limited electro-chemical applications. For EFCs, great promise exists for combinatorial testing of gradient materials, in which different enzyme bioelectrode surfaces are simulated on a substrate for the purpose of optimization [46,47,68]. SECM has been used in the screening of Pt–Co, Pt–Ru alloy, and Pt–C materials as cathode catalysts for polymer electrolyte fuel cells [57,69–72]. A similar application of SECM to screen and evaluate BFC catalysts is possible. SECM offers the possibility of automated, unattended library characterization, which is appealing for use in combinatorial material science. The electrochemical feedback mechanism in SECM provides the ability to visualize the spatial distribution of reaction rates across a materials library, thus aiding in the search for new and better catalysts.

For rapid evaluation, catalysts deposited along a line in an array electrode could be used for characterization. Detailed information can also be obtained from the approach curves of each library member in the array. From these data, reaction rates of each library member can be estimated. The region with the highest tip current value may correspond to the optimal architecture or activity, which can be used as a starting point for further optimization. In this way, the time-consuming testing of preparation protocols with a multitude of variable parameters could be shortened. A significant obstacle to library characterization can result from tip fouling, which can be due to organic contaminates, spilling of particulates from oxidizing surface, or catalytic poisoning of the UME tip. This can be avoided by periodically cleaning the tip by applying a voltage cycle sweep.

Array electrodes for SECM can be prepared using a variety of photolithography tools or using inkjet printing for preparing graded catalyst layers [47]. In 2005, Black et al. tried characterizing thin film combinatorial libraries of fuel cell catalysts materials using SECM. [69] In their work, a highly ordered, fine-resolved combinatorial array of electrodes were prepared using a combinatorial sputtering system that was used as a library for evaluating binary Pt–Ru catalytic materials. In 2003, Wilhelm et al. developed a sensor array with defined arrangement of GOx and HRP by a combination of microcontact printing and SECM-based modification [73]. A pattern was formed in which one region was modified by GOx and surrounded by a periodic grid of immobilized enzyme. GOx converts glucose and O_2 into gluconolactone and H_2O_2. The latter served as a substrate in the HRP-catalyzed oxidation of ferrocene–methanol. A GC image of the emerging ferrocenium derivative shows the combined action of both enzymes. The different signals in the resulting image can be traced back to the different availability of H_2O_2 and can be distinguished clearly from local variation of immobilized HRP activity [73].

14.8   SECM FOR MEMBRANES

To date, there has not been a unified consensus as to whether or not a membrane will be used to separate the anode and cathode chambers in an EFC. However, to avoid the thermodynamic losses caused by crossover of reactants from either electrode, the use of an ion-selective membrane becomes necessary and important. There are currently no commercially available membranes for EFCs. Modified proton exchange membranes such as Nafion, however, have been custom developed for EFCs, where the pH of the membrane when fully hydrated will stay close to 7. The EFCs are not yet "engineering-ready"; hence, the research on membranes is very limited. Nevertheless, SECM provides the ideal analytical setup to evaluate local mass transport processes across the membranes and tissue.

SECM has been used in conjunction with potentiometry techniques to monitor the transport of ions through ion-selective polymer membranes [74]. In their 2007 review, Wittstock et al. discussed a methodology to investigate transport processes in biological membranes, which in principle can be used to study any ion transport membrane [16]. To evaluate a membrane, it must be used as a separator to separate the donor and acceptor compartments. The UME tip moves in the acceptor compartment and measures the local concentrations of the substance emerging from the pores of the membrane using SG/TC mode. The use of ion-selective UME tips enables the determination of transport quantities for individual species. As shown in Figure 14.11, the transport processes can be driven by concentration, pressure, or potential gradients across the membrane. When an ionic species of interest emerges from a pore, the transport in the acceptor compartment occurs mainly by diffusion and is independent of the prevailing transport mechanism inside the pore. Steady-state concentration

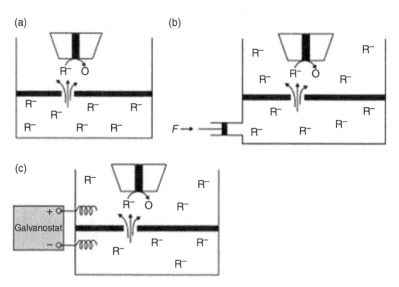

FIGURE 14.11 Schematic of SECM setup to study the transport processes in BFC membranes. The driving force for the transport can be (a) a concentration gradient between a donor and an acceptor compartment, (b) a pressure gradient between the compartments, or (c) a potential gradient between the two compartments. (Reprinted with permission from Ref. [16]. Copyright 2007, Wiley-VCH Verlag GmbH.)

gradients are formed above the pore opening in the acceptor compartment if the pore radius is smaller than 50 μm. The SECM IT mode experiments discussed earlier in this chapter will also be a relevant tool for the analysis of membrane separators.

14.9 PROBING SINGLE ENZYME MOLECULES USING SECM

Even though SECM is a potent tool for the investigation of microscale processes, its resolution is not high enough to probe a single molecule, primarily because of the nature of the amperometric measurements and the limitation of the electronics and instrumentation. In 2009, a new variant of SECM termed *scanning electrochemical potential microscopy* (SECPM) was developed and discussed by Baier and Stimming [75]. In SECPM, the potential difference between the tip and the substrate is measured at zero current (using a high-impedance potential amplifier), and this potential difference serves as a feedback signal in the (x–y) scanning mode. Due to the electrical double-layer effects, the potential of the tip–solution interface decreases with the distance from the substrate, thereby enabling mapping of the potential distribution of the interface in the (x–z) direction. SECPM can reveal the morphology and reactivity of the enzymes down to single-molecule level and has proven to be suitable for the investigation of organic and biological molecules adsorbed on electrode surfaces under electrochemical conditions.

SECPM is also a promising tool for mapping the charge distribution of adsorbed molecules. Since the local potential is measured in the electrical double-layer region, and no current flows between the tip and the sample, this technique has a great potential for the investigation of single enzyme molecules in their native environment as well as under reaction conditions. The ability to perform local reactivity measurements, depending on the surface potential, may open up new perspectives for studying not only electrocatalysis involving redox enzymes but also a variety of electrochemical surface science problems under *in situ* conditions. In their representative 2009 work, Baier and Stimming imaged HRP adsorbed on a highly ordered pyrolytic graphite surface to obtain a high-resolution image of a single HRP enzyme molecule using SECPM (Figure 14.12) [75].

14.10 COMBINING SECM WITH OTHER TECHNIQUES

Scanning force microscopy (SFM) and electrochemical scanning tunneling microscopy (ECSTM) routinely allow imaging with a lateral and vertical resolution that is clearly superior to that of SECM. However, neither method directly delivers chemically specific information. In such a situation, SECM comes in handy, as it is both robust and versatile to be integrated with other probing techniques. SECM can be coupled with other probe microscopy tools by integration of a UME into the sensors used for scanning probe techniques. The modification of UMEs for this purpose is robust enough to allow further modification of the integrated electrodes to enzyme electrodes, thereby increasing the chemical selectivity of the measurements.

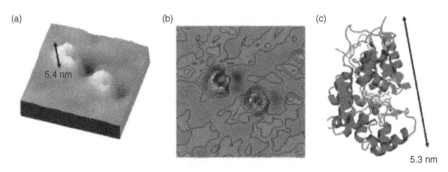

FIGURE 14.12 Imaging single HRP molecules using SECM. (a) 3D SECPM image of HRP adsorbed on highly ordered pyrolytic graphite (HOPG) electrode surface. (b) 2D contour plot of the same sample. (c) Crystal structure of HRP with an approximate dimension of 5.3 nm in agreement with the SECPM images. (Reprinted with permission from Ref. [75]. Copyright 2009, Wiley-VCH Verlag GmbH.) (Please see the color version of this figure in Color Plates section.)

However, each probe must be fabricated individually for this purpose [76,77]. Recent instrumental developments in the field of SPM are aimed at combined techniques, which provide simultaneous information on the surface topography and on the surface physical and chemical properties.

14.10.1 Atomic Force Microscopy

Atomic force microscopy (AFM) in tapping mode is particularly useful for imaging soft sample surfaces such as biological specimens, membranes, and enzyme layers. The high-resolution imaging capability of an AFM and the chemical information imaging capability of SECM can be combined into a single instrument to use the advantages of both instrumental techniques. The combination of AFM–SECM is a particularly attractive strategy for obtaining complementary electrochemical and topographical information with high lateral resolution in a single time- and space-correlated measurement [77,78]. The combination of AFM with SECM allows position-specific probing of electrochemical surface properties of modified electrodes. An AFM probe with integrated UMEs promises to be a powerful electroanalytical tool. The integration of an electroactive area at an exactly defined distance above the apex of the AFM tip allows the distance between the electrode and the probe to remain constant, regardless of the surface topography, which enables simultaneous electrochemical and AFM imaging in contact and tapping modes [79]. Figure 14.13 shows the schematic of the AFM–SECM imaging of enzyme coated on a micropatterned polymer spot electrode [45]. Imaging enzyme activity in AFM tapping mode was performed at a micropattern of isolated enzyme-containing polymer spots. Electrochemical images recorded in tapping mode reveal a current response and lateral resolution comparable to the electrochemical images of the same sample recorded in contact mode [45].

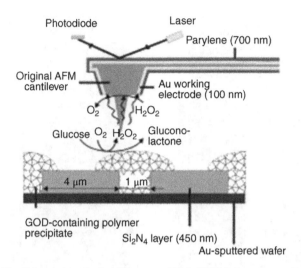

FIGURE 14.13 Schematic of simultaneous AFM–SECM imaging and the reactions involved at the surface of the micropatterned sample with integrated electrode operation in GC mode. (Reprinted with permission from Ref. [45]. Copyright 2003, Wiley-VCH Verlag GmbH.)

Photolithography and ion beam techniques have been used to fabricate and integrate the SECM UME and the AFM cantilever for high spatial resolution imaging [48,80,81]. The measurements by the integrated AFM–SECM were made in 2004 by Hirata et al. using dynamic force microscopy, which uses the resonance enhancement of the force sensitivity by oscillating the AFM cantilever at resonance frequency [48]. Imaging GOx activity using the integrated AFM–SECM showed a high degree of resolution that even showed grain, aggregation, and membrane defects on the GOx-modified enzyme electrode and agreed well with the topography image obtained using the AFM tip (Figure 14.14). The dynamic force microscopy-based imaging can be applied to both conducting and insulating surfaces. The resolutions of both current and topography were high enough to visualize the activity of enzyme molecules on electrode surfaces and accordingly can be used in EFC catalysis for the investigation of composite enzyme biocatalysts. Continued improvements in optimizing the experimental parameters such as cantilever spring constant, resonance frequency, and oscillation amplitude for minimizing the noise level in current are being realized [48].

14.10.2 Confocal Laser Scanning Microscopy

For electrochemical applications, confocal laser scanning microscopy (CLSM) allows a 3D investigation of diffusion layers at a spatial resolution that is compatible with the lateral resolution of SECM without any modification [82]. In 2008, Nogala et al. combined SECM with CLSM and SPM to image laccase embedded in sol–gel film

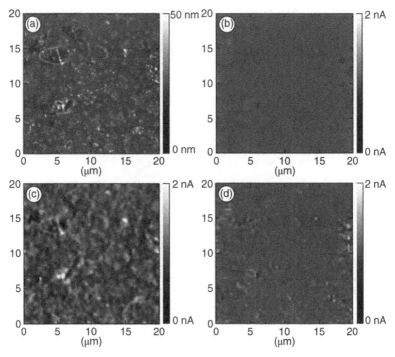

FIGURE 14.14 Simultaneously recorded topography and SECM current images of a polyion complex layer containing GOx-modified graphite electrode using DFM mode SECM–AFM. (a) Topographic image; (b) SECM current image (probe potential: +0.9 V) taken without glucose; (c) SECM current image taken at 10 mM glucose; (d) same conditions as (c) but base electrode potential was set to +0.9 V. (Reprinted with permission from Ref. [48]. Copyright 2004, Elsevier.) (Please see the color version of this figure in Color Plates section.)

and yielded a high degree of correlation between the optically detected features and the electrochemical signals as shown in Figure 14.15 [66]. Combinations of CLSM and SECM have been used to characterize diffusion layers in front of microelectrodes using fluorescent dyes [82–84].

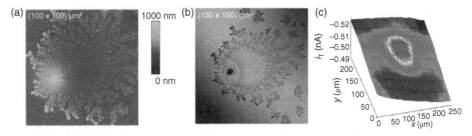

FIGURE 14.15 Combined SPM–CLSM–SECM imaging of laccase aggregate. (a) SFM image; (b) CLSM reflection mode image (50× objective); (c) SG/TC mode SECM image. (Reprinted with permission from Ref. [66]. Copyright 2008, Elsevier.) (Please see the color version of this figure in Color Plates section.)

14.11 OUTLOOK

Since its introduction, SECM has been used in a variety of electrochemical applications and has been proven to be particularly useful in bioelectroanalytical characterization including imaging of enzyme-modified electrodes, live cell imaging, and small-scale characterization of biological membranes. As a stand-alone technique, SECM is a powerful tool for analytical applications, but, beyond this, when combined with other imaging tools such as SPM and confocal microscopy, the power and versatility of this technique can be significantly enhanced. SECM has not been specially used in BFC studies, but a significant portion of the SECM literature discusses the use of this tool for the investigation of enzyme-modified electrodes, which are very relevant to EFC research. Using this tool, it is possible to locate immobilized entities on modified electrode surfaces, determine their spatial resolution, investigate their kinetic activity, and also study the transport of substrate and products in and around the functionalized surface on a microscopic scale. There is a strong trend toward the exploitation of developments in micro- and nanomanufacturing techniques to allow the miniaturization of biological devices such as biosensors and lab-on-a-chip devices, and it is only the beginning of an era when SECM will find an important role in the characterization of biological systems.

LIST OF ABBREVIATIONS

ABTS	2,2′-azinobis(3-ethylbenzothiazoline-6-sulfonate)
AFM	atomic force microscopy
BFC	biological fuel cell
CLSM	confocal laser scanning microscopy
DNA–CNT	deoxyribonucleic acid–carbon nanotube
ECSTM	electrochemical scanning tunneling microscopy
EFC	enzymatic fuel cell
FB	feedback
GC	generation–collection
GOx	glucose oxidase
HOPG	highly ordered pyrolytic graphite
HRP	horseradish peroxidase
IT	induced transfer
LYZ–CNT	lysozyme–carbon nanotube
ORR	oxygen reduction reaction
RC	redox competition
SECM	scanning electrochemical microscopy
SECPM	scanning electrochemical potential microscopy
SFM	scanning force microscopy
SG/TC	substrate generation–tip collection
SPM	scanning probe microscopy
TG/SC	tip generation–substrate collection
UME	ultramicroelectrode

REFERENCES

1. Unwin PR, Bard AJ. Scanning electrochemical microscopy. 9. Theory and application of the feedback mode to the measurement of following chemical reaction rates in electrode processes. *J Phys Chem* 1991;95:7814–7824.

2. Unwin PR, Bard AJ. Scanning electrochemical microscopy. 14. Scanning electrochemical microscope induced desorption: a new technique for the measurement of adsorption/desorption kinetics and surface diffusion rates at the solid/liquid interface. *J Phys Chem* 1992;96:5035–5045.

3. Wipf DO, Bard AJ. Scanning electrochemical microscopy. 7. Effect of heterogeneous electron-transfer rate at the substrate on the tip feedback current. *J Electrochem Soc* 1991;138:469–474.

4. Bard AJ, Mirkin MV, Unwin PR, Wipf DO. Scanning electrochemical microscopy. 12. Theory and experiment of the feedback mode with finite heterogeneous electron-transfer kinetics and arbitrary substrate size. *J Phys Chem* 1992;96:1861–1868.

5. Shiku H, Shiraishi T, Ohya H, Matsue T, Abe H, Hoshi H, Kobayashi M. Oxygen consumption of single bovine embryos probed by scanning electrochemical microscopy. *Anal Chem* 2001;73:3751–3758.

6. Zhang MQ, Girault HH. Fingerprint imaging by scanning electrochemical microscopy. *Electrochem Commun* 2007;9:1778–1782.

7. Bard AJ, Fan FRF, Kwak J, Lev O. Scanning electrochemical microscopy: introduction and principles. *Anal Chem* 1989;61:132–138.

8. Kwak J, Bard AJ. Scanning electrochemical microscopy: apparatus and two-dimensional scans of conductive and insulating substrates. *Anal Chem* 1989;61:1794–1799.

9. Kwak J, Bard AJ. Scanning electrochemical microscopy: theory of the feedback mode. *Anal Chem* 1989;61:1221–1227.

10. Bard AJ, Denuault G, Lee C, Mandler D, Wipf DO. Scanning electrochemical microscopy: a new technique for the characterization and modification of surfaces. *Acc Chem Res* 1990;23:357–363.

11. Bard AJ, Denault G, Friesner RA, Dornblaser BC, Tuckerman LS. Scanning electrochemical microscopy: theory and application of the transient (chronoamperometric) SECM response. *Anal Chem* 1991;63:1282–1288.

12. Sun P, Laforge FO, Mirkin MV. Scanning electrochemical microscopy in the 21st century. *Phys Chem Chem Phys* 2007;9:802–823.

13. Roberts WS, Lonsdale DJ, Griffiths J, Higson SPJ. Advances in the application of scanning electrochemical microscopy to bioanalytical systems. *Biosens Bioelectron* 2007;23:301–318.

14. Fernandez JL, Bard AJ. Scanning electrochemical microscopy. 50. Kinetic study of electrode reactions by the tip generation–substrate collection mode. *Anal Chem* 2004;76:2281–2289.

15. Amemiya S, Guo JD, Xiong H, Gross DA. Biological applications of scanning electrochemical microscopy: chemical imaging of single living cells and beyond. *Anal Bioanal Chem* 2006;386:458–471.

16. Wittstock G, Burchardt M, Pust SE, Shen Y, Zhao C. Scanning electrochemical microscopy for direct imaging of reaction rates. *Angew Chem Int Ed* 2007;46:1584–1617.

17. Pedrosa VA, Gnanaprakasa T, Balasubramanian S, Olsen EV, Davis VA, Simonian AL. Electrochemical properties of interface formed by interlaced layers of DNA- and lysozyme-coated single-walled carbon nanotubes. *Electrochem Commun* 2009;11:1401–1404.

18. Horrocks BR, Mirkin MV, Pierce DT, Bard AJ, Nagy G, Toth K. Scanning electrochemical microscopy. 19. Ion-selective potentiometric microscopy. *Anal Chem* 1993;65:1213–1224.

19. Horrocks BR, Schmidtke D, Heller A, Bard AJ. Scanning electrochemical microscopy. 24. Enzyme ultramicroelectrodes for the measurement of hydrogen peroxide at surfaces. *Anal Chem* 1993;65:3605–3614.

20. Lee C, Kwak JY, Anson FC. Application of scanning electrochemical microscopy to generation/collection experiments with high collection efficiency. *Anal Chem* 1991;63:1501–1504.

21. Sklyar O, Trauble M, Zhao CA, Wittstock G. Modeling steady-state experiments with a scanning electrochemical microscope involving several independent diffusing species using the boundary element method. *J Phys Chem B* 2006;110:15869–15877.

22. Barker AL, Gonsalves M, Macpherson JV, Slevin CJ, Unwin PR. Scanning electrochemical microscopy: beyond the solid/liquid interface. *Anal Chim Acta* 1999;385:223–240.

23. Yamada H, Matsue T, Uchida I. A microvoltammetric study of permeation of ferrocene derivatives through a planer bilayer lipid membrane. *Biochem Biophys Res Commun* 1991;180:1330–1334.

24. Barker AL, Macpherson JV, Slevin CJ, Unwin PR. Scanning electrochemical microscopy (SECM) as a probe of transfer processes in two-phase systems: theory and experimental applications of SECM-induced transfer with arbitrary partition coefficients, diffusion coefficients, and interfacial kinetics. *J Phys Chem B* 1998;102:1586–1598.

25. Barker AL, Unwin PR. Measurement of solute partitioning across liquid/liquid interfaces using scanning electrochemical microscopy–double potential step chronoamperometry (SECM–DPSC): principles, theory, and application to ferrocenium ion transfer across the 1,2-dichloroethane/aqueous interface. *J Phys Chem B* 2001;105:12019–12031.

26. Barker AL, Unwin PR, Zhang J. Measurement of the forward and back rate constants for electron transfer at the interface between two immiscible electrolyte solutions using scanning electrochemical microscopy (SECM): theory and experiment. *Electrochem Commun* 2001;3:372–378.

27. Zhou FM, Unwin PR, Bard AJ. Scanning electrochemical microscopy. 16. Study of second-order homogeneous chemical reactions via the feedback and generation collection modes. *J Phys Chem* 1992;96:4917–4924.

28. Pierce DT, Unwin PR, Bard AJ. Scanning electrochemical microscopy. 17. Studies of enzyme mediator kinetics for membrane-immobilized and surface-immobilized glucose oxidase. *Anal Chem* 1992;64:1795–1804.

29. Zhao CA, Wittstock G. Scanning electrochemical microscopy of quinoprotein glucose dehydrogenase. *Anal Chem* 2004;76:3145–3154.

30. Wittstock G. Modification and characterization of artificially patterned enzymatically active surfaces by scanning electrochemical microscopy. *Fresenius J Anal Chem* 2001;370:303–315.

31. Wei C, Bard AJ, Mirkin MV. Scanning electrochemical microscopy. 31. Application of SECM to the study of charge-transfer processes at the liquid–liquid interface. *J Phys Chem* 1995;99:16033–16042.

32. Zaumseil J, Wittstock G, Bahrs S, Steinrucke P. Imaging the activity of nitrate reductase by means of a scanning electrochemical microscope. *Fresenius J Anal Chem* 2000;367:352–355.

33. Wittstock G, Wilhelm T, Bahrs S, Steinrucke P. SECM feedback imaging of enzymatic activity on agglomerated microbeads. *Electroanalysis* 2001;13:669–675.

34. Shiku H, Matsue T, Uchida I. Detection of microspotted carcinoembryonic antigen on a glass substrate by scanning electrochemical microscopy. *Anal Chem* 1996;68:1276–1278.

35. Pierce DT, Bard AJ. Scanning electrochemical microscopy. 23. Reaction localization of artificially patterned and tissue-bound enzymes. *Anal Chem* 1993;65:3598–3604.

36. Kranz C, Wittstock G, Wohlschlager H, Schuhmann W. Imaging of microstructured biochemically active surfaces by means of scanning electrochemical microscopy. *Electrochim Acta* 1997;42:3105–3111.

37. Zhao C, Sinha JK, Wijayawardhana CA, Wittstock G. Monitoring beta-galactosidase activity by means of scanning electrochemical microscopy. *J Electroanal Chem* 2004;561:83–91.

38. Zhao C, Wittstock G. Scanning electrochemical microscopy for detection of biosensor and biochip surfaces with immobilized pyrroloquinoline quinone (PQQ)-dependent glucose dehydrogenase as enzyme label. *Biosens Bioelectron* 2005;20:1277–1284.

39. Niculescu M, Gaspar S, Schulte A, Csoregi E, Schuhmann W. Visualization of micropatterned complex biosensor sensing chemistries by means of scanning electrochemical microscopy. *Biosens Bioelectron* 2004;19:1175–1184.

40. Parra A, Casero E, Vazquez L, Jin J, Pariente F, Lorenzo E. Microscopic and voltammetric characterization of bioanalytical platforms based on lactate oxidase. *Langmuir* 2006;22:5443–5450.

41. Holt KB. Using scanning electrochemical microscopy (SECM) to measure the electron-transfer kinetics of cytochrome *c* immobilized on a COOH-terminated alkanethiol monolayer on a gold electrode. *Langmuir* 2006;22:4298–4304.

42. Wijayawardhana CA, Wittstock G, Halsall HB, Heineman WR. Spatially addressed deposition and imaging of biochemically active bead microstructures by scanning electrochemical microscopy. *Anal Chem* 2000;72:333–338.

43. Kranz C, Lotzbeyer T, Schmidt HL, Schuhmann W. Lateral visualization of direct electron transfer between microperoxidase and electrodes by means of scanning electrochemical microscopy. *Biosens Bioelectron* 1997;12:257–266.

44. Oyamatsu D, Hirano Y, Kanaya N, Mase Y, Nishizawa M, Matsue T. Imaging of enzyme activity by scanning electrochemical microscope equipped with a feedback control for substrate–probe distance. *Bioelectrochemistry* 2003;60:115–121.

45. Kueng A, Kranz C, Lugstein A, Bertagnolli E, Mizaikoff B. Integrated AFM–SECM in tapping mode: simultaneous topographical and electrochemical imaging of enzyme activity. *Angew Chem Int Ed* 2003;42:3238–3240.

46. Gaspar S, Mosbach M, Wallman L, Laurell T, Csoregi E, Schuhmann W. A method for the design and study of enzyme microstructures formed by means of a flow-through microdispenser. *Anal Chem* 2001;73:4254–4261.

47. Turcu F, Hartwich G, Schafer D, Schuhmann W. Ink-jet microdispensing for the formation of gradients of immobilised enzyme activity. *Macromol Rapid Commun* 2005;26:325–330.

48. Hirata Y, Yabuki S, Mizutani F. Application of integrated SECM ultra-microelectrode and AFM force probe to biosensor surfaces. *Bioelectrochemistry* 2004;63:217–224.

49. Wilhelm T, Wittstock G. Generation of periodic enzyme patterns by soft lithography and activity imaging by scanning electrochemical microscopy. *Langmuir* 2002;18:9485–9493.

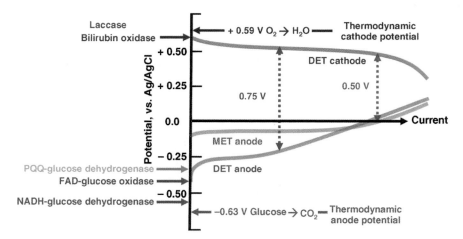

FIGURE 2.2 Conceptual drawing of polarization curves that illustrates deviations from ideality for a bioanode that is catalyzing oxidation of glucose by enzymes with several different cofactors and a biocathode that is reducing molecular oxygen by a copper oxidase.

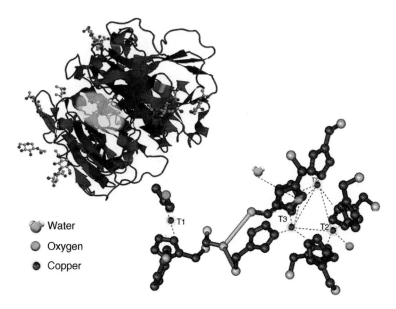

FIGURE 3.1 Ball-and-stick representation of the T1 and T2/T3 copper binding sites of laccase from *T. versicolor* (NCBI Protein Data Bank: 1GYC from *T. versicolor*) viewed using Cn3D ver. 4.1 (NCBI) showing coordination of oxygen. (Adapted with permission from Ref. [27]. Copyright 2010, Elsevier.) Protein structure of *T. versicolor* laccase is shown (RCSB Protein Data Bank: 1GYB); copper atoms in the active site are highlighted in yellow.

Enzymatic Fuel Cells: From Fundamentals to Applications, First Edition. Edited by Heather R. Luckarift, Plamen B. Atanassov, and Glenn R. Johnson.
© 2014 John Wiley & Sons, Inc. Published 2014 by John Wiley & Sons, Inc.

FIGURE 4.8 Soluble quinoprotein glucose dehydrogenase from *A. calcoaceticus* in complex with PQQH$_2$ and glucose. (Reprinted with permission from Ref. [110]. Copyright 2003, Elsevier.) Structure retrieved from NCBI (PDB ID: 1CQ1), viewed in Cn3D 4.3. Protein homodimer colored in blue/purple tube worm structure with calcium ions colored yellow for clarity; PQQ cofactor represented as space fill model.

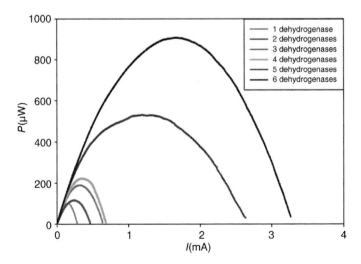

FIGURE 5.6 Representative power curves for lactate/air biological fuel cells in 500 mM sodium lactate at room temperature with different degrees of oxidation of the fuel. (Reproduced with permission from Ref. [93]. Copyright 2011, Elsevier.)

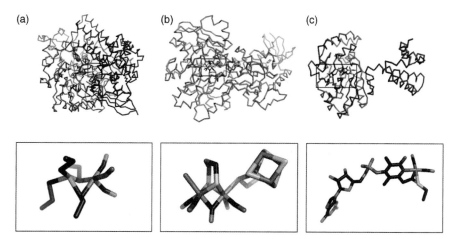

FIGURE 6.1 Representative structures of hydrogenases. Crystal structures of (a) the [NiFe]-hydrogenase from *D. vulgaris* Miyazaki F (PDB ID: 1H2R, 1.40 Å), (b) the [FeFe]-hydrogenase from *C. pasteurianum* (PDB ID: 3C8Y, 1.39 Å), and (c) the [Fe]-hydrogenase from *Methanocaldococcus jannaschii* (PDB ID: 3F47, 1.75 Å). The top row shows the holoprotein structure and the bottom row shows the hydrogen-activating site in stick representation. Atoms are labeled as carbon (black), oxygen (red), nitrogen (blue), sulfur (orange), and iron (rust). All protein structure figures were created in PyMOL [144].

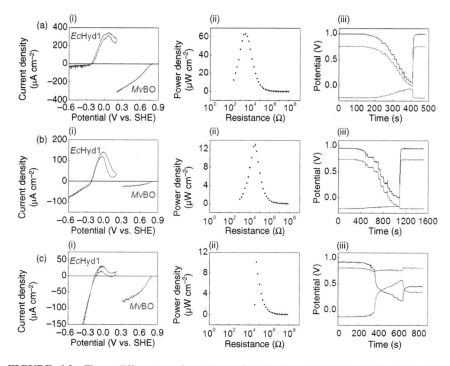

FIGURE 6.3 Three different configurations of *Ec*Hyd1 and *Mv*BOx H_2/O_2 fuel cells operating at 25 °C using stationary electrodes of 1.25 cm^2: (a) PEM, (b) 96% H_2/4% O_2, and (c) 4% H_2/96% air. (*See text for full caption.*)

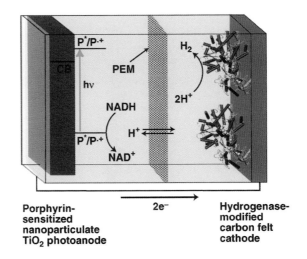

FIGURE 6.4 Schematic drawing of the photoelectrochemical biological fuel cell developed by Hambourger et al. for the production of hydrogen from biomass using solar energy. (Reprinted with permission from Ref. [72]. Copyright 2008, American Chemical Society.)

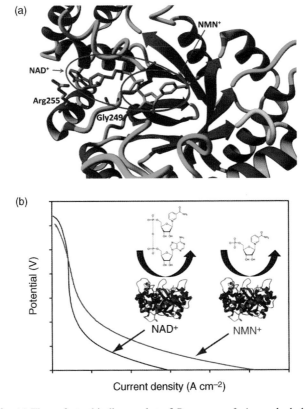

FIGURE 7.2 (a) The cofactor binding pocket of *Pyrococcus furiosus* alcohol dehydrogenase D K249G/H255R double mutant, indicating the position of bound NMN⁺ (orange) and NAD⁺ (purple), as well as the mutated residues. (*See text for full caption.*)

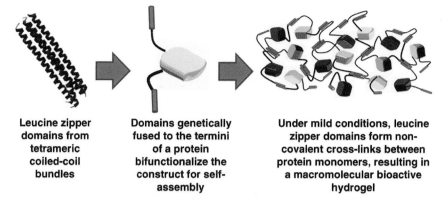

| Leucine zipper domains from tetrameric coiled-coil bundles | Domains genetically fused to the termini of a protein bifunctionalize the construct for self-assembly | Under mild conditions, leucine zipper domains form non-covalent cross-links between protein monomers, resulting in a macromolecular bioactive hydrogel |

FIGURE 7.3 Schematic of leucine zipper fusions to enable self-assembly of various proteins of interest into a robust macromolecular hydrogel.

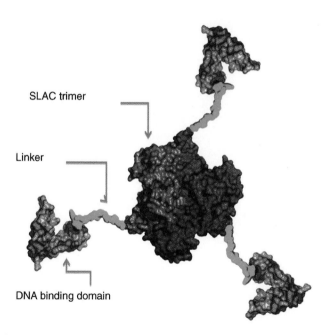

SLAC trimer

Linker

DNA binding domain

FIGURE 7.4 Schematic diagram of the SLAC–zinc finger fusion protein. The SLAC enzyme is trimeric and the C-terminus of each monomer (blue, bluish gray, purple) is fused to a DNA binding zinc finger domain (brown) via a flexible peptide linker (green). The DNA binding domain is a three zinc finger module from a mouse transcription factor Zif268 that specifically binds double-stranded DNA having the sequence 5′-GCGTGGGCG-3′. (Reproduced with permission from Ref. [66]. Copyright 2011, Royal Society of Chemistry.)

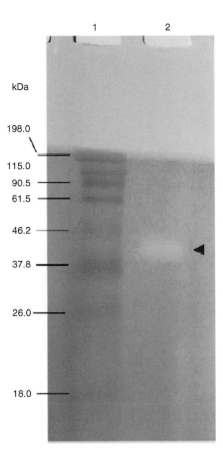

FIGURE 8.1 *Ganoderma lucidum* laccase activity after SDS-PAGE. Gel was incubated in Remazol Brilliant Blue R solution and decolorization of dye is seen surrounding the laccase band (arrow, lane 2). Molecular weight markers are in lane 1. (Adapted with permission from Ref. [57]. Copyright 2007, Elsevier.)

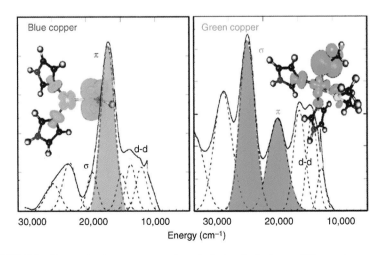

FIGURE 8.3 Representative spectroscopic and electronic features of blue and green copper centers. (Adapted with permission from Ref. [80]. Copyright 2006, American Chemical Society.)

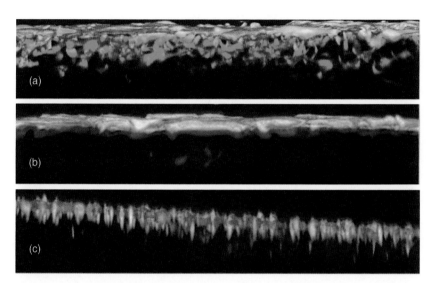

FIGURE 12.4 (a) Enzyme distribution in native chitosan scaffolds stained with fluorescein (green) combined with fluorescent (Alexa Fluor 546)-stained malate dehydrogenase (purple). (b) Enzyme distribution in butyl-modified chitosan scaffolds (same staining). (c) Enzyme distribution in ALA-modified chitosan scaffolds (same staining). (Reproduced with permission from Ref. [27]. Copyright 2009, American Chemical Society.)

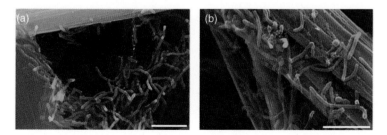

FIGURE 13.6 (a, b) *S. oneidensis* cells on Toray paper electrodes. Three-dimensional images prepared with Anaglyph Maker, ver. 1.08. Scale bars = 5 μm. (Unpublished images taken by author KEF.) 3D red cyan glasses are recommended to view this image correctly.

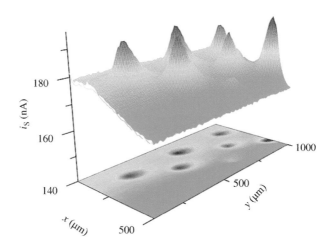

FIGURE 14.8 ORR activity imaging by TG/SC mode of an array of Pt spots supported on glassy carbon. (Reprinted with permission from Ref. [60]. Copyright 2003, American Chemical Society.)

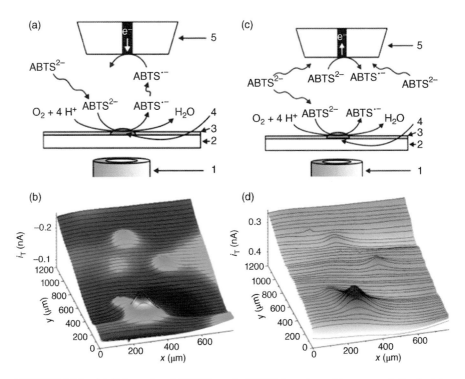

FIGURE 14.9 (a) Schematic of SG/TC mode SECM imaging of laccase activity (not to scale): (1) microscope objective; (2) glass slide; (3) laccase–silicate film; (4) laccase aggregate; (5) UME. (b) SG/TC mode SECM image obtained using 1 mM $ABTS^{2-}$ as mediator in 0.1 M phosphate buffer (pH 4.8) at $E_T = -0.1$ V. (c) Schematic of RC mode SECM imaging of laccase activity (not to scale). (d) RC mode SECM image: $E_T = +0.4$ V and other conditions as in (b). (Reprinted with permission from Ref. [66]. Copyright 2008, Elsevier.)

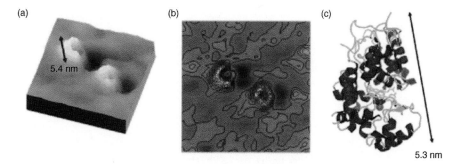

FIGURE 14.12 Imaging single HRP molecules using SECM. (a) 3D SECPM image of HRP adsorbed on highly ordered pyrolytic graphite (HOPG) electrode surface. (b) 2D contour plot of the same sample. (c) Crystal structure of HRP with an approximate dimension of 5.3 nm in agreement with the SECPM images. (Reprinted with permission from Ref. [75]. Copyright 2009, Wiley-VCH Verlag GmbH.)

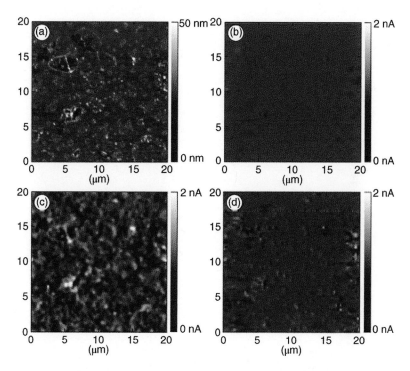

FIGURE 14.14 Simultaneously recorded topography and SECM current images of a polyion complex layer containing GOx-modified graphite electrode using DFM mode SECM–AFM. (a) Topographic image; (b) SECM current image (probe potential: +0.9 V) taken without glucose; (c) SECM current image taken at 10 mM glucose; (d) same conditions as (c) but base electrode potential was set to +0.9 V. (Reprinted with permission from Ref. [48]. Copyright 2004, Elsevier.)

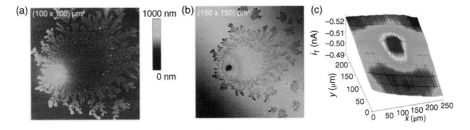

FIGURE 14.15 Combined SPM–CLSM–SECM imaging of laccase aggregate. (a) SFM image; (b) CLSM reflection mode image (50× objective); (c) SG/TC mode SECM image. (Reprinted with permission from Ref. [66]. Copyright 2008, Elsevier.)

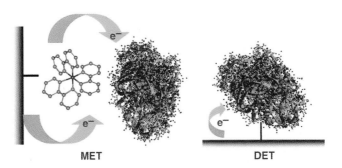

MET   **DET**

FIGURE 15.1 Schematic diagram of MET and DET illustrating the flow of electrons.

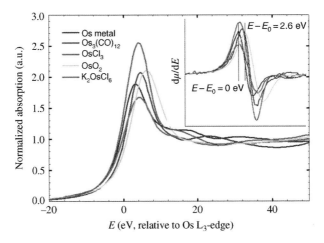

FIGURE 15.5 Normalized Os L_3-edge XANES spectra of the indicated Os reference compounds. The inset plots the first derivative spectrum of the XANES, which is the conventional method for determining E_0 by taking the energy value at the first maximum of the spectrum.

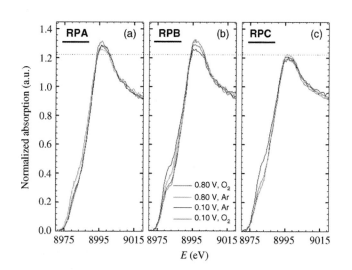

FIGURE 15.10 *In situ* Cu k-edge XANES spectra for MET laccase using (a) RPA, (b) RPB, and (c) RPC plots.

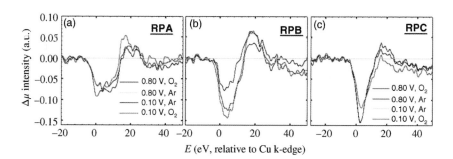

FIGURE 15.11 Cu k-edge $\Delta\mu$ curves calculated for the MET XANES shown in Figure 15.10. The RPA $\Delta\mu$ (a) was calculated using Equation 15.4, whereas RPB (b) and RPC (c) curves were calculated with Equation 15.5.

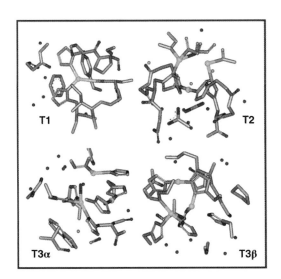

FIGURE 15.13 Drawings of the laccase TNC clusters used in the FEFF calculations showing each type of C as the central "absorber" atom. Each cluster consisted of ~90 atoms surrounding the absorber within a 7 Å radius. (Reproduced with permission from Ref. [73]. Copyright 2014, Wiley-VCH.)

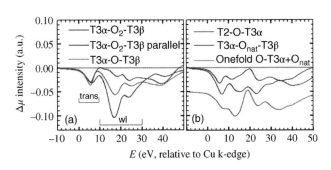

FIGURE 15.16 Simulated $\Delta\mu_t$ spectra calculated using Equation 15.7 for six of the structures depicted in Figure 15.15a–i. The onefold O-T3α + O$_{nat}$ curve (plot b, green line) was offset on the $\Delta\mu$ axis by −0.05 for illustrative purposes.

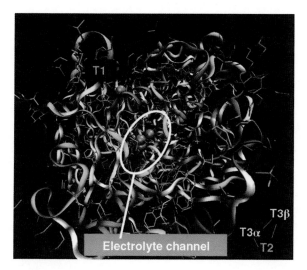

FIGURE 15.19 Diagram of laccase (*T. versicolor*) illustrating the close proximity of the T3 coppers to the larger electrolyte channel.

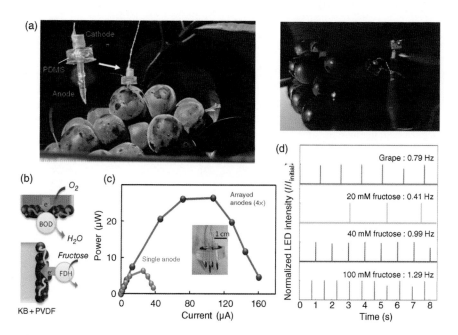

FIGURE 17.3 (a) Photograph of the assembled MEFC to be inserted into a grape. (b) Schemes of O_2 reduction at the enzymatic gas diffusion cathode (upper panel) and fructose oxidation at the enzymatic needle anode (lower panel). (c) Power output of the insertion MEFC using single anode and arrayed anodes (4×). (d) Monitoring of sugar level in a raw grape with a self-powered fructose-sensing device. The device consists of the insertion MEFC and an LED system, whose blink interval is correlated with the fructose concentration.

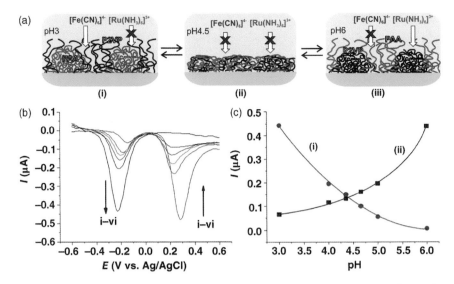

FIGURE 18.5 (a) The polymer brush permeability for differently charged redox probes controlled by a solution pH value: (i) the positively charged protonated P2VP domains allow the electrode access for the negatively charged redox species; (ii) the neutral hydrophobic polymer thin film inhibits the electrode access for all ionic species; (iii) the negatively charged dissociated PAA domains allow the electrode access for the positively charged redox species. (b) The differential pulse voltammograms obtained for the mixed-polymer brush in the presence of $[Fe(CN)_6]^{4-}$, 0.5 mM, and $[Ru(NH_3)_6]^{3+}$, 0.1 mM, at the variable pH of the solution: (i) 3.0, (ii) 4.0, (iii) 4.35, (iv) 4.65, (v) 5.0, and (vi) 6.0. The background solution was composed of 0.1 M phosphate buffer titrated to the specified pH values. (c) The peak current dependences on the pH value for the (i) anionic, $[Fe(CN)_6]^{4-}$, and (ii) cationic, $[Ru(NH_3)_6]^{3+}$, species, as derived from the differential pulse voltammograms measured at the variable pH values. (Adapted with permission from Ref. [85]. Copyright 2010, Wiley-VCH Verlag GmbH.)

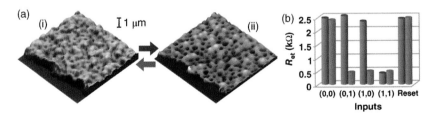

FIGURE 18.11 The signal-responsive membrane associated with an ITO electrode and coupled with the enzyme-based logic gates. (a) AFM topography images ($10 \times 10\ \mu m^2$) of the membrane with the closed (i) and open (ii) pores. (b) The electron transfer resistance, R_{et}, of the switchable interface derived from the impedance spectroscopy measurements obtained upon different combinations of the input signals. The blue and red bars correspond to the OR–Reset and AND–Reset systems, respectively. (Adapted with permission from Ref. [102]. Copyright 2009, American Chemical Society.)

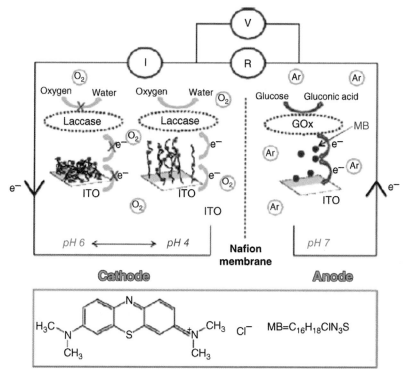

FIGURE 19.2 Glucose/O$_2$ EFC composed of a bioanode and a pH-switchable logically controlled biocathode. (Reprinted with permission from Ref. [119]. Copyright 2009, American Chemical Society.)

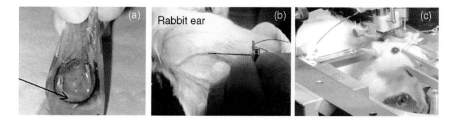

FIGURE 19.4 (a) Photograph of a GDH/PPO EFC based on soluble mediators encased by membranes implanted in a rat. (Reprinted with permission from Ref. [110]. Copyright 2010, PLoS.) (b) Photographs of the assembled BFC for power generation from a rabbit vein. (Reprinted with permission from Ref. [124]. Copyright 2011, Royal Society of Chemistry.) (c) Photograph of the experimental setup during *in vivo* tests of a CDH/BOx microscale EFC operating in the brain tissue of a living rat [125].

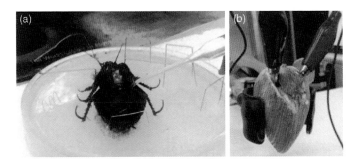

FIGURE 19.5 (a) Photograph of the experimental setup of a TOx–GOx/BOx EFC implanted into a cockroach. (Reprinted with permission from Ref. [67]. Copyright 2012, American Chemical Society.) (b) Photograph of the experimental setup of a clam with implanted (PQQ)-GDH/laccase EFC. (Reprinted with permission from Ref. [130]. Copyright 2012, Royal Society of Chemistry.)

50. Evans SAG, Brakha K, Billon M, Mailley P, Denuault G. Scanning electrochemical microscopy (SECM): localized glucose oxidase immobilization via the direct electrochemical microspotting of polypyrrole–biotin films. *Electrochem Commun* 2005;7:135–140.

51. Turyan I, Matsue T, Mandler D. Patterning and characterization of surfaces with organic and biological molecules by the scanning electrochemical microscope. *Anal Chem* 2000;72:3431–3435.

52. Strike DJ, Hengstenberg A, Quinto M, Kurzawa C, Koudelka-Hep M, Schuhmann W. Localized visualization of chemical cross-talk in microsensor arrays by using scanning electrochemical microscopy. *Mikrochim Acta* 1999;131:47–55.

53. Ge FY, Tenent RC, Wipf DO. Fabricating and imaging carbon-fiber immobilized enzyme ultramicroelectrodes with scanning electrochemical microscopy. *Anal Sci* 2001;17:27–35.

54. Kranz C, Kueng A, Lugstein A, Bertagnolli E, Mizaikoff B. Mapping of enzyme activity by detection of enzymatic products during AFM imaging with integrated SECM–AFM probes. *Ultramicroscopy* 2004;100:127–134.

55. Oyamatsu D, Kanaya N, Hirano Y, Nishizawa M, Matsue T. Area-selective immobilization of multi enzymes by using the reductive desorption of self-assembled monolayer. *Electrochemistry* 2003;71:439–441.

56. Glidle A, Yasukawa T, Hadyoon CS, Anicet N, Matsue T, Nomura M, Cooper JM. Analysis of protein adsorption and binding at biosensor polymer interfaces using X-ray photon spectroscopy and scanning electrochemical microscopy. *Anal Chem* 2003;75:2559–2570.

57. Fernandez JL, Raghuveer V, Manthiram A, Bard AJ. Pd–Ti and Pd–Co–Au electrocatalysts as a replacement for platinum for oxygen reduction in proton exchange membrane fuel cells. *J Am Chem Soc* 2005;127:13100–13101.

58. Fernandez JL, Walsh DA, Bard AJ. Thermodynamic guidelines for the design of bimetallic catalysts for oxygen electroreduction and rapid screening by scanning electrochemical microscopy. M–Co (M: Pd, Ag, Au). *J Am Chem Soc* 2005;127:357–365.

59. Weng YC, Fan FRF, Bard AJ. Combinatorial biomimetics. Optimization of a composition of copper(II) poly-L-histidine complex as an electrocatalyst for O_2 reduction by scanning electrochemical microscopy. *J Am Chem Soc* 2005;127:17576–17577.

60. Fernandez JL, Bard AJ. Scanning electrochemical microscopy. 47. Imaging electrocatalytic activity for oxygen reduction in an acidic medium by the tip generation–substrate collection mode. *Anal Chem* 2003;75:2967–2974.

61. Fernandez JL, Mano N, Heller A, Bard AJ. Optimization of "wired" enzyme O_2 electroreduction catalyst compositions by scanning electrochemical microscopy. *Angew Chem Int Ed* 2004;43:6355–6357.

62. Eckhard K, Chen XX, Turcu F, Schuhmann W. Redox competition mode of scanning electrochemical microscopy (RC-SECM) for visualisation of local catalytic activity. *Phys Chem Chem Phys* 2006;8:5359–5365.

63. Sklyar O, Ufheil J, Heinze J, Wittstock G. Application of the boundary element method numerical simulations for characterization of heptode ultramicroelectrodes in SECM experiments. *Electrochim Acta* 2003;49:117–128.

64. Ivnitski D, Artyushkova K, Rincón RA, Atanassov P, Luckarift HR, Johnson GR. Entrapment of enzymes and carbon nanotubes in biologically synthesized silica: glucose oxidase-catalyzed direct electron transfer. *Small* 2008;4:357–364.

65. Simkus RA, Laurinavicius V, Boguslavsky L, Skotheim T, Tanenbaum SW, Nakas JP, Slomczynski DJ. Laccase containing sol–gel based optical biosensors. *Anal Lett* 1996;29:1907–1919.

66. Nogala W, Burchardt M, Opallo M, Rogalski J, Wittstock G. Scanning electrochemical microscopy study of laccase within a sol–gel processed silicate film. *Bioelectrochemistry* 2008;72:174–182.

67. Karnicka K, Eckhard K, Guschin DA, Stoica L, Kulesza PJ, Schuhmann W. Visualisation of the local bio-electrocatalytic activity in biofuel cell cathodes by means of redox competition scanning electrochemical microscopy (RC-SECM). *Electrochem Commun* 2007;9:1998–2002.

68. Maciejewska M, Schafer D, Schuhmann W. SECM imaging of spatial variability in biosensor architectures. *Electrochem Commun* 2006;8:1119–1124.

69. Black M, Cooper J, McGinn P. Scanning electrochemical microscope characterization of thin film combinatorial libraries for fuel cell electrode applications. *Meas Sci Technol* 2005;16:174–182.

70. Jayaraman S, Hillier AC. Construction and reactivity mapping of a platinum catalyst gradient using the scanning electrochemical microscope. *Langmuir* 2001;17:7857–7864.

71. Jayaraman S, Hillier AC. Screening the reactivity of Pt_xRu_y and $Pt_xRu_yMo_z$ catalysts toward the hydrogen oxidation reaction with the scanning electrochemical microscope. *J Phys Chem B* 2003;107:5221–5230.

72. Jayaraman S, Hillier AC. Construction and reactivity screening of a surface composition gradient for combinatorial discovery of electro-oxidation catalysts. *J Comb Chem* 2004;6:27–31.

73. Wilhelm T, Wittstock G. Analysis of interaction in patterned multienzyme layers by using scanning electrochemical microscopy. *Angew Chem Int Ed* 2003;42:2247–2250.

74. Gyurcsanyi RE, Pergel E, Nagy R, Kapui I, Lan BTT, Toth K, Bitter I, Lindner E. Direct evidence of ionic fluxes across ion selective membranes: a scanning electrochemical microscopic and potentiometric study. *Anal Chem* 2001;73:2104–2111.

75. Baier C, Stimming U. Imaging single enzyme molecules under *in situ* conditions. *Angew Chem Int Ed* 2009;48:5542–5544.

76. Kueng A, Kranz C, Mizaikoff B. Scanning probe microscopy with integrated biosensors. *Sens Lett* 2003;1:2–15.

77. Kueng A, Kranz C, Lugstein A, Bertagnolli E, Mizaikoff B. AFM-tip-integrated amperometric microbiosensors: high-resolution imaging of membrane transport. *Angew Chem Int Ed* 2005;44:3419–3422.

78. Gardner CE, Macpherson JV. Atomic force microscopy probes go electrochemical. *Anal Chem* 2002;74:576A–584A.

79. Lugstein A, Bertagnolli E, Kranz C, Kueng A, Mizaikoff B. Integrating micro- and nanoelectrodes into atomic force microscopy cantilevers using focused ion beam techniques. *Appl Phys Lett* 2002;81:349–351.

80. Kranz C, Friedbacher G, Mizaikoff B. Integrating an ultramicroelectrode in an AFM cantilever: combined technology for enhanced information. *Anal Chem* 2001;73:2491–2500.

81. Lehrer C, Frey L, Petersen S, Sulzbach T, Ohlsson O, Dziomba T, Danzebrink HU, Ryssel H. Fabrication of silicon aperture probes for scanning near-field optical microscopy by focused ion beam nano machining. *Microelectron Eng* 2001; 57–8: 721–728.

82. Cannan S, Macklam ID, Unwin PR. Three-dimensional imaging of proton gradients at microelectrode surfaces using confocal laser scanning microscopy. *Electrochem Commun* 2002;4:886–892.

83. Rudd NC, Cannan S, Bitziou E, Ciani L, Whitworth AL, Unwin PR. Fluorescence confocal laser scanning microscopy as a probe of pH gradients in electrode reactions and surface activity. *Anal Chem* 2005;77:6205–6217.

84. Barker AL, Unwin PR, Gardner JW, Rieley H. A multi-electrode probe for parallel imaging in scanning electrochemical microscopy. *Electrochem Commun* 2004;6:91–97.

15

IN SITU X-RAY SPECTROSCOPY OF ENZYMATIC CATALYSIS: LACCASE-CATALYZED OXYGEN REDUCTION

SANJEEV MUKERJEE AND JOSEPH ZIEGELBAUER

Department of Chemistry and Chemical Biology, Northeastern University, Boston, MA, USA

THOMAS M. ARRUDA

Department of Chemistry, Salve Regina University, Newport, RI, USA

KATERYNA ARTYUSHKOVA AND PLAMEN ATANASSOV

Department of Chemical and Nuclear Engineering and Center for Emerging Energy Technologies, University of New Mexico, Albuquerque, NM, USA

15.1 INTRODUCTION

Biological fuel cells (BFCs) are energy-producing devices, akin to conventional fuel cells, except that they employ biological materials (typically enzymes or microbes) as catalysts on one or both of the electrodes. BFCs eliminate the need for precious metal catalysis and offer a unique advantage in high selectivity toward a single reactant or product that enables operation without the use of a compartment separator or membrane. There are examples of BFCs throughout the literature dating as far back as the 1960s. Initially, the goal was to produce an autonomous power source for an artificial heart [1–3]. To this end, the first successful BFC demonstration in 1964 by Yahiro et al. employed a variety of anode enzymes such as glucose oxidase (GOx), D-amino acid oxidase, and alcohol dehydrogenase in conjunction with an oxygen cathode [4]. Their investigations concluded that the redox mechanism of flavoproteins involves electron transfer from substrate to enzyme (contrasted with pyridinoproteins,

Enzymatic Fuel Cells: From Fundamentals to Applications, First Edition. Edited by Heather R. Luckarift, Plamen Atanassov, and Glenn R. Johnson.

which involve hydrogen transfer), but that power output was low. Although the low power output and short cell lifetime rendered BFCs inapplicable for the artificial heart project, the focus was shifted to powering implantable medical applications [5–7]. However, the energy requirements needed for intermittent charge pulses for nerve stimulators or drug delivery pumps could not be met by BFCs, and Li/SOCl$_2$ batteries were introduced as the alternative [8]. A resurgence of BFC research occurred in the 1980s and 1990s for the purpose of supplying energy to the grid, with limited commercial success [9,10]. Nonetheless, many significant advancements were achieved during this period in fuel cell technology, BFC mediator development, and membrane architecture [11–13]. For example, the incorporation of transition metal-based electron mediators for both cathodes and anodes became standard practice [14–16], and membrane-type BFCs operating at ~1 V were developed [17]. Earlier this decade, reports began to emerge describing BFCs that could operate under a wide range of conditions. Much of the success of these devices can be attributed to the development of "wired" metal complex mediators such as those developed by Heller and coworkers [10,18–21]. One particularly interesting demonstration was reported in 2003 by Mano et al., in which the enzyme catalysts were encapsulated into a redox polymer hydrogel matrix and immobilized directly onto carbon fibers and implanted into a grape [22]. The cell employed a GOx anode for the two-electron oxidation of glucose and bilirubin oxidase (BOx) for the four-electron reduction of dioxygen (Equations 15.1 and 15.2:)

$$\beta\text{-d-glucose} \xrightarrow{\text{GOx}} \delta\text{-gluconolactone} + 2H^+ + 2e^- \tag{15.1}$$

$$O_2 + 4H^+ + 4e^- \xrightarrow{\text{BOx}} 2H_2O \tag{15.2}$$

This BFC was important as it illustrated that a compartmentless BFC could be constructed and implanted into a physiological medium. The total volume of the electrode fibers was $0.0026\,\text{mm}^3$ and both reactants were readily available from the grape sap, that is, oxygen and glucose (>30 mM). The maximum power density achieved was $\sim 2.5\,\mu\text{W mm}^{-2}$ at 0.52 V when the cathode was implanted near the grape skin where the O_2 concentration was highest. This demonstrated principally that a BFC can be constructed smaller than any other power-generating device on the market today. As a result, similar devices are expected to find a niche for generating low power for implantable sensors, and may be inexpensive enough to be disposable [10].

15.2   DEFINING THE ENZYME/ELECTRODE INTERFACE

All BFCs and biological sensors—no matter the type of device—have one important aspect in common: the requirement for effective electronic communication between an electrode current collector and an enzyme. In the case of BFCs, it is obvious that efficient electron transfer must occur from enzyme to current collector (anode), and vice versa on the cathode, in order to ensure the highest possible cell efficiency. Any enzyme/substrate reaction that occurs without transferring the electron through the internal circuit of the cell would consume reactant without contributing power to the cell, and hence decrease the cell's overall power density. In the case of biosensors, any such inefficiency would result in incorrect sensing with potentially fatal consequences if the inefficiency was large enough. For example, a glucose sensor (such as one for monitoring blood glucose in diabetes patients) that is only 60% efficient could actually be consuming 40% more blood glucose than it reports, indicating the patient needs to take a glucose supplement. However, the excess glucose taken as a result could cause hyperglycemia in the patient, leading to organ damage and death. Considering such design constraints and the associated consequences, it is imperative to fully understand the link between enzyme and electrode.

15.3 DIRECT ELECTRON TRANSFER VERSUS MEDIATED ELECTRON TRANSFER

Direct electron transfer (DET), as the name suggests, insinuates that electron transfer occurs between enzyme and electrode without the use of any third-party mediators (Figure 15.1). For DET to occur, several criteria must be met: (i) the enzyme must be immobilized on the electrode with its active site within electron tunneling distance, (ii) the immobilization must be stable over time, and (iii) ideally, enzymes should be arranged in a monolayer (close-packed array) that eliminates residual enzymes on top of the initial monolayer that are not expected to be active and can, in fact, impede diffusion of reactants to the active enzymes.

DET is fundamentally the simplest method of enzyme association and can be achieved by one of the three methods: (a) physical adsorption, (b) electrostatic interaction, and (c) covalent bonding [23,24]. In 2006, Shleev et al. immobilized laccase on a functionalized carbon electrode, for example, via glutaraldehyde, to form a stable laccase film of one monolayer thickness [25]. The resulting enzyme association revealed bioelectrocatalytic activity for the reduction of dioxygen, and two distinct redox peaks were observed under anaerobic conditions. The redox peaks, at 99 and 530 mV versus Ag/AgCl, were attributed to the formal potentials of the type

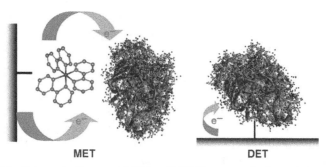

MET   DET

FIGURE 15.1 Schematic diagram of MET and DET illustrating the flow of electrons. (Please see the color version of this figure in Color Plates section.)

2/type 3 (T2/T3) and type 1 (T1) Cu centers, respectively. This observation is unique for the ability to determine a formal redox potential for the trinuclear Cu center potentiostatically, but, more importantly, the study also demonstrated the ability to observe electron tunneling directly to the trinuclear center, despite being deeply embedded within the protein structure.

Immobilization by covalent binding involves bonding the enzyme directly to the electrode surface using one or more residues of the enzyme or a chemical tether [23,24]. Such immobilization methods have been successfully employed for laccase using thiol-rich compounds such as α-lipoic acid [23,24] and 4-amino-thiophenol [26,27] on gold electrodes. In the former case, the enzyme was chemically tethered to a nanoporous gold electrode by using an Au−S bond at the gold interface. Following esterification, the N-hydroxysuccinimide groups were replaced by an amino group on the surface of the laccase. This results in an immobilized laccase that is chemically bonded to the gold in short enough proximity to allow for electron tunneling. Interestingly, this covalent coupling method may have been further enhanced by the physical adsorption of laccase directly onto the nanoporous gold. Laccase of *Trametes versicolor* contains eight lysine residues, four of which are on, or near, the surface and can conceivably supply a bridge to the nanoporous gold via the N-terminus of the side chain. Michaelis–Menten-type kinetic analysis by Qiu et al. [23,24] revealed that smaller (micrometer size) nanoporous gold particles improve the catalytic efficiency of laccase, possibly by exposing the active site to the substrate, and enhancing mass transfer.

15.3.1 Mediated Electron Transfer

Although DET is the simpler method for manufacturing BFC electrodes, there are limitations that may limit the utility of DET electrodes. Among these limitations are enzyme loading (which impacts the power density) and catalyst retention. The latter continues to be addressed by novel methods of immobilization, whereas the former remains an issue. As a result, DET to date has primarily found application in sensor development where current densities are sufficient to "sense" a chemical species. Increasing the surface area of the electrodes by using nanotubes and nanoparticle surfaces, however, does to some extent alleviate the limitation of loading [28,29]. Power densities, however, such as those obtained for a laccase cathode and carbon/ascorbate anode are routinely on the order of $\mu W\,cm^{-2}$ [29].

Thus, to realize the full potential of BFCs, investigators are now considering mediated electron transfer (MET) to increase the electronic association between the total enzyme on the electrode and, hence, the power density. By using a redox mediator to shuttle electrons to and from electrode to enzyme, it is possible to exploit the entire three-dimensional electrode area. These mediators, whether diffusion-based or wired [30], therefore capitalize on a catalyst layer that is much thicker (100 μm) than the monolayers in DET. Although MET has been shown to produce power densities several orders of magnitude higher than DET, the presence of an added electron transfer step, however, introduces another avenue for possible inefficiency.

During the electroreduction of O_2 to form H_2O (Equation 15.2), the initial step is transfer of an electron from the electrode to the mediator. The mediator is reduced and, in turn, an electron is transferred to the nearest enzyme; in doing so, the mediator is reoxidized whereas the enzyme is reduced (Figure 15.1). An ideal mediator provides electron transfer that is completely reversible, and a number of mediators have been synthesized that include a wide array of structures, valence states, or redox potentials, or are tailored to exhibit properties such as specific diffusion coefficients [18–21,30]. Mediators are typically classed into two categories: (i) diffusional or (ii) immobilized (or wired). An example of a common diffusional mediator is 2,2′-azinobis(3-ethyl-benzothiazoline-6-sulfonate) (ABTS), which is divalent and can serve as a cosubstrate with oxygen-reducing enzymes. ABTS works by accepting electrons from the electrode and diffusing through solution until it encounters an enzyme to which it can transfer its electrons [17]. Unfortunately, this type of electron transfer relies on incidental encounters between a solution species and an enzyme, which is inherently inefficient. Immobilized or wired mediators, in contrast, are fixed to the electrode in close proximity to the enzyme and provide relatively high electrode kinetics for a wide variety of enzymatic systems [18–22,31,32].

A variety of properties must be considered when selecting mediators for BFCs, including stability, toxicity, and, most importantly, redox potential. The mediator is expected to be oxidized primarily at electrode potentials greater than the redox potential of the mediator. Conversely, the mediator is expected to be reduced when the potential is below that of the mediator potential. The resulting mixed potential then represents the open-circuit potential of a cell employing two such mediated electrodes. Appropriate mediators must then be chosen with a redox potential close to the potential of the enzyme used. Although a wide variety of mediators have been considered, the current state-of-the-art mediators involve osmium-based organometallic moieties tethered to a polymer backbone such as polyvinylimidazole (PVI) [10,30]. The redox potential of these mediators can be tuned by changing the substituent on the organometallic osmium complex. A 2009 study by Hudak et al. demonstrated an H_2/O_2 fuel cell that operated with a maximum power density of $0.7\,\text{mW cm}^{-2}$ using osmium-based mediators on a laccase biocathode [33]. Further, oxygen reduction current densities were on the order of $1–10\,\text{mA cm}^{-2}$, several orders of magnitude higher that that obtained for DET biocathodes [34].

A detailed electrokinetic analysis of osmium mediator/laccase biocathodes was published in 2008 by Gallaway and Calabrese Barton [31]. In their work, the authors described the biomolecular rate constants for mediation to be between 250 and $9.4 \times 10^4\,\text{M}^{-1}\,\text{s}^{-1}$ when the redox potentials of mediator and laccase are close or far apart, respectively. Such values were used to determine the optimal mediator redox potential of $0.66\,\text{V}$ versus reversible hydrogen electrode (RHE), and a mediator structure was proposed to achieve this value.

15.4 THE BLUE COPPER OXIDASES

Blue copper oxidase (BCO) is a classification of oxidoreductase enzymes that contain at least one blue or T1 copper and a T2/T3 trinuclear cluster. These enzymes typically give rise to a characteristic blue color as a result of a strong absorption band around

600 nm due to the T1 Cu. Many BCOs are known and have been isolated from plant, animal, and fungal sources alike. Some of the more widely studied BCOs include laccase, ascorbate oxidase, ceruloplasmin, phenoxazinone synthase, BOx, dihydro-geodin oxidase, sulochrin oxidase, and FET3 (from *Saccharomyces cerevisiae*) [35,36]. These enzymes all have a common catalytic ability to oxidize a substrate that results in a reduced Cu^+ state, which subsequently reduces O_2. The native state of most BCOs is fully oxidized; that is, each of the four Cu ions is divalent. Therefore, for the full reduction of O_2 to H_2O (Equation 15.2), the enzyme must oxidize four substrate molecules.

BCOs exhibit variable substrate specificity; plant and fungal laccases, for example, typically exhibit wide substrate specificity and can oxidize a variety of aminophenols, diphenols, and aryl diamines. The mechanism for these enzymes is consistent with Marcus theory [37], and suggests that oxidation occurs in the outer sphere and there is likely no specific binding pocket. Conversely, some BCOs such as ascorbate oxidase (specificity toward L-ascorbate) [38,39] and ceruloplasmin (specificity toward Fe^{2+}) [40,41] are often stereospecific and highly substrate specific [42].

Substrate-specific BCOs, when reacting with substrates under natural conditions, follow Michaelis–Menten kinetics with reasonably high catalytic activity [42]. The substrate–enzyme reaction, or reduction of the Cu^{2+} sites, is the rate-determining step for the overall enzymatic reaction [37], whereas the oxygen reduction reaction (ORR) occurs much more rapidly. This type of enzymatic reaction would not participate in any electrocatalytic electron transfer unless the "substrate" is actually the electrode. Therefore, the goal for BFCs is to configure the enzyme by a method that allows the electrode to assume the role of substrate so that the electrons can travel through an external circuit.

15.4.1 Laccase

As with most BCOs, laccase is readily isolated from many species of plants, animals, and fungi [42]. The properties of laccase (formula weight, structure, redox potential, and the presence/absence of certain active site Cu ions), however, can vary significantly depending on the source. Although many forms of laccase have been studied [30,42], the white rot fungus (*T. versicolor, Coriolus versicolor*, or *Polyporus versicolor*) laccase catalyzes the $4e^-$ reduction of O_2 to H_2O at a high redox potential (780–800 mV vs. RHE at pH 4.0) and is the most widely used in BFC research due to commercial availability and a known sequence and crystal structure [43]. These fungal laccases exhibit a full complement of Cu ions (one T1, one T2, and two T3) and the T1 site is located in relatively close proximity to the enzyme surface.

Despite the broad substrate specificity of many laccases, the primary catalytic reaction of interest for BFC research is the ORR. Hence, the focus turns to the Cu sites of BCOs, such as laccase. Much is already known about the functionality of each of the types of Cu; for example, it is known that ORR is initiated by reducing the T1 Cu (near the surface) [25]. The resulting electron tunnels down a His–Cys–His tripeptide to the trinuclear Cu center (TNC) [44–46]. As the activation site for ORR, the redox potential of the T1 site determines the redox potential for ORR. The actual ORR, however, occurs at the TNC, which is embedded 12.5 Å inside the enzyme. ORR occurs when O_2 binds to the fully reduced TNC in the presence of protons, resulting in H_2O and reoxidized Cu ions. The TNC, which contains the T2/T3 Cu ions, is accessible to oxygen and protons via two electrolyte channels that connect the interior active site to the exterior of the enzyme [43]. Further, the oxidized forms of T1 and T2 Cu are electron paramagnetic resonance (EPR) active due to the unpaired $3d^9$ electron in a high spin state, whereas the two T3 Cu ions are EPR inactive due to antiferromagnetic coupling by a bridged oxide/hydroxide (referred to as O_{nat}). The reduced form (Cu^+) of all types of Cu has no unpaired electrons, and, hence, is EPR inactive.

In laccase, the T1 Cu is ligated to two imidazole moieties (via N) from the surrounding histidine residues as well a sulfur atom from the nearby cysteine residue. Both the T2 and T3β Cu ions are ligated to two imidazoles and the T3α Cu has three imidazole ligands. The crystal structure also reveals a bridged O between the T3α and T3β coppers with Cu−O bond distances of 2.08 and 2.19 Å, respectively. Despite all that is known about the structure of laccase, however, ambiguities still remain in respect to ORR mechanisms. Many ORR mechanisms have been proposed [42,47,48], but due to the complexity of the reaction center, the precise mechanism has yet to be identified [25]. In addition, many studies present rate constants for ORR and proposed intermediate steps that suggest that the reaction occurs quickly, rendering steady-state measurements ineffective [31,49,50]. One method to circumvent this issue is to freeze trap the quickly decaying (or reoxidizing) intermediate state and make measurements accordingly. Using this technique in 2002, Lee et al. measured X-ray absorption spectroscopy (XAS) and EPR of the native intermediate of laccase from *Rhus vernicifera* by chemically treating a solution of laccase with stoichiometric amounts of reductant and flash freezing the intermediates with liquid N_2 [49]. In doing so, they determined that the native intermediate is a fully oxidized active site with an oxygen radical bound to the TNC, whereas the resting oxidized form has the T2 site magnetically isolated from the T3 Cu ions. Additionally, it was suggested that the decay of the native intermediate to the resting oxidized form is slow due to the loss of bridging O_2 reduction products, whereas the re-reduction of the intermediate is rapid. Finally, the study indicated that both the native intermediate and resting oxidized states are fully oxidized Cu ions, differing only by structural changes associated with the presence or absence of bridging oxo by-products.

15.5 *IN SITU* XAS

The study described above provides fundamental accounts of enzyme behavior, albeit following chemical treatment, and at 77 K. Thus, although the findings are significant, it is unclear if the mechanism can be extrapolated to room temperature and physiologically relevant electrochemical conditions, such as those in a BFC. Thus, the primary benefit of the XAS studies described herein is the ability to investigate laccase *in situ*, under realistic electrochemical and BFC conditions, and correlated to various electrochemical and spectroscopic methods.

The *in situ* XAS analyses described herein were performed using X-ray absorption near-edge structure (XANES) analysis and extended X-ray absorption fine structure (EXAFS) analysis and analyzed in conjunction with FEFF8.0 modeling to investigate changes to the active site as a function of potential and presence of O_2. Fundamental electrochemical methods such as cyclic voltammetry (CV), chronoamperometry, and ORR polarization measurements were then compared to illustrate the electrocatalytic activity of laccase. Finally, an in-depth analysis of the $\Delta\mu$/FEFF8 results lead to a new proposed mechanism for the laccase-catalyzed ORR.

For the discussion presented herein, three representative redox polymers (poly $(VI_{15}[Os(tpy)(bpy)]^{2+/3+})$ (RPA), poly($VI_{20}[Os(tpy)(dm\text{-}bpy)]^{2+/3+}$ (RPB), and poly $(VI_{12}[Os(bpy)_2Cl]^{+/2+})$ (RPC)) were considered (Figure 15.2). A range of redox polymers can be prepared with varying activity and redox potential (E_0) and are typically based on PVI backbones (Table 15.1) [31,34,51,52].

Redox polymer (RP) films are cast from a mixture of RP, laccase solution, and polyethylene glycol diglycidyl ether (PEGDGE) (61% polymer, 32% laccase, 7% PEGDGE) onto either gold or glassy carbon rotating disk electrodes. Anaerobic CVs of the resulting redox polymer hydrogels show characteristic redox peaks associated with the immobilized laccase (Figure 15.3). Under anaerobic conditions, laccase is

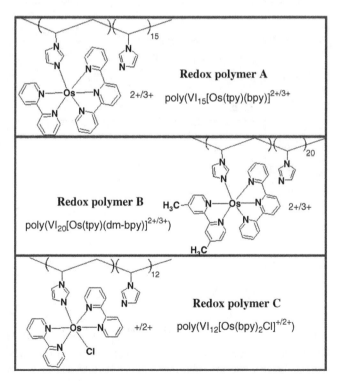

FIGURE 15.2 Drawings of the three osmium-based redox polymers employed in this study.

TABLE 15.1 Summary of Redox Polymers Described in This Chapter

Code	Redox Polymer	E_0 (V vs. RHE)	Os (wt%)	Reference
RPA	poly(VI$_{15}$[Os(tpy)(bpy)]$^{2+/3+}$)	0.82	9.2	[53]
RPB	poly(VI$_{20}$[Os(tpy)(dm-bpy)]$^{2+/3+}$)	0.77	6.6	[34]
RPC	poly(VI$_{12}$[Os(bpy)$_2$Cl]$^{+/2+}$)	0.43	11	[51]

Key: bpy, 2,2'-bipyridyl; dm-bpy, 4,4'-dimethyl-2,2'-bipyridine; tpy, 6',2''-terpyridine; VI, vinylimidazole.

essentially "inactive" due to the absence of O_2 as substrate. Thus, the CVs can be used to determine the redox profiles of the $Os^{2+/3+}$ (RPA) and $Os^{+/2+}$ (RPC) species alone by locating the midpoint between the two peaks, i_{pa} and i_{pc} (Table 15.2).

The redox event observed from CVs shows peak separation, $\Delta E_p \sim 59$ mV, that indicates an Os redox couple that is reversible with $n = 1$ electrons transferred, as

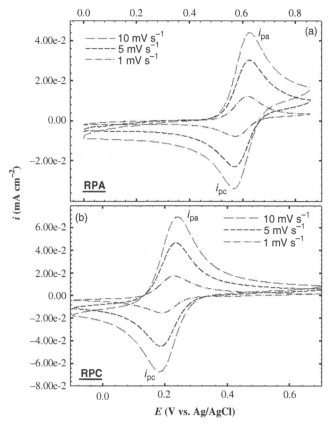

FIGURE 15.3 Cyclic voltammograms of (a) redox polymer A and (b) redox polymer C collected at 10 mV s^{-1} on a stationary 5 mm glassy carbon RDE tip and 20 °C. The electrolyte used was Ar purged, 100 mM citrate buffer at pH 4.00. Current density reports geometric surface area. CV traces for polymers A and B were identical; data not shown.

TABLE 15.2   Redox Profiles Derived from Cyclic Voltammograms of Various Redox Polymers with Immobilized Laccase

Redox Polymer	E_{exp}^0 (V vs. RHE)	ΔE_p (mV)	Randles–Sevcik Equation (see Equation 15.4)
RPA	0.81	59.8	$i_p = 0.465\nu^{1/2} - 2.40 \times 10^{-3}$
RPC	0.42	57.9	$i_p = 0.765\nu^{1/2} - 6.90 \times 10^{-3}$

expected for a Nernstian reaction, where R is the gas constant, T is temperature, n is the number of electrons transferred, and F is Faraday's constant (Equation 15.3).

$$\Delta E_p \approx \frac{2.3RT}{nF} \tag{15.3}$$

A further measure of reversibility can be observed from the CVs by considering the Randles–Sevcik equation (Equation 15.4):

$$i_p = 0.4463 \left(\frac{F^3}{RT}\right)^{1/2} n^{3/2} D_m^{1/2} c_m \nu^{1/2} \tag{15.4}$$

where i_p is peak current density, F is Faraday's constant, R is the gas constant, T is temperature, n is the number of electrons transferred, D_m is the apparent electron diffusion coefficient of the mediator, c_m is the concentration of mediator, and ν is the scan rate [31,54–56]. For a reversible system, a plot of i_p versus $\nu^{1/2}$ yields a linear relationship and allows for the determination of Randles–Sevcik parameters. When applied to the experimental data (Figure 15.3), linear relationships are observed, suggesting the reaction was highly reversible and yielding correlation coefficients >0.999 for both redox polymers. The analysis provides Randles– Sevcik parameters (D_m, c_m, etc.) as a comprehensive kinetic analysis of these mediators in agreement with prior results [31], but also confirms reversibility as proposed (Table 15.2).

ORRs determined by potentiostatic polarization for laccase with RPA and RPC hydrogels exhibit the typical sigmoidal reduction wave indicative of ORR (Figure 15.4). Note that the ORR onset for RPC shifts by approximately −400 mV as the redox potential of the reaction is determined by the redox potential of the mediator. During rotating disk electrode studies, a Levich-type relationship was observed at low rotations (between 0 and 100 rpm), but this does not hold true at rotations higher than 400 rpm (Figure 15.4, inset); that is, supplying more O_2 to the reaction center via rotation appears to have no effect on the diffusion-limiting current. There are two possible explanations for this behavior, both involving O_2 diffusion: (i) the redox polymer film is thicker than the diffusion layer causing mass transport limitations, or (ii) the mass transport limitations occur as a result of slow O_2 diffusion into the center of the enzyme (via two electrolyte channels) where the TNC resides. The latter of these two explanations is considered more plausible, as the polymer films have been measured to be on the order of 1/10th the thickness of the diffusion layer [31].

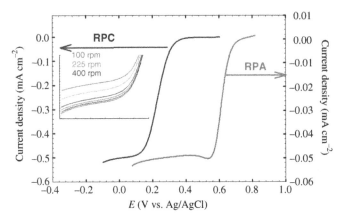

FIGURE 15.4 ORR polarization sweeps for the indicated laccase/RP hydrogels collected at $10\,mV\,s^{-1}$ in O_2-saturated $100\,mM$ citrate buffer at pH 4.00 at $20\,°C$ shown at 400 rpm. The inset shows RPC at various rotation rates as pertaining to the Levich analysis.

The electrochemical results provide a diagnostic measure to illustrate that MET works efficiently on these redox materials. However, it provides neither mechanistic insight into the ORR process nor structural information in regard to the laccase active site. In this respect, *in situ* $\Delta\mu$–XANES provides a useful tool to consider in parallel with experimental data to elucidate information on active site transformations during the reaction, specifically the O_x binding site at the TNC.

For the laccase from *T. versicolor* described herein, *in situ* XAS measurements were performed at both the Cu k-edge and the Os L_3-edge for each of the three Os-based redox mediators presented (Figure 15.2). The Os L_3-edge will be discussed briefly, followed by an in-depth examination of the Cu k-edge analyses for both MET and unmediated electron transfer (uMET). (For the purpose of discussion herein, uMET is considered to be fundamentally different from DET, because of the absence of any specific preferential orientation that ensures efficient DET.)

15.5.1 Os L_3-Edge

XANES can be useful in determining the valence state of a material, provided the appropriate reference spectra have been collected. This is achieved by comparing the edge energy changes of an unknown valence state sample with the edge energy of known standards. An analysis of this type on the osmium mediators, in parallel with experimental data from CVs, should theoretically corroborate that the osmium redox process is a one-electron transfer step.

XAS measurements of model osmium compounds (Os powder, $Os_3(CO)_{12}$, $OsCl_3$, OsO_2, and K_2OsCl_6) revealed inconsistent shifts in the electron binding energy ΔE_0 of the $2p^{3/2}$ electron (L_3 edge) (Figure 15.5). For example, trivalent Os ($OsCl_3$) exhibits an edge position nearly identical to that of the Os metal, whereas the zero-valent $Os_3(CO)_{12}$ material reveals an edge position nearly $+1.3\,eV$ above the metallic

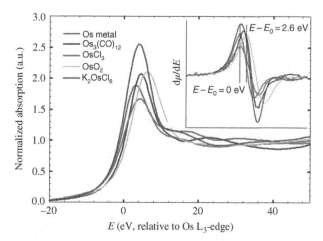

FIGURE 15.5 Normalized Os L_3-edge XANES spectra of the indicated Os reference compounds. The inset plots the first derivative spectrum of the XANES, which is the conventional method for determining E_0 by taking the energy value at the first maximum of the spectrum. (Please see the color version of this figure in Color Plates section.)

Os. Further, the two 4+ materials reveal edge positions that are different by 2.5 eV with OsO_2 shifting +2.6 eV and K_2OsCl_6 only increasing by approximately +0.1 eV. The reason for these discrepancies can largely be attributed to the electron-withdrawing capability of the ligands that surround Os as described by ligand field theory (LFT). Strong electron-withdrawing ligands such as CO and O are considered to be π-acceptors, effectively drawing electron density away from the Os, hence increasing the binding energy of the Os electrons. In contrast, ligands such as Cl^- are low-field ligands (π-donors) that lead to weaker electron-withdrawing capability and smaller ΔE_0. To further complicate matters, even standard compounds of the same ligand have been shown to exhibit significantly different E_0 values as symmetry can affect the binding energy. Thus, when attempting measurements of this type, compounds of the same ligand and similar geometries *must* be employed to ensure an accurate assessment. Unfortunately, it is practically impossible to determine the valence state changes of the osmium mediator samples using the standards employed here. Furthermore, the commercial availability of N-containing osmium compounds is extremely limited, making a comprehensive collection of appropriate standards difficult. Nonetheless, a qualitative assessment can still be made by comparing the E_0 values of the osmium mediators at various potentials such as lower and higher than the standard redox potential of the mediator.

Figure 15.6a–c shows the Os L_3-edge XANES spectra of each mediator (RPA, RPB, and RPC) with their corresponding derivative spectra (Figure 15.6d–f). The two aspects of the XANES that are expected to yield information are the white-line intensities and ΔE_0. White-line intensities generally increase in magnitude as the valence state increases as a result of more states available for an electronic transition to occur. In the examples provided, the white-line intensities do not change significantly,

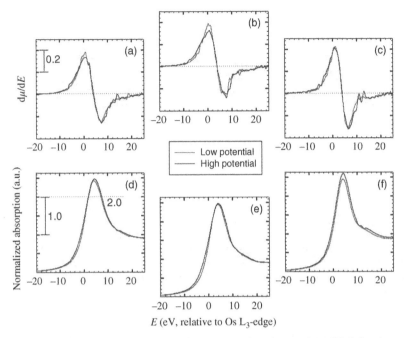

FIGURE 15.6   First derivative spectra for (a) RPA, (b) RPB, and (c) RPC for the corresponding normalized Os L_3-edge XANES spectra in (d) RPA, (e) RPB, and (f) RPC.

and there are also negligible ΔE_0 values obtained as a function of electrode potential. Such observations could be explained by one of the following three phenomena: (i) the osmium valence state remains unchanged, (ii) there is a substantial amount of unreactive osmium present, or (iii) the one-electron transfer is invisible to XANES due to electron shielding. Because the CVs have already shown that the mediators exhibit a reversible redox couple with $n = 1$, theory (i) seems unlikely, and, as such, explanations (ii) and (iii) will be considered in more detail.

When fabricated to the electrode surface, the polymer film dries and a thick redox polymer hydrogel crust forms around the edge of film. The crust is considerably thicker than the 1–3 μm active portion of the film, which makes the outermost Os (and laccase) susceptible to mass transport limitations. This Os would essentially be isolated from the active material in the catalyst layer, rendering it electrochemically unresponsive, and thus it would not change its oxidation state and contribute to an increase in ΔE_0. However, this portion of the electrode only represents a minimal portion, and care is taken to avoid irradiating the edges of the catalyst deposition, thus eliminating this phenomenon as the main cause of a negligible ΔE_0.

In lieu of the above assumptions, theory (iii) is considered further to explain the lack of shift in E_0. First consider an atom from the first-row transition elements, such as iron, as many studies have been published in which Fe valence states are reported as determined by XANES edge shifts [57–60]. Oxidation of Fe is often accompanied by an edge shift of 8–10 eV (from $0 \rightarrow 3^+$) depending on the ligand. However, it is important to mention that these measurements are often carried out at the Fe k-edge, which will in fact be influenced by the nuclear charge to a greater extent than higher shell electrons due to proximity [61]. The end result being the net positive effective nuclear charge (Z_{eff}) will exert enough influence on the k-shell electrons to require substantially more energy to remove it, especially as electrons are being removed from the valence shell. However, this is not the case with the Os samples where the L_3-edge is probed. The removal of valence electrons is expected to increase the binding energy via increasing Z_{eff}, but Os still has many more electrons that assist in shielding the Z_{eff}, thus resulting in an undetectable ΔE_0. Similar effects have been observed for platinum, where XANES were measured at oxidizing potentials (>1 V) and no ΔE_0 was obtained.

15.5.2 uMET

Cu k-edge spectra for unmediated laccase under (a) anaerobic and (b) aerobic conditions are shown in Figure 15.7. Unlike the relatively featureless Os L_3-edge, the Cu k-edge reveals a dipole-allowed transition ($1s \rightarrow 4p$) in the region of E_0. This transition has been shown to occur when laccase is reduced to the Cu^+ state and is absent at the fully oxidized Cu^{2+} state [62]. Note that this feature appears to be present in both aerobic and anaerobic conditions, although it is subject to change with oxidation state. For example, Figure 15.7a reveals a small decrease in the $1s \rightarrow 4p$ transition as the potential is increased, which happens to coincide with a slightly increased white-line intensity. Even though this suggests a slight oxidation with potential, laccase is still mostly reduced in the absence of O_2, despite the fact that the difficulty in "trapping" reduced laccase has been well documented [42,48,62]. The

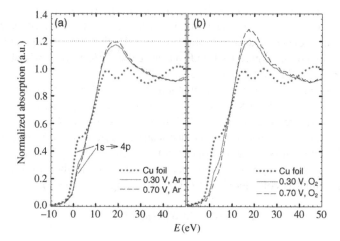

FIGURE 15.7 *In situ* Cu k-edge XANES spectra for uMET laccase collected as a function of potential in 100 mM citrate buffer at pH 4.00 under (a) aerobic and (b) anaerobic conditions.

observations indicate that under anaerobic conditions, electrode potential has less of an effect on the oxidation state of Cu in laccase than does the removal of its substrate. This is explained by either poor reversibility, that is, the electrode cannot reoxidize the active site in the absence of O_2, or photoreduction of the oxidized active site. Laccase, however, has not been shown to undergo photoreduction [62], even though some BCOs have [63,64]. Therefore, it is suggested that under these conditions electron transfer from the electrode to the enzyme occurs readily, whereas electron transfer from the enzyme back to the electrode occurs poorly or not at all. In contrast, under aerobic conditions (Figure 15.7b) and oxidizing potential, the XANES spectrum exhibits a much larger white line and almost no transition at low potential than without O_2. The O_2 here clearly reoxidizes the laccase, however, with a potential preference. It is surprising that at low potential a fully reduced laccase is observed where others have needed to freeze quench the reduced state under anaerobic conditions [62]. Nevertheless, this experiment demonstrates that the oxidation state of Cu can be altered electrochemically for an uMET laccase system.

In addition to the above discussion, it should be noted that that spectra in Figure 15.7 represent mixed-valent Cu under all conditions, with the highest percentage of Cu^{2+} at 0.7 V (aerobic conditions) and highest percentage of Cu^+ at 0.3 V (anaerobic conditions). This is evident as the $1s \rightarrow 4p$ transition is never entirely complete. Similar findings were observed in a comprehensive Cu k-edge study of laccase (*R. vernicifera*) published in 1987 by Kau et al. [65], using a wide variety of Cu standards (40 Cu^{2+} and 19 Cu^+). In the previous study, H_2O_2 was used to chemically reduce the T3 coppers (using T2-depleted enzyme), resulting in a large transition at 8949 eV. This reduction was attributed only to the T3 coppers as EPR revealed no changes to the T1 spectrum. It is also important to remember that these measurements were made at steady state (~ 1.5 h for six scans) and the reaction occurs rapidly. This means that the measured XANES spectrum actually represents an average state in which all coppers exist for the duration of the measurements. For example, the O_2-saturated laccase at 0.7 V appears mostly oxidized as suggested above; however, this does not mean that it never exists in the reduced state. It only means that the reduced state is very short lived under these oxidizing conditions, as the ORR polarizations (Figure 15.4) indicate an active ORR process, which would not be the case if Cu did not become reduced.

To further analyze the data, the $\Delta\mu$ method was considered to determine where the O_x species resides in the TNC active site. The difference spectra for the XANES (Figure 15.7) are presented in Figure 15.8 using the indicated XANES scans as the reference scan. Note that different reference scans are used to amplify either the effect of electrode potential (same gas), effect of O_2 presence (same potential; E), or both (combination of gas and potential). For example, the situation where *only* the effect of electrode potential is considered was calculated using Equation 15.5:

$$\Delta\mu = \mu(0.7 \text{ V, Ar}) - \mu(0.3 \text{ V, Ar}) \qquad (15.5)$$

where $\mu(0.7 \text{ V, Ar})$ is the XANES at 0.7 V under anaerobic conditions and $\mu(0.3 \text{ V, Ar})$ is the XANES at 0.3 V, also under anaerobic conditions. Each $\Delta\mu$ reveals a

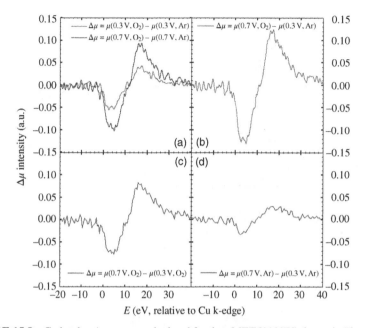

FIGURE 15.8 Cu k-edge $\Delta\mu$ curves calculated for the uMET XANES shown in Figure 15.7. Each curve was calculated via equations in respective plots. (Reproduced with permission from Ref. [74]. Copyright 2014, Wiley-VCH.)

minimum near 4 eV followed by a maximum at 16 eV, all of which are symmetrical (i.e., the magnitude of the minimum of a given peak is almost equal to that of the maximum). All curves also exhibit very similar line shapes, differing only in the magnitude of the peaks that appear to change systematically. The trend in $\Delta\mu$ magnitude is such that the largest magnitude $\Delta\mu$ occurs (Figure 15.8a) where both E and O_2 are considered, the lowest magnitude (Figure 15.8d) where E is considered in Ar (anaerobic), and the mixed E and gas curves in between. Although it is not possible to determine the O_x binding position by this qualitative assessment alone (FEFF calculations are also needed), these $\Delta\mu$ data confirm that the most active situation occurs under aerobic conditions for uMET and least active under anaerobic conditions. The line shape assessment is further discussed in considering FEFF analysis below.

15.5.3 Mediated Electron Transfer

The plots shown in Figure 15.9 reveal the normalized Cu k-edge XANES data collected at two potentials, under O_2 and Ar saturation for MET using the redox polymers indicated previously. At first glance, the data appear similar, and unlike the XANES obtained for uMET, there is no 1s → 4p transition observed at either low or high potential, or on the introduction of O_2. This suggests that all four coppers exist in the divalent state predominantly, or at least spend a high percentage of the time fully

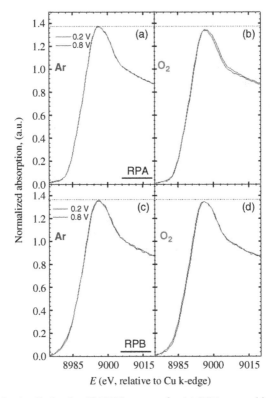

FIGURE 15.9   *In situ* Cu k-edge XANES spectra for (a) RPA anaerobic, (b) RPA aerobic, (c) RPB anaerobic, and (d) RPB aerobic conditions.

oxidized, thus not revealing any transition indicative of Cu^+. Surprisingly, the XANES collected under aerobic conditions (Figures 15.9b and d) yield a white-line intensity that is slightly lower than that of the anaerobic scans. This result is unexpected as more oxidized samples tend to *increase* white lines of first-row transition elements due to the availability of d-states (in conjunction with some p–d hybridization). Recall that XANES signal is a result of not only local electronic transitions but also multiple scattering of low-energy photoelectrons [66,67]. It is possible that the decrease in white-line intensity is a result of an adsorbed O_x species that could block or impede the backscattering amplitude of the low-energy photoelectrons, resulting in decreased white-line intensity. Either way, the data are inconclusive, suggesting that it is only possible to resolve divalent Cu. This is the likely explanation for the lack of features in the spectra considering the timescale of ORR in this enzyme. For instance, in 2008, Gallaway and Calabrese Barton published bimolecular rate constants on the order of $10^5 \, s^{-1} \, M^{-1}$ [31], suggesting the turnover from Cu^{2+}/Cu^+ occurs in subsecond timescales. As observed for uMET, the reduced state may be more difficult to resolve due to the fast turnover; in fact, it would be even more difficult to resolve for MET as the electron transfer efficiency is expected to be much higher.

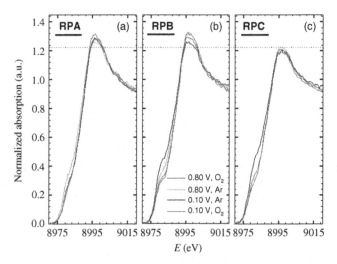

FIGURE 15.10 *In situ* Cu k-edge XANES spectra for MET laccase using (a) RPA, (b) RPB, and (c) RPC plots. (Please see the color version of this figure in Color Plates section.)

Another possible explanation for the lack of features in Figure 15.9 is poor electrical contact between the redox polymer hydrogel and the electrode. This would cause the enzyme to appear "dead" in the beam, thus remaining oxidized and featureless. To address this phenomenon, experiments were conducted in which the electrode preparation included the incorporation of carbon paper into the redox polymer hydrogel. This increased the active surface area and electronic conductivity by increasing the contact between the hydrogel and electrode. The resulting XANES for redox polymers RPA, RPB, and RPC are presented in Figure 15.10 and reveal features that were not previously present in the near-edge spectrum (Figure 15.9). Each polymer produces a transition in the edge jump region described above as the $1s \rightarrow 4p$ transition, in addition to changing white-line intensities. The guideline is provided to assist in comparison of the white-line intensities (Figure 15.10), and it is clear that all of the white-line intensities for RPC are lower in magnitude than the intensities for RPA and RPB. This indicates that there is a significant difference in the electronic structure of the enzyme Cu ions as a function of redox polymer alone, by providing either a different electron transfer mechanism or an alteration due to the RP itself. There are significant differences between the structures of RPC and those of RPA and RPB (Figure 15.2) including the valence state of the Os, which if in close enough proximity could provide ligand field effects that could cause the intensity to decrease. For example, consider the T2 Cu, which is ligated linearly to two nitrogen atoms. The ligand field splitting is such that the p_x and p_y orbitals are degenerate, whereas p_z is at higher energy. However, supplying a third ligand field would destroy the degeneracy of the $p_{x,y}$ level, requiring more energy to populate the p_y states [65]. Of course, this would only affect one-quarter of the total XANES signal, as there are three other Cu ions not undergoing the same ligand field splitting. Further, some

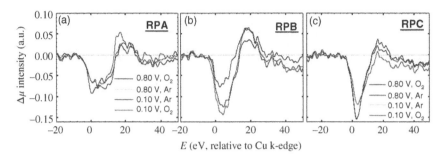

FIGURE 15.11   Cu k-edge $\Delta\mu$ curves calculated for the MET XANES shown in Figure 15.10. The RPA $\Delta\mu$ (a) was calculated using Equation 15.4, whereas RPB (b) and RPC (c) curves were calculated with Equation 15.5. (Please see the color version of this figure in Color Plates section.)

discrepancies are revealed even between the scans of RPA and RPB, which are very similar in structure. The white-line intensities at 0.1 V under O_2 (red lines), for example, are of approximately equal magnitude, but RPB exhibits a much smaller white line at 0.1 V Ar (blue lines) than RPA, indicating that it is more reduced. RPC, in contrast, exhibits similar near-edge features to those of RPB in that its s $\rightarrow$ p transition is largest at 0.1 V and under anaerobic conditions. It seems that the observed behavior of RPB and RPC more closely resembles the uMET situation where Cu is reduced at low potentials, whereas RPA actually appears to be reduced more at high potentials. Nonetheless, the results are still ambiguous and require further study; by removal of the T1/T2 coppers, for example, a definitive explanation may be obtained.

Finally, the $\Delta\mu$ curves for the XANES in Figure 15.10 are replotted in Figure 15.11 to clearly illustrate the changes discussed above. Note that RPA and RPB result in very similar $\Delta\mu$ curves, resembling those shown in Figure 15.8 (uMET), white RPC yields $\Delta\mu$ lines that have a strong negative peak near 2 eV and only a very small positive peak to follow. The $\Delta\mu$ were calculated using the relationship (Equation 15.6) for RPA:

$$\Delta\mu = \mu(E, \text{gas}) - \mu(0.8 \text{ V}, \text{Ar}) \tag{15.6}$$

and the relationship (Equation 15.7) for RPB and RPC:

$$\Delta\mu = \mu(E, \text{gas}) - \mu(0.1 \text{ V}, \text{Ar}) \tag{15.7}$$

The reason for using these reference scans is that RPA appears to be the most reduced (and likely free of O_x) at this potential, where RPB and RPC appear most reduced at 0.1 V. The idea is to obtain a signature line shape and then compare it with signatures generated by FEFF. There appears to be no significant trends in the $\Delta\mu$ for all three of the redox polymers in terms of changes with electrode potential, thus making further analysis confusing. Therefore, the $\Delta\mu$ shall be discussed in terms of their respective line shapes following a FEFF8 analysis described below.

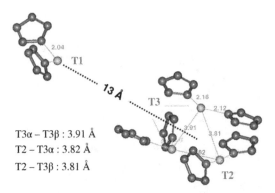

T3α – T3β : 3.91 Å
T2 – T3α : 3.82 Å
T2 – T3β : 3.81 Å

FIGURE 15.12 Graphical representation of the active site of laccase (*T. versicolor*).

15.5.4 FEFF8.0 Analysis

Prior to FEFF analysis, it is important to understand the structure of the active site. The schematic in Figure 15.12 illustrates the local structure of the T1 and TNC active sites of laccase as they are relevant to FEFF analysis. Note that the T1 site is located near the surface of the enzyme (~4–5 Å), whereas the TNC is more deeply embedded. The T1 Cu is directly ligated to two histidine nitrogen atoms and a sulfur atom (not shown) from the nearby cysteine residue, whereas the T2 Cu is nearly collinear with two histidine N atoms and each of the T3 Cu ions are coordinated with three histidine N atoms. Also not shown in this schematic (Figure 15.12) is a bridged oxo-species, typically shown as a single oxygen atom, which is responsible for antiferromagnetically coupling the two T3 coppers. During the course of the FEFF analysis, O atoms and O_2 molecules are moved about the TNC in all plausible configurations to determine the location of the binding site. (*Note*: Although three different cluster sizes were investigated, the large cluster size (7 Å from absorbing Cu, ~90 atoms) was used as it was deemed to provide more meaningful results.)

The base structures for which the large cluster calculations were performed are shown in Figure 15.13. The structures were isolated from the native protein by selecting all atoms within a 7 Å radius from the excited-state Cu, including the bridged oxygen as depicted. Each of these structures contains ~90 atoms for which FEFF excludes the outermost atoms such that the total atoms in the calculation are 87. A representative set of simulated XANES is presented in Figure 15.14 for each of the "components" of the blank cluster (without the bridged O) with the equally weighted "merged" scan included. Each of the individual components exhibit distinct characteristics, such as double peaks in the white-line area (T1) and a large pre-edge peak followed by a narrow but intense white line (T2), whereas both T3 coppers give rise to a broad white line. The merged line is calculated using the relationship (Equation 15.8)

$$\mu_{mer} = \frac{\sum \chi_c(E)}{n}$$

(15.8)

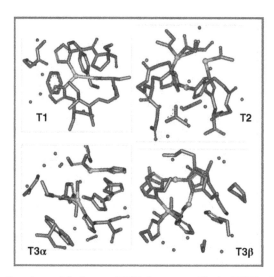

FIGURE 15.13 Drawings of the laccase TNC clusters used in the FEFF calculations showing each type of C as the central "absorber" atom. Each cluster consisted of ~90 atoms surrounding the absorber within a 7 Å radius. (Reproduced with permission from Ref. [74]. Copyright 2014, Wiley-VCH.) (Please see the color version of this figure in Color Plates section.)

where $\chi_c(E)$ is the calculated XANES of an individual component and n is the total number of components (in this case, 4). The individual simulations, however, are not necessarily useful for comparison with experimental data as the latter are comprised of an average of the individual components; thus, a consideration of the merged components may provide more insight. For most calculations, significant changes to

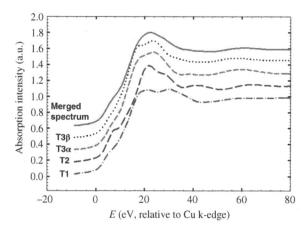

FIGURE 15.14 Representative FEFF8 XANES simulations for each of the four Cu components along with the convoluted spectrum. Note the simulation shown here was for the blank laccase cluster.

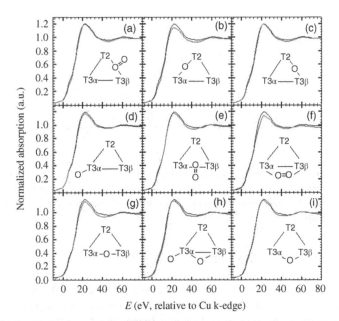

FIGURE 15.15 FEFF8 generated XANES spectra for the 90-atom "large" clusters with oxygen bound to the TNC as depicted in each plot. In each plot, the upper line represents the O-containing species whereas the lower line plots the blank cluster. (Reproduced with permission from Ref. [74]. Copyright 2014, Wiley-VCH.)

only one or two of the individual component spectra will also in turn lead to the changes in the merged spectrum.

Figure 15.15 shows the FEFF8 simulated XANES spectra for each of the large clusters considered (pictured as insets). Each graph overlays the simulated XANES for the "clean" cluster to illustrate the effect of O on the TNC. The simulations in Figure 15.15a, c, d, and e exhibit only modest changes with the addition of O/O_2, whereas plots in Figure 15.15b, f, g, h, and i show reasonably large changes to the white-line area. Surprisingly, the changes to the white lines for each case involve a *decrease* in white-line intensity. The experimental results discussed earlier for XANES collected under aerobic conditions (MET) underwent a similar decrease in white-line intensity and were attributed to diminishing absorption intensity due to interference from the O_x; in other words, the added O_x impedes multiple photo-electron backscattering paths causing the intensity to decrease.

For further analysis, the $\Delta\mu_t$ curves are presented in Figure 15.16 for each of the configurations and reveal a significant difference from the blank calculation. Each $\Delta\mu$ was calculated by using the relationship (Equation 15.9)

$$\Delta\mu_t = \mu(\text{TNC-O, site}) - \mu(\text{blank}) \tag{15.9}$$

where $\mu(\text{TNC-O, site})$ represents the simulated XANES as a function of O/O_2-binding site and $\mu(\text{blank})$ is the simulation without any O. The general shape of the $\Delta\mu_t$ curves

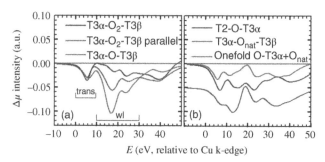

FIGURE 15.16 Simulated $\Delta\mu_t$ spectra calculated using Equation 15.7 for six of the structures depicted in Figure 15.15a–i. The onefold O-T3α + O$_{nat}$ curve (plot b, green line) was offset on the $\Delta\mu$ axis by -0.05 for illustrative purposes. (Please see the color version of this figure in Color Plates section.)

is such that there is an intense dip between 0 and 40 eV with several peaks and dips within. The region from 0 to 10 eV represents the energy range where the 1s → 4p transition occurs and 10–30 eV represents the white-line area in the XANES. The dip at 5 eV is a result of the decrease in the pre-edge transition. It was previously shown that the intensity of this transition decreases (hence the dip in the $\Delta\mu$) when laccase becomes oxidized. In fact, Kau et al. [65] published similar difference spectra in 1987 that reveal the same phenomenon, except that the difference was calculated by subtracting the oxidized laccase from the reduced, resulting in spectra that reveal maxima rather than minima in this region. At higher energy, between 10 and 30 eV, a split peak is observed with local minima at 17 and 24 eV, respectively. Two of the three lines in Figure 15.16a (red and green lines) follow a similar shape and trend, only differing in magnitude, whereas the remaining $\Delta\mu$ curves are significantly different. These $\Delta\mu_t$ signatures represent the most likely O$_x$-binding configurations and, as such, were selected for comparison with experimental results.

First, if we consider the uMET $\Delta\mu$ curves in Figure 15.8, each $\Delta\mu$ measurement exhibited nearly identical line shapes, only differing in their intensities. This line shape consistency suggests the O$_x$ binding site is the same for all curves, but reveals nothing about the actual structure. When a representative of the uMET $\Delta\mu$ curve is overlaid with the indicated $\Delta\mu_t$ signature for comparison (Figure 15.17), the signature exhibits a dip around 18 eV that the experimental data do not, and the uMET $\Delta\mu$ bears a reasonable likeness to the theoretical curve. Also, in comparison to the remaining $\Delta\mu_t$ signatures (Figure 15.16), this simulation seems to represent the best fit to the uMET $\Delta\mu$ data suggesting that it is the most likely O-binding configuration.

For MET XANES for RPA and RPB, very little change was observed as a function of potential, yet there was an overall *decrease* in the white-line intensity. Figure 15.18a and b shows the prior XANES data replotted as a function of purge gas, and the overlaid XANES in each case reveals a shift in E_0 as well as a small decrease in white-line intensity, consistent with some of the modeled XANES in Figure 15.15. Further, Figure 15.18c overlays the $\Delta\mu$ lines for the above XANES, with $\Delta\mu_t$ for the two clusters (depicted to the right of the plot; Figure 15.18). Reasonable qualitative

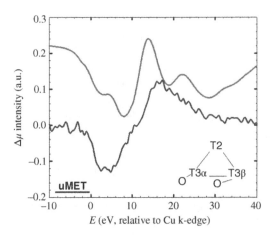

FIGURE 15.17 Comparison of uMET $\Delta\mu = \mu(0.7\,V, O_2) - \mu(0.3\,V, Ar)$ (upper line) with simulated $\Delta\mu_t$ for the structure shown. $\Delta\mu_t$ curve (lower line) was offset by $(\Delta\mu \times 4) + 0.2$ and shifted $-5\,eV$ for visualization. (Reproduced with permission from Ref. [74]. Copyright 2014, Wiley-VCH.)

agreement is observed between theoretical and experimental $\Delta\mu$ curves, with RPA matching well with the $\Delta\mu_t$ curve and the O_2 parallel structure, and with RPB resembling the cluster containing onefold O on the T3α plus the bridged O. This does not confirm that RPA (or RPB with its matched structure) is exclusively represented by the parallel O_2 structure, even though those signatures fit the best. It does, however, establish a link between the experimental $\Delta\mu$ for MET and uMET and two very closely related structures, which form the main components of the proposed ORR mechanism.

15.6 PROPOSED ORR MECHANISM

The ultimate goal of the analysis described was to understand the ORR mechanism for laccase as it occurs *in situ*, as a BFC electrocatalyst. The $\Delta\mu$ measurements shown, along with the FEFF8 modeling analysis, reveal two possible structures that could readily be modeled into a mechanism scheme. Laccase has long been known to reside in a "resting oxidized" state [42,47], which contains all four coppers in the divalent state and a bridged O/OH between the two T3 coppers. In order for ORR to occur, dioxygen must enter the interior of the enzyme to access the TNC. This is possible via one of the two electrolyte channels (Figure 15.19). The larger of the two is speculated to be the more direct route for dioxygen to access the TNC. The two coppers immediately exposed to the channel are the two T3 coppers, further indicating that a plausible ORR mechanism may begin here at this location.

The proposed mechanism herein is depicted in Figure 15.20. Notice that in order to initiate ORR at the two T3 coppers, the bridged O/OH group must be removed first. This is achieved by first reducing two of the three TNC coppers (Figure 15.20, structure II). Exactly which of the three structures shown is the correct one is

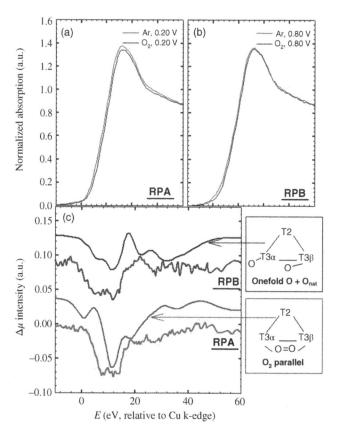

FIGURE 15.18 Cu k-edge XANES of laccase and (a) RPA and (b) RPB plotted as a function of electrolyte saturation gas at the indicated fixed potential. (c) Comparison plots of $\Delta\mu$ spectra calculated using the XANES in (a) and (b), $\Delta\mu = \mu(O_2) - \mu(Ar)$, with the $\Delta\mu_t$ simulations of the clusters depicted on the right. Visualization offsets used in (c) were $\Delta\mu + 0.13$ and $E - 1.5$ eV (onefold $O + O_{nat}$) (top curves $\Delta\mu_t$), $\Delta\mu + 0.09$ (RPB, bottom curves), and $\Delta\mu + 0.09$ and $E - 5.5$ eV for O_2 parallel simulation. (Reproduced with permission from Ref. [74]. Copyright 2014, Wiley-VCH.)

impossible to distinguish by XAS/FEFF8 as the structures themselves are identical; only the valence of the individual coppers has changed. The electrons are then transferred to the bridged O/OH in the presence of $2H^+$ to form a fully oxidized intermediate with no bridged-O (III) and H_2O. Dioxygen from the electrolyte channel then has the opportunity to bind to the T3 coppers in a parallel fashion, with each O atom coordinated to one T3 copper (Figure 15.20, structure IV). This structure is favored over the perpendicular analog as suggested by the XANES (Figure 15.15e) and $\Delta\mu_t$ (Figure 15.16a) line shapes and is likely more energetically favorable with both O atoms being coordinated. Also, the fact that this structure was even detected using XAS suggests that it is a stable, long-lived intermediate, as the quicker decaying intermediates would be averaged out of the total signal. For example, consider a

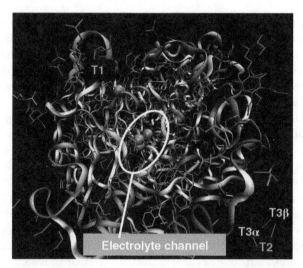

FIGURE 15.19 Diagram of laccase (*T. versicolor*) illustrating the close proximity of the T3 coppers to the larger electrolyte channel. (Please see the color version of this figure in Color Plates section.)

simple hypothetical reaction where sample A reacts with sample B to produce an intermediate AB*, which quickly decomposes into product C (Equation 15.10).

$$A + B \rightarrow AB^* \rightarrow C \qquad (15.10)$$

If XAS were employed to track the reaction progress (assuming A contains the element of interest for XAS characterization), AB* will not be detected as XAS is bulk

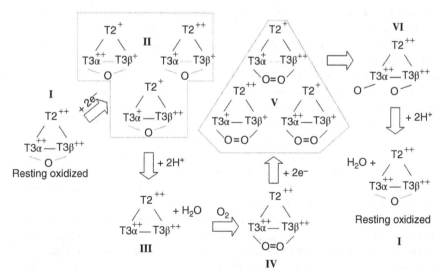

FIGURE 15.20 Proposed ORR mechanism. (Reproduced with permission from Ref. [74]. Copyright 2014, Wiley-VCH.)

averaging. Instead, the XAS spectrum will resolve either A (for relatively slow reactions) or the product C, as both of those species will have spent a high percentage of the time of measurement in those states. This is not to suggest that the intermediate AB^* does not impart *any* influence on the spectrum, it only suggests that the component of the spectrum that is due to AB^* is such a small percentage of the overall signal that it would not be visible.

The next step is a further two-electron reduction of the TNC, again with three nondistinguishable possibilities (structures V). Notice the structure here is identical to structure IV (with the exception of the Cu valence state again), which means that the two processes could occur simultaneously, or the dioxygen adsorption step could occur quickly with the electron transfer step happening much slower. Either way, both structures are consistent with the experimental findings. Structure VI forms when the two reduced coppers transfer their electrons to dioxygen, effectively splitting the molecule with one O atom remaining bridged between the two T3 coppers and the other one becoming singly coordinated with T3α. The $\Delta\mu_t$ signature from this structure, much like structures IV and V, was also consistent with the experimental data as discussed above, indicating it was stable enough to be detected by XAS and thus included in this mechanism. Finally, the O singly coordinated to T3α copper detaches in the presence of H^+ to form H_2O and returns the TNC to the resting oxidized state.

The mechanism presented above represents a series of structures based on prior knowledge of the laccase system from the literature and the XAS/FEFF8 analysis presented in this chapter. It is to some extent still unclear in areas (particularly structures II and V), but the aim was to use *in situ* XAS to elucidate the mechanism of ORR in laccase as it occurs on a BFC cathode. As a result, the mechanism in Figure 15.20 describes the behavior of laccase under the constraints of the experimental conditions for which the measurements were made. It is important to note that many other mechanisms have been proposed. In Particular, the Solomon and Atanassov research groups have been instrumental in providing important and detailed information on the active sites of a variety of blue copper oxidases [42,45,47–50,62,65,68–73].

For example, in 1996, Solomon et al. used methods such as EPR, XAS, magnetic circular dichroism, multiple Raman techniques, and ultraviolet (UV)–visible absorption spectroscopy to study laccase and other multicopper oxidases (MCOs) [42]. The measurements were made on chemically altered (reduced or oxidized) laccase from *R. vernicifera*, often at cryogenic temperatures. Two comprehensive parallel mechanisms were presented, revealing a nine-step process from the fully oxidized "native intermediate" to fully reduced and back via reoxidation. A three-step side reaction was also shown involving a parasitic and slow decay of the native intermediate to the resting state. Interestingly, Solomon's mechanism underwent the same removal of the bridged O/OH, as shown in structure III, with the four subsequent steps showing laccase reacting with substrate to reduce all coppers. However, the remainder of the mechanism differs from that in Figure 15.20 in that dioxygen is bound at a T2/T3 site. It is worth noting that the T2/T3α ORR mechanism was attempted (see Figure 15.15a), but revealed little change from the blank cluster and thus was ruled out in favor of the parallel mechanism. Also, a T2–T3β mechanism was attempted, but it failed as another atom from the 7 Å active site cluster was too close in proximity to the dioxygen.

15.7 OUTLOOK

The ability to study *in situ* XAS provides unique insight that illustrates changes to a dynamic system. *In situ* XANES for uMET laccase, for example, revealed relatively significant changes under aerobic conditions. Subsequent FEFF8 modeling showed that the observed line shape was consistent with a trinuclear cluster containing a single onefold bonded oxygen atom on the T3α site plus a natural bridged oxygen atom (O_{nat}) between the two T3 coppers. In contrast, for MET, *in situ* XANES curves revealed significant changes as a function of both potential and purge gas. Each potential O-binding structure in the enzyme was considered and, by comparison with experimental data, used to elucidate likely binding sites.

Even though the active site transformations occur relatively quickly, by using *in situ* XAS it is possible to determine the binding site of dioxygen in laccase and propose a new mechanism for oxygen reduction. The proposed mechanism involves an initial two-electron reduction of the TNC followed by electron transfer to the O_{nat} (bridged O/OH group) between the two T3 coppers. This provides a vacancy for dioxygen to enter from the electrolyte channel of the enzyme and bind to the T3 coppers in a parallel fashion. Following a further two-electron step that re-reduces the TNC, the electrons are transferred to the O atom (in the presence of H^+) that is singly coordinated to the T3α, thus removing the O to form H_2O and converting the TNC back to the resting oxidized state.

ACKNOWLEDGMENTS

This project was a small portion of a multiuniversity research initiative focused on biological fuel cells, supported by the Air Force Office of Scientific Research under contract number FA9550-06-1-0264. Deboleena Chakraborty and Scott Calabrese Barton are gratefully acknowledged for supplying copious amounts of laccase and redox polymer. XAS data were collected at the Case Western Reserve University beamline X3-B at the National Synchrotron Light Source at Brookhaven National Laboratory (Upton, NY), which is supported by the Department of Energy, Office of Science, Office of Basic Energy Sciences under Contract No. DE-AC02-98CH10886.

LIST OF ABBREVIATIONS

ABTS	2,2′-azinobis(3-ethylbenzothiazoline-6-sulfonate)
BCO	blue copper oxidase
BFC	biological fuel cell

BOx	bilirubin oxidase
bpy	2,2'-bipyridyl
CV	cyclic voltammetry
DET	direct electron transfer
dm-bpy	4,4'-dimethyl-2,2'-bipyridine
EPR	electron paramagnetic resonance
EXAFS	extended X-ray absorption fine structure
GOx	glucose oxidase
LFT	ligand field theory
MCO	multicopper oxidase
MET	mediated electron transfer
ORR	oxygen reduction reaction
PEGDGE	polyethylene glycol diglycidyl ether
PVI	polyvinylimidazole
RHE	reversible hydrogen electrode
RP	redox polymer
RPA	poly(VI$_{15}$[Os(tpy)(bpy)]$^{2+/3+}$)
RPB	poly(VI$_{20}$[Os(tpy)(dm-bpy)]$^{2+/3+}$)
RPC	poly(VI$_{12}$[Os(bpy)$_2$Cl]$^{+/2+}$)
T1	type 1
T2	type 2
T3	type 3
TNC	trinuclear copper center
tpy	6',2''-terpyridine
uMET	unmediated electron transfer
UV	ultraviolet
VI	vinylimidazole
XANES	X-ray absorption near-edge structure
XAS	X-ray absorption spectroscopy

REFERENCES

1. Appleby AJ, Ng DYC, Weinstein H. Parametric study of the anode of an implantable biological fuel cell. *J Appl Electrochem* 1971;1:79–90.

2. Arzoumanidis GG, O'Connell JJ. Electrocatalytic oxidation of D-glucose in neutral media with electrodes catalyzed by 4,4',4'',4'''-tetrasulfophthalocyanine. *J Phys Chem* 1969;73:3508–3510.

3. Colton CK, Drake RF. Analysis of *in vivo* deoxygenation of human blood: a feasibility study for an implantable biological fuel cell. *Trans Am Soc Artif Intern Organs* 1969;15:187–198.

4. Yahiro AT, Lee SM, Kimble DO. Bioelectrochemistry. 1. Enzyme utilizing bio-fuel cell studies. *Biochim Biophys Acta* 1964;88:375–383.

5. Gebhardt U, Rao JR, Richter GJ. A special type of Raney-alloy catalyst used in compact biofuel cells. *J Appl Electrochem* 1976;6:127–134.

6. Rao JR, Richter GJ, Von Sterm F, Weidlich E. The performance of glucose electrodes and the characteristics of different biofuel cell constructions. *Bioelectrochem Bioenerg* 1976;3:139–150.

7. Rao JR, Richter GJ, Von Sterm F, Weidlich E, Wenzel M. Metal–oxygen and glucose–oxygen cells for implantable devices. *Biomed Eng* 1974;9:98–103.

8. Auborn JJ, French KW, Lieberman SI, Shah VK, Heller A. Lithium anode cells operating at room temperature in inorganic electrolytic solutions. *J Electrochem Soc* 1973;120:1613–1619.

9. Bath TD. *Alternative Fuels and the Environment.* Lewis Publishers, Boca Raton, FL, 1995.

10. Heller A. Miniature biofuel cells. *Phys Chem Chem Phys* 2004;6:209–216.

11. Karube IS, Matsunaga TK, Kuriyama S. Biochemical energy conversion by immobilized whole cells. *Ann NY Acad Sci* 1981;369:91–98.

12. Matsunaga TK, Suzuki S. Some observations on immobilized hydrogen-producing bacteria: behavior of hydrogen in gel membranes. *Biotechnol Bioeng* 1980;22:2607–2615.

13. Suzuki SK. Energy production with immobilized cells. *Appl Biochem Bioeng* 1983;4:281–310.

14. Persson B, Gorton L, Johansson G, Torstensson A. Biofuel anode based on D-glucose dehydrogenase, nicotinamide adenine dinucleotide and a modified electrode. *Enzyme Microb Technol* 1985;7:549–552.

15. Turner APF, Aston WJ, Higgins IJ, Davis G, Hill HAO. Applied aspects of bioelectrochemistry: fuel cell, sensors, and bioorganic synthesis *Biotechnol Bioeng Symp* 1982;12:401–412.

16. Turner APF, Ramsay G, Higgins IJ. Applications of electron transfer between biological systems and electrodes. *Biochem Soc Trans* 1983;11:445–448.

17. Palmore GTR, Kim H-H. Electro-enzymatic reduction of dioxygen to water in the cathode compartment of a biofuel cell. *J Electroanal Chem* 1999;464:110–117.

18. Kim H-H, Mano N, Zhang Y, Heller A. A miniature membrane-less biofuel cell operating under physiological conditions at 0.5 V. *J Electrochem Soc* 2003;150:A209–A213.

19. Mano N, Kim H-H, Heller A. On the relationship between the characteristics of bilirubin oxidases and O_2 cathodes based on their "wiring". *J Phys Chem B* 2002;106:8842–8848.

20. Mano N, Kim H-H, Zhang Y, Heller A. An oxygen cathode operating in a physiological solution. *J Am Chem Soc* 2002;124:6480–6486.

21. Mano N, Mao F, Heller A. A miniature biofuel cell operating in a physiological buffer. *J Am Chem Soc* 2002;124:12962–12963.

22. Mano N, Mao F, Heller A. Characteristics of a miniature compartment-less glucose–O_2 biofuel cell and its operation in a living plant. *J Am Chem Soc* 2003;125:6588–6594.

23. Qiu H, Xu C, Huang X, Ding Y, Qu Y, Gao P. Adsorption of laccase on the surface of nanoporous gold and the direct electron transfer between them. *J Phys Chem C* 2008;112:14781–14785.

24. Qiu H, Xu C, Huang X, Ding Y, Qu Y, Gao P. Immobilization of laccase on nanoporous gold: comparative studies on the immobilization strategies and the particle size effects. *J Phys Chem C* 2009;113:2521–2525.

25. Shleev S, Pita M, Yaropolov AI, Ruzgas T, Gorton L. Direct heterogeneous electron transfer reactions of *Trametes hirsuta* laccase at bare and thiol-modified gold electrodes. *Electroanalysis* 2006;18:1901–1908.

26. Shleev S, Christenson A, Serezhenkov V, Burbaev D, Yaropolov A, Gorton L, Ruzgas T. Electrochemical redox transformations of T1 and T2 copper sites in native *Trametes hirsuta* laccase at gold electrode. *Biochem J* 2005;385:745–754.

27. Ivnitski D, Atanassov P. Electrochemical studies of intramolecular electron transfer in laccase from *Trametes versicolor*. *Electroanalysis* 2007;19:2307–2313.

28. Rahman MA, Noh H-B, Shim Y-B. Direct electrochemistry of laccase immobilized on Au nanoparticles encapsulated-dendrimer bonded conducting polymer: application for a catechin sensor. *Anal Chem* 2008;80:8020–8027.

29. Zheng W, Zhou HM, Zheng YF, Wang N. A comparative study on electrochemistry of laccase at two kinds of carbon nanotubes and its application for biofuel cell. *Chem Phys Lett* 2008;457:381–385.

30. Calabrese Barton S, Gallaway J, Atanassov P. Enzymatic biofuel cells for implantable and microscale devices. *Chem Rev* 2004;104:4867–4886.

31. Gallaway J, Calabrese Barton S. Kinetics of redox polymer-mediated enzyme electrodes. *J Am Chem Soc* 2008;130:8527–8536.

32. Gallaway J, Wheeldon I, Rincón R, Atanassov P, Banta S, Calabrese Barton S. Oxygen-reducing enzyme cathodes produced from SLAC, a small laccase from *Streptomyces coelicolor*. *Biosens Bioelectron* 2008;23:1229–1235.

33. Hudak NS, Gallaway JW, Calabrese Barton S. Mediated biocatalytic cathodes operating on gas-phase air and oxygen in fuel cells. *J Electrochem Soc* 2009;156:B9–B15.

34. Calabrese Barton S, Kim H-H, Binyamin G, Zhang Y, Heller A. Electroreduction of O_2 to water on the "wired" laccase cathode. *J Phys Chem B* 2001;105:11917–11921.

35. Messerschmidt A, Huber R. The blue oxidases, ascorbate oxidase, laccase and ceruloplasmin. Modelling and structural relationships. *Eur J Biochem* 1990;187:341–352.

36. Sakurai T, Kataoka K. Basic and applied features of multicopper oxidases, CueO, bilirubin oxidase, and laccase. *Chem Rec* 2007;7:220–229.

37. Marcus RA, Sutin N. Electron transfers in chemistry and biology. *Biochim Biophys Acta* 1985;811:265–322.

38. Dodds ML. Comparative substrate specificity studies of ascorbic acid oxidase and copper ion catalysis. *Arch Biochem* 1948;18:51–58.

39. Wimalasena K, Dharmasena S. Substrate specificity of ascorbate oxidase: unexpected similarity to the reduction site of dopamine β-monooxygenase. *Biochem Biophys Res Commun* 1994;203:1471–1476.

40. Huber CT, Frieden E. Substrate activation and the kinetics of ferroxidase. *J Biol Chem* 1970;245:3973–3978.

41. Osaki S, Johnson DA, Frieden E. The possible significance of the ferrous oxidase activity of ceruloplasmin in normal human serum. *J Biol Chem* 1966;241:2746–2751.

42. Solomon EI, Sundaram UM, Machonkin TE. Multicopper oxidases and oxygenases. *Chem Rev* 1996;96:2563–2606.

43. Piontek K, Antorini M, Choinowski T. Crystal structure of a laccase from the fungus *Trametes versicolor* at 1.90-Å resolution containing a full complement of coppers. *J Biol Chem* 2002;277:37663–37669.

44. Lowery MD, Guckert JA, Gebhard MS, Solomon EI. Active-site electronic structure contributions to electron-transfer pathways in rubredoxin and plastocyanin: direct versus superexchange. *J Am Chem Soc* 2002;115:3012–3013.

45. Machonkin TE, Solomon EI. The thermodynamics, kinetics, and molecular mechanism of intramolecular electron transfer in human ceruloplasmin. *J Am Chem Soc* 2000;122:1 2547–12560.

46. Rorabacher DB. Electron transfer by copper centers. *Chem Rev* 2004;104:651–698.

47. Solomon EI, Machonkin TE, Sundaram UM. Spectroscopy of multi-copper oxidases. In: Messerschmidt A (ed.), Multi-Copper Oxidases. World Scientific, Singapore, 1997, pp. 103–108.

48. Yoon J, Liboiron BD, Sarangi R, Hodgson KO, Hedman B, Solomon EI. The two oxidized forms of the trinuclear Cu cluster in the multicopper oxidases and mechanism for the decay of the native intermediate. *Proc Natl Acad Sci USA* 2007;104:13609–13614.

49. Lee S-K, George SD, Antholine WE, Hedman B, Hodgson KO, Solomon EI. Nature of the intermediate formed in the reduction of O_2 to H_2O at the trinuclear copper cluster active site in native laccase. *J Am Chem Soc* 2002;124:6180–6193.

50. Palmer AE, Lee SK, Solomon EI. Decay of the peroxide intermediate in laccase: reductive cleavage of the O−O bond. *J Am Chem Soc* 2001;123:6591–6599.

51. Forster RJ, Vos JG. Synthesis, characterization, and properties of a series of osmium- and ruthenium-containing metallopolymers. *Macromolecules* 1990;23:4372–4377.

52. Ohara TJ, Rajagopalan R, Heller A. Glucose electrodes based on cross-linked bis(2,2′-bipyridine)chloroosmium(+/2+) complexed poly(1-vinylimidazole) films. *Anal Chem* 1993;65:3512–3517.

53. Gao ZQ, Binyamin G, Kim HH, Calabrese Barton S, Zhang YC, Heller A. Electro-deposition of redox polymers and co-electrodeposition of enzymes by coordinative crosslinking. *Angew Chem Int Ed* 2002;41:810–813.

54. Elmgren M, Lindquist SE, Sharp M. Charge propagation through a redox polymer film containing enzymes—effects of enzyme loading, pH and supporting electrolyte. *J Electroanal Chem* 1993;362:227–235.

55. Forster RJ, Vos JG, Lyons MEG. Controlling processes in the rate of charge transport through [Os(bipy)$_2$(PVP)$_n$Cl]Cl redox polymer-modified electrodes. *J Chem Soc, Faraday Trans* 1991;87:3761–3767.

56. Nakahama S, Murray RW. The effect of composition of a ferrocene-containing redox polymer on the electrochemistry of its thin film coatings on electrodes. *J Electroanal Chem* 1983;158:303–322.

57. Berry AJ, O'Neill HSC, Jayasuriya KD, Campbell SJ, Foran GJ. XANES calibrations for the oxidation state of iron in silicate glass. *Am Mineral* 2003;88:967–977.

58. Kwiatek WM, Galka M, Hanson AL, Paluszkiewicz C, Chichocki T. XANES as a tool for iron oxidation state determination in tissues. *J Alloys Compd* 2001;325:276–282.

59. Wilke M, Farges F, Petit P-E, Brown GE Jr., Martin F. Oxidation state and coordination of Fe in minerals: an Fe K-XANES spectroscopic study. *Am Mineral* 2001;86:714–730.

60. Wilke M, Hahn O, Woodland AB, Rickers K. The oxidation state of iron determined by Fe K-edge XANES—application to iron gall ink in historical manuscripts. *J Anal At Spectrom* 2009;24:1364–1372.

61. Shriver D, Atkins P. *Inorganic Chemistry*, 2nd edition. W.H. Freeman and Company, New York, 1999.

62. Cole JL, Tan GO, Yang EK, Hodgson KO, Solomon EI. Reactivity of the laccase trinuclear copper active site with dioxygen: an X-ray absorption edge study. *J Am Chem Soc* 1990;112:2243–2249.

63. Chance B, Angiolillo P, Yang EK, Powers L. Identification and assay of synchrotron radiation-induced alterations on metalloenzymes and proteins. *FEBS Lett* 1980;12: 178–182.

64. Penner-Hahn JE, Murata M, Hodgson KO, Freeman HC. Low-temperature X-ray absorption spectroscopy of plastocyanin: evidence for copper-site photoreduction at cryogenic temperatures. *Inorg Chem* 1989;28:1826–1832.

65. Kau LS, Spira-Soloman DJ, Penner-Hahn JE, Hodgson KO, Solomon EI. X-ray absorption edge determination of the oxidation state and coordination number of copper. Application to the type 3 site in *Rhus vernicifera* laccase and its reaction with oxygen. *J Am Chem Soc* 1987;109:6433–6442.

66. Meitzner G, Huang ES. Analysis of mixtures of compounds of copper using K-edge X-ray absorption spectroscopy. *Fresenius J Anal Chem* 1992;342:61–64.

67. Sipr O, Rocca F, Fornasini P. On the origin of the differences in the Cu K-edge XANES of isostructural and isoelectronic compounds. *J Phys: Condens Matter* 2009;21:255401.

68. Augustine AJ, Kragh ME, Sarangi R, Fujii S, Liboiron BD, Stoj CS, Kosman DJ, Hodgson KO, Hedman B, Solomon EI. Spectroscopic studies of perturbed T1 Cu sites in the multicopper oxidases *Saccharomyces cerevisiae* Fet3p and *Rhus vernicifera* laccase: allosteric coupling between the T1 and trinuclear Cu sites. *Biochemistry* 2008;47: 2036–2045.

69. Andersson KK, Schmidt PP, Katterle B, Strand KR, Palmer AE, Lee S-K, Solomon EI, Graslund A, Barra A-L. Examples of high-frequency EPR studies in bioinorganic chemistry. *J Biol Inorg Chem* 2003;8:235–247.

70. Quintanar L, Yoon J, Aznar CP, Palmer AE, Andersson KK, Britt RD, Solomon EI. Spectroscopic and electronic structure studies of the trinuclear Cu cluster active site of the multicopper oxidase laccase: nature of its coordination unsaturation. *J Am Chem Soc* 2005;127:13832–13845.

71. Sundaram UM, Zhang HH, Hedman B, Hodgson KO, Solomon EI. Spectroscopic investigation of peroxide binding to the trinuclear copper cluster site in laccase: correlation with the peroxy-level intermediate and relevance to catalysis. *J Am Chem Soc* 1997;119:12525–12540.

72. Xu F, Shin W, Brown SH, Wahleitner JA, Sundaram UM, Solomon EI. A study of a series of recombinant fungal laccases and bilirubin oxidase that exhibit significant differences in redox potential, substrate specificity, and stability. *Biochim Biophys Acta* 1996;1292:303–311.

73. Ivnitski D, Artyushkova K, Rincón RA, Atanassov P, Luckarift HR, Johnson GR. Entrapment of enzymes and carbon nanotubes in biologically synthesized silica: glucose oxidase-catalyzed direct electron transfer. *Small* 2008;4:357–364.

74. Arruda TA, Chakraborty L, Lawton JS, Calabrese-Barton S, Atanassov P, Mukerjee S. Direct observation of oxygen binding and reduction in laccase (*T. versicolor*) by *in situ* X-ray absorption spectroscopy. *ChemElectroChem* 2014;under review.

16

ENZYMATIC FUEL CELL DESIGN, OPERATION, AND APPLICATION

VOJTECH SVOBODA

School of Materials Science and Engineering, Georgia Institute of Technology, Atlanta, GA, USA

PLAMEN ATANASSOV

Department of Chemical and Nuclear Engineering and Center for Emerging Energy Technologies, University of New Mexico, Albuquerque, NM, USA

16.1 INTRODUCTION

Enzymatic fuel cells (EFCs) have been demonstrated to generate power from a variety of unconventional renewable fuels (e.g., sugars). Those are high energy density fuels and thus EFC is a theoretical solution for high energy density power generation that outperformes conventional electrochemical power sources. The main advantages of EFCs are as follows:

- Specialty fuels, for example, carbohydrates, alcohols, and fatty acids, can be used.
- There is the potential for complete oxidation and utilization of the "high-energy" fuels.
- They offer green technology, minimal environmental footprint, and biodegradability.
- They have biocompatible design for implantable medical applications.
- Enzymes are renewable and availability is unlimited (unlike precious metals or fossil fuels).

Enzymatic Fuel Cells: From Fundamentals to Applications, First Edition. Edited by Heather R. Luckarift, Plamen Atanassov, and Glenn R. Johnson.
© 2014 John Wiley & Sons, Inc. Published 2014 by John Wiley & Sons, Inc.

- With economy of scale, enzymes can be obtained at low cost.
- Safety concerns are limited with EFC and they are much lower to conventional electrochemical power sources.

EFCs typically generate low power, in comparison with conventional batteries and fuel cells, and with their initial prototypes energy density and energy utilization from source fuels tend to be low. With further development, however, the advantages of EFCs are expected to provide competitive benefits for certain applications over the conventional electrochemical power sources. In hybrid approach, EFCs may be *combined* with conventional batteries, fuel cells, and supercapacitors to form advanced power sources with improved performance for specific applications.

EFCs are electrochemical systems that consist of an anode, a cathode, and an electrolyte. Design of EFC prototypes was inspired by conventional batteries and fuel cells, but there are substantial differences that lead to completely new design concepts and requirements. Specifically, in contrast to conventional batteries, the oxidized substance in the EFC is not carried in the electrodes, but instead stored as a "fuel." In contrast to conventional fuel cells, EFCs use highly selective enzymes in the anode and cathode reactions and they can operate without any membrane separation, in neutral aqueous electrolyte, and at room temperature and are capable to provide deep, or complete, fuel oxidation.

16.2   BIOBATTERIES AND EFCs

In the field of enzymatic power sources, researchers talk about enzymatic biobatteries and EFCs. The terminology is randomly used, but some major differences between their respective design and operation might be defined. EFCs are devices that generate power as long as fuel is externally supplied to the system for instance with a continuous flow-though fuel system. In contrast, a biobattery is a device with a limited amount of fuel in the system. When the fuel is depleted, the device stops producing electrical power and the entire biobattery must be replaced. A gray area between both is a system designed with a replaceable fuel cartridge. From the biobattery perspective, the fuel cartridge change or other type of fuel refills by users might be classified as a chemical recharge, though conventional rechargeable batteries are only electrically rechargeable. From the EFC perspective, this is just an ordinary fuel tank replacement. Certain charge transfer reactions of EFCs are reversible under favorable conditions and, as a result, such biobatteries may allow also for electrical recharging. These systems are typically equipped with nonenzymatic reversible redox cathodes [1].

In this chapter, we describe design criteria for the individual components of EFCs and provide guidance for designing enzymatic power sources in general. The practical difference between biobatteries and EFCs in relation to current design is mainly in the sizing and operation mode of specific components. Herein, we do not differentiate between biobatteries and EFCs in respect to terminology, in order to fully reflect the breadth of enzymatic power sources.

16.3 COMPONENTS

16.3.1 Anodes

Enzymes in anodic reactions facilitate biocatalytic oxidation of a substrate, supplied as a fuel (see Chapter 2). The power generated is dictated by the number of catalytic sites in direct association with the electrode and the substrate availability for the charge transfer reaction. Thus, high enzyme loading in the electrode is required for effective current generation that, in turn, demands a high surface area for enzyme immobilization and open porosity for mass transport. An ideal electrode material, for example, provides multifunctional porosity: mesopores for enzyme immobilization and macropores for mass transport of fuel. If we consider the state of the art, anode electrode materials are typically synthesized from biopolymers doped with conductive additives [2,3]. Enzyme immobilization can be directly on a planar conductive surface or within multiple layers in a hydrogel or similar three-dimensional matrix that allows the additional integration of electron transfer (ET) mediators. Immobilized mediators enable ET to the conductive surface via electron hopping over neighboring redox moieties and exhibit a diffusion transfer mechanism [4]. The current state of the art for immobilized mediators is based primarily on osmium complexes (see Chapter 8) [5–7]. Implementation of such mediators to EFC is limited, however, by complicated synthesis, stability, and environmental problems, as well as safety and cost.

Diffusive mediators are small mobile redox molecules that, when supplied with a fuel, can facilitate ET from distant enzymes to the conductive electrode's surface. Diffusive mediators are advantageous in terms of electrode synthesis, safety, pollution, and cost but impose significant burden on design and operation. EFCs with diffusive mediators must be equipped with ion-selective membranes in order to avoid internal ionic short circuit. Furthermore, diffusive mediators must be supplied along with the fuel, which complicates user preparation and handling.

The electronically conductive network in the electrode functions as an internal current collector and can be formed from conductive nanoparticles integrated within a porous matrix or from a continuous rigid backbone. One example is the biopolymer chitosan fabricated with a conductive additive of carbon nanotubes [2,3,8]. Ideally, the electronically conductive network will also provide mechanical stability to the electrode and support the structure against vibrations, fuel flow-through pressures, and other mechanical shocks. The electronically conductive network is essentially an internal current corrector within the anode and, as such, must be electronically "wired" to the external current collector of the anode compartment with minimal contact and ohmic resistance. The electrode materials described are typically delicate and mechanically soft and must be carefully handled to maintain their open high-porosity structure.

One further consideration in anode design is that some oxidoreductase enzymes are oxygen sensitive and undergo side reactions in the presence of oxygen and substrates that compete with the desired charge transfer reaction. The side reactions do not require electron flux through an external electric circuit (i.e., do not generate power), and therefore continue as long as substrate and oxygen are present in the EFC. Such self-discharging processes consume fuel, thereby reducing the total generated power and overall efficiency. An example of such a process is observed when glucose oxidase (GOx) is used as an anodic catalyst for oxidation of glucose; where oxygen acts as an electron sink and the process consumes glucose and produces hydrogen peroxide in the fuel. This mechanism has been used in glucose sensors but must be avoided in EFC systems [9].

16.3.2 Cathodes

The cathode in EFCs facilitates charge transfer reduction in which hydrogen protons and electrons are consumed. Hydrogen protons and electrons are formed in oxidation reactions in the anode, whereas protons are transferred to the cathode through an aqueous-based electrolyte. Electrons are simultaneously transferred through an electric load in an outer circuit. The most preferred cathode reaction for EFCs is oxygen reduction. Air can be used as a practically limitless source of oxygen and this approach reduces the overall mass and volume of the cathode. Such "air-breathing" oxygen-reducing cathodes have been developed and perfected for conventional fuel cells (see Chapter 3). Most of the available (conventional) air-breathing cathodes are fabricated from platinum-based catalysts with typical Pt loading of $\sim 0.5\, \mathrm{mg\, cm^{-2}}$. Platinum, however, poses a significant cost factor that limits industrial scaling, and it is not a green technology (rather, it is a nonbiodegradable heavy metal pollutant). Therefore, non-platinum catalysts are preferred, and despite currently lower performance to platinum (in acid or in alkaline electrolyte), they can offer significant benefits at neutral pH. Catalysts for conventional fuel cells are designed to operate with alkaline or acidic electrolyte, and the neutral pH required for EFC operation actually lowers their catalytic performance. As a result, potential catalyst poisoning with EFC fuels, intermediates, or waste products must be verified before catalyst development. This verification is still useful even when the cathode is separated from the electrolyte by an ion selective membrane.

In addition, air-breathing cathodes must be integrated with suitable gas diffusion layers (GDLs) (and membrane or membrane-type layers) to prevent catalyst flooding from contact with the fuel. GDLs have a high contact angle with water (i.e., they are hydrophobic) that provides a supply of oxygen (or air) to the catalyst while expelling water generated as a product of the reaction at the cathode. GDLs also support the electronic conductivity of the cathode. Such cathode configuration in conventional fuel cell technology is called a membrane electrode assembly (MEA). For EFC applications, only the cathode is applied on the MEA, whereas the anodic side remains as a bare membrane or a membrane-type coating that is exposed to fuel; such assembly is called a *half-MEA* cathode.

Application of the conventional half-MEA air-breathing cathode in EFC can be problematic because of mechanical integrity of the MEA assembly. Sulfonated tetrafluoroethylene membranes (such as Nafion) work well as electrical separators between anode and cathode and can provide an effective seal to prevent fuel leakage. In an asymmetric MEA, the bare Nafion side is exposed to the aqueous fuel solution in the EFC, which can result in swelling of the membrane and high shear forces that can lead to fast structural disintegration of the cathode.

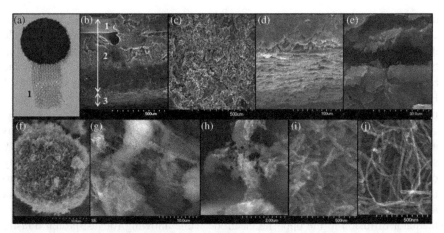

FIGURE 16.1 (a) Photograph of a gas diffusion electrode (GDE). Scanning electron microscopy (SEM) images of different magnification of GDE. (b) Cross section of GDE with (2) GDL (carbon black (CB) with 35 wt% polytetrafluoroethylene (PTFE)) and (3) MWCNT (multiwalled carbon nanotube) catalytic layer (3.5 wt% PTFE) with (1) nickel mesh as current collector. (c) Top view of GDL (CB with 35 wt% PTFE). (d, e) Higher magnification SEMs of MWCNT catalytic layer with 3.5 wt% PTFE as binder. (f) Micro-emulsion droplet of PTFE covered with CB (Vulcan 72R). SEM images of PTFE-modified MWCNTs with (g) 35 wt% PTFE, (h) 22 wt% PTFE, and (i) 3.5 wt% PTFE. (j) MWCNT paper. (Reproduced with permission from Ref. [15]. Copyright 2012, Wiley-VCH Verlag GmbH.)

Oxygen reduction can be catalyzed by enzymes, and air-breathing cathodes with laccase and bilirubin oxidase as enzymatic catalysts, for example, have been demonstrated [10–15]. Enzymatically catalyzed cathodes avoid many of the problems discussed for platinum and other precious metal catalysts, but they often provide lower power density and limited stability. Specifically with air-breathing cathodes, enzymes that catalyze oxygen reduction must be immobilized at the tripoint between hydrophilic (H^+ conductivity), hydrophobic (O_2 supply), and conductive (e^- conductivity) interfaces for effective catalysis (Figure 16.1) (see Chapter 3).

Research, development, and design of a high-power air-breathing oxygen reduction cathode for EFCs that is mechanically stable and supports enzyme stability have been poorly addressed in the literature, despite the fact that cathodes are typically the power-limiting electrode in EFC systems (see Chapter 3). An alternative design concept is application of a chemical redox cathode, such as Prussian blue, supported with an ionic liquid electrolyte that provides a reversible redox reaction and allows electric recharge of the cathode [1]. Such design does, however, increase the mass and volume of the EFC in comparison with air-breathing cathodes.

16.3.3 Separator and Membrane

As with all conventional electrochemical power sources, separation distance between the EFC anode and cathode should be minimal to reduce power limitation due to ion diffusion between the electrodes. At the same time, both electrodes must be electronically insulated to avoid internal short circuit. Both criteria are achieved with a simple thin separator that must provide maximal electronic resistance and maximum ion transport. Separators must also demonstrate mechanical and thermal endurance for the external operating conditions. In contrast to conventional fuel cells, one advantage of EFCs is the high selectivity of certain enzyme-catalyzed reactions that allow for a membraneless design (or the use of a very simple physical separator).

Ion-selective membranes must be used in EFCs when the following conditions apply: (i) diffusive mediators are used, (ii) charge transfer reactions are not completely selective, (iii) anode (or anolyte) and cathode (or catholyte) are not chemically compatible, or (iv) anode oxygen sensitivity combined with eventual oxygen leak into the cell from the cathode. For instance, in a membraneless system, diffusive mediators act as an internal ionic short circuit. Therefore, a membrane must be used to avoid mediator diffusion to the opposite electrode while allowing for fast proton transfer kinetics and thus uncompromised power output. The most typical membranes in commercial oxygen reduction cathodes are Nafion membranes that ensure fast proton conductivity in operating conditions specific to conventional fuel cells. Nafion is electronically nonconductive and provides excellent mechanical and thermal stability. In EFCs, however, Nafion membranes suffer from poor proton conductivity at neutral pH. The high initial ion conductivity decreases with a gradual ion exchange due to neutralization of the internal micelle structure.

In air-breathing cathodes, ion-selective membranes or coatings provide an additional important function in preventing flooding of the catalyst layer with fuel solution that would otherwise prevent supply of oxygen through the opened gas channels. Nafion seals the fuel solution in the EFC and prevents oxygen penetration to the anodic compartment: this is specifically important when oxygen-sensitive enzymes are used with the anode.

16.3.4   Reference Electrode

EFC power sources do not need a reference electrode to operate, but reference electrodes are important for the design and development of EFCs as a means to identify the performance and lifetime limiting electrode. Integration of EFCs with a reference electrode allows for measurement of the individual electrodes' voltage at open circuit and under electric loading this allows for identification whether the anode or cathode is power (and/or stability) limiting. As EFCs operate in aqueous, pH-neutral, buffered electrolyte, Ag/AgCl electrodes or saturated calomel electrodes (SCEs) are the typical reference electrodes of choice and are commercially available in the form of microelectrodes with diameters as low as 0.5 mm.

16.3.5   Fuel and Electrolyte

As discussed previously, the substrate oxidized within the EFC is not integrated within the electrodes but is supplied directly as fuel. The fuel is typically an aqueous solution of the substrate with sufficient salts to form an electrolyte. The kinetics and stability of enzymes immobilized to EFC electrodes are strongly dependent on pH and, therefore, fuel is buffered to maintain the required pH level. Enzyme kinetics and stability vary considerably with changes in pH; thus, fuel solutions must be optimized for each specific enzymatic system. The optimal pH determination becomes increasingly complex for cascades that use multiple enzymes. In this case, computational models combined with experimental characterization can be used to optimize fuel constituents.

Diffusive mediators that facilitate ET between enzymes and the conductive electrode (in both anode and cathode configurations) can be provided along with the fuel, although this does impart a burden on practical applications of such EFCs. EFC design criteria must address all aspects of operational constraints such as safety, stability, cost, and environmental regulations. For example, if a fuel and mediator mix (and its associated waste stream) is classified as a safety or environmental hazard, the fuel and waste must be handled in accordance with regulations. In contrast, some EFCs use simple sugar solutions, such as saps and juices, that require only water-based electrolytes, and their waste can be expelled to the ambient environment. In these systems, fuel and buffer in the form of dehydrated powder might be provided in a dry system that merely requires the user to add water.

Some enzymes are oxygen sensitive and thus the fuel must be deoxygenated before being supplied to the EFC. In laboratory conditions, this is typically provided by bubbling nitrogen or argon through the solution. For practical applications, however, the use of oxygen-sensitive enzymes and the associated fuel preparation is complicated. Supply of sealed deoxygenated fuel tanks is possible, but the use of oxygen-tolerant enzymes is preferred for practical EFC applications.

The rate of enzyme-catalyzed reactions follows Michaelis–Menten kinetics, defined as

$$v = \frac{V_{\text{max}}[S]}{K_{\text{m}} + [S]} \tag{16.1}$$

where v is the reaction rate, V_{max} represents the maximum rate at substrate saturated concentration, and the Michaelis constant K_{m} is the substrate concentration at which the reaction rate is equal to half of V_{max}. [S] is the substrate concentration. Figure 16.2 shows a Michaelis–Menten curve for $V_{\text{max}} = 4$ mol l^{-1} s^{-1} and $K_{\text{m}} = 1$ mol l^{-1} that represents reaction rate dependence on substrate concentration. The initial linear part of the curve is typically leveraged for biosensor applications, whereas EFCs operate optimally at the substrate saturated kinetics close to the maximal rate (V_{max}). The substrate saturated concentration for EFC is typically low (on the order of tens or hundreds millimolar).

Theoretically, full oxidation of glucose provides $24e^-$ per molecule, and 3.57 Ah g^{-1} of solid glucose. This is approximately 10 times higher charge capacity than the graphite electrode materials of conventional Li-ion batteries and is comparable to current development goals for modern silicon nanostructure-based materials. With an operating voltage of 0.35 V per cell, glucose as fuel can theoretically produce 1250 Wh kg^{-1} and 1924 Wh l^{-1} (from solid glucose). However, with lower concentration of substrates that is required for EFC operation, the parameters become less competitive. For example, a fuel of 100 mM glucose solution in water provides specific charge of only 64 mAh g^{-1}, specific energy of 22.4 Wh kg^{-1}, and energy density of 2.5 Wh l^{-1} (assuming complete fuel utilization). This does not account for the mass and volume of electrodes, packaging, and all other parts of the system.

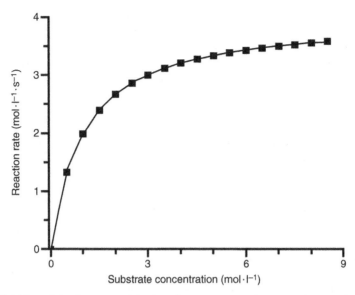

FIGURE 16.2 Michaelis–Menten kinetics of an enzyme-catalyzed reaction rate dependence on substrate concentration. The example is plot for $V_{max} = 4$ mol l^{-1} s^{-1} and $K_m = 1$ mol l^{-1}.

Supplying higher concentrations of fuel (greater than the saturated value of the Michaelis–Menten kinetics) leads to substrate inhibition with certain enzymes. Additionally, the increased substrate viscosity of a highly concentrated fuel slows the mass transport and substrate diffusion toward the catalytically active surface, particularly for high surface area, tortuous, and highly porous electrodes. In fact, power generation of the current EFC systems with high enzyme loading and highly porous electrodes is typically limited by fuel mass transport. The substrate mass transport, therefore, is significantly enhanced when fuel is provided as a continuous flow through the electrodes. A further problem of highly concentrated fuels is the accumulation of reaction products that inhibit the charge transfer reaction kinetics by steric hindrance. The accumulation of reaction products can also lower the stability of enzymes at the electrode. For instance, gluconolactone is the by-product of single-step glucose oxidation when using GOx as the enzyme catalyst. Gluconolactone can be hydrolyzed to gluconic acid, which lowers the pH of the electrolyte specifically in pores of the electrode, and the pH shift reduces the kinetics of enzyme catalysis, and ultimately denatures the immobilized enzyme catalyst. Ideally, complete substrate oxidation reduces, or partially eliminates, issues with by-product formation. Practically in flow-through EFCs, by-product removal is facilitated by feeding fresh fuel into the system.

In summary, optimal saturated substrate concentrations, the need for by-product removal, and the need for flow-through agitated hydrodynamic conditions are the main aspects for optimal fuel selection for each specific EFC design. Future design may address recycling water from the reaction waste for on-board formation of fresh fuel, from a concentrated solution or solid phase stock. To our knowledge, such system has yet to be built and demonstrated with EFCs.

16.4 SINGLE-CELL DESIGN

A single-cell EFC is the keystone for development of a complete EFC-based power source. It is of much simpler design and construction in comparison with a stack of EFC cells. On the other hand, it has a very limited potential for practical applications due to low power and low operating voltage (typically in the range of 0.25–0.35 V at maximum power generation). The total output power of a single-cell EFC can be increased by scaling (i.e., increasing the size of the electrode increases the total generated current), whereas the operating voltage range stays the same. The size increase of a single-cell EFC has design limits, and such systems must be equipped with a direct current DC/DC converter (voltage booster). All DC/DC converters require certain minimal power input for proper operation and can consume considerable power to operate. Currently, there are no commercial converters that can boost such a low operating voltage of a single-cell EFC that also consume reasonably low power with respect to the EFC power generation. Despite these limitations, single-cell EFCs are important for system development, characterization, and optimization as they eliminate the impact of cell-to-cell variability in a stack when examining reproducibility and performance characteristics. Design of a single-cell EFC is therefore an essential step before design and optimization of a complete EFC stack.

A single-cell EFC can be designed as a system with an anode and cathode compartment that is akin to conventional electrochemical power sources or downsized to microfluidic devices. Both concepts will be discussed herein as they offer distinct opportunities in future development. Applications with extremely low power demand that require cheap and disposable power supplies are addressed by microfluidic-type EFCs. Higher power demand applications and more robust systems, with potentially replaceable electrodes, are more amenable to compartment EFC designs. A further area of development is the fabrication of implantable medical devices, in which design constraints can be addressed by merging both microfluidic and compartment-type EFCs.

16.4.1 Design of Single-Cell EFC Compartment

The compartment of a single-cell EFC consists of the following components:

- *Anode*: Typically high surface area, open porosity, with immobilized enzymes. The anode might contain an integrated internal electronically conductive network (internal current collector) and an internal mechanical support structure (or both integrated in one).
- *Cathode*: Similar construction to the anode, or in the form of an air-breathing electrode, or construction akin to the air-breathing MEA of conventional proton exchange membrane fuel cells.
- The anode and cathode compartments are formed by electrode packaging that also serves as an external current collector and encloses the fuel in the cell. For the case of oxygen-sensitive enzymes, the electrode packaging and sealing must be hermetically tight.

- A membrane or other simple separator that electronically separates the anode and cathode, prevents internal electronic short circuit, and allows minimal separation distance between anode and cathode ($\ll 1$ mm). The small separation distance reduces the diffusion length for protons. EFCs can operate with only a simple separator, such as cellophane, that is wetted in aqueous fuel. If ion-selective membranes must be used, they must provide fast proton conductivity in pH-neutral aqueous electrolytes.

- A seal between the packaged electrode components prevents fuel leakage, oxygen penetration from ambient atmosphere (in the case of oxygen-sensitive enzymes), and electronic insulation. Nafion membranes can also function as seals.

- A mechanism that holds all the components together is typically provided with set screws integrated around the design of the electrode packaging so as to avoid electronic short circuit of the anode and cathode.

- A fuel port is provided for filling the EFC with fresh fuel, along with an exit port for consumed fuel and waste products. For oxygen-sensitive EFCs, the ports must be reclosable or equipped with self-closing valves.

- Contactors for electrical connection to external electric circuits are included.

- A single-cell EFC specifically for research purposes might have integrated ports for a reference electrode for operation in the standard three-electrodes configuration.

Low ohmic resistance, electronic connections between anode and cathode, and the external current collectors integrated with the housing are very important for high-power operation. In reality, the entire "wired" electronic pathway from enzymes to external electric load must be of minimal ohmic resistance. Anode materials possess high surface area and open porosity for high enzyme loading and are designed to be hydrophilic to the aqueous fuels, which can impose problems with contact resistance between the electrode and the anodic current collector. Contact might be facilitated by mechanically compressing the current collector into the anode compartment, but this, in turn, can cause problems during assembly, when force compacting the anode material in the hardware and closing the system. In addition, the electrode material may become dislodged due to moisture or vibration, resulting in a loose electronic contact in the cell. Any means of firm mechanical attachment of the anode material to the anode compartment for electronic contact purposes is thus beneficial. Notably, the anode compression force must be low enough to avoid any significant reduction of porosity that would lower mass transport of the fuel to the bulk and structural damage of the material, for example, breaking of the internal current collector that would increase ohmic resistance. Alternatively, a spring-loaded contact might be applied to provide balanced force and to compensate for variations of mechanical properties, size, and volume changes with electrode wetting and swelling in the fuel, as well as vibrations and mechanical impacts imposed with portable applications.

A large contact area between the electrode material and the current collector of the electrode compartment (housing) is preferred. The current collector of the housing may be designed as the entire inner conductive walls of the electrode compartment, fabricated from conductive materials or a surface-deposited conductive coating. Material for the current collector must be selected according to chemical and electrochemical compatibility. Specifically, the materials must be chemically stable in the fuel, reaction intermediates, generated by-products, and eventually comply with biocompatibility requirements. Electrochemical compatibility of the electrodes should address corrosion, specifically at cathodes. Enzymes are sensitive to direct contact with materials of high electrochemical potential as it leads to their fast denaturing. The best known structural materials are carbons; noble metal contacts of gold or platinum are impractical for EFCs due to high costs. Alternatively, gold-coated contacts for higher conductivity, electrochemical stability, and low contact resistance are a viable option. Biocatalytic electrodes have been demonstrated in a simple design with titanium wire current collectors. On scaling, however, anode compartments of titanium are hard to machine, their cost is high, and they are not environment friendly. Thus, current collectors or entire electrode packaging made of graphite seems to be the material of choice. Graphite is easy to machine and is relatively cheap. There are different graphite grades and various graphite composites that vary in electronic conductivity, mechanical stability, and porosity. Some low-density graphite materials are permeable for gases and liquids between the individual graphite layers, even at very low pressure.

Current EFCs are typically cathode limited even with efficient oxygen supply. Oxygen-reducing cathodes in EFCs supplied directly with oxygen gas provide highest power configuration. Additionally, in conventional fuel cells, the oxygen feed is sometimes humidified for the best catalytic performance. However, feeding oxygen to EFCs imposes critical technical, safety, and cost burdens on the EFC power system. As a result, an energetically passive air supply leveraged from other systems provides better technical justification. A flooded cathode, which uses dissolved oxygen in the fuel solution, should be avoided due to low oxygen supply and thus low power performance.

An exploded view schematic image of a modular EFC research cell is shown in Figure 16.3 (top left) and features (i) an-air breathing oxygen reduction half-MEA

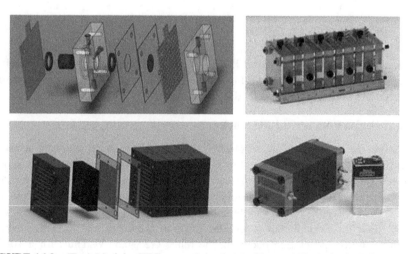

FIGURE 16.3 (Top) Modular EFC research hardware. (Bottom) Optimized EFC prototype with bipolar electrodes. (Left) Exploded drawing. (Right) Photograph of EFC stack assembly. (Reproduced with permission from Ref. [16]. Copyright 2010, The Electrochemical Society.)

cathode, (ii) a GOx-based anode fabricated on a carbon felt internal current collector, and (iii) a stainless steel plate anode and cathode external current collector. The EFC is shown in a three-electrode configuration with a port for the reference electrode and a pair of ports for filling the anode with fuel.

16.5   MICROFLUIDIC EFC DESIGN

In microfluidics-based EFCs, the fuel and oxidizer are supplied in two parallel colaminar streams in a single channel, thus avoiding the need for a membrane. This is particularly important for EFCs with diffusive mediators or for systems with less selective enzyme-catalyzed charge transfer reactions. Avoiding membranes in this manner significantly improves power performance, which is further facilitated with hydrodynamic fuel conditions in the flow that improves the substrate and by-products mass transport.

   In 2006, Kjeang et al. analyzed a microfluidic EFC using a 2D computational approach and concluded that microfluidic EFCs are power limited by the kinetics of the enzyme-catalyzed reaction as oppose to the typical mass transport limitation [17]. In their example, EFCs with a one-step oxidation of ethanol generate higher power than an EFC that catalyzes full oxidation of methanol, but the latter provides significantly higher energy density (approximately four to five times). By specific enzyme patterning, the authors confirmed that diffusion-limited operation using mixed enzyme patterning provides nearly complete fuel use. Microfluidic EFCs generate higher power density in advection-dominated regimes, but they become significantly reduced at lower fuel velocity due to back diffusion and fuel starvation. The overall power generation in the microfluidic EFCs was identified as being limited by the kinetically slowest enzyme. This is in contrast to the compartment-designed EFCs, which are typically power limited by performance on the system level at the cathode and by substrate mass transport at the anode.

16.6   STACKED CELL DESIGN

Stacking an assembly of individual EFCs is the most direct system-level approach to increasing power generation and is more flexible and universal than simply increasing the electrode size of a single-cell EFC.

16.6.1   Series-Connected EFC Stack

Because of the low voltage of a single EFC, cells in a stack are typically connected in series for increased total voltage. For higher stack voltages, and thus more series-connected cells, a practical problem occurs in respect to a large difference between the system open-circuit voltage (OCV) and voltage at the maximal power loading. In this case, a DC/DC converter or simple circuitry to limit OCV becomes necessary. A single EFC, for example, may provide OCV in the range of 0.7–1 V, whereas maximum power generation occurs in the range 0.25–0.35 V. Thus, for a 10-cell stack, the operating voltage at maximum power is between 2.5 and 3.5 V, equivalent to two series-connected primary alkaline batteries (e.g., Zn/MnO_2 with KOH electrolyte) predominantly used for powering a wide range of small consumer electronics. However, the OCV voltage of such a stack is between 7 and 10 V, and those existing electronics are not designed to handle such an input voltage. As a result, the application circuitry must be custom designed to withstand high initial voltage with EFC power supply or, alternatively, the EFC stack must be equipped with a DC/DC converter or a maximal voltage cutoff power conditioning system. In the latter case, the converter or the power cutoff circuitry should not impose any load on the EFC stack during no-power demand cycles. When an application starts to demand power, only then should the following sequence proceed: (1) the EFC is connected and becomes essentially electrically loaded with the power conditioning circuitry, or the DC/DC converter is activated, and both regulate (lower) the EFC stack voltage to predefined voltage input for the application; (2) only then the EFC stack is connected to the application load and begins to generate power and at this point the conditioning circuitry may be disengaged. Conventional alkaline primary batteries have end-of-discharge cutoff voltages of 0.9 V per cell and that results in 1.8 V for a 2-cells conventional alkaline power source and corresponds to 180 mV per cell in the 10-cells EFC stack. In a highly dynamic operation with pulse-type loading, the EFC stack must withstand the low voltage cutoff limit with short time peak pulses. If the frequency of peak power pulses is relatively low, then some EFC systems can withstand peak voltages lower than 50 mV per cell.

The influence of degradation mechanisms is specific to each EFC design. Degradation can occur with electrodes, enzymes, and mediators alike. Therefore, understanding degradation mechanisms and their dependence on voltage is an important aspect of EFC design. System design, electrode sizing, and performance balance between the anode and cathode must be leveraged to exclude critical operating voltage for specific electrodes. As described, a reference electrode might be integrated with the EFC so that a cutoff voltage limit can be defined precisely and independently for both electrodes. In general, the application circuitry can be modified to accommodate a lower cutoff voltage, and the initial voltage conditioning and the cutoff limit may be handled and controlled with a DC/DC converter at the penalty of increased system cost and a substantial power loss.

An EFC stack composed of five modular research hardware single-cell EFCs is shown in Figure 16.3 (top left). The anode and cathode contactors of each cell are visible on the top and the individual cells are electrically insulated; the intercell electric connection is provided with external U-shape leads. This allows for series or parallel connection of the cells in the stack as well as ease of monitoring of the individual cell voltage.

16.6.2 Parallel-Connected EFC Stack

The individual cells in an EFC stack can also be electrically connected in parallel. This may seem counterintuitive in relation to the low operating voltage of a single EFC cell, and to our knowledge no parallel-connected EFC stack has been reported as of this writing. There are specific reasons, however, why this option may become applicable: (1) there is a problem with a single manifold series-type fuel supply and flow-through (discussed later in this chapter), (2) high cell-to-cell performance variation, (3) high cell-to-cell variation in operational and shelf-life stability, and (4) potential application requirement for a high system reliability. The drawback of this approach is the necessity for using custom-designed DC/DC converters that provide voltage boost and control the application requirements, as this is associated with significant power losses in the converter and adds cost to the system. The architecture and requirements of DC/DC converters for EFC applications are discussed later in this chapter.

16.7 BIPOLAR ELECTRODES

A minimal mass and volume of an EFC stack are the essential design criteria of the power source, but simple stacking of the individually designed EFC cells does not comply with such criteria. The weight and volume minimization of the stack might be inspired by conventional micro fuel cell stacks. The most direct approach is a design of bipolar electrodes for series-connected cells. A bipolar electrode is a single-body electrode that functions on one side as an anode of the cell (i) and on the other side as a cathode of the adjacent cell ($i + 1$) in the stack. The intercell electronic connection is provided directly through the conductive bulk of the bipolar plate. This allows for minimal separation distance between the cells and minimal material mass and volume. The bipolar electrode typically features holes for the stack screws that hold the entire stack together and facilitate alignment of the individual electrodes and cells. The bipolar electrodes can feature integrated channels and openings to supply fuel and oxidizer. Those fuel and oxidizer lines must be designed to allow for homogenous distribution and must facilitate filling and degassing of dry EFC with fuel. Forcing fuel solution into the EFC must push out the gas (air) from the system during filling. The bipolar plates must also have machined settings for incorporation of the seals between the bipolar plates (between electrodes). The most convenient material for the bipolar electrodes is graphite or a graphite composite that allow for high electronic conductivity, chemical and electrochemical stability, and ease of mechanical and machining properties. Various grades of graphite and graphite composites are available; feedings of liquid-phase fuels and gas-phase oxidizer require a high-density and low-porosity graphite composite, which is impermeable for liquids and gases at the relatively low pressure applied with EFCs.

A specific arrangement/design might be applied for the first and the last cell in the stack. The first and the last electrode do not need to be bipolar as they are serving only one cell, but they do need to provide electronic contacts to the external circuit contactor. Additionally, those plates must have integrated ports for fuel input and output to and from the stack. The external circuit contactors are typically made of mechanically rigid and stable materials that can handle the force imposed with connection of electrical leads. Even though the contactors are not in direct contact with the fuel/electrolyte, fuel contamination might occur (or any aerosol contamination from the external environment), leading to corrosion at the interface between the contactor and the bipolar plate, which in turn increases the impedance of the stack cell. Graphite composite materials provide high electronic conductivity, chemical stability, and sufficient mechanical properties at certain thickness for the contactor application. Alternatively, metal contactors (stainless steel or aluminum) might provide a thin and lightweight alternative for the contactors.

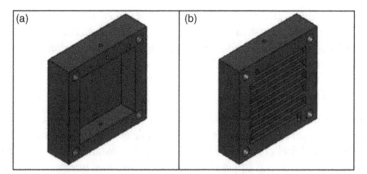

FIGURE 16.4 Schematic image of a bipolar electrode: view of anode (a) and cathode (b) sides. (Reproduced with permission from Ref. [16]. Copyright 2010, The Electrochemical Society.)

A schematic image of a bipolar electrode is shown in Figure 16.4. The cathodic side has an integrated serpentine channel for feeding and distributing oxygen (oxidizer), and the anodic side has a rectangular compartment for anode integration. Two independent feed channels are also visible through the bipolar plate that allows for fuel filling and oxidized feeding in series throughout the entire stack. A schematic image and photograph of a five-EFC stack with bipolar electrodes is shown in Figure 16.3 (bottom). The stack has two integrated lines for fuel and oxidizer embedded in the bipolar plates and two pairs of ports for the lines are visible at the end plates. Design optimization and application of bipolar electrodes provides significant volumetric reduction and increased electrode size in comparison with an EFC stack composed of five research modular single-cell EFCs (refer to the top image in Figure 16.3). The EFC stack with bipolar plates, however, does not allow for cells with parallel connections and does not provide ports for reference electrode integration.

16.8 AIR/OXYGEN SUPPLY

As described previously, despite the maximum power performance for a cathode fed directly with oxygen, the design is not practical because of its technical requirements, safety, cost, and its need for the user to service the feed system. It may be more feasible, however, to supply a low-pressure, low flow rate compressed air feed leveraged from other parts of the system, in a passive form. Oxidizer feed supply may be necessary for special applications in which not enough ambient oxygen or air is available. In general, for a low power generating EFC system, it is envisioned that passive diffusion of ambient air to the cathode is the most technically viable approach.

16.9 FUEL SUPPLY

Because of much higher enzyme stability in the dried (lyophilized) state in comparison with hydrated enzymes, it is perceived that EFCs will be delivered dry and users will activate them by filling with fuel just before required operation. This has to be done safely, simply, quickly, and efficiently. Fuel must be completely filled and typically gas/air must be pushed out by fuel and removed from the system for the best wetting and substrate/ion blockage-free operation. Air must be removed from the system to avoid oxygen poisoning for oxygen-sensitive enzymes.

16.9.1 Fuel Flow-Through

EFC anodes in the compartment configuration and operation are typically limited by diffusion of fuel (substrate) to enzymes in the bulk of the porous tortuous electrode material. Increased fuel concentration might help, but maximal substrate concentration is relatively low as it is limited by substrate inhibition, increased fuel viscosity, and accumulation of highly concentrated waste products in the fuel that all hinder the charge transfer reaction.

The mass transport of fuel is significantly accelerated with forced fuel flow through the electrode. Fuel flow during operation is also beneficial for removing waste products from the EFC. The drawback of fuel flow, however, is the additional power and energy consumption required for pumping. The minimal fuel flow rate f (m^3 s^{-1}) to compensate the fuel consumption in a single-cell EFC during the EFC operation can be calculated using Equation 16.2:

$$f = \frac{P \cdot S}{V \cdot F \cdot n \cdot c \cdot \eta} \qquad (16.2)$$

where P is power density (W m^{-2}) per projected geometrical surface area of the electrode S(m^2), V is the single-cell operating voltage (V) at power generation P, F is Faraday's constant (96,485), n is the number of electrons in the charge transfer reaction, c is concentration of fuel (mol m^{-3}), and η is fuel utilization efficiency. For example, a single-cell glucose-based EFC with electrode geometric surface area of 6.45 cm^2 (1 in.^{2}) that operates at power density of 5 mW cm^{-2} (total 32.3 mW) at 0.35 V per cell and with 0.1 mol l^{-1} glucose concentration in the fuel and fuel utilization efficiency of 75% requires minimal flow rate of 382 µl min^{-1} for a single-type enzyme EFC that provides a single oxidation step (2e$^-$ reaction). For the EFC with enzyme cascades that provides complete oxidation (24e$^-$ charge transfer per molecule of glucose), the minimal flow rate requirement drops to 32 µl min^{-1}. Such flow rate is already in the microfluidic range.

There are two viable approaches to inducing fuel flow through an EFC. One is to integrate an extremely low power consumption pump. The consumed power for pumping must be significantly low with respect to the total generated power by the entire EFC system. The other is to use a passive pumping system that does not consume electric power.

In a circular profile tubing of diameter D (m) and length l (m), pressure drop of ∇p (Pa) is required to maintain a laminar flow of a fluid with dynamic viscosity μ (Pa s) and velocity u (m s^{-1}). In this case, the pressure drop is dominated by the fluid flow friction in the tubing and can be simplified by Equation 16.3 [18]:

$$\nabla p = \frac{32 \cdot \mu \cdot l \cdot u}{D^2} \qquad (16.3)$$

Then the pumping power W (W) can be calculated for flow rate f ($m^3 s^{-1}$) (Equation 16.4):

$$W = \nabla p \cdot f = \frac{32 \cdot \mu \cdot l \cdot u \cdot f}{D^2} \qquad (16.4)$$

The pressure drop in the fuel line is only one part of the entire fuel circuit, but as is explained later in this chapter, the fuel line diameter is specifically designed to be very small and thus this part of the circuit might become dominating. The pressure drop through the bulk of the electrodes strongly depends on porosity and tortuosity of the material as well as on the electrode geometry and the design of the flow path.

Micropumps for extremely low flow rates, based on electrical, magnetic, light, thermal, and other actuated mechanisms, have been proposed and demonstrated. Micropumps can be fabricated on various substrates such as glass, Si wafers, or polymers and thus might be easily integrated with the EFC hardware. There are two groups of micropumps: (1) reciprocating displacement micropumps that have rotating or oscillating elements (e.g., piston, or peristaltic mechanism) that exert pressure and displace the working liquid, and (2) continuous dynamic flow micropumps that actuate continuous flow of the working liquid with direct transformation of mechanical or nonmechanical energy [19,20]. Micropumps include electrokinetic, electrohydrodynamic, magnetohydrodynamic, electrochemical, electroosmotic, and electrophoretic actuation. Electrostatic micropumps have low power consumption (1 mW), a very short stroke (5 µm), and provide flow rates of 70–850 µl min^{-1} with pressure heads of 2.5–31 kPa; however, they typically require around 200 V for actuation, as demonstrated by Nisar et al. in 2008 [21]. In 2005, Teymoori and Sani reported an electrostatic peristaltic pump of size 7 mm × 4 mm × 1 mm that requires low actuation voltage of 18.5 V and provides a flow rate of 9.5 µl min^{-1} [22]. Piezoelectric micropumps require a relatively high actuation voltage and provide a short actuation stroke and a wide range of flow rates of 1–2400 µl min^{-1}. They consume low power (around 3 mW) and require actuation voltage of 80 V. In 2005, Feng and Kim demonstrated a piezoelectric micropump with a footprint of 10 mm × 10 mm × 1.6 mm and fabricated with biocompatible materials that provided a flow rate of 3.2 µl min^{-1} and maximum back pressure of 0.12 kPa [23]. Such micropumps could be integrated with implantable EFC systems; however, the actuation voltage might be critical for implantable applications. Pan et al. showed that larger electromagnetic micropumps (10 mm × 10 mm × 8 mm) provide flow rates of 40 ml min^{-1} for air and 2.1 ml min^{-1} for water, with power consumption of 500 mW [24]. Their lowest power consumption design (25 mm × 10 mm × 10 mm), equipped with a polydimethylsiloxane membrane and a permanent magnet, was reported to provide 200–800 µl min^{-1} with 7.5 kPa back pressure at 13 mW power consumption. The latest development of low power consumption micropumps indicates potential for integration with EFC systems, but little is known regarding the feasibility of combining an EFC with a self-powered, pump-forced, fuel flow-through system.

Passive flow-through systems might use external mechanical vibrations induced with mobile applications, chemical energy, physical principles of capillary effects, suction of an aqueous solution with dehydrated hydrophilic material, or induction of fuel flow by feeding the fuel into a dehydrated hydrophilic material (e.g., a cellulose-based) or other hydrophilic polymers. An example of such a passive fuel flow-through system integrated with a microfluidic EFC has been demonstrated and is described later in this chapter.

16.9.2 Fuel Flow-Through System

The fuel flow-through system should feed fuel from a reservoir to the EFC and output consumed fuel (waste solution) to a waste reservoir or simply dispose the waste to the ambient environment in a liquid form or in a gas form by evaporation, depending on the fuel composition. All this must be provided with minimal mass and volume of the flow-through system and with minimal power consumption. For a single-cell EFC, this is a straightforward design. For a series-connected EFC, it is complicated due to electrical short-circuiting through the fuel circuit lines that reduces the total generated power. As discussed, EFC fuel typically contains a buffered salt solution to provide ionic conductivity and maintain the pH balance that is critical for kinetics of enzyme-catalyzed reactions as well as for enzyme stability. An example of specific conduct-ance κ (S m^{-1}) dependence on ionic strength of pH 7 potassium phosphate is shown in Figure 16.5. Now, consider a fuel with 200 mM potassium phosphate buffer of pH 7 supplied through a single line manifold (diameter 1 mm) to a stack of two series-connected cells. The length of the line between the adjacent anodes is 10 mm. The fuel-like resistance R (Ω) is then calculated with Equation 16.5:

$$R = \frac{l}{A} \cdot \frac{1}{\kappa} \tag{16.5}$$

where l (m) is the length of the fuel line and A (m^2) is its cross-sectional area. The resulting fuel line resistance in the example is 4.86 kΩ. The power losses due to the short-circuiting of two cells through the fuel line during maximal power operation

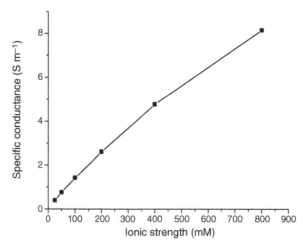

FIGURE 16.5 Dependence of specific conductance on ionic strength of potassium phosphate buffer, pH 7.

(0.35 V) are 0.025 mW. The ionic short circuit between cells through the fuel line may become significant for a micro-EFC stack, especially for a stack design with minimal cell separation distance (a typical design requirement) and large-diameter fuel channel (selected for minimal pressure head and minimal pumping power). To reduce the losses in the single manifold fed fuel system, the intercell fuel line should have a small diameter and be as long as possible. This, however, needs to be balanced with pressure requirements and power consumption of an integrated micropump. This also lowers the performance of the passive flow-through system, and slows down the initial fuel filling that might be performed manually by the user to attain complete filling and degassing (bubble removal) of the entire system. The initial filling can hardly be provided in a reasonable time period with an integrated micropump.

An additional problem of a single manifold fuel filling is decreasing fuel concentration and increasing by-product concentration after each cell in the series. Therefore, the cells closer to the end of the series-connected stack are supplied with lower fuel concentration and increased waste concentration. It is important to realize that even without any actual flow of fuel, ionic bridges between the individual cells of the stack are established through the fuel line. Then, power and fuel are continuously consumed by the losses, even during periods without any external electrical loading.

Individual fuel lines to each cell of the stack (parallel fuel lines) might avoid some of those problems, but fuel must be ionically insulated for each cell in both the fuel source tank/reservoir and the waste container. Moreover, such parallel fuel lines are impractical and difficult to integrate in the stack hardware (e.g., bipolar plates); in this case, the design of the entire stack assembly becomes much more complicated. External tubing connections to each cell are not a good design solution, and parallel fuel supply lines make the system complicated, heavier, and bulkier.

The operating flow-through system should avoid any air bubble introduction into the EFC and should be designed to push fuel into the EFC, rather than suck it out with introduced under-pressure to the fuel output line. Ideally, the fuel feed system should facilitate removing gas bubbles or accumulating them in designed area of the system without having to pass the bubbles through the entire fuel line and connected cells.

16.9.3   Fuel Flow-Through Operation and Fuel Waste Management

The fuel flow-through system should be activated only in the event of EFC power generation, or in regular intervals to refresh the fuel in the system and potentially perform self-system checks such as cell voltage at the pump electric loading.

Fuel flow-through operating regimes remain to be optimized for EFCs and will depend on the EFC design and operating algorithm of specific applications. In some cases, it might be beneficial to actuate the flow-through system for discrete periods and follow up with steady-state operation. The flow-though actuation might be triggered with time intervals, monitored drop of power generation, decrease of individual cell voltage, reached limit of individual electrodes' voltage (systems with reference electrode), or with monitored waste or by-product accumulation in the EFC.

As discussed, the fuel composition might define the fuel waste management. For instance, if hazardous diffusive mediators are supplied with the fuel, then the waste must be treated as a hazardous chemical waste and must be collected in a waste fuel reservoir and cannot be simply disposed to the ambient environment. In future development of the EFC technology, there might be a need for on-board reprocessing of the waste to a fresh fuel. What may be particularly important is extracting water from the waste for reuse, as water poses a significant contribution to the system mass and volume. In the case of EFC low fuel utilization (e.g., high flow rates are required for high power generation), the spent fuel might be reconditioned by dissolving solid substrate to retain the required substrate concentration and directed back to the fuel supply reservoir or provide flow-through in the reverse direction. The number of the fuel regeneration cycles would depend on the reaction product concentration and its effect on the reaction kinetics and enzyme stability. Reaction products may accumulate from incomplete catalysis and may require treatment or their capture from the regenerated fuel. EFCs with complete substrate oxidation mitigate this issue, as accumulation of by-products are reduced or eliminated entirely by generating only carbon dioxide that can be simply extracted from the system.

16.10 STORAGE AND SHELF LIFE

Stability of enzymes immobilized in electrodes of EFC power sources remains a critical parameter of the technology. Enzymes dispersed in buffer solutions typically have lower stability compared with frozen or lyophilized (freeze-dried) enzyme preparations. Therefore, it is expected that EFC power systems will be delivered and stored in dry or frozen form. For both cases, recommended and maximum allowed storage temperature ranges must be defined. Despite enzyme immobilization protocols (see Chapter 11) that improve enzyme stability, long storage periods at temperatures above 60 °C remain a challenge for most EFC systems.

The following are general recommendations for EFC storage and operation:

- A dry EFC should be filled with fuel just before intended operation.
- If an activated (already fuel filled) EFC requires longer period storage without any power generation need, it is recommended to flush the system with a buffer solution (including enzyme stabilizers, if available). Hydrated EFCs might be then potentially frozen and stored. This method must be considered during system design to account for fuel expansion due to ice formation (e.g., by implementation of expandable zones in the cells).

16.11 EFC OPERATION, CONTROL, AND INTEGRATION WITH OTHER POWER SOURCES

16.11.1 Activation

If the EFC is delivered and stored dry or hydrated with stabilizer and buffer solution but without substrate, then the EFC needs to be activated by filling it with fuel. It is expected that a user will initially fill the system manually with a syringe or pipette or with an external pump that might guarantee proper pressure and gradual filling. Depending on the design, altering the EFC position while filling might be necessary to ensure proper degassing. Note that even EFCs with an excellent design of fuel supply system must still be filled and operated without gas bubbles that significantly impair the performance.

Similarly, frozen systems are activated by melting the fuel in the EFC—gradually and completely—and the entire system must be tempered to the ambient operating temperature. After filling, the system might require several minutes for wetting and diffusion of substrate (and mediator) to the bulk of the porous electrode.

With current developments, it is anticipated that activated EFCs filled with fresh fuel will be immediately used for power generation. Due to fuel consumption with side reactions and internal losses even without external electric loading, the fuel-wetted EFC will begin to degrade. Additionally, mediator instability, and possible mediator internal short circuit will also consume fuel and lower power generation.

16.12   EFC CONTROL

Current EFCs generate low power density and thus impose only very minimal safety risk in terms of electricity generation. Therefore, EFC control can be very simple and mostly focused on preserving the system stability and lifetime. For disposable EFCs, there is practically no need for any control system, except for overvoltage protection and minimal voltage cutoff on the application side. For future EFCs with higher power generation and longer term stability requirements, a power management system will be necessary. An overloaded EFC stack might overpolarize and fast degrade the weakest cell, which then increases ohmic losses and reduces power generation. In general, high electrode polarization accelerates enzyme denaturing and might be destructive also for mediators, and thus shortening the system lifetime. Therefore, the stack—and potentially individual cell—voltage should be monitored, and power generation should be controlled within defined voltage limits. When tight control of specifically an anode or cathode is required by design, then a reference electrode might be integrated with selected cell/s or with each cell to provide three electrodes configuration for precise measurement and voltage control. This solution does not seem practical for a low-power and low-cost EFC system, however a simple quasi reference electrode (e.g. chloridized silver wire) might be applied for this utility. With the current development, power management control that measures and excludes the weakest cell from the stack seems to be unlikely to be realized in most EFC applications. As of this writing, any current development or demonstration of a power management system for EFC applications has yet to be realized.

16.13   POWER CONDITIONING

As described previously, certain applications and configurations require a DC/DC converter or a DC/alternating current (AC) inverter to combine EFCs with application hardware. Such a requirement covers single-cell EFC systems and EFC stacks with parallel-connected cells. In those cases, the DC/DC converter must boost the EFC voltage to the required application level that is typically higher than the single-cell voltage. The voltage supply requirement for a microcontroller- or microprocessor-based electronic load, with low-voltage complementary oxide semiconductor components, is typically 3 V. Some low-voltage microcontrollers, like the Freescale Semiconductor MC9S08LL16, can operate in a restricted low-energy demand mode with only 1.9 V. Very efficient DC/DC voltage boosters for EFC applications must be custom designed to address both the specific application and the specific EFC power source. EFCs vary in design as well as in the optimal operating conditions. Commercially available DC/DC converters are not currently applicable to EFC power systems due to their high power consumption in relation to the EFC low power generation. In 2011, Wu et al. designed and tested a low-power DC/DC voltage booster based on electromagnetic inductance for a microbial fuel cell [25]. The induced voltage in the inductor V_{ind} (V) is proportional to the inductance L (H) and the rate of current i (A) change in time t (s) (Equation 16.6):

$$V_{ind} = -L \cdot \frac{di}{dt} \tag{16.6}$$

Wu et al. used a field-effect transistor (FET) modulated with an oscillator's square wave signal to switch the current on and off in the frequency range 1 kHz to 1 MHz [25]. Their DC/DC converter was capable of boosting the input voltage from 0.2 to >3 V on the output. They further demonstrated that the converter with inductance >5 mH cannot produce 3 V with frequencies over 10 kHz. On the other hand, low inductance requires higher frequencies at which the booster power efficiency drops. Additionally, reducing duty cycles from 50 to 18% improved efficiency by 25%. As we already indicated, DC/DC voltage boosters must be custom designed to the specific load and EFC system. Wu et al. recognized three specific criteria for development: (1) minimal booster power consumption in the oscillator and FET, (2) maximal power production from the power source, and (3) a minimum output voltage of 3 V. Their demonstrated optimized voltage booster operating at 20 kHz amplified input voltage of 0.2–0.4 to >3 V and consumed around 20 µW. The booster was functional with an electric load greater than kiloohms that makes the electric load of 0.9 mW and of which the losses are only 2.22%. The maximal power of this specific voltage booster is inadequate for current EFC applications. Nevertheless, custom development of DC/DC converters and voltage boosters will have to continue parallel with EFC development and for specific applications requirements. In the future, low-power electronics might be developed specifically for EFC needs. Furthermore, EFC performance and characteristics will be standardized and better understood. This will allow for development of an EFC computation model, and DC/DC converter design will use modeling to achieve optimal design for application-specific operating conditions.

16.14   OUTLOOK

This chapter provides an overview of current EFC designs and summarizes constrains and limitations with the individual components of the systems. EFC power sources differ from conventional batteries and fuel cells in terms of design and operation. The intrinsic EFC characteristics require novel materials and design of new components.

The ongoing research provides new functional materials that must be tailored and designed to be integrated in the EFC with respect to other components and system operation. Thus, the design effort must proceed in parallel to the development of materials and enzyme catalysts.

The power performance of EFCs is typically limited with substrate and oxidizer mass transport towards the enzyme catalysts. The EFCs low power generation, however, prevents any active high energy demand fuel and oxidizer supply solution. In this chapter, we discusse multiple options and solutions to minimize such limitations.

Operation and control of EFC power sources also differs from the conventional power systems. Specifically, we discuss the advantage and drawbacks of EFC single-cells versus series- and parallel-connected cells in EFC stacks. The need for a custom designed DC/DC converter for EFC applications is explained, as well as the problem of high voltage difference between OCV and operating voltage under electric load. A special effort must be dedicated to storage (shelf life) and operating conditions that affect the enzyme stability and, thus, lifetime of the EFCs. As explained, this directly influences the system design and operating procedures.

LIST OF ABBREVIATIONS

AC	alternating current
CB	carbon black
DC	direct current
EFC	enzymatic fuel cell
ET	electron transfer
FET	field-effect transistor
GDE	gas diffusion electrode
GDL	gas diffusion layer
GOx	glucose oxidase
MEA	membrane electrode assembly
MWCNT	multiwalled carbon nanotube
OCV	open-circuit voltage
PTFE	polytetrafluoroethylene
SCE	saturated calomel electrode
SEM	scanning electron microscopy

REFERENCES

1. Addo PK, Arechederra RL, Minteer SD. Towards a rechargeable alcohol biobattery. *J Power Sources* 2011;196:3448–3451.

2. Cooney MJ, Lau C, Windmeisser M, Liaw BY, Klotzbach T, Minteer SD. Design of chitosan gel pore structure: towards enzyme catalyzed flow-through electrodes. *J Mater Chem* 2008;18:667–674.

3. Konash A, Cooney MJ, Liaw BY, Jameson DM. Characterization of enzyme–polymer interaction using fluorescence. *J Mater Chem* 2006;16:4107–4109.

4. Gallaway JW, Calabrese Barton SA. Kinetics of redox polymer-mediated enzyme electrodes. *J Am Chem Soc* 2008;130:8527–8536.

5. Calabrese Barton S, Kim H-H, Binyamin G, Zhang Y, Heller A. The "wired" laccase cathode: high current density electroreduction of O_2 to water at +0.7 V (NHE) at pH 5. *J Am Chem Soc* 2001;123:5802–5803.

6. Kim H-H, Mano N, Zhang Y, Heller A. A miniature membrane-less biofuel cell operating under physiological conditions at 0.5 V. *J Electrochem Soc* 2003;150:A209–A213.

7. Mano N, Kim H-H, Zhang Y, Heller A. An oxygen cathode operating in a physiological solution. *J Am Chem Soc* 2002;124:6480–6486.

8. Cooney MJ. Design of macropore structure for enzyme fuel cells operation. *Abstr Pap Am Chem Soc* 2005;230:U1672–U1672.

9. Chen W, Cai S, Ren Q-Q, Wen W, Zhao Y-D. Recent advances in electrochemical sensing for hydrogen peroxide: a review. *Analyst* 2012;137:49–58.

10. Farneth WE, D'Amore MB. Encapsulated laccase electrodes for fuel cell cathodes. *J Electroanal Chem* 2005;581:197–205.

11. Ghindilis AL. Direct electron transfer catalysed by enzymes: application for biosensor development. *Biochem Soc Trans* 2000;28:84–89.

12. Hudak NS, Gallaway JW, Calabrese Barton S. Formation of mediated biocatalytic cathodes by electrodeposition of a redox polymer and laccase. *J Electroanal Chem* 2009;629:57–62.

13. Palmore GTR, Kim H-H. Electro-enzymatic reduction of dioxygen to water in the cathode compartment of a biofuel cell. *Electroanal Chem* 1999;464:110–117.

14. Sun Y, Calabrese Barton S. Methanol tolerance of a mediated, biocatalytic oxygen cathode. *J Electroanal Chem* 2006;590:57–65.

15. Lau C, Adkins ER, Ramasamy RP, Luckarift HR, Johnson GR, Atanassov P. Design of carbon-nanotube-based gas diffusion cathode for O_2 reduction by multicopper oxidase. *Adv Energy Mater* 2012;2:162–168.

16. Svoboda V, Lindstrom U, Singhal S, Lau C, Atanassov P. *Advanced glucose–air enzymatic fuel cell for portable applications.* 217th ECS Meeting, Vancouver, 2010.

17. Kjeang E, Sinton D, Harrington DA. Strategic enzyme patterning for microfluidic biofuel cells. *J Power Sources* 2006;158:1–12.

18. Crowe, CT, Elger DF, Roberson JA. *Engineering Fluid Mechanics.* John Wiley & Sons, Inc., New York, 2001.

19. Chen L, Lee S, Choo J, Lee EK. Continuous dynamic flow micropumps for microfluid manipulation. *J Micromech Microeng* 2008;18:013001.

20. Laser DJ, Santiago JG. A review of micropumps. *J Micromech Microeng* 2004;14:R35–R64.

21. Nisar A, Afzulpurkar N, Mahaisavariya B, Tuantranont A. MEMS-based micropumps in drug delivery and biomedical applications. *Sens Actuators B: Chem* 2008;130:917–942.

22. Teymoori MM, Sani AA. Design and simulation of a novel electrostatic micromachined pump for drug delivery applications. *Sens Actuators A: Phys* 2005;117:222–229.

23. Feng GH, Kim ES. Piezoelectrically actuated dome-shaped diaphragm micropump. *J Microelectromech Syst* 2005;14:192–199.

24. Pan T, McDonald SJ, Kai EM, Ziaie B. A magnetically driven PDMS micropump with ball check-valves. *J Micromech Microeng* 2005;15:1021–1026.

25. Wu PK, Biffinger JC, Fitzgerald LA, Ringeisen BR. A low power DC/DC booster circuit designed for microbial fuel cells. *Process Biochem* 2012;47:1620–1626.

17

MINIATURE ENZYMATIC FUEL CELLS

TAKEO MIYAKE AND MATSUHIKO NISHIZAWA

Department of Bioengineering and Robotics, Tohoku University, Sendai, Japan;
Core Research for Evolutional Science and Technology (CREST), Japan Science and
Technology Agency, Tokyo, Japan

17.1 INTRODUCTION

Enzymatic fuel cells (EFCs) are power devices in which enzymes are used as electrocatalysts to convert biochemical energy directly into electricity, in contrast to metallic catalysts commonly used in fuel cells [1,2]. The extremely high reaction selectivity of enzymes, as compared with metallic catalysts such as platinum, eliminates the need for fuel purification and allows a separator-free design that consists of a single pair of anode and cathode electrodes exposed to solutions containing both fuel and oxygen. In 2003, Mano et al. effectively demonstrated the ultimate simplicity of EFC systems for power generation using a pair of enzyme-modified carbon fibers (CFs) inserted into a peeled grape (Figure 17.1) [3]. Such a simple EFC system provides high flexibility in structural design and miniaturization and, as such, miniature enzymatic fuel cells (MEFCs) could be an attractive power source for future portable and implantable micro-devices [4,5]. In this chapter, we describe three types of MEFCs fabricated using a series of microelectromechanical system (MEMS)-related techniques. (1) An insertion MEFC is a type of cell that generates electricity from sugars in living organisms [6]. For example, a needle-shaped bioanode was inserted into a grape to drive a self-powered sugar monitor. The development of a carbon nanotube

Enzymatic Fuel Cells: From Fundamentals to Applications, First Edition. Edited by Heather R. Luckarift, Plamen Atanassov, and Glenn R. Johnson.

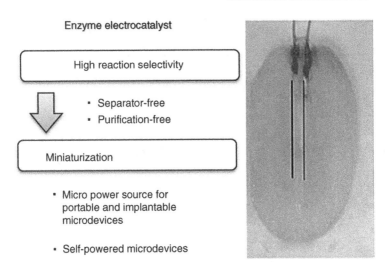

FIGURE 17.1 Miniature enzymatic fuel cells: points of merit for miniaturization and possible applications. The photograph represents an MEFC embedded in a raw grape using a pair of enzyme-modified carbon fibers. (Reproduced with permission from Ref. [3]. Copyright 2003, American Chemical Society.)

(CNT)-based flexible bioanode film further contributed to the miniaturization of the insertion MEFC devices [7]. (2) A microfluidic MEFC consists of microchannels for continuous fuel supply, in which the power generation and performance depend on several fluidic parameters including the flow velocity, electrode configuration, and channel dimensions [8]. The series and parallel stacking of microfluidic MEFCs increase the level of output voltage [9] and net lifetime [10], respectively. Finally, (3) a sheet-shaped MEFC is described that could be combined with wearable electronics of the future [11].

17.2 INSERTION MEFC

The high reaction selectivity of enzyme catalysts provides unique advantages of MEFCs, including the possibility of direct power generation from carbohydrates in biofluids such as juices or blood [3,12,13]. Such harvesting of biochemical ambient energy has attracted attention as a "green" technology. Power generation from real living organisms, however, entails the consideration that natural organisms are generally covered by skin. One possible solution is to use a needle-shaped insertion anode to approach fuels through the skin. Another important consideration is that oxygen is limited in many organisms to a lower concentration than that of sugars. In addition, biofluids may contain reaction inhibitors for cathodic enzymes, such as ascorbic acid and urate [14]. Taking these considerations into account, an insertion MEFC with a needle anode and a gas diffusion cathode designed to be exposed to atmospheric air was assembled, as illustrated in Figure 17.2.

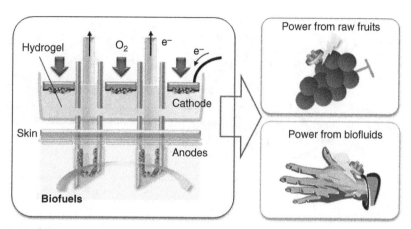

FIGURE 17.2 Schematic structure of an insertion MEFC designed to use biochemical energy in living organisms.

17.2.1 Insertion MEFC with Needle Anode and Gas Diffusion Cathode

The insertion MEFC device described above (Figure 17.3) consists of a fructose dehydrogenase (FDH)-modified needle anode for fructose oxidation and a bilirubin oxidase (BOx)-modified carbon paper cathode for reduction of oxygen in ambient air (Figure 17.3). The anode and cathode are assembled using a polydimethylsiloxane (PDMS) chamber and an ion-conducting agarose hydrogel (pH 5.0) as the inner matrix. This structural design allows the use of fructose in grapes and also protects the cathode from reaction inhibitors in the grape juice. The cell performance using a raw grape at room temperature is shown in Figure 17.3c. The device with a single anode generated $111 \mu W \, cm^{-2}$ ($6.3 \mu W$) of electrical power with a current density of $442 \mu A \, cm^{-2}$ at 0.25 V. The total performance could be amplified by connecting an array of needle anodes in parallel. A device with an array of four needle anodes, for example, demonstrates a fourfold increase in power, $\sim 26.5 \mu W$ (Figure 17.3c). The insertion MEFC was combined with a light-emitting diode (LED) device consisting of a charge pump IC, a $1 \mu F$ ceramic capacitor, and a red LED. The blink interval of the LED is inversely proportional to the power of the fuel cell, which is roughly proportional to the concentration of the fuel [6,15]. In practice, the LED blinks at a higher frequency with an increase in the fructose concentration (Figure 17.3d), and as a result, the concentration of fructose within the grape was estimated to be roughly 20–40 mM. In fact, a separate measurement revealed that the true concentration of fructose in the grape was 35 mM.

In 2011, Miyake et al. developed an insertion MEFC device for power generation from blood with a needle glucose anode prepared by co-immobilization of glucose dehydrogenase (GDH), diaphorase (Dp), and vitamin K_3 (VK_3)-pendant poly-L-lysine [6]. The fuel cell performance was evaluated by inserting the device into a vein of a rabbit ear and reached $130 \mu W \, cm^{-2}$ at 0.56 V. An anti-biofouling modification of the needle tip with 2-methacryloyloxyethyl phosphorylcholine (MPC) polymer was required, however, to stabilize the cell performance [6].

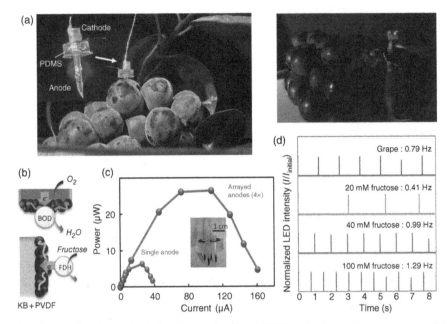

FIGURE 17.3 (a) Photograph of the assembled MEFC to be inserted into a grape. (b) Schemes of O_2 reduction at the enzymatic gas diffusion cathode (upper panel) and fructose oxidation at the enzymatic needle anode (lower panel). (c) Power output of the insertion MEFC using single anode and arrayed anodes (4×). (d) Monitoring of sugar level in a raw grape with a self-powered fructose-sensing device. The device consists of the insertion MEFC and an LED system, whose blink interval is correlated with the fructose concentration. (Please see the color version of this figure in Color Plates section.)

17.2.2 Windable, Replaceable Enzyme Electrode Films

Also in 2011, Miyake et al. developed a flexible enzyme–CNT ensemble film and wound it onto a metal needle for direct oxidation of fuel from inside a living organism [7]. Such an enzyme electrode can be easily mounted onto the insertion MEFC and readily replaced when the original is degraded.

Nanostructured carbons have been widely used for fabricating enzyme-modified electrodes because of their large surface area. However, because they are random aggregates of particulate or tubular nanocarbons, the postmodification of enzymes in the intrananospace is generally hard to control, and the brittle nature of such aggregated electrodes limits a miniaturized design. As an alternative, we have used a carbon nanotube forest (CNTF) consisting of extremely long (~1 mm) single-walled CNTs, which can be handled with tweezers and used as a 100% binder-free carbon film. When liquids are introduced into the as-grown CNTF (CNTs with a pitch of 16 nm) and dried, the CNTF shrinks to a near-hexagonal close-packed structure (CNTs with a pitch of 3.7 nm) because of the surface tension of the liquids. By using an enzyme solution as the liquid, the CNTF is expected to entrap the enzymes dynamically during the shrinkage, as illustrated in Figure 17.4a. The *in situ* monitoring of electrocatalytic activity of the

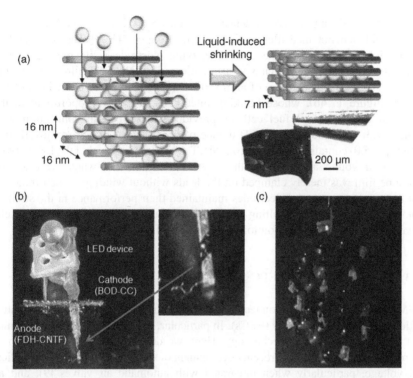

FIGURE 17.4 (a) Schematic diagram of enzyme entrapment inside a CNTF by liquid-induced shrinkage. The photograph shows a freestanding enzyme–CNTF ensemble film that can be manipulated with tweezers. (b) A stand-alone, self-powered miniature sugar monitor, at the tip of which an FDH–CNTF ensemble film was wound. (c) A number of the self-powered monitors are inserted into grapes.

CNTF films on soaking in a buffer containing FDH and D-fructose demonstrates currents that correlate with thickness (80 μA for 12 μm thick film and 25 μA for 4 μm thick film), indicating that the FDH molecule can entirely penetrate inside the CNTF films. The content of FDH inside a CNTF film was controllable by the concentration of FDH solution in which the as-grown CNTF films were incubated. The FDH content increased toward the theoretical limiting value (6.2 μg), at which the void volume of as-grown CNTF is fully occupied by FDH. Such controlled entrapment of enzymes can also be examined via the degree of CNTF shrinkage. Typically, the CNTF film without enzymes shrank to one-quarter of the original area; the CNTF film treated with FDH solution, by comparison, shrank to one-half of the original area. The degree of shrinkage also depended on the size of enzyme; a smaller enzyme such as laccase, for example, led to shrinkage equivalent to one-third of the original area.

These results support our methodology, which induces *in situ* regulation of intrananospace of the CNTF by the quantity and size of entrapped enzymes [7]. By connecting the FDH-based anode with a laccase-based cathode, the fuel cell performance was evaluated in an O_2-saturated McIlvaine buffer containing 200 mM fructose. With stirring, the power density reached $1.8 \, \text{mW cm}^{-2}$ (at 0.45 V), 84% of which could be maintained after continuous operation for 24 h [7]. The "freestanding and flexible" character of the present enzyme electrode is its most attractive advantage from a practical viewpoint. A piece of FDH–CNT film was wound onto the electric leads (needle) of an insertion MEFC and combined with an LED device (Figure 17.4b), whose blinking interval is inversely proportional to the power of the biological fuel cell, as previously described in connection with Figure 17.3. The leads of the LED device were immersed in a stirred McIlvaine buffer (pH 5.0) containing fructose (0.2 M). The blinking interval of the LED driven by the film-wound anode and cathode was similar to that when the enzyme–nanotube film was merely clamped on the leads without winding, which indicates that the enzyme–nanotube electrodes maintained their performance in the strained wound conditions. The resulting devices were put into grape berries as "self-powered" miniature sugar monitors (Figure 17.4c).

17.3   MICROFLUIDIC MEFC

MEMS-related microfabrication techniques have contributed greatly to the miniaturization of biological fuel cells (BFCs). In particular, microfluidic chips have significant overlap with fuel cell technology. Here, we describe the structural design of microfluidic MEFCs [8], how recent development of stacked cells provides a practical cell voltage, particularly when integrated with automatic air valves [9], and an automatically switchable fuel flow system that prolongs the operational lifetime [10].

17.3.1   Effects of Structural Design on Cell Performances

The performance of a microfluidic MEFC depends on several structural parameters, including the fuel flow velocity, electrode configuration, and channel dimensions. In 2008, Togo et al. constructed an enzymatic glucose/O_2 fuel cell within a microfluidic channel, as shown in Figure 17.5a, for studying the influence of electrode configuration and fluidic channel height on cell performances [8]. The cell was composed of a BOx-adsorbed O_2-reducing cathode and an enzymatic glucose-oxidizing anode prepared by co-immobilization of GDH, Dp, and VK_3-pendant poly-L-lysine. The effects of pre-electrolysis of O_2 at the upstream cathode on the performance of the downstream anode can also be evaluated using this system. Figure 17.5b shows that the V–I curve obtained with the cathode in an upstream configuration was superior to that with a downstream cathode configuration, especially in the higher current region, where the maximum cell current (I_{max}) increased ~10% by placing the cathode upstream. This result proved that the cell design with an upstream cathode is effective at protecting the anode from oxidation. The cell performance was also affected by the channel height, as shown in Figure 17.5c. This experiment used microfluidic BFCs having electrodes on both lower and upper walls of the channel, which were constructed by sandwiching a silicone rubber spacer of 0.1 or 1 mm thickness between two electrode-patterned glass slides. Both glass slides have an upstream

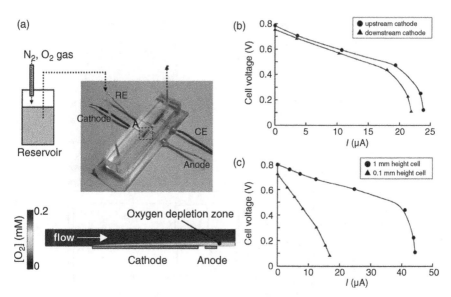

FIGURE 17.5 (a) Microfluidic MEFC. (b) *V–I* curves of microfluidic cells operating with air-saturated phosphate buffer (pH 7) containing 0.1 M NaCl, 10 mM glucose, and 1 mM nicotinamide adenine dinucleotide (NAD$^+$) at room temperature, with a flow rate of 0.3 ml min^{-1}. The cells were operated with an (●) upstream cathode or (▲) downstream cathode. Channel height: 1 mm. (c) *V–I* curves of sandwich-type microfluidic cells with channel height of (●) 1 mm and (▲) 0.1 mm.

cathode and a downstream anode. The I_{max} of the 1 mm high cell was almost twice that of a single set of electrodes and directly corresponds to the increased electrode area. In contrast, the I_{max} of the 0.1 mm high cell composed of two sets of electrodes was significantly smaller, mainly because of depletion of O_2 in the narrower flow channel. From the viewpoint of the volumetric density, however, the 0.1 mm high narrower cell is superior to the 1 mm high cell; the volume density in respect to I_{max} and maximum cell power (P_{max}) of the 0.1 mm high cell was 3.8 and 2 times greater, respectively, than that for the 1 mm high cell. The optimum efficiency in operation, with the highest density output, would presumably occur when the flow condition formed a depletion layer comparable to the channel height.

17.3.2 Automatic Air Valve System

The possible output voltage of a single EFC is determined thermodynamically, and is lower than 1 V. Therefore, many applications require cell stacking (via a series connection), which can be problematic because of short-circuiting of cells through the ion-conductive fuel solution. The series connection of BFCs, therefore, requires a system for ionic isolation between each cell. In 2009, the Nishizawa research group devised one possible strategy for making such ionic isolation systems, based on air trapping by the superhydrophobic area produced between fuel cells arrayed in a

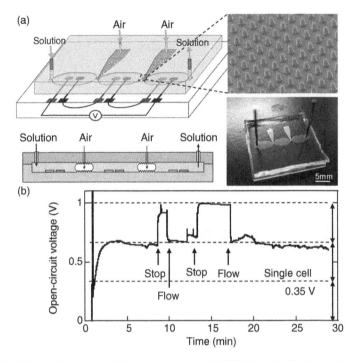

FIGURE 17.6   (a) Structures of the series-connected MEFCs in a fluidic chip, which has lotus leaf-like superhydrophobic valves at the gap between each MEFC. (b) The time course of the open-circuit voltage measured between the two ends of the electrode circuit during the charge and recharge of a 0.1 M NaCl and 0.1 M glucose-containing 50 mM phosphate buffer solution (pH 7).

microfluidic channel [9]. As shown in Figure 17.6a, the glucose/O_2 EFC is composed of three sets of enzyme-modified anode/cathode pairs, each of which has a separately measured open-circuit voltage (OCV) of ~0.35 V. Importantly, a lotus leaf-like micropillar array, which was designed to be 20 μm in height, 15 μm in diameter, and spaced 15 μm apart, was prepared on the wall of the gap between each couple. The bottle-shaped superhydrophobic air reservoirs are open at their ends in order to take in air from the outside. The OCV between the two ends of the circuit was measured during the regulation of fuel flow (Figure 17.6b). The OCV was ~0.65 V even under solution flow, a value roughly twice that of a single cell. This indicates that one of the two gates can be closed (ionically insulated) during solution flow, and that small droplets or a mist of fuel solution would transfer across the "closed" gate. By using a further hydrophobic material such as fluorocarbon polymers and/or a more sophisticated cell design, it is possible that a stable series connection in a flow system can be achieved. When the solution flow was stopped, the OCV quickly increased to ~1 V, suggesting that both gates were truly closed by the air valves, creating a series connection of the three cells. Such a change in OCV was reversible for successive stop and flow operations. The effect of channel height was also studied and it was found that a channel that was sufficiently narrow (~40 μm), with a concomitantly high surface area to volume ratio, was required for effective utilization of the surface hydrophobicity in order to function as a valve.

17.3.3 SPG System

The lifetime of an operating EFC is limited typically to 1 week because of enzyme deactivation and/or leaching. This longevity has been gradually improved by both protein engineering [16,17] and enzyme immobilization techniques [18]. As an alternative strategy, we describe here a sequential power generation (SPG) system, an automatic backup system with a stored fresh cell stack in which the power of each cell decreases gradually but the net output can be maintained by automatic replacement with a newly activated cell [10]. As shown in Figure 17.7, the SPG device is

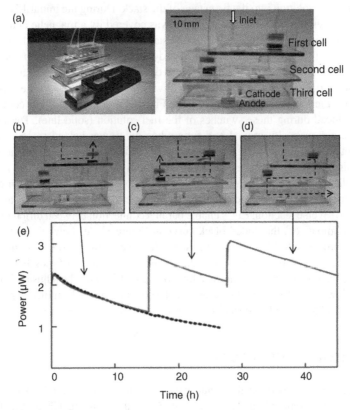

FIGURE 17.7 (a) Photograph and schematic of the sequential electric power generation system for prolonging the net lifetime of a miniature biofuel cell stack. Photographs of the biofuel cell stack at (b) 0 h, (c) 20 h, and (f) 30 h after introducing fuel solution into the first stage of the stack. (e) Time course of the power measured in the phosphate buffer solution (pH 7) containing 50 mM glucose, 5 mM NAD^+, and 0.1 M NaCl. The cells were connected in parallel with a 100 kΩ load. The net lifetime of the system (solid line) is prolonged as compared with a single biological fuel cell (dashed line).

composed of PDMS microchannels and electrodes fabricated on three glass slides that are layered and connected through holes covered initially by magnetic chips and a poly(lactic-*co*-glycolic acid) (PLGA) film. The binding of PLGA was achieved using pressurized CO_2 [19]. When a fuel solution flows into the microchannel of the top stage of the stack, the cell starts generating electric power and the PLGA polymer used as glue starts to dissolve. At a specified time interval, when the glue has sufficiently dissolved, the magnetic cover is lifted up by way of a neodymium magnet set in the roof of the microchannel. After opening the magnetic chip cover, the fuel solution then flows down into the second stage microchannel to activate and operate the fresh cell in the second stage. By repeating this cycle, the total power output is maintained at a stable level, and the net lifetime of the stacked BFCs is extended.

Figure 17.7b–d shows photographs of a SPG system in which three cells are stacked and connected in parallel with a 100 kΩ load, at (a) 0 h, (b) 20 h, and (c) 30 h after introducing fuel solution into the first stage of the stack. During the initial 15 h, because the hole between the first and second stages was covered by a magnetic chip, the fuel solution flows out from the outlet of the first stage (Figure 17.7b). After the cover had opened, the magnetic chip moved up to the neodymium magnet at the roof of the channel and closed the outlet. Then the solution flowed down into the second stage and out from the second outlet (Figure 17.7c). The next switching of the flow to the third stage proceeded in the same manner (Figure 17.7d). Figure 17.7e shows the time course of the power produced during these switches of the fuel solution (solid line). The magnetic chips were automatically moved at a time when the power halved ($\sim$15 h), and the subsequent power generation by the freshly fueled cell was superimposed on the decay of the old cells at this interval. In general, an electrical cell cannot be connected in parallel to a cell of different voltage because of reverse reaction at each electrode. In contrast, because enzyme-modified electrodes have intrinsic reaction selectivity and show resistance to reverse reaction, EFCs enable parallel connection between different voltage cells. For comparison, the dotted black curve in Figure 17.7e shows the power decay observed from a single BFC without the stack. In the case of the SPG system, a power level of 2 ± 1 μW was maintained for extended periods (about three times that of a single cell). Since the time interval for the switching of the fuel flow can be adjusted from a few hours to a few weeks by the Fe_3O_4 content in the cover chip and the molecular weight of the PLGA sealant, the SPG system concept is applicable to other EFCs.

17.4   FLEXIBLE SHEET MEFC

A flexible sheet-shaped MEFC that can be combined with advanced flexible electronics was prepared in 2012 by Haneda et al. by using CF as the flexible, conductive base for the enzyme electrodes [11]. This research group modified such a CF strip with (i) CNTs and FDH for the oxidation of fructose and (ii) Ketjenblack (KB) and BOx for the reduction of oxygen from ambient air. The premodifications with CNT or KB increase the specific surface area of the CF electrodes, resulting in effective enzyme immobilization and, ultimately, higher power. The FDH anode strip and the BOx cathode strip were attached to both sides of an agarose hydrogel (3 mm

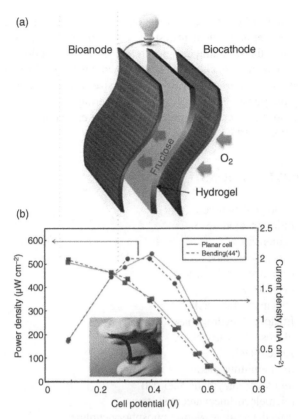

FIGURE 17.8 (a) A sheet-shaped EFC constructed by stacking enzyme-modified nano-engineered carbon fabric strips with a hydrogel film that retains electrolyte solutions and fructose fuel. (b) Performance of the sheet-shaped biofuel cell (1 cm × 0.2 cm) with and without bending. The internal agarose layer was made with 150 mM McIlvaine buffer solution (pH 5.0) containing 200 mM fructose.

thick) prepared with fructose (150 mM McIlvaine buffer solution containing 200 mM fructose; pH 5.0), as shown in Figure 17.8a. Even as a stand-alone assembly with the fructose-containing gel sheet, the maximum power density reached 550 μW cm^{-2} at 0.4 V. Importantly, this device could be bent to a 44° angle without loss of output power (Figure 17.8b). Bending in excess of this value caused fracture of the agarose hydrogel sheet; the device could, however, be made resistant to mechanical stress by using more elastic hydrogels such as those made from polyvinyl alcohol.

17.5   OUTLOOK

This chapter describes three types of MEFCs fabricated with a variety of MEMS-related techniques: insertion, microfluidic, and sheet. The insertion MEFC could harvest ambient biochemical energy *in situ* from inside fruit, such as grapes. Incorporating microfluidic elements, such as automatic air valves and automatic flow switches, into the structural design should theoretically improve the cell voltage and lifetime to practical levels. The sheet-shaped MEFC could be combined with advanced electronics in the future. Because the size of electronic devices is usually dominated by the power sources [20], the engineering advances focused on miniaturization described here can be expected to promote early practical applications and commercialization of EFCs.

LIST OF ABBREVIATIONS

BFC	biological fuel cell
BOx	bilirubin oxidase
CF	carbon fiber
CNT	carbon nanotube
CNTF	carbon nanotube forest
Dp	diaphorase
EFC	enzymatic fuel cell
FDH	fructose dehydrogenase
GDH	glucose dehydrogenase
KB	Ketjenblack
LED	light-emitting diode
MEFC	miniature enzymatic fuel cell
MEMS	microelectromechanical systems
MPC	2-methacryloyloxyethyl phosphorylcholine
NAD^+	nicotinamide adenine dinucleotide
OCV	open-circuit voltage
PDMS	polydimethylsiloxane
PLGA	poly(lactic-*co*-glycolic acid)
SPG	sequential power generation
VK_3	vitamin K_3

REFERENCES

1. Willner I, Katz E. *Bioelectronics*. Wiley-VCH Verlag GmbH, Weinheim, 2005.
2. Moehlenbrock MJ, Minteer SD. Extended lifetime biofuel cells. *Chem Soc Rev* 2008;37:1188–1196.
3. Mano N, Mao F, Heller A. Characteristics of miniature compartment-less glucose–O_2 biofuel cell and its operation in living plant. *J Am Chem Soc* 2003;125:6588–6594.
4. Gellett W, Kesmez M, Schumacher J, Akers N, Minteer SD. Biofuel cells for portable power. *Electroanalysis* 2010;22:727–731.
5. Calabrese Barton S, Gallaway J, Atanassov P. Enzymatic biofuel cell for implantable and microscale devices. *Chem Rev* 2004;104:4867–4886.

6. Miyake T, Haneda K, Nagai N, Yatagawa Y, Onami H, Yoshino S, Abe T, Nishizawa M. Enzymatic biofuel cells designed for direct power generation from biofluids in living organisms. *Energy Environ Sci* 2011;4:5008–5012.

7. Miyake T, Yoshino S, Yamada T, Hata K, Nishizawa M. Self-regulating enzyme–nanotube ensemble films and their application as flexible electrodes for biofuel cells. *J Am Chem Soc* 2011;133:5129–5134.

8. Togo M, Takamura A, Asai T, Kaji H, Nishizawa M. Structural studies of enzyme-based microfluidic biofuel cells. *J Power Sources* 2008;178:53–58.

9. Togo M, Morimoto K, Abe T, Kaji H, Nishizawa M. *Microfluidic biofuel cells: series-connection with superhydrophobic air valves*. Digest of Technical Papers, Transducers 09, 2009, pp. 2102–2105.

10. Miyake T, Oike M, Yoshino S, Yatagawa Y, Haneda K, Nishizawa M. Automatic, sequential power generation for prolonging the net lifetime of a miniature biofuel cell stack. *Lab Chip* 2010;10:2574–2578.

11. Haneda K, Yoshino S, Ofuji T, Miyake T, Nishizawa M. Sheet-shaped biofuel cell constructed from enzyme-modified nanoengineered carbon fabric. *Electrochim Acta* 2012;82:175–178.

12. Wen D, Xu X, Dong S. A single-walled carbon nanohorn-based miniature glucose/air biofuel cell for harvesting energy from soft drinks. *Energy Environ Sci* 2011;4:1358–1363.

13. Wei X, Liu J. Power sources and electrical recharging strategies for implantable medical devices. *Front Energy Power Eng China* 2008;2:1–13.

14. Kang C, Shin H, Heller A. On the stability of the "wired" bilirubin oxidase oxygen cathode in serum. *Bioelectrochemistry* 2006;68:22–26.

15. Hanashi T, Yamazaki T, Tsugawa T, Ferri S, Nakayama D, Tomiyama M, Ikebukuro K, Sode K. Biocapacitor—a novel category of biosensor. *Biosens Bioelectron* 2009;24:1837–1842.

16. Seng T, Schwaneberg U. Protein engineering in bioelectrocatalysis. *Curr Opin Biotechnol* 2003;14:590–596.

17. Todd-Holland J, Lau C, Brozik S, Atanassov P, Banta S. Engineering of glucose oxidase for direct electron transfer via site-specific gold nanoparticle conjugation. *J Am Chem Soc* 2011;133:19262–19265.

18. Rubenwolf S, Kerzenmacher S, Zengerle R, Stetten FV. Strategies to extend the lifetime of bioelectrochemical enzyme electrodes for biosensing and biofuel cell applications. *Appl Microbiol Biotechnol* 2011;89:1315–1322.

19. Yang Y, Xie Y, Kang X, Lee LJ, Kniss DA. Assembly of three-dimensional polymeric constructs containing cells/biomolecules using carbon dioxide. *J Am Chem Soc* 2006;128:14040–14041.

20. Heller A. Potentially implantable miniature batteries. *Anal Bioanal Chem* 2006;385:469–473.

18

SWITCHABLE ELECTRODES AND BIOLOGICAL FUEL CELLS

EVGENY KATZ, VERA BOCHAROVA, AND JAN HALÁMEK

Department of Chemistry and Biomolecular Science, and NanoBio Laboratory (NABLAB), Clarkson University, Potsdam, NY, USA

18.1 INTRODUCTION

Rapid development of novel electrochemical systems was achieved in the 1970s and 1980s when a new concept of chemically modified electrodes was introduced [1,2]. Application of organic chemistry methods to the functionalization of electrode surfaces, and later pioneering of novel self-assembly methods, fostered the development of numerous modified electrodes with properties unusual for bare conducting surfaces [3–6]. While attempting to harness enhanced catalytic properties of electrodes and their selective responses to different redox species, the modified electrodes rapidly became imminent components of various electroanalytical systems and fuel cells [1–8]. Novel bioelectrochemical systems [9,10], particularly used in biosensors and biological fuel cells (BFCs) [11–14], have emerged upon introduction of modified electrodes with bioelectrocatalytic properties. To continue the remarkable success, electrodes functionalized with various signal-responsive materials (including molecular, supramolecular, and polymeric species) attached to electrode surfaces as monolayers or thin films were pioneered to allow switchable/tunable properties of the functional interfaces controlled by external signals [15–18].

Over the past two decades, sustained advances in chemical modification of the electrodes have given us a large variety of electrodes, switchable by various physical and/or chemical signals between electrochemically active and inactive

Enzymatic Fuel Cells: From Fundamentals to Applications, First Edition. Edited by Heather R. Luckarift, Plamen Atanassov, and Glenn R. Johnson.
© 2014 John Wiley & Sons, Inc. Published 2014 by John Wiley & Sons, Inc.

states [15–18]. However, very few of them were used in BFCs [19,20]. Different mechanisms were involved in the transition of the electrode interfaces between the active and inactive states depending on the properties of the modified surfaces and the nature of the applied signals. The activity of the switchable electrodes was usually controlled by physical signals (optical, electrical, or magnetic) that failed to provide direct communication between the electrodes and their biochemical environment in BFCs [21–29]. Switchable electrodes controlled by biochemical rather than physical signals are needed to design a BFC adjustable to its biochemical environment according to the presence or absence of biochemical substances. A new approach became possible when a polymer-modified electrode switchable between ON/OFF states by pH values was coupled to biochemical reactions generating pH changes *in situ* [30,31]. This allowed transduction of biochemical input signals (e.g., glucose concentration) to the pH changes governing the electrochemical activity of the switchable electrode [32]. Using a new concept of Boolean logic gates and their biocomputing networks based on enzymatic reactions [33–38], the complexity of the enzyme system controlling the electrode activity was scaled up. Finally, the switchable electrodes controlled by complex multienzyme systems, being reversibly activated–inactivated by various patterns of different biochemicals [39,40], were assembled in a BFC producing electrical power depending on the biochemical environment [20,41–43].

This chapter gives an overview of different signal-responsive electrochemical interfaces, particularly emphasizing the importance of scaling up the complexity of the signal processing systems by the application of biomolecular logic systems integrated with signal-responsive interfaces and glimpses of the diverse challenges and opportunities in the near future. The chapter uncovers a new area of fundamental research activity in BFC studies—the coupling of BFCs with biocomputing systems to yield "smart" bioelectronic devices generating electrical power on demand upon logic processing of biochemical signals.

18.2   SWITCHABLE ELECTRODES FOR BIOELECTRONIC APPLICATIONS

Functionalization of electrode surfaces with various signal-responsive materials stimulates development of the systems with switchable properties on demand, remotely controlled by single or mixed external signals of different nature (electrical potential, magnetic field, light, chemical/biochemical inputs). Further increase in complexity of modified electrodes has been achieved by coupling them with unconventional biomolecular computing systems logically processing multiple biochemical signals. This approach resulted in the formation of various interfaces capable of interconnecting complex variations of biomarkers corresponding to different physiological conditions with electronic devices. As a result, the switchable electrodes were integrated with various smart biosensing and signal processing systems as well as used to assemble BFCs producing power on demand.

18.3   LIGHT-SWITCHABLE MODIFIED ELECTRODES BASED ON PHOTOISOMERIZABLE MATERIALS

Electrodes functionalized with self-assembled monolayers or thin films of photochemically isomerizable species (e.g., spiropyran [44,45], azobenzene [22,23], diarylethene, and phenoxynaphthacenequinone derivatives [46–48]) were developed to switch reversibly interfacial properties upon irradiation with different light signals. Sophisticated photoisomerizable supra-molecular complexes were self-assembled on electrode surfaces to operate as light-activated molecular "machines" switching interfacial properties and facilitating electron transport at their specific conformations [49]. Optoelectronic devices for information processing (e.g. flip-flop memory units) were engineered using switchable electrode-interfaces modified with photoisomerizable species [50].

Mechanisms responsible for switchable properties of the photoisomerizable interfaces are different and depend on photochemical reactions of the immobilized organic species. For example, immobilized diarylethene and phenoxynaphthacenequinone derivatives directly change their electrochemical properties upon photoisomerization, being redox inactive in one of the isomeric states whereas demonstrating reversible redox transformations in another state, thereby allowing light-induced switching of the modified electrodes between electrochemically mute and active states; the latter were used to mediate electrocatalytic cascades amplifying the interfacial redox transformations [46–48]. Another mechanism controlling switchable properties of electrode interfaces is based on the changing shape or charge of the photoisomerizable molecules. Photochemically activated reversible changes of azobenzene moieties associated with electrode interfaces between *cis*- and *trans*-isomeric states affected access of diffusional redox species to the conducting support or activated electron-transporting molecular "machinery" associated with the photoisomerizable entities [22,49]. Photoinduced reversible isomerization process of surface-confined spiropyran units was accompanied by protonation–deprotonation of the isomeric states, thus affecting the surface charge, and resulted in different electrochemical behavior of charged diffusional redox species [44,45]. Figure 18.1a exemplifies the photoswitchable electrochemical reaction of cytochrome c (Cyt) controlled by the charge interactions at a spiropyran photoisomerizable monolayer [21]. Note that pyridine units in the functional monolayer provided Cyt c orientation favorable for the electron transfer (ET) process, whereas differently charged spiropyran (SP) / merocyanine (MRH$^+$) photoisomerizable species controlled the distance between positively charged Cyt c and the electrode surface, enabling the photoswitchable electrochemical process followed by cyclic voltammetry (Figure 18.1b). Application of this light-controlled interface for biocatalytic reactions following the primary Cyt c electrochemical process allowed photochemical control over complex biocatalytic cascades [21].

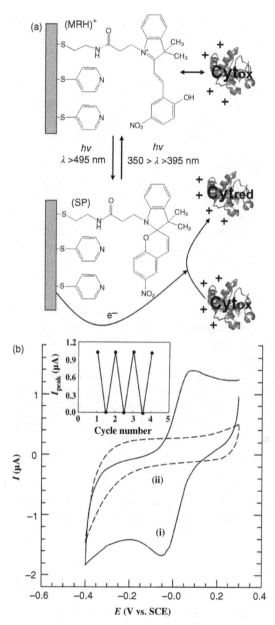

FIGURE 18.1 (a) The electrochemical process of Cyt controlled by the charge at the electrode surface functionalized with pyridine units and SP/MRH$^+$ photoisomerizable species (Cyt$_{ox}$ and Cyt$_{red}$ denote the initial oxidized state and electrochemically generated reduced state of Cyt, respectively). (b) CVs obtained for Cyt at different isomeric states of the electrode interface: (i) SP produced by visible light; (ii) MRH$^+$ generated by ultraviolet irradiation. *Inset*: Reversible switching between the activated and inhibited electrochemical processes of Cyt modulated by light signals. (Adapted with permission from Ref. [21]. Copyright 1995, American Chemical Society.)

18.4 MAGNETOSWITCHABLE ELECTROCHEMICAL REACTIONS CONTROLLED BY MAGNETIC SPECIES ASSOCIATED WITH ELECTRODE INTERFACES

Magnetic beads and nanorods chemically functionalized with organic shells were associated with electrode interfaces to control their properties and affect electrochemical reactions due to rearrangement of the magnetic species upon application of an external magnetic field [26–29,51–55]. Magnetic microparticles functionalized with chemically attached redox species were reversibly moved by an external magnetic field from solution to the electrode interface and back [26,27,51,52]. While being suspended in the solution, the redox-functionalized magnetic particles were unable to interact directly with the electrode surface, thus demonstrating no electrochemical response. On magneto-induced translocation to the electrode surface, they were integrated with the conducting support and demonstrated their redox activity. In addition to the magneto-induced reversible activation-inactivation of the redox reactions for the species chemically attached to the particles, magnetic particles were proven to be valuable for triggering biocatalytic cascades mediated by the redox species [51]. Adaptive multifunctional magnetic nanowires were designed to control the operation of electrochemical sensors and microfluidic devices, providing external control over electrocatalytic and bioelectrocatalytic processes upon application of a magnetic field [28,29,54,55]. This approach is ready for applications in BFCs, particularly based on microfluidic systems.

Strong inhibition of electrochemical reactions was demonstrated upon magneto-induced association of hydrophobic magnetic nanoparticles with electrode interfaces [52,56,57]. Magnetic nanoparticles functionalized with hydrophobic organic shells were originally dispersed in a nonaqueous liquid phase immiscible with water (e.g., toluene), which was the second liquid phase above an aqueous solution. The nano-particles were then magnetically attracted to the electrode surface located in the aqueous electrolyte solution, thus generating a hydrophobic thin film on the conducting interface isolating the electrode surface from the aqueous solution [52]. This resulted in many interesting nontrivial changes in the electrode properties. The generated hydrophobic film obviously inhibited access of water-soluble redox species to the electrode surface, thereby preventing their electrochemical reactions and causing a huge increase in the interfacial ET resistance observed in faradaic imped-ance spectra [52]. An interesting effect was observed for electrodes chemically modified with redox species—the electrochemical response for them was not inhibited upon magneto-induced deposition of the hydrophobic particles atop the modified electrode interface [56,57]. This allowed separation of the diffusional and surface-confined electrochemical reactions when the former were observed only in the absence of the hydrophobic magnetic nanoparticles on the electrode surface, whereas the latter were observed in the absence and presence of the nanoparticle film on the electrode interface. However, the mechanism of redox reactions of the surface-confined species was different in the presence and absence of the hydrophobic thin film generated on the surface [56,57]. When the hydrophobic magnetic nanoparticles were suspended in the nonaqueous solution and the redox-functionalized electrode surface was facing an aqueous electrolyte solution, the surface-confined redox species demonstrated electrochemical behavior typical for the aqueous environment. On the contrary, when the hydrophobic film composed of the attracted magnetic nanoparticles was formed on the electrode surface, the surface-confined species became immersed in the nonaqueous microenvironment demonstrating a redox process typical for the conditions when ET cannot be accompanied with proton transfer. In addition to that, the hydrophobic magnetic nanoparticles were applied to transport water-insoluble redox species to the electrode surface and use them for electrocatalytic reactions that are not possible in aqueous solutions [56,57]. It should be noted that the changes induced by the hydrophobic magnetic nanoparticles were reversible and the electrode surface was exposed to the aqueous solution upon magneto-induced lifting of the nanoparticles from the surface. The separation of diffusional and surface-confined electrochemical reactions as well as the alteration of electrochemical mechanisms induced by the external magnetic field applied on the electrode interface allowed switching between different bioelectrocatalytic and photoelectrocatalytic reactions [56–58]. The magnetoswitchable electrode interfaces operating in the presence and absence of the hydrophobic magnetic nanoparticles were applied in biosensing systems switchable between different analytes [59], in electronic systems switchable between single-electron and multielectron transfer properties [60], and in electrochemical information processing systems (e.g., for the realization of "write–read–erase" memory devices) [61,62].

Application of magnetoswitchable electrode interfaces in fuel/BFCs is still awaiting research efforts—only fragmentary studies were performed in this fascinating direction. For example [62], Pt/Ru-functionalized magnetic spheres were used for a magnetic-field stimulated methanol oxidation and oxygen reduction processes. The electrocatalytic alloy magnetic particles were prepared by a galvanostatic codeposition of platinum and ruthenium onto nickel spheres. The electrocatalytic oxidation of methanol and reduction of oxygen was triggered by switching the position of an external magnet below the surface of the carbon electrode to confine the Pt/Ru-coated particles. The magnetic stimulation of the redox processes of methanol and oxygen allowed the reversible activation and deactivation of the operation of direct methanol fuel cells. Such switching of fuel cells would enable on-demand power generation, for meeting the specific needs of power consuming units. A similar approach is even more promising in BFCs activated by an external magnetic field.

Large-scale aligned arrays of conductive nanowires formed on the electrode surface in the presence of an external magnetic field are beneficial for improvement of carrier collection and overall efficiency of diffusional electrochemical and bioelectrocatalytic reactions. Au-shell/$CoFe_2O_4$-magnetic-core nanoparticles were self-assembled in the presence of a magnetic field to yield a "forest" of standing nanowires increasing the electrode surface area, Figure 18.2a, and thus amplifying electrochemical responses of a diffusional redox probe by ca. 6-fold followed by cyclic voltammetry, Figure 18.2b [63]. The process was reversed when the magnetic field was switched off resulting in disaggregation of the conducting nanowires. The primary electrochemical reaction of the electron relay was coupled with the bioelectrocatalytic oxidation of glucose in the presence of soluble glucose oxidase

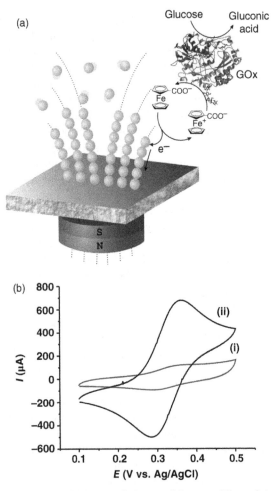

FIGURE 18.2 (a) Magnetic field-controlled reversible assembling of Au-coated magnetic nanoparticles and their use as a nanostructured electrode with the enhanced ability of electro-chemical oxidation of ferrocene monocarboxylic acid coupled with glucose oxidation bio-catalyzed by GOx. (b) CVs of ferrocene monocarboxylic acid (0.1 mM) obtained in the absence (i) and presence (ii) of the magnetic field. (Adapted with permission from Ref. [63]. Copyright 2008, American Chemical Society.)

(GOx) resulting in the amplification of the biocatalytic cascade controlled by the nanostructured assembly on the electrode surface (Figure 18.2a). The studied nano-electrode array was suggested as a general platform for electrochemical biosensors and BFCs with the enhanced current outputs controlled by the structure of the self-assembled nanowires. The most important advantage of the present system was an easy control over the generated nanoelectrodes assembly on the electrode surface by varying the time intervals for the nanowires growth and by switching the magnetic field ON and OFF, thus resulting in the nanowires self-assembling and disassembling, respectively.

The recently emerged magnetoswitchable electrochemical and bioelectrochemical systems have inspired the design and construction of very sophisticated adaptive bioelectronic devices controlled by external magnetic signals [54]. These devices will find numerous applications in different areas of bioelectronics and particularly in switchable/tunable BFCs.

18.5   MODIFIED ELECTRODES SWITCHABLE BY APPLIED POTENTIALS RESULTING IN ELECTROCHEMICAL TRANSFORMATIONS AT FUNCTIONAL INTERFACES

Chemical transformations at electrode interfaces induced by applied potentials can substantially change electrochemical properties of electrodes [64]. Even bare metal surfaces can significantly change properties when oxide thin films are generated at positive potentials [65]. Moreover, modification of electrode surfaces with redox polymers, which can be electrochemically converted to different oxidation states, significantly amplifies the potential-induced surface changes [66]. Electron-transfer, optical and wettability properties of redox-polymer functionalized electrodes were reversibly changed by potential application [67,68]. Incorporation of metallic species into polymer matrices or their self-assembling in nanostructured ensembles on electrode interfaces can amplify the potential-induced surface changes even more [24,25,69–71]. For example, Cu^{2+} cations associated with a polymer on an electrode surface can be reversibly transformed between an ionic state when an oxidative potential is applied and metallic clusters upon application of reductive potentials [24,25]. The metal clusters, when they are electrochemically produced in a polymer thin film, convert it into a conducting medium, whereas cationic species do not provide electronic conductivity in the polymer. This resulted in dramatic changes in the ET resistance at the modified electrode surface modulated by reversible potential applications and was used to control bioelectrocatalytic reactions at the interface, particularly allowing switchable behavior of a BFC [19,24,25].

Functionalization of electrode surfaces with signal-responsive materials aided to establish an entirely new electrode switching behavior [17,72,73]. The most frequently used approach is based on surface-confined pH-responsive polyelectrolytes reversibly switchable between a charged hydrophilic form permeable for redox species of the opposite charge and a neutral hydrophobic state that is not permeable for ionic species [17,74,75]. These polyelectrolytes can be switched between permeable and nonpermeable forms upon local interfacial pH changes generated by electrochemical reactions, thus being switchable by potentials applied on the modified electrode surface [76,77]. For example, a poly(4-vinyl pyridine) (P4VP)-brush-modified indium-tin oxide (ITO) electrode was used to reversibly switch the interfacial activity upon electrochemical signals [76]. Application of an external potential electrochemically reducing O_2 resulted in the concomitant consumption of hydrogen ions at the electrode interface, thus yielding a higher pH value and triggering deprotonation and restructuring of the P4VP brush on the electrode surface (Figure 18.3a). The initial swollen state of the protonated P4VP brush (pH 4.4)

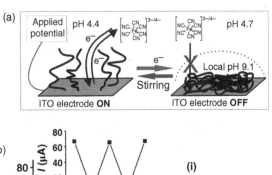

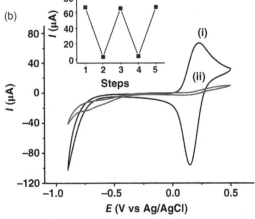

FIGURE 18.3 (a) pH-controlled reversible switching of the P4VP brush between ON (left) and OFF (right) states allowing and restricting the anionic species penetration to the electrode surface, thus activating and inhibiting their redox process. (b) CVs obtained on the P4VP brush-modified ITO electrode in the presence of 0.5 mM $K_4[Fe(CN)_6]$: (i) prior to the application of the potential on the electrode; (ii) after application of -0.85 V to the electrode for 20 min. The background electrolyte was composed of 1 mM lactic buffer (pH 4.4) and 100 mM sodium sulfate saturated with air. *Inset*: The reversible switching of the peak current value upon "closing" the interface by the electrochemical signal and restoring the electrode activity by the solution stirring. (Adapted with permission from Ref. [76]. Copyright 2010, American Chemical Society.)

was permeable for anionic $[Fe(CN)_6]^{4-}$ redox species (Figure 18.3b, curve (i)), whereas the electrochemically produced local pH of 9.1 resulted in the deprotonation of the polymer brush. The produced hydrophobic shrunken state of the polymer brush was impermeable for the anionic redox species, thus fully inhibiting its redox process at the electrode surface (Figure 18.3b, curve (ii)). It should be noted that the pH change was generated locally at the modified interface, whereas the bulk pH value had very little change. The return of the interface to the electrochemically active state was achieved by disconnecting the applied potential followed by stirring the electrolyte solution or by slow diffusional exchange of the electrode-adjacent thin layer with the bulk solution. The developed approach allowed the electrochemically triggered reversible inhibition ("closing") of the electrode interface (Figure 18.3b, inset). The opposite "opening" process was also electrochemically triggered when a mixed polymeric brush composed of poly(2-vinyl pyridine) (P2VP) and polyacrylic acid (PAA) was associated with an electrode surface [77]. Similarly to the previous example, the local interfacial pH value was increased upon electrochemical reduction of O_2, resulting in the transition of the polyelectrolyte thin film from a neutral state to a negatively charged form due to dissociation of the PAA component. This resulted in the switch of the modified electrode surface from a hydrophobic inactive state to a hydrophilic negatively charged state permeable for cationic redox species (e.g., $[Ru(NH_3)_6]^{3+}$). Application of this approach to different interfacial systems will allow a vast range of switchable electrodes with externally controlled activity useful for application in biosensors and BFCs.

18.6   CHEMICALLY/BIOCHEMICALLY SWITCHABLE ELECTRODES

Many different systems were designed to prepare modified electrodes with surfaces sensitive to various chemical and biochemical signals. Particularly, electrode surfaces functionalized with receptor units were used to bind selectively chemical inputs resulting in changes of interfacial properties (electrical, optical, piezoelectric, etc.) [78–80]. Recent advances in the development of novel signal-responsive materials [81,82] allowed their immobilization on electrode surfaces to activate and inactivate electrochemical processes reversibly by chemical or biochemical signals. In one of the examples [83], a nanostructured porous membrane prepared from a polyelectrolyte, quarternized poly(2-vinylpyridine), qP2VP, cross-linked with 1,4-diiodobutane, was deposited on an ITO electrode surface and the membrane pores were reversibly closed and opened upon reacting the membrane with cholesterol and washing it out, respectively, Figure 18.4a and b. Cholesterol binding to the polymer chains through the formation of hydrogen bonds resulted in the swelling of the membrane and closing of the pores (Figure 18.4b), whereas washing of cholesterol from the membrane restored its initial state with the open pores (Figure 18.4a). For example, swelling of the membrane caused by the uptake of cholesterol from 0.13 M solution resulted in the double increase of the dry membrane thickness (from 192 to 369 nm), whereas the pore diameter decreased by half (from 588 to 290 nm), resulting in the triple decrease of the membrane porosity from 51.7 to 17.3% [83]. The reversible opening and closing of the membrane pores was visualized by atomic force microscopy (AFM) and followed by measuring the electrode ET resistance for a diffusional redox probe using faradaic impedance spectroscopy (Figure 18.4c). The cyclic addition and removal of cholesterol to and from the membrane resulted in the modulated interfacial resistance controlled by chemical signals (Figure 18.4c, inset).

pH-sensitive polymer brushes tethered to electrode surfaces demonstrated pH-switchable interfacial behavior when a neutral state of the polyelectrolyte was impermeable for ionic redox species keeping the electrode mute [81,82], whereas the ionized state of the polymer brush was permeable for ionic redox species of the opposite charge allowing their access to the conducting support and activating the electrochemical process [17,84,85]. Protonation–deprotonation of the polymer brush was controlled by the pH value, which could be changed in the bulk solution or varied locally at the interface by means of electrochemistry (see Section 18.5) [17,76,77,84,85]. Further sophistication of the pH-switchable interfaces was achieved by electrode modification with mixed-polymer systems [17,84,85]. For example, P2VP and PAA, tethered to an ITO electrode

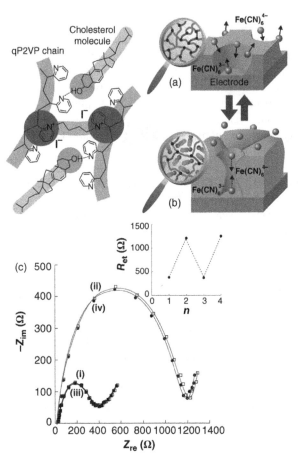

FIGURE 18.4 Structure of the cross-linked polyelectrolyte gel membrane and the reversible change of the chemical stimuli-responsive membrane in the absence (a) and presence (b) of cholesterol. (c) Faradaic impedance spectra obtained for the membrane-modified electrode in the absence (i, iii) and presence (ii, iv) of cholesterol corresponding to the open and closed pores, respectively. *Inset* (c): The reversible changes of the electron transfer resistance, R_{et}, derived from the impedance spectra upon addition and removal of cholesterol. (Adapted with permission from Ref. [83]. Copyright 2007, American Chemical Society.)

as a mixed brush, demonstrated three differently charged states controlled by an external pH value: positively charged due to protonation of the P2VP component (pH 3), neutral when the charges of P2VP and PAA are compensated (pH 4.5), and negatively charged when the PAA component is dissociated (pH 6) (Figure 18.5a) [84,85]. The pH-controlled switching between the different charges allowed discrimination between electrochemical reactions of oppositely charged redox species: $[Fe(CN)_6]^{4-}$ and $[Ru(NH_3)_6]^{3+}$. The negatively charged species were allowed to access the conducting support and demonstrate their electrochemical activity only when the mixed-polymer brush was positively charged. On the contrary, the positively charged species accessed

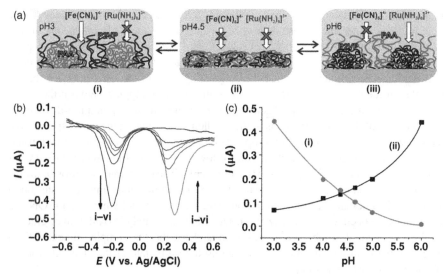

FIGURE 18.5 (a) The polymer brush permeability for differently charged redox probes controlled by a solution pH value: (i) the positively charged protonated P2VP domains allow the electrode access for the negatively charged redox species; (ii) the neutral hydrophobic polymer thin film inhibits the electrode access for all ionic species; (iii) the negatively charged dissociated PAA domains allow the electrode access for the positively charged redox species. (b) The differential pulse voltammograms obtained for the mixed-polymer brush in the presence of $[Fe(CN)_6]^{4-}$, 0.5 mM, and $[Ru(NH_3)_6]^{3+}$, 0.1 mM, at the variable pH of the solution: (i) 3.0, (ii) 4.0, (iii) 4.35, (iv) 4.65, (v) 5.0, and (vi) 6.0. The background solution was composed of 0.1 M phosphate buffer titrated to the specified pH values. (c) The peak current dependences on the pH value for the (i) anionic, $[Fe(CN)_6]^{4-}$, and (ii) cationic, $[Ru(NH_3)_6]^{3+}$, species, as derived from the differential pulse voltammograms measured at the variable pH values. (Adapted with permission from Ref. [85]. Copyright 2010, Wiley-VCH Verlag GmbH.) (Please see the color version of this figure in Color Plates section.)

the electrode being electrochemically active only when the modified interface was negatively charged, and the neutral state of the modified interface was not permeable for all ionic species keeping the electrode mute for all electrochemical reactions. Gradual pH changes in the supporting electrolyte solution demonstrated a reversible transition from the electrochemical reaction of $[Fe(CN)_6]^{4-}$ to the redox process of $[Ru(NH_3)_6]^{3+}$ and back (Figure 18.5b and c) [84,85]. This switchable/tunable behavior of the modified interface was considered as a prototype of future on-demand drug releasing systems and as a 2-to-1 multiplexer for unconventional chemical information processing systems [84,85]. It will also be a powerful tool for regulating bioelectrocatalytic reactions at electrodes in BFCs.

Even more interesting electrochemical properties were discovered for the polymer brush tethered to an electrode surface and functionalized with bound redox species [30]. $Os(dmo-bpy)_2$ (dmo-bpy = 4,4'-dimethoxy-2,2'-bipyridine) redox groups were covalently bound to the P4VP brush chains grafted on an ITO electrode. The polymer-bound redox species were found to be electrochemically active at pH <5 when the polymer is protonated, swollen, and the chains are flexible. On changing pH to the values higher than 6, the polymer chains lose their charges and the produced shrunken state of the polymer does not show electrochemical activity. This was explained by the poor mobility of the polymer chains in the shrunken state restricting the direct contact between the Os complex units and the conducting support. A low density of the Os complex in the polymer film does not allow the electron hopping between the redox species, and their electrochemical activity can be achieved only upon quasi-diffusional translocation of the polymer chains in the swollen state bringing the redox species to a short distance from the conducting support (Figure 18.6a). The reversible activation and deactivation of the modified electrode was followed by cyclic voltammetry measured at pH 4 (ON state) and pH 6 (OFF state) (Figure 18.6b). The reversible transition of the Os complex-functionalized polymer brush between the swollen and shrunken states upon varying pH values allowed the modulation of the electrode activity between the ON and OFF states, respectively (Figure 18.6b, inset). Bioelectrocatalytic oxidation of glucose in the presence of soluble GOx was mediated by the Os complex-functionalized electrode being in the ON state, while muting the OFF state of the electrode for the bioelectrochemical reaction [30].

In 2008, the Katz research group used the pH-controlled reversible transition of the Os complex-functionalized electrode between the electrochemically active and

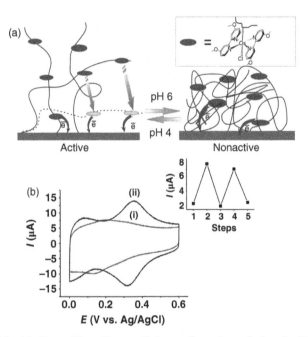

FIGURE 18.6 (a) Reversible pH-controlled transformation of the Os–P4VP polymer brush on the electrode surface between electrochemically active and inactive states. (b) CVs of the Os–P4VP-modified electrode obtained upon measurements performed in neutral and acidic solutions: (i) pH 7.0; (ii) pH 3.0. *Inset*: Reversible switching of the Os–P4VP-modified electrode activity. (Adapted with permission from Ref. [30]. Copyright 2008, American Chemical Society.)

inactive states to activate/deactivate the bioelectrocatalytic glucose oxidation upon performing enzymatic reactions changing pH values [30,32]. Two soluble hydrolytic enzymes, esterase (Est) and urease, were used to change the pH value *in situ*, thus affecting the bioelectrocatalytic activity of the modified electrode for glucose oxidation in the presence of soluble GOx. Addition of ethyl butyrate to the solution resulted in the formation of butyric acid biocatalyzed by Est. The generated butyric acid yielded pH ~3.8 when the Os complex-functionalized electrode was electrochemically active, thus enabling the bioelectrocatalytic oxidation of glucose (Figure 18.7a). Then the addition of urea to the solution resulted in the formation of

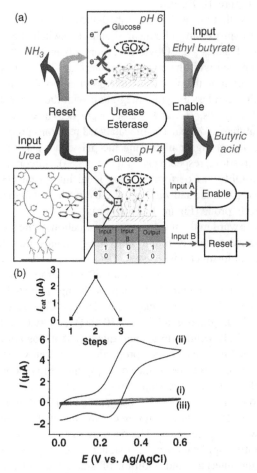

FIGURE 18.7 (a) Switchable bioelectrocatalytic oxidation of glucose controlled by external enzymatic reactions. (b) CVs obtained for the switchable bioelectrocatalytic glucose oxidation when the system is (i) in the initial OFF state, pH ~6.5; (ii) enabled by the ethyl butyrate input signal, pH ~3.8; and (iii) inhibited by the urea reset signal, pH ~7.5. *Inset*: Switchable bioelectrocatalytic current ON and OFF by the enzyme-processed biochemical signals. (Adapted with permission from Ref. [32]. Copyright 2008, American Chemical Society.)

ammonia biocatalyzed by urease restoring the initial pH ~6.5 and disabling the glucose oxidation at the modified electrode. The switching between electrode ON/OFF states upon the addition of two chemical input signals could be presented as a system mimicking electronic enable–reset circuitry operating in the digital form, where the absence/presence of the input signals is digitally presented as 0/1 inputs and the resulting electrode OFF/ON states are considered as 0/1 states [32]. The bioelectrocatalytically active and inactive states of the modified electrode switched *in situ* by the enzymatic reactions were followed by cyclic voltammetry (Figure 18.7b). The reversible transition between the active and inactive states demonstrated the possibility to modulate the bioelectrocatalytic activity by biochemical signals (Figure 18.7b, inset).

It should be noted that the system described above has the disadvantage of using soluble enzymes for processing chemical signals to switch the electrode interface between ON and OFF states [32]. Obviously, more advanced systems should include these enzymes integrated with the switchable interface. This was achieved upon the enzymes immobilization on the pH-switchable polymer-brush using Au nanoparticles as a platform for the seamless enzyme integration with the switchable interface [86]. Two immobilized enzymes, Est and urease, were activated by the corresponding substrates, ethyl butyrate and urea. The biocatalyzed reactions produced *in situ* butyric acid or ammonia, decreasing or increasing the pH value, resulting in the electrode activation or inactivation, respectively (Figure 18.8a). Importantly, the electrode switching ON and OFF was performed upon local interfacial pH changes without alteration of the bulk pH value and was followed by cyclic voltammetry in the presence of a diffusional redox probe (Figure 18.8b and c).

One more step forward in the increasing sophistication of the switchable electrode systems was attained when different activating signals described above were integrated in one multi-functional system [87]. The novel system included pH-switchable polymer brush associated with an ITO electrode surface and magnetic nanoparticles functionalized with GOx enzyme being activated with the glucose substrate. When the GOx-modified magnetic nanoparticles were suspended in the solution $(0.3\,mg\,ml^{-1})$, glucose addition $(10\,mM)$ resulted in the formation of gluconic acid in the bulk volume in a small quantity that was not enough to change the pH value for switching the electrode properties. However, when the nanoparticles were concentrated on the electrode surface by applying the magnetic field of 2.5 T, the local interfacial pH decrease generated upon biocatalytic oxidation of glucose and formation of gluconic acid was already enough for the P4VP brush protonation and for opening of the electrode interface for an electrochemical reaction of a diffusional redox probe, shown in Figure 18.9a. The opening of the electrode surface occurred only upon cooperative application of two signals: the magnetic field to collect the nanoparticles on the interface and addition of glucose to activate the biocatalytic reaction yielding the pH change. The interfacial changes were followed by faradaic impedance spectroscopy that demonstrated the ET resistance, R_{et}, decreased corresponding to the open state of the interface only after both activating signals (magnetic field and glucose) were applied (Figure 18.9b). Thus, the system was mimicking Boolean AND logic operation with two signals of different nature: magnetic and

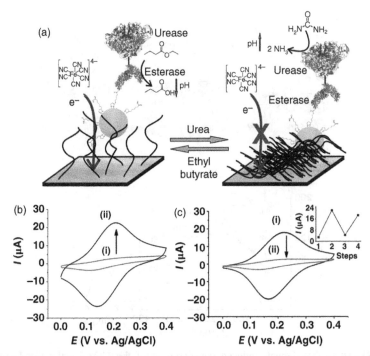

FIGURE 18.8 (a) Switching the electrode ON and OFF upon biocatalyzed reactions of the immobilized enzymes. (b, c) CVs obtained on the enzyme–P4VP-modified ITO electrode in the presence of 0.2 mM $K_4[Fe(CN)_6]$ upon application of chemical signals. (b) 0.1 M Na_2SO_4 solution with the bulk pH 6: (i) before and (ii) 30 min after the application of ethyl butyrate, 10 mM. (c) 0.1 M Na_2SO_4 solution with the bulk pH 4.5: (i) before and (ii) 30 min after the application of urea, 2 mM. Arrows show the direction of the biocatalytically induced changes. *Inset* (c): The reversible changes of the peak current upon cyclic ON–OFF transformations induced by the biochemical signals. (Adapted with permission from Ref. [86]. Copyright 2010, Royal Society of Chemistry.)

chemical. The logically processed signals governing switchable properties of modified electrodes are discussed in detail in the next section.

18.7 COUPLING OF SWITCHABLE ELECTRODES WITH BIOMOLECULAR COMPUTING SYSTEMS

Most of the above-discussed signal-responsive switchable electrodes have very limited abilities for scaling up the complexity of the signal processing systems. All electrodes responding to the physical signals, such as light, magnetic field, or electrical potential, cannot be easily assembled in networks with concatenated units processing multiple signals. Some of the reported electrochemical systems were able to respond to more than one input signal, performing Boolean logic operations and memory (read–write–erase or flip-flop) functions [50,61,87]. However, even these

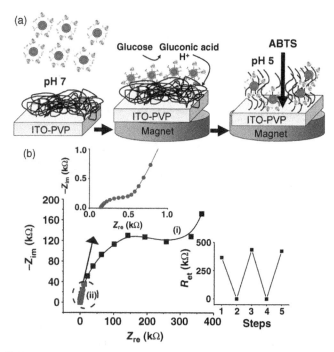

FIGURE 18.9 (a) Magneto-induced concentration of the GOx-functionalized Au shell/CoFe$_2$O$_4$ magnetic core nanoparticles on the electrode surface modified with P4VP brush to open the interface for a diffusional redox probe upon local pH changes produced in course of a biocatalytic reaction. (b) Impedance spectra (Nyquist plots) obtained on the P4VP-modified electrode with the GOx nanoparticles magneto-confined at the electrode: (i) in the absence of glucose; (ii) in the presence of 10 mM glucose (also shown at a smaller scale). *Inset*: Reversible switching of the R_{et} by adding and removing glucose. The impedance spectra were recorded in the presence of ABTS, 0.1 mM, used as a soluble redox probe. (Adapted with permission from Ref. [87]. Copyright 2009, American Chemical Society.)

electrodes can hardly be further scaled up to more sophisticated information processing systems.

Advances achieved in unconventional molecular and biomolecular computing allowed the formulation of complex chemical systems processing information and mimicking Boolean logic gates and their networks responding to many chemical input signals [36,88–92]. Recently pioneered enzyme-based logic systems were shown to be scalable to complex networks composed of several concatenated logic gates processing multiple chemical input signals in the programmed way [33,37,38,93–100]. Different logic gates (AND, OR, XOR, INHIB, NOR, etc.) [33,94,96–98], and their various combinations performing arithmetic operations [95], and complex logic algorithms (e.g., IMPLICATION, when not only the values but also the correct order of the input signals affect the final output result) [37,38], were designed using multienzyme systems. These systems have been connected to signal responding materials associated with electrode surfaces, thus allowing electrode responses to many chemical signals in a preprogrammed way [31,101]. Particularly, the pH-switchable polymer membranes and brushes described above reached the next sophistication level upon their functional integration with enzyme-based logic gates and networks.

Because many enzymatic reactions consume or yield hydrogen ions and many polymer-based signal-responsive systems are sensitive to pH changes [81,82], enzyme systems performing AND/OR Boolean logic operations and producing pH changes upon biocatalytic reactions have been designed and coupled with nanostructured signal-responsive materials associated with electrode interfaces [102]. For example, an AND logic gate was composed of an aqueous solution containing dissolved sucrose, O_2 and urea, while the enzymes, glucose oxidase (GOx) and invertase (Inv), operated as input signals, Figure 18.10a [102]. The absence of each enzyme in the system was considered as the input signal 0, whereas the presence of the enzyme (in a specific optimized concentration) was considered as the input signal 1. The whole reaction chain included conversion of sucrose to glucose catalyzed by INV, followed by the oxidation of glucose catalyzed by GOx and resulting in the formation of gluconic acid, thus yielding acidic pH values. The reaction chain proceeds only in the presence of both enzymes (input signals 1,1), whereas the absence of either or both of them (input signals 0,0; 0,1; and 1,0) inhibits the formation of the acidic medium (Figure 18.10b). The output signal produced by the biochemical system was considered as 0 when the pH changes were small (ΔpH < 0.2) and as 1 when ΔpH > 1 (Figure 18.10b, inset). The system demonstrated AND logic behavior with the characteristic truth table (Figure 18.10c). After the reaction was completed, another enzyme input of urease was used to catalyze the hydrolysis of urea and to reset the pH value to the original neutral value due to formation of ammonia. The whole AND–Reset cycle mimics the performance of the respective electronic circuitry (Figure 18.10d). Similarly, the OR logic gate was composed of ethyl butyrate, glucose, O_2, and urea dissolved in an aqueous solution, whereas two enzymes (GOx and Est) were used as input signals (Figure 18.10e). Both enzymes activated biocatalytic reactions independently; GOx catalytically oxidized glucose and Est catalytically hydrolyzed ethyl butyrate, both resulting in acidification of the solution (Figure 18.10f). Thus, the system preserved the initial neutral pH (ΔpH < 0.2; the output signal 0) only in the absence of both enzymes (input signals 0,0), whereas the reactions (either or both together) yielded the acidic media (ΔpH > 1; the output signal 1) upon input signals 0,1; 1,0; and 1,1 (Figure 18.10f, inset), demonstrating behavior typical for the OR gate with the respective truth table (Figure 18.10g). The logic operation resulting in the acidification of the solution was followed by the addition of the reset enzyme urease, returning the system to the original pH value. The whole reaction set could be expressed in terms of the equivalent electronic system: OR–Reset (Figure 18.10h).

The pH-changing AND/OR logic systems with the Reset function were used to control porosity of a polymer membrane deposited on an ITO electrode surface [102]. The pores were open (Figure 18.11a, image (ii)), allowing access of a diffusional redox probe to the conducting support and resulting in a small ET resistance, R_{et}, measured by impedance spectroscopy only when the pH was decreased upon

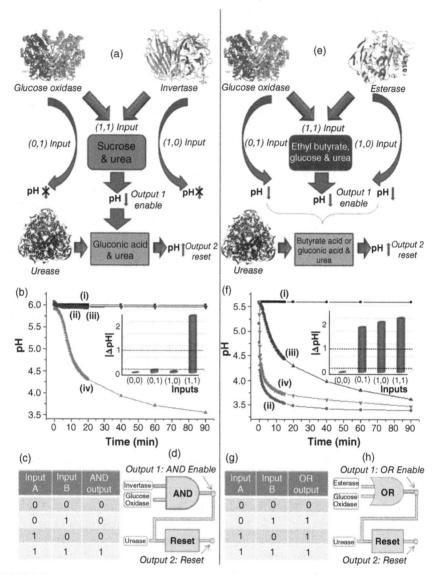

FIGURE 18.10 The biochemical logic gates with the enzymes used as input signals to activate the gate operation followed by the Reset function. (a) The AND gate based on GOx- and INV-catalyzed reactions. (b) pH changes generated *in situ* by the AND gate upon different combinations of the input signals: (i) 0,0; (ii) 0,1; (iii) 1,0; and (iv) 1,1. *Inset*: Bar diagram showing the pH changes as the output signals of the AND gate. (c) The truth table of the AND gate showing the output signals in the form of pH changes generated upon different combinations of the input signals. (d) Equivalent electronic circuit for the biochemical AND–Reset logic operations. (e) The OR gate based on GOx- and Est-catalyzed reactions. The sketches (f–h) of the OR gate are analogous to those (b–d) of the AND gate. (Adapted with permission from Ref. [102]. Copyright 2009, American Chemical Society.)

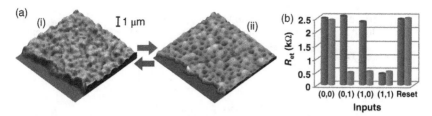

FIGURE 18.11 The signal-responsive membrane associated with an ITO electrode and coupled with the enzyme-based logic gates. (a) AFM topography images ($10 \times 10 \, \mu m^2$) of the membrane with the closed (i) and open (ii) pores. (b) The electron transfer resistance, R_{et}, of the switchable interface derived from the impedance spectroscopy measurements obtained upon different combinations of the input signals. The blue and red bars correspond to the OR–Reset and AND–Reset systems, respectively. (Adapted with permission from Ref. [102]. Copyright 2009, American Chemical Society.) (Please see the color version of this figure in Color Plates section.)

successful completion of the biocatalytic reactions (Figure 18.11b). Activation of the biocatalytic Reset function resulted in a pH increase and closing of the membrane (Figure 18.11a, image (i)), thus restoring the initial high R_{et} value (Figure 18.11b). Similarly, the enzyme-based AND/OR logic gates were used to open an electrode interface modified with a pH-sensitive P4VP-brush by applying various combinations of chemical input signals [40].

Further increase of the signal-processing system complexity was achieved upon concatenation of enzyme logic gates [39]. The logic network composed of three enzymes (alcohol dehydrogenase (ADH), glucose dehydrogenase (GDH), and GOx; Figure 18.12a), operating in concert as four concatenated logic gates (AND/OR; Figure 18.12b), was designed to process four different chemical input signals (reduced form of nicotinamide adenine dinucleotide (NADH), acetaldehyde, glucose, and oxygen). The cascade of biochemical reactions culminated in pH changes controlled by the pattern of the applied chemical input signals. The pH changes produced *in situ* were coupled with a pH-sensitive P4VP brush-functionalized electrode, resulting in the interface switching from the OFF state, when the electrochemical reactions are inhibited, to the ON state, when the interface is electrochemically active. Soluble $[Fe(CN)_6]^{3-/4-}$ was used as an external redox probe to analyze the state of the interface and to follow the changes produced *in situ* by the enzyme logic network, depending on the pattern of the applied biochemical signals. The chemical signals processed by the enzyme logic system and transduced by the sensing interface were read out by electrochemical means (Figure 18.12c). The whole set of the input signal combinations included 16 variants. Those that corresponded to the logic output 1, according to the Boolean logic encoded in the logic circuitry, resulted in the acidic medium and thus in the electrode activation (Figure 18.12d).

The logic gates and networks controlling states of switchable electrode interfaces were initially developed to demonstrate the concept of the coupling between biomolecular computing systems and modified electrodes. However, their practical

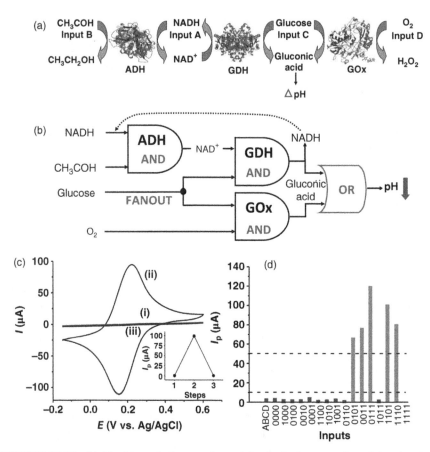

FIGURE 18.12 (a) The biocatalytic cascade used for the logic processing of the chemical input signals and producing *in situ* pH changes as the output signal. (b) The equivalent logic circuitry for the biocatalytic cascade. (c) CVs obtained for the ITO electrode modified with the P4VP brush in (i) the initial OFF state, pH ~6.7, (ii) ON state enabled by the logic system at pH ~4.3, and (iii) *in situ* reset to the OFF state. *Inset*: Reversible switching of the electrode activity. (d) Anodic peak currents, I_p, for the 16 possible input combinations. The dotted lines show threshold values separating logic 1, undefined, and logic 0 output signals. (Adapted with permission from Ref. [39]. Copyright 2009, American Chemical Society.)

use is highly feasible in the area of smart multisignal processing biosensors, actuators, and BFCs, particularly for biomedical applications [103,104]. Recently designed logic gates by the Katz research group for the analysis of pathophysiological conditions corresponding to different injuries are activated by various injury biomarkers [105–107]. The logic analysis of many biomarker signals significantly increases fidelity of the biomedical conclusion, particularly when biomarkers with limited specificity are used in the analysis. It should be noted that the biggest challenge in realization of these logic systems for biomedical applications is a relatively small difference between logic 0 and logic 1 levels of the digitized input

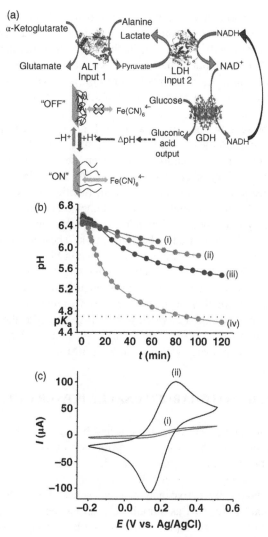

FIGURE 18.13 (a) The biocatalytic cascade used for the logic processing of the biomarkers characteristic of liver injury, resulting in *in situ* pH changes and activation of the electrode interface. (b) pH changes generated *in situ* by the biocatalytic cascade activated with various combinations of the two biomarker input signals (ALT, LDH): (i) 0,0; (ii) 0,1; (iii) 1,0; (iv) 1,1. The dotted line corresponds to the pK_a value of the P4VP brush. (c) CVs obtained for the ITO electrode modified with the P4VP polymer brush in (i) the initial OFF state, pH 6.3, and (ii) the ON state enabled by the (ALT, LDH) input combination 1,1, pH 4.75. (Adapted with permission from Ref. [108]. Copyright 2011, American Chemical Society.)

signals. Whereas convenient arbitrary concentrations of chemical inputs were used for the biocomputing concept demonstration (usually logic 0 was represented by the absence of the corresponding species), in the biomedical applications the logic levels of the biochemical input signals are defined based on their medical meaning: logic 0 and logic 1 input levels correspond to normal physiological and elevated pathological concentrations of biomarkers, respectively, which may appear with a small difference.

A system representing the first example of an integrated sensing-actuating chemical device with Boolean logic for processing natural biomarkers at their physiologically relevant concentrations has been developed [108]. Biomarkers characteristic of liver injury, alanine transaminase (ALT) and lactate dehydrogenase (LDH), were processed by a biocatalytic system functioning as a logic AND gate (Figure 18.13a). The nicotinamide adenine dinucleotide (NAD^+) output signal produced by the system upon its activation in the presence of both biomarkers was then biocatalytically converted to the pH decrease. The acidic pH value produced by the system as a response to the biomarkers, shown in Figure 18.13b, triggered the restructuring of a polymer-modified electrode interface. This allowed soluble redox species to approach the electrode surface, thus switching the electrochemical reaction ON (Figure 18.13c). Small concentration changes of the $NADH/NAD^+$ system (0.3 mM) were converted into a large current corresponding to the electrochemical process of the $[Fe(CN)_6]^{4-}$ redox probe (10 mM), thus amplifying the output signal generated by the enzyme logic system by at least 30-fold.

18.8   BFCs WITH SWITCHABLE/TUNABLE POWER OUTPUT

BFCs with a switchable power output controlled by external physical or chemical signals were developed. The switchable properties of BFCs were achieved by modification of biocatalytic electrodes with signal-responsive materials sensitive to various stimuli-triggering reversible transitions of the electrode interfaces between electrochemically active and inactive states. Electrical, magnetic, and biochemical signals were used as stimuli controlling the electrode activity. Biochemical systems of various complexity mimicking Boolean logic operations AND/OR as well as multienzyme/multilogic gate networks were functionally coupled with the pH-sensitive electrodes in the BFC. This allowed switching the power production ON and OFF in the BFC upon application of many different biochemical signals logically processed by the enzyme systems according to the built-in logic "program." The studied switchable BFCs controlled by electric, magnetic, and biochemical signals exemplify novel bioelectronic devices activated by external signals on demand. They open a new facet in the multidisciplinary field of bioelectronics and keep a big promise for practical applications, particularly as implantable sustainable energy sources for biomedical applications. This section provides an overview of various switchable/tunable BFCs with the power production controlled by external signals applied to the stimuli-responsive electrodes discussed in the previous sections.

18.8.1   Switchable/Tunable BFCs Controlled by Electrical Signals

The system with the electrochemically switchable interfacial resistance [25] was used as a platform for enzyme-biocatalytic electrodes and their assembly in a switchable biological fuel cell [19]. A PAA thin film loaded with Cu^{2+} cations was modified with iso-2-cytochrome *c* (from *Saccharomyces cerevisiae*), which can serve as a specific ET mediator for O_2-reducing cytochrome oxidase (COx; isolated from a Keilin–Hartree heart muscle) [109]. In order to immobilize Cyt c with the alignment favorable for its communication with COx, the redox protein was attached to the polymer film via a single-cysteine residue providing orientation of the active heme site to the external direction. The affinity complex between the immobilized Cyt c and COx was electrocatalytically active for O_2 reduction and the COx–Cyt c-modified electrode was used as a cathode in a BFC by Katz and Willner in 2003 (Figure 18.14a) [19]. Another Cu^{2+}-loaded PAA-modified electrode was covalently modified with pyrroloquinoline quinone (PQQ) operating as an ET mediator and an amino-derivatized synthetic analog of the flavin adenine dinucleotide (FAD) cofactor, (2-aminoethyl)-flavin adenine dinucleotide. The redox dyad composed of PQQ-FAD was used to reconstitute apo-glucose oxidase (apo-GOx; EC 1.1.3.4, from *Aspergillus niger*) on the electrode surface [110,111], which was then used as an anode in the BFC (Figure 18.14b) [19]. Both bioelectrocatalytic electrodes had switchable interfacial resistance controlled by the potential applied on them and resulting in conductive (−0.5 V vs. saturated calomel electrode (SCE)) and

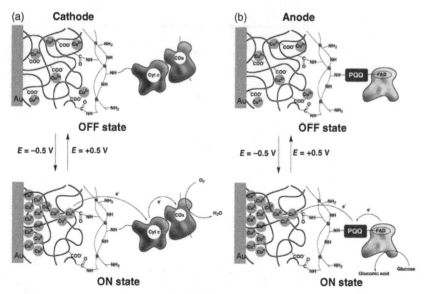

FIGURE 18.14 Reversible transformations of the PAA polymer film loaded with Cu^{2+} cations upon application of reductive (−0.5 V vs. SCE) and oxidative (+0.5 V) potentials to the modified electrode. Electrochemical switching of the cathode (a) and anode (b) biocatalytic interfaces between active (−0.5 V) and inactive (+0.5 V) states. (Adapted with permission from Ref. [19]. Copyright 2003, American Chemical Society.)

nonconductive (+0.5 V) states of the copper-loaded polymer film (Figure 18.14). The conductive state of the polymer support, containing aggregated Cu^0 clusters, allowed the ET between the biocatalytic enzymes and the electrode resulting in the ON state of the bioelectrocatalytic electrode, whereas the nonconductive state, which included only ionic Cu^{2+}, disconnected the enzymes from the electrode support and produced the OFF state of the electrode. The switchable electrodes combined in the BFC produced high and low values of the current and voltage in the cell upon reversible switching of the electrodes ON and OFF (Figure 18.15). The electrochemical transformation of the electrically nonconductive ionic state of

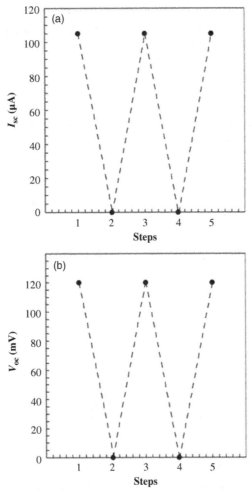

FIGURE 18.15 Reversible switching of I_{sc} (a) and V_{oc} (b) generated in the switched OFF state (after application of +0.5 V; steps 2 and 4) and switched ON state (after application of −0.5 V; steps 1, 3, and 5) of the biological fuel cell. (Adapted with permission from Ref. [19]. Copyright 2003, American Chemical Society.)

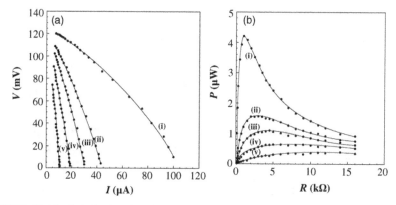

FIGURE 18.16 Polarization functions (a) and power release functions (b) obtained after application of the reductive potential of −0.5 V to the biocatalytic electrodes for different time intervals: (i) 200 s; (ii) 400 s; (iii) 600 s; (iv) 800 s; (v) 1000 s. (Adapted with permission from Ref. [19]. Copyright 2003, American Chemical Society.)

copper (Cu^{2+}) to the electronically conductive copper clusters (Cu^0) was a relatively slow process, taking ~1000 s for the complete conversion, which might be stopped prior to the complete reduction of Cu^{2+} ions and their full transformation to the Cu^0 clusters. This allowed the time-controlled tuning of the interfacial resistance. Indeed, partial transformation of copper to the electronically conductive clusters resulted in the fractional decrease of the interfacial resistance. Variable resistance at the bioelectrocatalytic electrodes resulted in different voltage–current output of the BFC, observed in the form of different polarization curves and power production (Figure 18.16a and b, respectively) upon application of the reductive potential (−0.5 V) to the electrodes for different time intervals. Therefore, the present approach allowed not only switchable but also tunable power production by the BFC [19].

18.8.2 Switchable/Tunable BFCs Controlled by Magnetic Signals

Application of magnetic micro/nanoparticles and nanorods to achieve switchable electrode interfacial properties was described with examples above (see Section 18.4). This approach is still waiting to reach its pinnacle of future use in fuel/BFCs. Another approach to the magnetic control of electrochemical reactions is based on the magnetohydrodynamic effect, which is well known for simple inorganic electrochemical processes [112–116], but it is rarely applied in bioelectrochemistry [117]. Recent development of the theory quantitatively describing the mechanism of the magnetohydrodynamic effect allowed its effective application in complex bioelectrocatalytic systems and BFCs [117–120].

A depletion layer produced at an electrode interface upon electrochemical consumption of redox species generates a concentration gradient and results in the diffusion of the reactive species from the bulk solution to the electrode surface, Figure 18.17a. Application of a permanent magnetic field in the direction perpendicular to

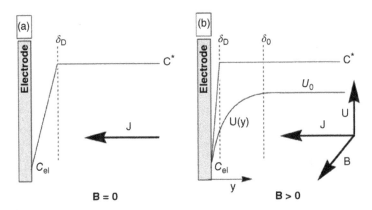

FIGURE 18.17 Depletion layer at an electrode surface in the absence (a) and presence (b) of external magnetic field applied perpendicular to the diffusional flux. J is the diffusional flux of the substrate, C_{el} is the substrate concentrations at the electrode surface, δ_D and δ_0 are the Nernst diffusion layer thickness and hydrodynamic boundary layer thickness, respectively, and U_0 is the fluid velocity on the outer edge of the hydrodynamic boundary layer. (Adapted with permission from Ref. [118]. Copyright 2004, American Chemical Society.)

the diffusion flux of the charge-transporting species results in their rotation in the magnetic field, yielding the rotation of the entire liquid phase. This phenomenon results in shrinking of the depletion layer at the electrode surface and enhances the mass transport and generated current at the electrode (Figure 18.17b). Application of a magnetic field to a BFC (in the direction perpendicular to the current between the cathode and anode) results in a convective mass transport, thus resulting in the enhancement of the current generated by the cell.

A biofuel cell composed of bioelectrocatalytic electrodes based on reconstituted enzymes was applied to demonstrate the phenomenon of the current enhancement in the presence of a constant magnetic field [120]. The BFC included a lactate-oxidizing anode based on LDH (EC 1.1.1.27, from rabbit muscle, type II) [121,122], whereas the oxygen-reducing cathode was based on COx integrated with Cyt c (Figure 18.18) [109]. Both bioelectrocatalytic electrodes included redox enzymes interacting with the corresponding cofactors/natural mediators NAD^+ and Cyt c for LDH and COx, respectively, immobilized in an aligned monolayer configuration. An additional mediator (PQQ) facilitated the ET from the enzymatically reduced NADH to the electrode conducting support. The BFC produced an open-circuit voltage (OCV), V_{oc}, of ~120 mV and a short-circuit current density, i_{sc}, of ~100 $\mu A\,cm^{-2}$ in the absence of magnetic field. Upon application of the magnetic field, $B = 0.92\,T$, the voltage generated by the BFC was almost unchanged, but the current output was increased by approximately threefold reflecting the mass transport enhancement originating from the magnetohydrodynamic effect. The magneto-induced current output increase was observed at different loading resistances resulting in the shift of the polarization curve in the presence of the magnetic field (Figure 18.19a, curve (ii)), compared with the polarization curve recorded in the absence of the external magnet (Figure 18.19a, curve (i)). The current output was dependent on the intensity of

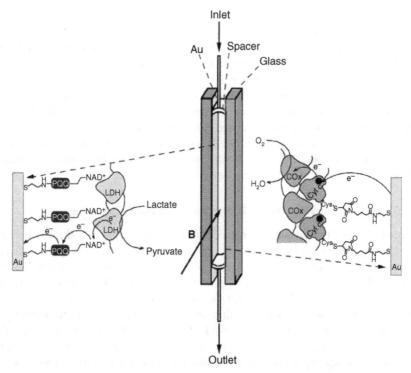

FIGURE 18.18 Magnetically controlled biological fuel cell with the biocatalytic anode based on the reconstituted NAD^+-dependent LDH and biocatalytic cathode based on the reconstituted COx/Cyt c system. Vector **B** shows the direction of the applied magnetic field. (Adapted with permission from Ref. [120]. Copyright 2005, American Chemical Society.)

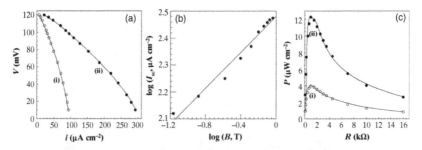

FIGURE 18.19 (a) Polarization curves obtained in the absence (i) and presence (ii) of the external magnetic field (0.92 T). (b) Dependence of I_{sc} generated by the biological fuel cell on the intensity of the external magnetic field. (c) Power released by the biological fuel cell as a function of the load resistance in the absence (i) and presence (ii) of the magnetic field (0.92 T). (Adapted with permission from Ref. [120]. Copyright 2005, American Chemical Society.)

magnetic flux density, B, revealing the slope of log i_{sc} versus log B plot (Figure 18.19b), which corresponds to 0.332 as predicted by the theoretical model ($i \propto B^{1/3}$) [118]. Thus, variation of the applied magnetic field allowed tuning of the current output generated by the BFC. The electrical power density output, $P = i \cdot V$, of the BFC was analyzed as a function of the external load resistance, R (Figure 18.19c). The power produced by the BFC at the optimum external resistance ($R = 1.2\,k\Omega$) was threefold higher upon application of the magnetic field of $B = 0.92\,T$. Cyclic switching ON/OFF of the magnetic field resulted in the reversible changes of the BFC activity reflected in the polarization curves and power release functions (Figure 18.19a and b, respectively). The switchable/tunable BFC was controlled by external magnetic signals at a distance without direct contact with the electrodes and solutions.

18.8.3   BFCs Controlled by Logically Processed Biochemical Signals

As soon as the pH-switchable bioelectrocatalytic electrodes and the pH change-producing enzyme logic systems were formulated, BFCs controlled by the logically processed biochemical signals became feasible [41–43]. Only one switchable bio-electrocatalytic electrode is needed in order to control the BFC activity, because the power production in the BFCs requires simultaneous operation of both electrodes: the cathode and anode for the oxygen reduction and the fuel oxidation, respectively. It should be noted that in most of the known BFCs the cathode reaction is the same, that is, bioelectrocatalytic oxygen reduction, whereas the anodic process might differ substantially depending on the used fuel and the applied biocatalyst. This makes it more practical to control the activity of the bioelectrocatalytic oxygen reduction at the modified cathode being switched between the ON and OFF states by the enzyme-induced pH changes (Figure 18.20).

The above-described (see Section 18.6) polyvinylpyridine (PVP) brush functionalized with the Os(dmo-bpy)$_2$Cl redox centers was used to generate the pH-switchable interface capable of mediating electron transport with enzymes only being in the ON state at pH < 4.5 [30]. Laccase is a well-known enzyme frequently

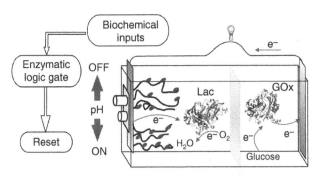

FIGURE 18.20   The switchable biological fuel cell controlled by pH changes produced *in situ* by the enzyme logic systems processing biochemical inputs. (Adapted with permission from Ref. [20]. Copyright 2009, Wiley-VCH Verlag GmbH.)

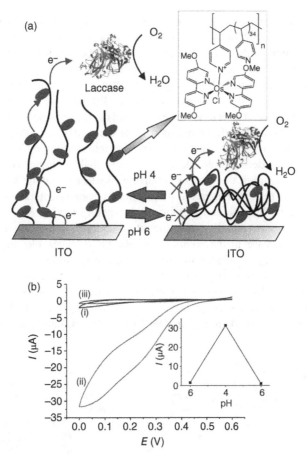

FIGURE 18.21 (a) The bioelectrocatalytic cathode for O_2 reduction in the presence of laccase controlled by pH. (b) The cyclic voltammograms obtained for the modified electrode in the presence of O_2 at different pH values: (i) ~6; (ii) ~4; (iii) ~6. *Inset*: Reversible changes of the electrocatalytic current in the cyclic voltammograms upon stepwise pH changes. (Adapted with permission from Ref. [41]. Copyright 2009, American Chemical Society.)

used for bioelectrocatalytic reduction of oxygen in BFCs [123]; thus, it was selected to operate at the modified pH-switchable cathode (Figure 18.21a). The Os(dmo-bpy)$_2$Cl redox centers covalently attached to the polymer brush tethered to the electrode surface operated as the electron-transporting mediators between the conducting electrode support and laccase. The mediating process was possible when the polymer brush was in its swollen state at pH < 4.5 and was fully inhibited at pH > 5.5 when the polymer was shrunken. The corresponding cyclic voltammograms obtained in the presence of laccase and oxygen demonstrated no electrocatalytic current at pH ~ 6 when the electrode is in the OFF state (Figure 18.21b, curve (i)) and a well-defined cathodic electrocatalytic current for the O_2 reduction at pH ~ 4 when the electrode is active (Figure 18.21b, curve (ii)). The electrocatalytic current was inhibited again after the pH value returned to the initial value (Figure 18.21b, curve (iii)), thus demonstrating the reversible switching of the electrocatalytic process (Figure 18.21b, inset). This reversible switching of the electrocatalytic process can be coupled with the pH changes generated *in situ* by the enzyme logic systems described above (see Section 18.6).

The logically controlled O_2-reducing bioelectrode was coupled with a glucose-oxidizing bioelectrocatalytic system to yield a BFC (Figure 18.20) [41,42]. The bioelectrochemical oxidation of glucose (100 mM) in the presence of soluble glucose oxidase (GOx, 250 units·mL^{-1}) and methylene blue (MB, 0.1 mM) operating as a diffusional electron transfer mediator [124] was selected as the simplest example of an anodic reaction to demonstrate the logically switchable biofuel cell. The anodic compartment containing the glucose-oxidizing bioelectrocatalytic system in 0.1 M phosphate buffer (pH 7; under argon) was separated from the cathodic switchable O_2-reducing electrode by a Nafion membrane, preventing mixing of the cathodic and anodic solutions. The cathodic compartment, in addition to the O_2-reducing system (112 units ml^{-1} laccase in the solution and the Os complex electron relay at the electrode surface), also included the enzyme logic/reset system.

Enzyme systems producing *in situ* pH changes as the logic output signals were developed specifically for controlling the electrochemical activity of the polymer-brush-modified electrodes [41]. In a simple example, two parallel biocatalytic reactions were applied to mimic an OR logic gate, as shown in Figure 18.22a.

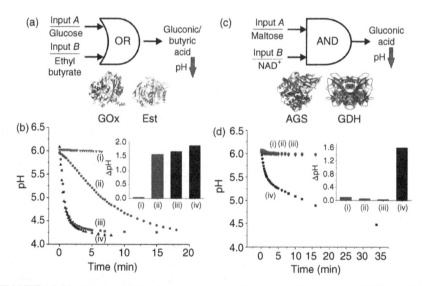

FIGURE 18.22 (a) The OR logic gate based on the concerted operation of GOx and Est activated by the inputs of glucose and ethyl butyrate. (b) pH changes generated *in situ* by the OR gate upon different combinations of the input signals: (i) 0,0; (ii) 0,1; (iii) 1,0; (iv) 1,1. (c) The AND logic gate based on the concerted operation of AGS and GDH activated by the inputs of maltose and NAD$^+$. (d) pH changes generated *in situ* by the AND gate upon different combinations of the input signals: (i) 0,0; (ii) 0,1; (iii) 1,0; (iv) 1,1. (Adapted with permission from Ref. [41]. Copyright 2009, American Chemical Society.)

Two enzymes, GOx (5 units ml^{-1}) and Est (5 units ml^{-1}), biocatalyzed the oxidation of glucose, input A (2 mM), and the hydrolysis of ethyl butyrate, input B (4 mM), respectively. The absence of glucose or ethyl butyrate was considered as the digital input signal 0, whereas their presence in the operational concentrations was considered as the input signal 1. The biocatalytic reactions in the presence of glucose and ethyl butyrate resulted in the formation of gluconic acid and butyric acid, respectively, both producing low pH values in the nonbuffered solution. Thus, in the presence of any substrate (input signals 0,1 or 1,0) or both of them together (input signals 1,1), one or both reactions proceeded and resulted in the acidification of the solution reaching pH ~ 4.2 (Figure 18.22b). The pH value was unchanged keeping the original value of pH ~ 6 only in the absence of the both substrates (input signals 0,0). Therefore, the features of the system corresponded to the OR logic operation (Figure 18.22b, inset). The AND logic gate was also composed of two enzymes operating cooperatively: amyloglucosidase (AGS, 5 units ml^{-1}) and GDH (5 units ml^{-1}) (Figure 18.22c). The enzyme system was activated by two biochemical input signals: maltose, input A (50 mM), and NAD^{+}, input B (0.5 mM), participating in a two-step chain reaction. In the first step, maltose was hydrolyzed by AGS to glucose and then glucose was oxidized by GDH to gluconic acid. The second reaction required NAD^{+} as an electron acceptor, and the chain reaction cannot be completed in the absence of NAD^{+} even though glucose was produced in the first step. The absence of maltose or NAD^{+} was considered as the digital input signal 0, whereas their presence in the operational concentrations was considered as the input signal 1. In the absence of any or both substrates (input signals 0,0; 0,1; or 1,0), the two-step reaction chain was not completed and the pH value was not changed. In the presence of the both substrates (input signals 1,1), the reaction proceeded until the very end, resulting in the formation of gluconic acid and acidification of the solution reaching pH ~ 4.3 (Figure 18.22d). Therefore, the features of the system corresponded to the AND logic operation (Figure 18.22d, inset).

The V–i (voltage vs. current density) polarization function of the BFC was obtained upon application of variable load resistances and by measuring the current and voltage generated on them (Figure 18.23a) . The power density produced by the BFC upon connecting to the variable resistances was derived from the V–i measurements (Figure 18.23b). The V–i function of the BFC in its nonactive state, pH ~ 6, revealed 80 mV for the OCV, V_{oc}, and 0.3 μA cm^{-2} for the short-circuit current density, i_{sc} (Figure 18.23a; curve (i)). The maximum released power density was $P_{max} = 6$ nW cm^{-2} (Figure 18.23b, curve (i)). The BFC was activated when the OR logic gate was used and 0,1; 1,0; or 1,1 input signals were applied (for the AND logic gate the "successful" combination of the input signals was 1,1). When the BFC was activated (through *in situ* pH changes enabling the biocatalytic cathodic process), the V–i function was changed demonstrating ~380 mV for the open-circuit voltage, V_{oc}, and 3 μA cm^{-2} for the short-circuit current density, i_{sc}, at pH ~4 (Figure 18.23a, curve (ii)). The maximum power density was significantly increased reaching 700 nW cm^{-2} (Figure 18.23b, curve (ii)). The patterns of the biochemical signals resulting in the BFC switching ON resembled the AND/OR logic functions (Figure 18.23c). After the BFC was activated by the set of the biochemical inputs processed by the enzyme logic gates, another Reset signal (urease and urea) was applied to bring

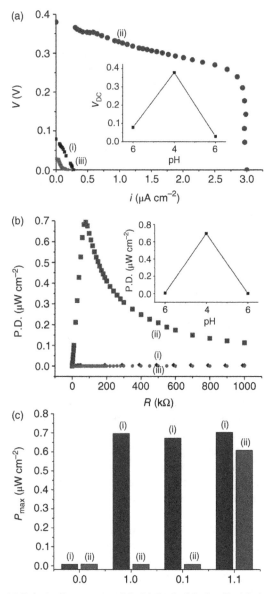

FIGURE 18.23 (a) Polarization curves of the biological fuel cell with the pH-switchable O_2 cathode obtained at different pH values generated *in situ* by the OR/AND enzyme logic gates: (i) pH ~6; (ii) pH ~4; (iii) pH ~6. *Inset*: The reversible changes of the open-circuit voltage produced by the cell at the variable pH. (b) Electrical power density generated by the biological fuel cell on different load resistances at different pH values generated *in situ* by the OR/AND enzyme logic gates: (i) pH ~6; (ii) pH ~4; (iii) pH ~6. *Inset*: The reversible changes of the maximum power density produced by the biological fuel cell at the variable pH. (c) The maximum power density produced by the biological fuel cell as a function of different combinations of the input signals: (i) the OR logic gate; (ii) the AND logic gate. (Adapted with permission from Ref. [41]. Copyright 2009, American Chemical Society.)

the pH value back to its initial value and to inactivate the BFC (Figure 18.23a and b; curve (iii)). Thus, the reversible cyclic activation–inhibition of the BFC was achieved by the application of various biochemical signals (Figure 18.23a and b, insets).

A similar BFC, as shown in Figure 18.20, was switched ON by much more sophisticated combination of the biochemical input signals when the bioelectrocatalytic cathodic process was controlled by the enzyme logic network composed of several concatenated gates (Figure 18.24) [42]. A sequence of biochemical reactions was designed to produce *in situ* pH changes as the final output of the biochemical cascade (Figure 18.24a). The reactions were activated by four chemical input signals,

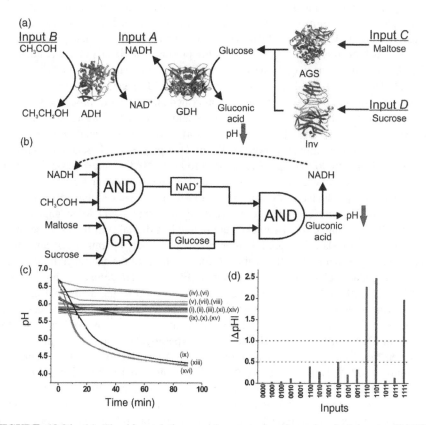

FIGURE 18.24 (a) The biocatalytic cascade processing four chemical inputs, NADH, acetaldehyde, maltose, and sucrose, upon concerted operation of four enzymes, ADH, AGS, INV, and GDH. (b) The logic network composed of three gates corresponding to the biocatalytic process. (c) The pH changes produced *in situ* by the biocatalytic system upon different combinations of the inputs: (i) 0,0,0,0; (ii) 1,0,0,0; (iii) 0,1,0,0; (iv) 0,0,1,0; (v) 0,0,0,1; (vi) 1,1,0,0; (vii) 1,0,1,0; (viii) 1,0,0,1; (ix) 0,1,1,0; (x) 0,1,0,1; (xi) 0,0,1,1; (xii) 1,1,1,0; (xiii) 1,1,0,1; (xiv) 1,0,1,1; (xv) 0,1,1,1; (xvi) 1,1,1,1. (d) The bar diagram showing pH changes as the output signals generated upon application of different patterns of the chemical inputs. Dashed lines show thresholds separating digital 0, undefined, and digital 1 output signals produced by the system. (Adapted with permission from Ref. [42]. Copyright 2009, Elsevier.)

NADH (0.5 mM), acetaldehyde (10 mM), maltose (100 mM), and sucrose (300 mM) (input signals A, B, C, and D, respectively), processed by four enzymes, ADH (5 units ml^{-1}), AGS (100 units ml^{-1}), INV (40 units ml^{-1}), and GDH (20 units ml^{-1}). The reactions were started from the production of NAD$^+$ and ethanol biocatalyzed by ADH in the presence of NADH and acetaldehyde being primary input signals. Another reaction chain resulted in the formation of glucose from sucrose and maltose (another couple of the primary inputs) biocatalyzed by INV and AGS, respectively. The *in situ* produced NAD$^+$ and glucose were reacted with GDH yielding gluconic acid as the final chemical product, which resulted in the formation of the acidic pH $\sim$ 4.3. The biochemical set of the reactions can be reformulated in the terms of the logic operations performed by the network composed of three logic gates, as shown in Figure 18.24b. Indeed, the production of NAD$^+$ being an intermediate product in the system was biocatalyzed by ADH only in the presence of NADH and acetaldehyde. The absence of either or both the inputs inhibited the reaction and NAD$^+$ was not formed. Thus, this part of the system operates as a Boolean AND logic gate. The production of glucose, also being an intermediate product, proceeded in two independent reactions biocatalyzed by AGS and INV in the presence of maltose and sucrose, respectively. Thus, the glucose formation was activated in the presence of any of the input signals or both of them. This part of the biochemical cascade represented the OR logic gate. The *in situ* produced NAD$^+$ and glucose reacted to yield gluconic acid and decrease the solution pH value upon the process biocatalyzed by GDH. The product and the respective pH decrease were obtained only in the presence of the both reactants, thus featuring the final AND logic gate (Figure 18.24b). The chemical input signals were defined as digital 0 in the absence of the respective chemicals and digital 1 upon addition of their operating concentrations optimized experimentally. All possible 16 combinations of the chemical input signals were examined for changing pH in the system (Figure 18.24c), and only three of them (1,1,0,1; 1,1,1,0; and 1,1,1,1) resulted in the production of the acidic pH of $\sim$4.3, which corresponds to ΔpH of $\sim$2.5 (Figure 18.24d). This result was consistent with the expected output of the logic network (Figure 18.24b), as well as with the biochemical scheme (Figure 18.24a). The experimental features of the system followed the Boolean logic function: O = (A AND B) AND (C OR D), where O is the output, and A, B, C, and D are the input signals. After completing the enzyme reactions resulting in the logically controlled acidic medium (pH = 4.1–4.5), the pH value was reset to the initial value (pH $\sim$ 6) by the formation of ammonia upon hydrolysis of urea (5 mM) biocatalyzed by urease (10 units ml^{-1}), thus performing the Reset function.

Similarly to the switchable BFC described above, the BFC was characterized by the polarization curve (Figure 18.25a) and the power release dependence (Figure 18.25b), obtained in the inactive and active states (curves (i) and (ii), respectively). However, the cell operation was controlled by four different biochemical signals applied in 16 combinations, only 3 of them being "successful" for switching the BFC ON according to the built-in logic "program" in the enzyme network (Figure 18.25c). After the BFC activation, another cycle can be started with the reset to the initial "mute" state by the application of the urease/urea signal.

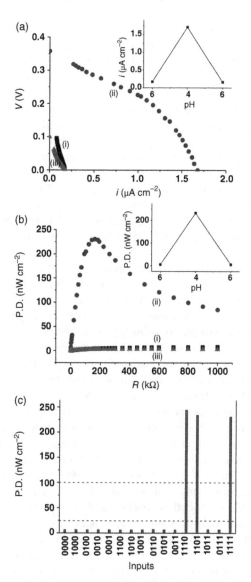

FIGURE 18.25   (a) Polarization curves of the biological fuel cell with the pH-switchable O$_2$ cathode obtained at different pH values generated *in situ* by the enzyme logic network: (i) pH ~6; (ii) pH ~4; (iii) pH ~6. *Inset*: The reversible changes of the short-circuit current density produced by the cell at the variable pH. (b) Electrical power density generated by the biological fuel cell on different load resistances at different pH values generated *in situ* by the enzyme logic network: (i) pH ~6; (ii) pH ~4; (iii) pH ~6. *Inset*: The reversible changes of the maximum power density produced by the biological fuel cell at the variable pH. (c) The maximum power density produced by the biological fuel cell as a function of different combinations of the input signals. (Adapted with permission from Ref. [42]. Copyright 2009, Elsevier.)

The biochemical networks demonstrate robust error-free processing of bio-chemical signals upon appropriate optimization of their components and intercon-nections [99,125,126]. However, the limit of the biocomputing network complexity is set by the cross-reactivity of the enzyme-catalyzed reactions. Only enzymes belonging to different biocatalytic classes (oxidases, dehydrogenases, peroxidases, hydrolases, etc.) could operate in a homogeneous system without significant cross-reactivity. If chemical reasons require the use of cross-reacting enzymes in the system, they must be compartmentalized using pattering on surfaces or applied in microfluidic devices. Application of more selective biomolecular interactions would be an advantage to make biocomputing systems more specific to various input signals and less cross-reactive in the chemical signal processing. This aim can be achieved by the application of highly selective biorecognition (e.g. immune) interactions for biocomputing [127]. One of the novel immune-based biocomputing systems was already applied for switching the BFC activity by the logically processed antibody signals [43].

A surface functionalized with a mixed monolayer of two different antigens: 2,4-dinitrophenyl (DNP) and 3-nitro-L-tyrosine (NT) loaded on human serum albumin (HSA) and bovine serum albumin (BSA), respectively, was used to analyze the input signals of the corresponding antibodies: anti-DNP (anti-dinitrophenyl IgG polyclonal from goat) and anti-NT (anti-nitrotyrosine IgG from rabbit) [127]. After binding to the surface, the primary antibodies were reacted with the secondary antibodies (anti-goat IgG-HRP (horseradish peroxidase) and anti-rabbit IgG-HRP (mouse origin IgG specific to goat immunoglobulin and mouse origin IgG specific to rabbit IgG), both labeled with HRP; 0.05 mg ml^{-1} each antibody) to attach the biocatalytic HRP tag to the immune complexes generated on the surfaces (Figure 18.26a). The primary anti-DNP and anti-NT antibodies were applied in four different combinations: 0,0; 0,1; 1,0; and 1,1, where the digital value 0 corresponded to the absence of the antibody and value 1 corresponded to their presence in the optimized concentrations: 8 µg ml^{-1} for anti-DNP and 0.2 µg ml^{-1} for anti-NT. The secondary antibody labeled with the HRP biocatalytic tag was bound to the surface only if the respective primary antibody was already there. Since the both secondary antibodies were labeled with HRP, the biocatalytic entity appeared on the surface upon application of 0,1; 1,0; and 1,1 signal combinations. Only in the absence of the both primary antibodies (signals 0,0) the secondary antibodies were not bound to the surface and the HRP biocatalyst did not appear there, thus resembling the OR logic operation. The assembled functional interface was reacted with 2,2′-azinobis(3-ethylbenzothiazoline-6-sulfonic acid) (ABTS, 0.5 mM) and H_2O_2 (0.5 mM). The biocatalytic oxidation of ABTS and concomitant reduction of H_2O_2 resulted in the increase of the solution pH only when the biocatalytic HRP tag was present on the surface (Figure 18.26b). This happened when the primary antibody signals were applied in the combinations 0,1; 1,0; and 1,1. The pH increase generated *in situ* by the enzyme reaction coupled with the immune recognition system yielded the inactive shrunken state of the polymer brush-modified electrode, thus deactivating the entire BFC. It should be noted that, for simplicity, the cathode was represented by a model redox system with a ferricyanide solution instead of the oxygen system (Figure 18.26c). The BFC being

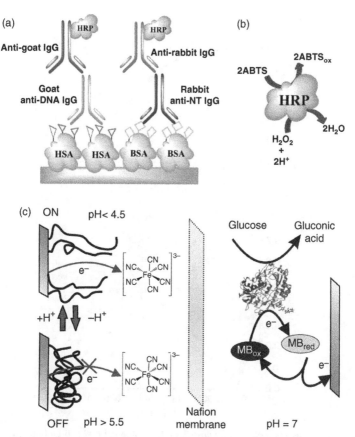

FIGURE 18.26 (a) The immune system composed of two antigens, two primary antibodies, and two secondary antibodies labeled with HRP biocatalytic tag used for the OR logic gate. (b) The biocatalytic reaction producing pH changes to control the biological fuel cell performance. (c) The biological fuel cell controlled by the immune-OR-logic gate due to the pH-switchable $[Fe(CN)_6]^{3-}$-reducing cathode. MB_{ox} and MB_{red} are oxidized and reduced states of the mediator methylene blue. (Adapted with permission from Ref. [43]. Copyright 2009, American Chemical Society.)

active at pH 4.5 (Figure 18.27a and b, curve (i)) was partially inactivated (Figure 18.27a and b, curve (ii)) by the pH increase up to 5.8 generated by the immune-based logic system. Since the output signal 1 from the logic system resulted in the inactivation of the BFC (operating as the inverter producing 0 output for input 1 and vice versa), the system modeled a NOR logic gate (Figure 18.27b, inset). After the BFC inactivation, the next cycle was started by the reset to the initial pH value activating the switchable electrode again. To activate the BFC, GOx (14.3 units ml^{-1}) and glucose (10 mM) were added to the cathodic compartment, resulting in the pH decrease to $\sim$4.2 due to the biocatalytic oxidation of glucose and formation of gluconic acid.

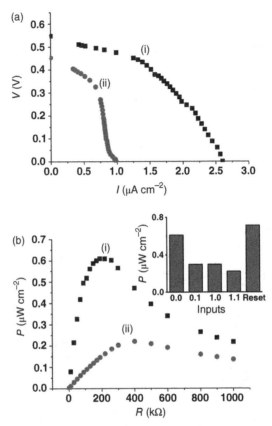

FIGURE 18.27 (a) The polarization curves of the biological fuel cell with the pH-switchable cathode obtained at different pH values generated *in situ* by the immune-OR-logic gate: (i) pH 4.5; (ii) pH 5.8. (b) Electrical power density generated by the biological fuel cell on different load resistances at different pH values generated *in situ* by the immune-OR-logic gate: (i) pH 4.5; (ii) pH 5.8. *Inset*: The maximum electrical power density produced by the biological fuel cell upon different combinations of the immune input signals. (Adapted with permission from Ref. [43]. Copyright 2009, American Chemical Society.)

18.9 OUTLOOK

Extensive work performed in the area of switchable electrode interfaces has led to numerous systems controlled by a large variety of physical and chemical signals. Particularly important results are expected when the research advances in signal-responsive materials, modified electrodes, and chemical computing are integrated into a new research area. Further scaling up the complexity of the chemical information processing systems will allow the next level of sophistication when switchable electrodes are controlled by various chemical processes. Particularly interesting would be combinations of biomolecular computing systems and switchable modified electrodes. Coupling of smart switchable electrodes with sophisticated multistep biochemical pathways could be envisaged in the continuing research. Integration of the signal-responsive electrodes with the information processing systems might be used to develop smart multisignal responsive biosensors and BFCs controlled by complex biochemical environment. The biochemically/physiologically controlled switchable electrodes will operate as an interface between biological and electronic systems in future micro/nanorobotic devices.

Switchable electrodes controlled by signal processing enzyme-based logic systems were already integrated in smart BFCs producing electrical power dependent on complex variations of biochemical signals [20,128]. Enzyme biocatalytic and immune biorecognition systems have been developed to control performance of switchable BFCs [41–43]. Future implantable BFCs producing electrical power on demand depending on physiological conditions are feasible as the result of the present research. Further development of sophisticated enzyme-based biocomputing networks will be an important phase in the development of smart bioelectronic devices. Scaling up the complexity of biocomputing system controlling BFC activity will be achieved by networking immune- and enzyme-based logic gates responding to a large variety of biochemical signals. BFCs switchable by enzyme-based or immunosystem-based keypad lock systems have been designed to operate as self-powered biomolecular information security systems [129,130]. The correct biomolecular "password" introduced into the keypad lock as a sequence of biomolecular input signals resulted in the activation of the BFC, whereas all other "wrong" permutations of the molecular inputs preserved the OFF state of the biological fuel cell. Further research directed to the increasing stability and robustness of the information processing biocatalytic electrodes could result in many practical applications, including, for example, bioelectrocatalytic barcode generation using electrodes characteristic of BFCs [131].

The present developments and future expectations are based on the application of a multidisciplinary approach that will require further collaborative contribution from electrochemists, specialists in materials science and unconventional molecular and biomolecular computing.

ACKNOWLEDGMENTS

This work was supported by NSF (Award No. CBET-1066397) and by the Semiconductor Research Corporation (Award 2008-RJ-1839G). The authors thank all students and collaborators who made this interesting research possible.

LIST OF ABBREVIATIONS

ABTS	2,2′-azinobis(3-ethylbenzothiazoline-6-sulfonate)
ADH	alcohol dehydrogenase
AFM	atomic force microscopy
AGS	aminoglucosidase

ALT	alanine transaminase
apo-GOx	apo-glucose oxidase
BFC	biological fuel cell
BSA	bovine serum albumin
COx	cytochrome oxidase
Cyt c	cytochrome c
dmo-bpy	4,4″-dimethoxy-2,2″-bipyridine
DNP	2,4-dinitrophenyl
Est	esterase
ET	electron transfer
FAD	flavin adenine dinucleotide
GDH	glucose dehydrogenase
GOx	glucose oxidase
HRP	horseradish peroxidase
HSA	human serum albumin
IgG	immunoglobulin G
INV	invertase
ITO	indium tin oxide
LDH	lactate dehydrogenase
MRH^+	merocyanine
NAD^+	nicotinamide adenine dinucleotide
NADH	nicotinamide adenine dinucleotide, reduced form
NT	3-nitro-L-tyrosine
OCV	open-circuit voltage
P2VP	poly(2-vinylpyridine)
P4VP	poly(4-vinylpyridine)
PAA	polyacrylic acid
PQQ	pyrroloquinoline quinone
PVP	polyvinylpyridine
qP2VP	quaternized poly(2-vinylpyridine)
SCE	saturated calomel electrode

REFERENCES

1. Murray RW. Chemically modified electrodes. *Acc Chem Res* 1980;13:135–141.

2. Wrighton MS. Surface functionalization of electrodes with molecular reagents. *Science* 1986;231:32–37.

3. Abruña HD. Coordination chemistry in two dimensions: chemically modified electrodes. *Coord Chem Rev* 1988;86:135–189.

4. Bain CD, Troughton EB, Tao Y-T, Eval J, Whitesides GM, Nuzzo RG. Formation of monolayer films by the spontaneous assembly of organic thiols from solution onto gold. *J Am Chem Soc* 1989;111:321–335.

5. Nuzzo RG, Fusco FA, Allara DL. Spontaneously organized molecular assemblies. 3. Preparation and properties of solution adsorbed monolayers of organic disulfides on gold surfaces. *J Am Chem Soc* 1987;109:2358–2368.

6. Chen D, Li JH. Interfacial design and functionization on metal electrodes through self-assembled monolayers. *Surf Sci Rep* 2006;61:445–463.

7. Zen JM, Kumar AS, Tsai D-M. Recent updates of chemically modified electrodes in analytical chemistry. *Electroanalysis* 2003;15:1073–1087.

8. Vielstich W, Gasteiger H, Lamm A (eds), *Handbook of Fuel Cells: Fundamentals, Technology, Applications.* John Wiley & Sons, Ltd, Chichester, UK, 2003.

9. Rusling JF, Forster RJ. Electrochemical catalysis with redox polymer and polyion–protein films. *J Colloid Interface Sci* 2003;262:1–15.

10. Willner I, Katz E. Integration of layered redox proteins and conductive supports for bioelectronic applications. *Angew Chem Int Ed* 2000;39:1180–1218.

11. Wang J. *In vivo* glucose monitoring: towards 'Sense and Act' feedback-loop individualized medical systems. *Talanta* 2008;75:636–641.

12. Gooding JJ. Advances in interfacial design for electrochemical biosensors and sensors: aryl diazonium salts for modifying carbon and metal electrodes. *Electroanalysis* 2008;20:573–582.

13. Moehlenbrock MJ, Minteer SD. Extended lifetime biofuel cells. *Chem Soc Rev* 2008;37: 1188–1196.

14. Calabrese Barton S, Gallaway J, Atanassov P. Enzymatic biofuel cells for implantable and microscale devices. *Chem Rev* 2004;104:4867–4886.

15. Katz E, Willner B, Willner I. Light-controlled electron transfer reactions at photoisomerizable monolayer electrodes by means of electrostatic interactions: active interfaces for the amperometric transduction of recorded optical signals. *Biosens Bioelectron* 1997;12:703–719.

16. Flood AH, Ramirez RJA, Deng WQ, Muller RP, Goddard WA, Stoddart JF. Meccano on the nanoscale—a blueprint for making some of the world's tiniest machines. *Aust J Chem* 2004;57:301–322.

17. Motornov M, Sheparovych R, Katz E, Minko S. Chemical gating with nanostructured responsive polymer brushes: mixed brush versus homopolymer brush. *ACS Nano* 2008;2:41–52.

18. Pita M, Katz E. Switchable electrodes: how can the system complexity be scaled up? *Electroanalysis* 2009;21:252–260.

19. Katz E, Willner I. A biofuel cell with electrochemically switchable and tunable power output. *J Am Chem Soc* 2003;125:6803–6813.

20. Katz E, Pita M. Biofuel cells controlled by logically processed biochemical signals: towards physiologically regulated bioelectronic devices. *Chem Eur J* 2009;15:12554–12564.

21. Willner I, Lion-Dagan M, Marx-Tibbon S, Katz E. Bioelectrocatalyzed amperometric transduction of optical signals using monolayer-modified Au-electrodes. *J Am Chem Soc* 1995;117:6581–6592.

22. Liu NG, Dunphy DR, Atanassov P, Bunge SD, Chen Z, Lopez GP, Boyle TJ, Brinker CJ. Photoregulation of mass transport through a photoresponsive azobenzene-modified nanoporous membrane. *Nano Lett* 2004;4:551–554.

23. Liu ZF, Hashimoto K, Fujishima A. Photoelectrochemical information storage using an azobenzene derivative. *Nature* 1990;347:658–660.

24. Zheng L, Xiong L. Layer-by-layer assembly of PA-EDTA/PAH multilayer films and their potential-switchable electrochemistry. *Colloids Surf A* 2006;289:179–184.

25. Chegel VI, Raitman OA, Lioubashevski O, Shirshov Y, Katz E, Willner I. Redox-switching of electrorefractive, electrochromic and conducting functions of Cu^{2+}/poly-acrylic acid films associated with electrodes. *Adv Mater* 2002;14:1549–1553.

26. Katz E, Baron R, Willner I. Magnetoswitchable electrochemistry gated by alkyl-chain-functionalized magnetic nanoparticles: controlling of diffusional and surface-confined electrochemical process. *J Am Chem Soc* 2005;127:4060–4070.

27. Katz E, Sheeney-Haj-Ichia L, Willner I. Magneto-switchable electrocatalytic and bio-electrocatalytic transformations. *Chem Eur J* 2002;8:4138–4148.

28. Laocharoensuk R, Bulbarello A, Mannino S, Wang J. Adaptive nanowire–nanotube bioelectronic system for on-demand bioelectrocatalytic transformations. *Chem Commun* 2007; 3362–3364.

29. Loaiza OA, Laocharoensuk R, Burdick J, Rodriguez MC, Pingarron JM, Pedrero M, Wang J. Adaptive orientation of multifunctional nanowires for magnetic control of bioelectrocatalytic processes. *Angew Chem Int Ed* 2007;46:1508–1511.

30. Tam TK, Ornatska M, Pita M, Minko S, Katz E. Polymer brush-modified electrode with switchable and tunable redox activity for bioelectronic applications. *J Phys Chem C* 2008;112:8438–8445.

31. Pita M, Minko S, Katz E. Enzyme-based logic systems and their applications for novel multi-signal-responsive materials. *J Mater Sci: Mater Med* 2009;20:457–462.

32. Tam TK, Zhou J, Pita M, Ornatska M, Minko S, Katz E. Biochemically controlled bioelectrocatalytic interface. *J Am Chem Soc* 2008;130:10888–10889.

33. Strack G, Pita M, Ornatska M, Katz E. Boolean logic gates using enzymes as input signals. *ChemBioChem* 2008;9:1260–1266.

34. Baron R, Lioubashevski O, Katz E, Niazov T, Willner I. Logic gates and elementary computing by enzymes. *J Phys Chem A* 2006;110:8548–8553.

35. Baron R, Lioubashevski O, Katz E, Niazov T, Willner I. Elementary arithmetic operations by enzymes: a model for metabolic pathway based computing. *Angew Chem Int Ed* 2006;45:1572–1576.

36. Katz E, Privman V. Enzyme-based logic systems for information processing. *Chem Soc Rev* 2010;39:1835–1857.

37. Strack G, Ornatska M, Pita M, Katz E. Biocomputing security system: concatenated enzyme-based logic operating as a biomolecular keypad lock. *J Am Chem Soc* 2008;130:4234–4235.

38. Niazov T, Baron R, Katz E, Lioubashevski O, Willner I. Concatenated logic gates using four coupled biocatalysts operating in series. *Proc Natl Acad Sci USA* 2006;103: 17160–17163.

39. Privman M, Tam TK, Pita M, Katz E. Switchable electrode controlled by enzyme logic network system: approaching physiologically regulated bioelectronics. *J Am Chem Soc* 2009;131:1314–1321.

40. Wang X, Zhou J, Tam TK, Katz E, Pita M. Switchable electrode controlled by Boolean logic gates using enzymes as input signals. *Bioelectrochemistry* 2009;77:69–73.

41. Amir L, Tam TK, Pita M, Meijler MM, Alfonta L, Katz E. Biofuel cell controlled by enzyme logic systems. *J Am Chem Soc* 2009;131:826–832.

42. Tam TK, Pita M, Ornatska M, Katz E. Biofuel cell controlled by enzyme logic network— approaching physiologically regulated devices. *Bioelectrochemistry* 2009;76:4–9.

43. Tam TK, Strack G, Pita M, Katz E. Biofuel cell logically controlled by antigen–antibody recognition: towards immune regulated bioelectronic devices. *J Am Chem Soc* 2009;131:11670–11671.

44. Lion-Dagan M, Katz E, Willner I. Amperometric transduction of optical signals recorded by organized monolayers of photoisomerizable biomaterials on Au-electrodes. *J Am Chem Soc* 1994;116:7913–7914.

45. Willner I, Lion-Dagan M, Katz E. Photostimulation of dinitrospiropyran-modified glucose oxidase in the presence of DNP-antibody. A biphase switch for the amperometric transduction of recorded optical signals. *Chem Commun* 1996; 623–624.

46. Wesenhagen P, Areephong J, Landaluce TF, Heureux N, Katsonis N, Hjelm J, Rudolf P, Browne WR, Feringa BL. Photochromism and electrochemistry of a dithienylcyclopentene electroactive polymer. *Langmuir* 2008;24:6334–6342.

47. Browne WR, Kudernac T, Katsonis N, Areephong J, Hielm J, Feringa BL. Electro- and photochemical switching of dithienylethene self-assembled monolayers on gold electrodes. *J Phys Chem C* 2008;112:1183–1190.

48. Doron A, Portnoy M, Lion-Dagan M, Katz E, Willner I. Amperometric transduction and amplification of optical signals recorded by a phenoxynaphthacenequinone monolayer electrode: photochemical and pH-gated electron transfer. *J Am Chem Soc* 1996;118:8937–8944.

49. Willner I, Pardo-Yissar V, Katz E, Ranjit KT. A photoactivated "molecular train" for optoelectronic applications: light-stimulated translocation of a β-cyclodextrin receptor within a stoppered azobenzene-alkyl chain supramolecular monolayer assembly on a Au-electrode. *J Electroanal Chem* 2001;497:172–177.

50. Baron R, Onopriyenko A, Katz E, Lioubashevski O, Willner I, Wang S, Tian H. An electrochemical/photochemical information processing system using a monolayer-functionalized electrode. *Chem Commun* 2006; 2147–2149.

51. Willner I, Katz E. Magnetic control of electrocatalytic and bioelectrocatalytic processes. *Angew Chem Int Ed* 2003;42:4576–4588.

52. Katz E, Sheeney-Haj-Ichia L, Basnar B, Felner I, Willner I. Magnetoswitchable controlled hydrophilicity/hydrophobicity of electrode surfaces using alkyl-chain-functionalized magnetic particles: application for switchable electrochemistry. *Langmuir* 2004;20:9714–9719.

53. Wang J, Kawde AN. Magnetic-field stimulated DNA oxidation. *Electrochem Commun* 2002;4:349–352.

54. Wang J. Adaptive nanowires for on-demand control of electrochemical microsystems. *Electroanalysis* 2008;20:611–615.

55. Wang J, Scampicchio M, Laocharoensuk R, Valentini F, González-García O, Burdick J. Magnetic tuning of the electrochemical reactivity through controlled surface orientation of catalytic nanowires. *J Am Chem Soc* 2006;128:4562–4563.

56. Katz E, Baron R, Willner I. Magnetoswitchable electrochemistry gated by alkyl-chain-functionalized magnetic nanoparticles: controlling of diffusional and surface-confined electrochemical process. *J Am Chem Soc* 2005;127:4060–4070.

57. Willner I, Katz E. Controlling chemical reactivity at solid–solution interfaces by means of hydrophobic magnetic nanoparticles. *Langmuir* 2006;22:1409–1419.

58. Katz E, Willner I. Switching of directions of bioelectrocatalytic currents and photocurrents at electrode surfaces by using hydrophobic magnetic nanoparticles. *Angew Chem Int Ed* 2005;44:4791–4794.

59. Katz E, Willner I. Hydrophobic magnetic nanoparticles induce selective bioelectrocatalysis. *Chem Commun* 2005; 4089–4091.

60. Katz E, Lioubashevski O, Willner I. Magnetoswitchable single-electron charging of Au-nanoparticles using hydrophobic magnetic nanoparticles. *Chem Commun* 2006; 1109–1111.

61. Katz E, Willner I. A quinone-functionalized monolayer electrode in conjunction with hydrophobic magnetic nanoparticles acts as a "write–read–erase" information storage system. *Chem Commun* 2005; 5641–5643.

62. Wang J, Musameh M, Laocharoensuk R, Gonzalez-Garcia O, Oni J, Gervasio D. Pt/Ru-functionalized magnetic spheres for a magnetic-field stimulated methanol and oxygen redox processes: towards on-demand activation of fuel cells. *Electrochem Commun* 2006;8:1106–1110.

63. Jimenez J, Sheparovych R, Pita M, Narvaez Garcia A, Dominguez E, Minko S, Katz E. Magneto-induced self-assembling of conductive nanowires for biosensor applications. *J Phys Chem C* 2008;112:7337–7344.

64. Vetter KJ. *Electrochemical Kinetics: Theoretical Aspects*. Academic Press, New York, 1967.

65. Woods R. Chemisorption at electrodes. Hydrogen and oxygen on noble metals and their alloys. In: Bard AJ (ed.), *Electroanalytical Chemistry*. Marcel Dekker, New York, 1976, pp. 1–162.

66. Bard AJ, Stratmann M (eds), *Encyclopedia of Electrochemistry. Vol. 10. Modified Electrodes*. Wiley-VCH Verlag GmbH, Weinheim, 2007.

67. Raitman OA, Katz E, Willner I, Chegel VI, Popova GV. Photonic transduction of a three-state electronic memory and of electrochemical sensing of NADH using surface plasmon resonance spectroscopy. *Angew Chem Int Ed* 2001;40:3649–3652.

68. Chegel V, Raitman O, Katz E, Gabai R, Willner I. Photonic transduction of electro-chemically-triggered redox-functions of polyaniline films using surface plasmon resonance spectroscopy. *Chem Commun* 2001; 883–884.

69. Riskin M, Basnar B, Chegel VI, Katz E, Willner I, Shi F, Zhang X. Switchable surface properties through the electrochemical or biocatalytic generation of Ag^0-nanoclusters on monolayer-supported electrodes. *J Am Chem Soc* 2006;128:1253–1260.

70. Riskin M, Basnar B, Katz E, Willner I. Cyclic control of the surface properties of a monolayer-functionalized electrode by the electrochemical generation of Hg nanoclusters. *Chem Eur J* 2006;12:8549–8557.

71. Riskin M, Katz E, Willner I. Photochemically-controlled electrochemical deposition and dissolution of Ag^0 nanoclusters. *Langmuir* 2006;22:10483–10489.

72. Combellas C, Kanoufi F, Sanjuan S, Slim C, Tran Y. Electrochemical and spectroscopic investigation of counterions exchange in polyelectrolyte brushes. *Langmuir* 2009;25:5360–5370.

73. Choi EY, Azzaroni O, Cheng N, Zhou F, Kelby T, Huck WTS. Electrochemical characteristics of polyelectrolyte brushes with electroactive counterions. *Langmuir* 2007;23:10389–10394.

74. Harris JJ, Bruening ML. Electrochemical and *in situ* ellipsometric investigation of the permeability and stability of layered polyelectrolyte films. *Langmuir* 2000;16:2006–2013.

75. Park MK, Deng SX, Advincula RC. pH-sensitive bipolar ion-permselective ultrathin films. *J Am Chem Soc* 2004;126:13723–13731.

76. Tam TK, Pita M, Trotsenko O, Motornov M, Tokarev I, Halámek J, Minko S, Katz E. Reversible "closing" of an electrode interface functionalized with a polymer brush by an electrochemical signal. *Langmuir* 2010;26:4506–4513.

77. Tam TK, Pita M, Motornov M, Tokarev I, Minko S, Katz E. Electrochemical nano-transistor from mixed polymer brush. *Adv Mater* 2010;22:1863–1866.

78. Diamond D, McKervey MA. Calixarene-based sensing agents. *Chem Soc Rev* 1996;25:15–24.

79. Yang DH, Ju M-J, Maeda A, Hayashi K, Toko K, Lee S-W, Kunitake T. Design of highly efficient receptor sites by combination of cyclodextrin units and molecular cavity in TiO_2 ultrathin layer. *Biosens Bioelectron* 2006;22:388–392.

80. Gabai R, Sallacan N, Chegel V, Bourenko T, Katz E, Willner I. Characterization of the swelling of acrylamidophenylboronic acid–acrylamide hydrogels upon interaction with glucose by Faradaic impedance spectroscopy, chronopotentiometry, quartz-crystal microbalance (QCM), and surface plasmon resonance (SPR) experiments. *J Phys Chem B* 2001;105:8196–8202.

81. Minko S. Responsive polymer brushes. *Polym Rev* 2006;46:397–420.

82. Luzinov I, Minko S, Tsukruk V. Adaptive and responsive surfaces through controlled reorganization of interfacial polymer layers. *Prog Polym Sci* 2004;29: 635–698.

83. Tokarev I, Orlov M, Katz E, Minko S. An electrochemical gate based on a stimuli-responsive membrane associated with an electrode surface. *J Phys Chem B* 2007;111: 12141–12145.

84. Motornov M, Tam TK, Pita M, Tokarev I, Katz E, Minko S. Switchable selectivity for gating ion transport with mixed polyelectrolyte brushes: approaching "smart" drug delivery systems. *Nanotechnology* 2009;20:43.

85. Tam TK, Pita M, Motornov M, Tokarev I, Minko S, Katz E. Modified electrodes with switchable selectivity for cationic and anionic redox species. *Electroanalysis* 2010;22: 35–40.

86. Bocharova V, Tam TK, Halámek J, Pita M, Katz E. Reversible gating controlled by enzymes at nanostructured interface. *Chem Commun* 2010;46:2088–2090.

87. Pita M, Tam TK, Minko S, Katz E. Dual magneto-biochemical logic control of electrochemical processes based on local interfacial pH changes. *ACS Appl Mater Interfaces* 2009;1:1166–1168.

88. de Silva AP, Uchiyama S. Molecular logic and computing. *Nat Nanotechnol* 2007;2:399–410.

89. Szacilowski K. Digital information processing in molecular systems. *Chem Rev* 2008;108:3481–3548.

90. Credi A. Molecules that make decisions. *Angew Chem Int Ed* 2007;46:5472–5475.

91. Pischel U. Chemical approaches to molecular logic elements for addition and subtraction. *Angew Chem Int Ed* 2007;46:4026–4040.

92. Andreasson J, Pischel U. Smart molecules at work—mimicking advanced logic operations. *Chem Soc Rev* 2010;39:174–188.

93. Baron R, Lioubashevski O, Katz E, Niazov T, Willner I. Two coupled enzymes perform in parallel the "AND" and "InhibAND" logic gates operations. *Org Biomol Chem* 2006;4: 989–991.

94. Baron R, Lioubashevski O, Katz E, Niazov T, Willner I. Logic gates and elementary computing by enzymes. *J Phys Chem A* 2006;110:8548–8553.

95. Baron R, Lioubashevski O, Katz E, Niazov T, Willner I. Elementary arithmetic operations by enzymes: a model for metabolic pathway based computing. *Angew Chem Int Ed* 2006;45:1572–1576.

96. Privman V, Zhou J, Halámek J, Katz E. Realization and properties of biochemical-computing biocatalytic XOR gate based on signal change. *J Phys Chem B* 2010;114: 13601–13608.

97. Melnikov D, Strack G, Zhou J, Windmiller JR, Halámek J, Bocharova V, Chuang M-C, Santhosh P, Privman V, Wang J, Katz E. Enzymatic AND logic gates operated under

conditions characteristic of biomedical applications. *J Phys Chem B* 2010;114:12166–12174.

98. Zhou J, Arugula MA, Halámek J, Pita M, Katz E. Enzyme-based NAND and NOR logic gates with modular design. *J Phys Chem B* 2009;113:16065–16070.

99. Privman V, Arugula MA, Halámek J, Pita M, Katz E. Network analysis of biochemical logic for noise reduction and stability: a system of three coupled enzymatic AND gates. *J Phys Chem B* 2009;113:5301–5310.

100. Tam TK, Pita M, Katz E. Enzyme logic network analyzing combinations of biochemical inputs and producing fluorescent output signals: towards multi-signal digital biosensors. *Sens Actuators B: Chem* 2009;140:1–4.

101. Minko S, Katz E, Motornov M, Tokarev I, Pita M. Materials with built-in logic. *J Comput Theor Nanosci* 2011;8:356–364.

102. Tokarev I, Gopishetty V, Zhou J, Pita M, Motornov M, Katz E, Minko S. Stimuli-responsive hydrogel membranes coupled with biocatalytic processes. *ACS Appl Mater Interfaces* 2009;1:532–536.

103. Wang J, Katz E. Digital biosensors with built-in logic for biomedical applications. *Isr J Chem* 2011;51:141–150.

104. Wang J, Katz E. Digital biosensors with built-in logic for biomedical applications—biosensors based on biocomputing concept. *Anal Bioanal Chem* 2010;398:1591–1603.

105. Zhou J, Halámek J, Bocharova V, Wang J, Katz E. Bio-logic analysis of injury biomarker patterns in human serum samples. *Talanta* 2011;83:955–959.

106. Halámek J, Bocharova V, Chinnapareddy S, Windmiller JR, Strack G, Chuang M-C, Zhou J, Santhosh P, Ramirez GV, Arugula MA, Wang J, Katz E. Multi-enzyme logic network architectures for assessing injuries: digital processing of biomarkers. *Mol BioSyst* 2010;6:2554–2560.

107. Halámek J, Windmiller JR, Zhou J, Chuang M-C, Santhosh P, Strack G, Arugula MA, Chinnapareddy S, Bocharova V, Wang J, Katz E. Multiplexing of injury codes for the parallel operation of enzyme logic gates. *Analyst* 2010;135:2249–2259.

108. Privman M, Tam TK, Bocharova V, Halámek J, Wang J, Katz E. Responsive interface switchable by logically processed physiological signals—towards "smart" actuators for signal amplification and drug delivery. *ACS Appl Mater Interfaces* 2011;3:1620–1623.

109. Pardo-Yissar V, Katz E, Willner I, Kotlyar AB, Sanders C, Lill H. Biomaterial engineered electrodes for bioelectronics. *Faraday Discuss* 2000;116:119–134.

110. Katz E, Riklin A, Heleg-Shabtai V, Willner I, Bückmann AF. Glucose oxidase electrodes via reconstitution of the apo-enzyme: tailoring of novel glucose biosensors. *Anal Chim Acta* 1999;385:45–58.

111. Willner I, Heleg-Shabtai V, Blonder R, Katz E, Tao G, Bückmann AF, Heller A. Electrical wiring of glucose oxidase by reconstitution of FAD-modified monolayers assembled onto Au-electrodes. *J Am Chem Soc* 1996;118:10321–10322.

112. Bund A, Ispas A, Mutschke G. Magnetic field effects on electrochemical metal depositions. *Sci Technol Adv Mater* 2008;9:2.

113. Anderson EC, Fritsch I. Factors influencing redox magnetohydrodynamic-induced convection for enhancement of stripping analysis. *Anal Chem* 2006;78:3745–3751.

114. Legeai S, Chatelut M, Vittori O, Chopart JP, Aaboubi O. Magnetic field influence on mass transport phenomena. *Electrochim Acta* 2004;50:51–57.

115. Leventis N, Gao XR. Steady-state voltammetry with stationary disk millielectrodes in magnetic fields: nonlinear dependence of the mass-transfer limited current on the electron balance of the Faradaic process. *J Phys Chem B* 1999;103:5832–5840.

116. Leventis N, Gao XR. Magnetohydrodynamic electrochemistry in the field of Nd–Fe–B magnets. Theory, experiment, and application in self-powered flow delivery systems. *Anal Chem* 2001;73:3981–3992.

117. Katz E, Lioubashevski O, Willner I. Magnetic field effect on cytochrome *c*-mediated bioelectrocatalytic transformations. *J Am Chem Soc* 2004;126:11088–11092.

118. Lioubashevski O, Katz E, Willner I. Magnetic field effects on electrochemical processes: a theoretical hydrodynamic model. *J Phys Chem B* 2004;108:5778–5784.

119. Lioubashevski O, Katz E, Willner I. Effects of magnetic field directed orthogonally to surfaces on electrochemical processes. *J Phys Chem C* 2007;111:6024–6032.

120. Katz E, Lioubashevski O, Willner I. Magnetic field effects on bioelectrocatalytic reactions of surface-confined enzyme systems: enhanced performance of biofuel cells. *J Am Chem Soc* 2005;127:3979–3988.

121. Katz E, Heleg-Shabtai V, Bardea A, Willner I, Rau HK, Haehnel W. Fully integrated biocatalytic electrodes based on bioaffinity interactions. *Biosens Bioelectron* 1998;13: 741–756.

122. Bardea A, Katz E, Bückmann AF, Willner I. NAD$^+$-dependent enzyme electrodes: electrical contact of cofactor-dependent enzymes and electrodes. *J Am Chem Soc* 1997;119:9114–9119.

123. Gallaway J, Wheeldon I, Rincón R, Atanassov P, Banta S, Calabrese Barton S. Oxygen-reducing enzyme cathodes produced from SLAC, a small laccase from *Streptomyces coelicolor*. *Biosens Bioelectron* 2008;23:1229–1235.

124. Bartlett PN, Tebbutt P, Whitaker RC. Kinetic aspects of the use of modified electrodes and mediators in bioelectrochemistry. *Prog React Kinet* 1991;16:55–155.

125. Melnikov D, Strack G, Pita M, Privman V, Katz E. Analog noise reduction in enzymatic logic gates. *J Phys Chem B* 2009;113:10472–10479.

126. Privman V, Strack G, Solenov D, Pita M, Katz E. Optimization of enzymatic biochemical logic for noise reduction and scalability: how many biocomputing gates can be interconnected in a circuit? *J Phys Chem B* 2008;112:11777–11784.

127. Strack G, Chinnapareddy S, Volkov D, Halámek J, Pita M, Sokolov I, Katz E. Logic networks based on immunorecognition processes. *J Phys Chem B* 2009;113:12154–12159.

128. Katz E. Biofuel cells with switchable power output. *Electroanalysis* 2010;22: 744–756.

129. Halámek J, Tam TK, Strack G, Bocharova V, Pita M, Katz E. Self-powered biomolecular keypad lock security system based on a biofuel cell. *Chem Commun* 2010;46:2405–2407.

130. Halámek J, Tam TK, Chinnapareddy S, Bocharova V, Katz E. Keypad lock security system based on immune-affinity recognition integrated with a switchable biofuel cell. *J Phys Chem Lett* 2010;1:973–977.

131. Strack G, Luckarift HR, Nichols R, Cozart K, Katz E, Johnson GR. Bioelectrocatalytic generation of directly readable code: harnessing cathodic current for long-term information relay. *Chem Commun* 2011;47:7662–7664.

19

BIOLOGICAL FUEL CELLS FOR BIOMEDICAL APPLICATIONS*

MAGNUS FALK AND SERGEY SHLEEV

Department of Biomedical Sciences, Malmö University, Malmö, Sweden

CLAUDIA W. NARVÁEZ VILLARRUBIA, SOFIA BABANOVA, AND PLAMEN ATANASSOV

Department of Chemical and Nuclear Engineering and Center for Emerging Energy Technologies, University of New Mexico, Albuquerque, NM, USA

19.1 INTRODUCTION

An implantable medical device (IMD) operates inside the body to monitor certain physiological parameters. Based on the power requirements of the device, IMDs are divided into two categories: (i) active (the devices that need power) and (ii) passive (those that do not require power). Examples of passive IMDs are artificial joints and vascular grafts, whereas active IMDs may include cardiac defibrillators, cardiac pacemakers, neurostimulators, drug pumps, cochlear implants, and retinal implants. The need to broaden the assortment of effectively active IMDs is constantly increasing.

Current IMDs require energy sources that can be either internal or external. External power can be provided from radio frequency pulses or external wiring from a

---

*The content of this chapter was previously published in ChemPhysChem and is reproduced here in its entirety: Falk M, Narváez Villarrubia CW, Babanova S, Atanassov P, Shleev S. Biological fuel cells for biomedical applications: colonizing the animal kingdom. *ChemPhysChem* 2013;14(10):2045–2058. Copyright 2013, Wiley-VCH Verlag GmbH. Reproduced with permission.

Enzymatic Fuel Cells: From Fundamentals to Applications, First Edition. Edited by Heather R. Luckarift, Plamen Atanassov, and Glenn R. Johnson.
© 2014 John Wiley & Sons, Inc. Published 2014 by John Wiley & Sons, Inc.

direct electrical supply [1]. Internal power can be provided via integrated batteries. These internal power supplies must supply reliable and sufficient energy for up to 10 years of continuous performance, be made up of biocompatible materials, and demonstrate long-term stability [1].

Research on power devices capable of harvesting energy *in vivo* from physiologically available organic compounds is striving to eliminate the need for an external recharging process. This recharging process is currently unavoidable, due to the use of batteries as power supply devices [2–5]. Advances in technology in the development of IMDs for health monitoring, organ replacements (or their simulations), drug delivery systems, and so on, have encouraged the research for alternative energy sources [2,4,6–8].

A fuel cell capable of harvesting electrical energy while operating *in vivo* is a promising alternative electric power source. Implantable fuel cells must ensure the battery's reliability criteria, such as continuous power supply, low self-discharging time, high reliability and efficiency, biocompatibility, miniature size, lightweight, and operation under physiological conditions [6]. In order to achieve these reliability requirements, the fuel cell design must overcome certain criteria inherent in its nature, such as catalyst stability, efficient fuel and oxidant mass transport, biocompatibility, and stability of the products generated within the cell. In contrast to a battery, fuel cells rely on reactants being introduced to the system, reactions occurring in the cell compartment, and products being released from it, which adds complexity to the biocompatibility criteria.

In the 1960s, feasibility studies of a glucose/O_2 fuel cell to power pacemakers and parametric studies on glucose anode for implantable applications were performed [9]. In this area, the Artificial Heart Program of the U.S. National Heart, Lung, and Blood Institute was the crucial investment [6]. The conversion of glucose to gluconic acid, with a two-electron transfer (ET) process, was demonstrated to be acceptable to power a prosthetic heart. The extraction of O_2 required to power a 12.5 mW implantable fuel cell was found to be dependent on the diffusion coefficient of molecular oxygen in blood; furthermore, the process was considered to be marginally feasible for this application [9].

In 1967, Jacobson presented a glucose/O_2 fuel cell prototype with an air-breathing cathode having a stable electrochemical output, generating a power density of $165\,\mu W\,cm^{-2}$ within the initial 24 h of operation (total 240 h) [10]. Later experiments showed differences in behavior of the device under physiological conditions when it was tested in pleural fluids and plant saps, in all likelihood due to protein adsorption on the electrode surface.

Both *in vitro* and *in vivo* studies, as well as animal trials, showed the feasibility of the implantable power device to work using glucose, which is naturally present in the blood. Four years after the introduction of glucose and the oxidation of other carbohydrates for power generation by Bockris et al. in 1964 [11], Wolfson et al. proposed the idea of fueling catalytically active electrodes by carbohydrates and O_2 occurring in interstitial fluids [8]. In 1967, Talaat et al. demonstrated the current output generation from electrodes immersed in the bloodstream [12].

In 1970, Drake et al. proposed another design to overcome the limitation of the physiological working conditions and performed 6-month *in vitro* and 30-day animal testing of their prototypes [6]. In animal tests, a maximum power output of 6.4 μW cm^{-2} for a short period of time and a power density of 2.2 μW cm^{-2} for continuous operation were reported. In the same study, a new fuel cell design was introduced, with dialysis membrane to limit the flow of high molecular weight compounds to the cell, an external biocompatible coating of the device to avoid unfavorable immunological responses by the host organism, and isolation of the cathode by O_2 and CO_2 phase separation within the cell [6].

Beginning in 1968, the implantable prototypes of fuel cells intended to power a pacemaker were developed by the American Hospital Supply Corporation [6], the Michael Reese Hospital [8], Siemens [13], and Tyco [14]. Industries also invested in the organochemical redox systems in order to develop devices that are able to oxidize not only glucose but also other fuels, such as amino acids [15]. However, the introduction of lithium iodine batteries as a power supply for pacemakers, and the improvement in its lifetime, led to a change in the direction of the application of glucose/O_2 fuel cells toward sensor technology [16,17].

Most of the implantable cardiac pacemakers need to be replaced because of the battery exhaustion within 5–8 years. The longevity of the IMDs is determined by the battery life. When the service life of the battery ends, it needs to be replaced, causing the patient to undergo painful surgery and incur enormous expenses. In contrast, the main advantage of fuel cells for IMD is its theoretical potential to function as long as the individual is alive. The devices can use substrates that already exist in the body—no artificial fuel and oxidants are required.

The basic challenge for the power generators designed for IMD is that they have to operate at physiological conditions, that is, under conditions that are a problem for the traditional fuel cells, in which reactions take place in acidic or alkaline media above 150 °C. This is combined with the low buffer capacitance of the intestinal fluids and the presence of substances, such as humic acids, which can cause electrode poisoning. Another issue is the simultaneous presence of both reactants (for the anode and cathode reactions), which requires high specificity of catalysts to avoid electrode depolarization. All of these drawbacks can be overcome by the application of biocatalysts and the construction of biological fuel cells (BFCs). Thus, among many different types of fuel cells, BFCs are promising devices for *in vivo* applications [2,4,7,18,19]. BFCs can operate under physiological conditions (neutral pH, temperatures between 25 and 50 °C, atmospheric pressure, etc.), converting the naturally present substrates into products that are tolerable to the host.

19.2   DEFINITION AND CLASSIFICATION OF BFCs

At this point, what is the exact definition of a BFC? According to some authors, BFCs are devices that use redox proteins and enzymes, microorganisms, or mitochondria, in order to assimilate a "fuel" and generate electric power [20]. At the same time, there is another broader definition in the literature: BFCs are fuel cells that convert organic and/or inorganic substrates, which are defined as biofuels and bio-oxidants [19,21]. The principles of these two types of biodevices are significantly different. The first type of device is developed on the basis of naturally occurring processes in the living cells and uses biological catalysts. The second type of device uses more "extravagant" fuel sources and explores nonbiological catalysts. In this chapter, the term BFCs will be used to describe the first type, which is more accurate in the view of operation and composition.

To design practically applicable BFCs for powering or for use as an IMD, several technologies have to be developed. The biodevices are aimed for implementation in different parts of the body, covering different needs with different applications, including micro- and nanoelectrodes to transfer the signal between nerves and muscles, sensors to monitor specific parameters in the body, power sources for implants—generating power during the patient's entire life, data communication devices between the body and an external unit, and sensing systems to provide trigger mechanisms for other implants. Thus, we need to discuss existing classifications of biodevices and decide what types of BFC designs are feasible to be used for powering IMDs. Different types of BFCs based on different biocatalysts, such as whole living cells, organelles, or isolated redox proteins and enzymes, have their own advantages and disadvantages, as described below.

19.2.1 Cell- and Organelle-Based Fuel Cells

In microbial fuel cells (MFCs), whole microorganisms are used to create the bioelectrodes, granting a high stability and long lifetime to the constructed biodevices [22]. By employing whole microorganisms as catalysts, the fuel cell is capable of fully oxidizing different organic fuels, such as sugars, fatty acids, and alcohols. This gives MFCs an advantage over other types of BFCs, having higher efficiency and longer lifetimes. However, a whole cell is significantly larger than organelles or isolated proteins, making MFCs generally rather large with small current and power densities. MFCs also suffer from slow mass transport of the fuel across the cell membranes and a low electron–electrode surface interaction due to biofilm formation [23]. A major issue for using MFCs to power IMDs is the high risk of infection, resulting in MFCs generally not being considered for implantable applications.

In 2010, Han et al. proposed to use an MFC implanted in the human colon [24]. Because of the large number of anaerobic microorganisms naturally present in the intestinal mucosa, and the measurable population of aerobic microorganisms in the lumen flowing with fecal mass, the proposed idea is intriguing and the risk of infection would not be a major concern. Specifically, a tubular MFC was designed, with a microbial anode intended to adhere to the colon mucosa and a platinum-containing cathode intended to be located in the center of the lumen. The rather large MFC (with a volume of 125 ml for each compartment) was investigated in a fluid designed to simulate the intestinal fluid. The biodevice generated a stable power generation for 2 months with an open-circuit voltage (OCV) of 552 mV and a maximum power density of $73 \, \mathrm{mW \, m^{-2}}$. The power output was limited by the cathode, which was attributed to the small surface area of the electrode and the low O_2 concentration.

Instead of using bacterial cells, Justin et al. proposed a new innovative approach employing white blood cells to generate current in 2011 [25]. The authors called the new biodevice a "metabolic biological fuel cell," which describes the nature of the power production very well—metabolic conversion of the substrate from living cells that are not microorganisms. The white blood cells were isolated from human blood and used as bioreactors at the anode. The two-chamber BFC separated by a Nafion proton exchange membrane was designed by combining the bioanode with a carbon fiber electrode placed in the cathodic compartment containing potassium ferricyanide. Although the obtained current densities were rather low and a long way from any practical application, with current output between 0.9 and $1.6\,\mu A\,cm^{-2}$, the research shows the possibility of using human cells and fluids to generate power. Such a design would overcome many of the problems connected with BFCs operating *in vivo*. The same group of researchers theorize how such a BFC could be used for powering neural implants [26].

Beginning in 2008, as an alternative to employing whole cells in the design of BFCs, Minteer and coworkers explored the possibility of using mitochondria as biocatalysts [27,28]. Mitochondria offer a compromise between the high efficiency of MFCs and the high volumetric catalytic activity of enzymatic fuel cells (EFCs). They contain all of the enzymes needed for certain fuel oxidation; moreover, researchers have shown that pyruvate, succinate, and fatty acids can be used as fuel with experimental evidence indicating the possibility of using amino acids [27]. Reported mitochondrial fuel cells have shown better performance ($\sim 0.2\,mW\,cm^{-2}$) compared with MFCs (~ 0.001–$0.1\,mW\,cm^{-2}$) [28], which can be attributed to several reasons. First, mitochondrial fuel cells have lower fuel transport limitations compared with MFCs; the lack of a cell membrane in the mitochondrial-based biodevice allows for small diffusion lengths. Second, use of mitochondria allows for a higher load of biocatalysts on the electrode surfaces. Third, mitochondria do not form biofilms that interfere in the electrode-active site–electrolytic solution interaction, which can decrease the actual efficiency of MFCs.

19.2.2 Enzymatic Fuel Cells

Enzymes are, in general, exceptional catalysts with a high substrate specificity, and due to the smaller size of enzymes compared with whole cells and organelles, enzymes can at least in theory be used to create fuel cells with higher power density compared with other types of BFCs. Indeed, EFCs with a power density of more than $1\,mW\,cm^{-2}$ with an OCV close to 1 V have been reported in the literature [29].

EFCs are very promising, when considering biocompatibility, selectivity, efficiency, and sensitivity criteria. In addition, EFCs could be employed to power nano- and microelectronic portable devices, drug delivery systems, biosensors, and IMDs [2,4,7]. The high selectivity of enzymes makes their use as natural catalysts in fuel cell applications very beneficial by eliminating problems of cross-reactions and poisoning of the electrodes. This allows for membraneless single-compartment fuel cells and casings to be designed, while still mitigating voltage loss that otherwise could arise. Enzymes can potentially be produced at a very low cost and display activity at neutral pH and room temperature; that is, operational conditions of IMD. However, it should be noted that when comparing EFCs with other types of BFCs, the efficiency of these biodevices is limited by incomplete fuel oxidation and shorter lifetime because of limited enzyme stability.

Due to the many advantages compared with other catalysts, as new enzymes and ways to implement them in electrode designs have emerged, the research interest of EFCs has grown significantly. For example, in 2004, Calabrese Barton et al. published a thorough comparison of the strengths and weaknesses of BFCs, listing possible applications [2]. In the same year, the advances in miniature BFCs were reviewed by Heller [4]. In 2008, Armstrong and coworkers published a detailed account of the use of enzymes in fuel cell technology [30]. Significant progress has been made in the last decade, and when considering mainly the implantable applications, the use of sugar/O_2 BFCs has received a lot of attention. Therefore, we chose to focus the remainder of this chapter on the design and development of EFCs, with special attention given to their biomedical applications.

19.3 DESIGN ASPECTS OF EFCs

Many aspects should be considered in the design of an EFC, where the most suitable design is dependent on the intended application. Some medical products that could potentially benefit from incorporation with EFCs are cochlear and retinal implants, functional electrical stimulation and intracranial pressure sensor implants, glaucoma and sphincter sensors, and artificial sphincters. Each of these applications imposes different demands on the IMD; for example, brain fluid has different characteristics (fuel concentration, pH, temperature, free O_2 content, etc.) compared with blood, saliva, tears, urine, and other physiological fluids. Enzyme choice is determined by the available fuel and oxidant in the host organism, which then is combined with a suitable electrode material. In order for the biodevice to function, an electrical connection between the biocatalyst (redox enzyme) and the electrode surface needs to be achieved. Furthermore, systems for mass transport of a substrate from the body fluid to the electrodes, as well as size and shape of the BFC, need to be considered based on the requirements of the intended IMDs, be it intravascular, extravascular, or transcutaneous.

19.3.1 Electron Transfer

For an EFC to operate, the enzyme needs to be electrically connected to the electrode surface. The ET between enzymes and electrodes has been reviewed in detail, for example, by Habermüller et al. in 2000 and Katz et al. in 2002 [31,32]. Electronic connection can be achieved via direct electron transfer (DET) or mediated electron transfer (MET) mechanisms.

In the first case, redox species, either in solution or immobilized in redox polymers attached to the electrode surface (second-generation biodevices), shuttle electrons between the enzyme and the electrode. Pioneering work in this research field applied to BFCs was performed by Heller in 2006 with many significant and important contributions, particularly regarding mechanisms to electrically connect to the redox sites of different enzymes with redox hydrogels [33]. To facilitate the electrical connection between enzymes and electrodes, Heller and coworkers attached enzymes to ferrocene groups [34]; additionally, they were the first to use Os-based redox polymers to efficiently "wire" enzymes to electrodes [35–38]. By using this design to create EFCs, a significantly higher power output and simpler construction could be achieved compared with earlier reported BFCs, by immobilizing the mediator together with the enzyme on the electrode surface [39]. The development of electrically "wired" enzymatic electrodes has had important impacts on both enzyme-based biosensors and BFCs [40,41].

In the case of DET, the active center of the enzyme is directly electrically connected to the electrode surface. In 2012, Shleev and coworkers reviewed the specifics of different DET-based BFCs [42]. However, only a minority of known redox enzymes are capable of DET [43]. The catalytic site is often buried deep within the protein matrix; consequently, the active center is insulated.

These two approaches have different advantages and disadvantages. MET-based biodevices usually have very efficient ET between the enzyme and the electrode surface, which leads to comparatively larger current output. However, to facilitate the wiring of enzymes, the redox potential of the redox polymer needs to be slightly positive toward the anodic enzyme to create the bioanode, and, vice versa, slightly negative to create the biocathode. This translates to a small voltage loss for MET-based devices compared with DET-based BFCs. The use of mediators in implantable devices could also potentially generate issues with toxicity of fabricated biodevices.

19.3.2 Enzymes

Due to the abundance of sugars and O_2 available in most implantable environments and generally nontoxic by-products, sugar-oxidizing and O_2-reducing enzymes are very attractive biocatalysts for designing EFCs.

Blue multicopper oxidases (BMCOs) such as laccase, ceruloplasmin, bilirubin oxidase (BOx), and ascorbate oxidase (AOx) have been extensively investigated as cathodic biocatalysts for DET-based biodevices [44]. BMCOs have a catalytic center consisting of four coppers: a type 1 (T1) Cu site, which accepts electrons from the substrate and from the electrode surface, and a type 2/type 3 (T2/T3) cluster, where O_2 is reduced directly to water. High redox potential laccases and BOx, with redox potential up to 780 and 670 mV versus normal hydrogen electrode (NHE), respectively [44,45], can be used to create efficient biocathodes with current densities up to a few mA cm^{-2}. In 2012, Shleev and coworkers used the DET ability of these enzymes to create several completely DET-based BFCs [42]. The enzymes have also been used in different MET-based approaches [46,47]; specifically, Heller and coworkers demonstrated the ability to create powerful biocathodes by "wiring" laccase or BOx using electrically conducting Os-containing hydrogels [48,49].

The performance of different biological components under operating conditions should also be considered. Laccases have optimum activity in acidic media and are inhibited by chloride ions, whereas BOx is chloride resistant and very active at neutral pH at the cost of having slightly lower redox potential (100 mV lower than that of laccases) [50]. For EFCs intended to operate in vertebrates, which have blood containing high chloride concentrations and with a neutral pH, BOx is the natural choice [51]. However, as noted by Shin et al. in 2007 and Wong et al. in 2008, deactivation of the enzyme occurs in the presence of urate, a compound present in serum [52,53]. For operation in organisms where the pH is lower, for example, grapes, high redox potential laccases are most suitable.

The most usual choice for anodic enzyme has been glucose oxidase (GOx) [53], which, when under mediated ET conditions, can effectively electro-oxidize glucose. The enzyme carries a flavin (flavin adenine dinucleotide (FAD)) buried deep within the enzyme, which makes DET difficult. Although DET with GOx has been reported in many different studies [54–57], there is an ongoing debate as to whether true DET is achieved, or whether the observed bioelectrocatalytic currents are due to naturally mediated glucose oxidation by free FAD, which has diffused out from the active centers of some partly denatured enzyme molecules.

A problem with using GOx in implantable conditions is that O_2 is the natural electron acceptor, and GOx produces hydrogen peroxide, a toxic by-product, in the presence of O_2. Hence, additional modifications of biodevices are required for their operation *in vivo*. In addition to the undesired biocompatibility issues, hydrogen peroxide can be both reduced and oxidized, and thus can create parasitic (crossover) currents, particularly in long-term, continuous operation of a biodevice. To avoid the influence of O_2 on bioanodes, other sugar-oxidizing enzymes can be used. One of the examples is the reduced form of nicotinamide adenine dinucleotide (NADH)-dependent glucose dehydrogenase (GDH). However, no reports of DET-based bioelectrocatalysis for this enzyme exist so far, and the ET in GDH-based bioanodes is generally performed using a nicotinamide adenine dinucleotide (NAD$^+$) cofactor [58]. To facilitate this, different catalysts for oxidation of NADH can be used as free mediators (in solution) or polymerized on the surface of the electrode [59–61]. However, NAD$^+$/NADH has not been successfully immobilized on an electrode surface, but rather needs to be fed externally to the cell, limiting the utility of such enzymes in implantable applications. To achieve DET, pyrroloquinoline quinone (PQQ)-dependent GDH can be used where the PQQ cofactor is part of the enzymatic structures [62–64]. Alternatively, cellobiose dehydrogenase (CDH) has been used to create DET-based bioanodes [65]. CDH consists of both FAD and heme domains, and the heme facilitates the electrical coupling to the electrode surface. A downside with using CDH is that glucose is not the natural substrate of the enzyme; thus, the catalytic efficiency is significantly reduced compared with GOx-based biodevices.

When considering EFCs operating in organisms other than vertebrates, such as plants and mollusks, other fuels apart from glucose are available. This opens up the possibility of using a large variety of enzymes, where the suitability depends on the organism in which they are intended to operate. Fructose dehydrogenase (FDH), an enzyme carrying one FAD and one heme c domain, which enables DET [66], can be used in plants to oxidize fructose. Trehalase can be used in insects to oxidize the sugar trehalose [67], which is hydrolyzed to glucose by the enzyme.

19.3.3   Electrodes and Electrode Materials

Development of new nanomaterials gives a basis to design very efficient and stable EFCs [68,69]. Many different strategies have been developed to incorporate or immobilize enzymes on different nanostructured electrodes based on mesoporous materials, nanoparticles, nanotubes, and nanocomposites [70]. Two examples of different nanostructured electrodes are shown in Figure 19.1.

Ideally, electrodes should have a high electrical conductivity, provide stabilizing interactions with enzymes, have a high surface to volume ratio to allow for a high enzyme load, and contain sufficiently large pore size to facilitate efficient mass transport of fuel(s) and oxidant(s) [2,71,72].

Electrodes made from spectrographic graphite are attractive to use due to their low cost, naturally high porosity, and uncomplicated nature [73]. Electrodes with a very high surface roughness can also be created via modification with metal or carbon nanoparticles, such as gold nanoparticles (AuNPs) or Ketjenblack (KB), respectively, and use of carbon nanotubes (CNTs). The development of composite carbon nanomaterials, such as single-walled carbon nanotubes (SWCNTs) and multiwalled carbon nanotubes (MWCNTs) has had a large impact on the development of EFCs. CNTs have many attractive properties, such as high electrical conductivity, thermal and chemical stability, and high mechanical strength, while displaying favorable interactions with enzymes [63,74–77]. CNT-based buckypapers are commercially available, and other carbon fiber papers such as Toray papers have also been employed as electrodes, displaying an enhanced ET due to high conductivity and porosity [48,78–80].

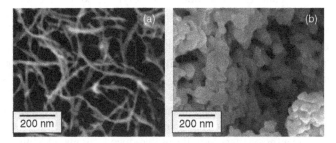

FIGURE 19.1   Scanning electron microscopy (SEM) images of nanostructured surfaces: (a) buckypaper electrodes fabricated using MWCNTs; (b) nanostructured gold electrode based on AuNPs.

19.3.4 Biodevice Design

The early developed implantable abiotic fuel cells, researched in the 1970s and implanted in rats, sheep, and dogs, were all rather bulky devices [6,13,81]. This minimum size restriction is still true even for modern batteries; because they require a casing to contain the electrolyte, they are therefore excluded for use in truly miniature applications. One of the salient points of EFCs is the ability for miniaturization and can, as remarked by Professor Heller, be made significantly smaller than batteries and other types of fuel cells [4,82]. Due to the high specificity of enzyme catalysts, no separator membranes are generally needed. By employing either electrode-immobilized mediators or a DET-based approach, and by having the enzymes immobilized on the electrodes, truly miniature devices for implantation can be created by excluding all membranes and compartments. Several miniature BFCs have been reported in the literature (see above), and were operated under physiological conditions both *in vitro* and *in vivo*.

When an O_2-sensitive enzyme is used in the design of the bioanode, membranes for fuel separation are necessary. The presence of a separator leads to increased resistance (ohmic losses) and, therefore, reduced power output, making miniaturization of the device very difficult. A possible solution is a nanoporous silica coating, working as a functional membrane [83]. Membranes are also necessary when soluble species are used in BFC construction. However, the need for membranes can also be circumvented by using enzymes, which oxidize glucose without subsequent O_2 reduction combined with enzyme-modified electrodes based on immobilized compounds without soluble species, thereby avoiding cross-reactions between the anode and cathode. This type of design is well suited for implantable BFCs and can be employed either by electrically connecting the enzymes to the electrodes via redox hydrogels or by using a DET-based mechanism.

Operation *in vivo* compared with *in vitro* poses many additional difficulties. First, *in vivo* conditions can be quite harsh, especially in vertebrates. These organisms have an aggressive immune system in which implanted BFCs will be subjected to a multitude of different proteins and cells present in the blood that will interact with the implant [84]. Biocompatibility is a major issue for long-term implantation. An implant placed into a blood vessel or a tissue will be recognized by the body as a foreign element, which can trigger an inflammation reaction. To avoid this, the materials used in the BFC design should be bioinert and biocompatible, and should preferably avoid triggering the host's immune system. Physiological fluids are also very complex liquids containing a multitude of low molecular weight compounds, which can affect the performance of the BFC. Second, the amount of sugar and O_2 available in the living organism is also quite limited. Free O_2 is significantly lower inside organisms compared with air-saturated solutions *in vitro*; approximately five times lower in human blood, for example [85]. The sugar of choice for many implantable applications is glucose; for example, ~5 mM being present in humans and 30 mM in grapes [85,86]. In organisms other than vertebrates, other sugars are present, such as trehalose in insects or fructose in plants, and can potentially be used as fuel by use of suitable anodic enzymes. Third, slow diffusion of carbohydrates and O_2 inside the living organism can severely reduce the performance of implanted BFCs.

The problem with a limited amount of fuel available for the bioanode can be lessened by designing bioanodes with better fuel efficiency often associated with "deep" and ultimately complete fuel oxidation. Glucose/O_2 EFCs generally make rather inefficient use of the fuel, where only two electrons are obtained from each glucose molecule. Since 2008, substantial effort has been made by the Minteer research group to introduce more efficient oxidation of biofuels by enzyme cascades, which are sets of enzymes that catalyze consecutive processes in which the product of the previous enzymatic oxidation is the substrate of the next one [87–92]. In 2013, Shao et al. demonstrated a more efficient glucose bioanode, where six electrons per molecule of glucose could be obtained via co-immobilization of different glucose-oxidizing enzymes, significantly increasing the current density and coulombic efficiency of the bioanode [93]. This strategy can be especially important for applications where the concentration of glucose is limited and where the bioanode limits the overall performance of the BFC. The approach of using multiple enzymes and enzyme cascades is powerful, but as of this writing it has yet to be demonstrated in any *in vivo* application.

By designing transcutaneous implants, many of the issues with the biocathode can be mitigated. A significant increase in effectiveness can be achieved by employing a gas diffusion biocathode, where O_2 is supplied to the electrode surface from the air, and protons are supplied from the electrolyte solution. Such a biocathode can be designed by immobilizing an O_2-reducing enzyme in a catalytic layer being supplied with O_2 via diffusion through a hydrophobic layer from the surrounding air, such as those described in previous studies [94–96]. This drastically increases the supply of O_2 to the biocathode and prevents contact with enzyme inhibitors present in physiological fluids [97]. Air-breathing devices can in theory be made very powerful (up to $100\,mA\,cm^{-2}$), where the porosity, thickness, hydrophobicity, and ionic conductivity of the liquid phase are all important factors, which should be taken into consideration when the biodevice is designed [98].

The limited amount of fuel and oxidant, as well as the slow diffusion of these compounds in the tissue, can be resolved for some applications by shifting the implant site from the physiological or extravascular tissue to an intravascular implant. However, good biocompatibility is of utmost importance, since blood coagulation can lead to thrombosis in the host organism, with potentially lethal consequences.

An enzyme's rather low stability and potential for deactivation in the presence of physiological fluids are significant hurdles for the practical realization of implanted EFCs. When considering IMDs requiring surgery, yearlong stability or more is required for viability of the biodevice. However, by using different rational immobilization strategies, enhanced stability and efficiency of the created biodevices can be realized. In 2012, Gutierrez-Sanchez et al. reported on a laccase-based biocathode with enhanced DET and tolerance to chloride anions, created by using AuNPs functionalized with diazonium salts [99]. Furthermore, in 2010, Ivnitski et al. showed that by encapsulating enzymes in a CNT/Nafion membrane or silica gel/CNT matrix, stability could be increased [100]. Stability for 30 days or more has been demonstrated by using silica gel matrices [80,95], chitosan [101–104], modified Nafion [101,105], and by tethering compounds that link the enzyme to the electrode surface [37,106,107], with yearlong stability shown for enzymes physically confined in micelles [105,108]. The creation of a matrix to entrap enzymes on the electrode surface is the best approach for the enzyme to gain structural stability outside its natural environment, although mass transfer problems might arise.

19.4 *IN VITRO* AND *IN VIVO* BFC STUDIES

During the past decades, researchers have focused on identifying different bio-catalytic species and their mediators and cofactors, isolating redox enzymes, under-standing their mechanisms of action, stabilizing biocatalysts *ex vivo*, and developing new materials and feasible electrode designs. Besides investigations *in vitro*, designs also need to be evaluated *in vivo* in order to validate the biocompatibility of materials and the performance of the devices. Evaluation of BFCs, the materials from which they are constructed, and the products of bioelectrocatalytic reactions have to satisfy safety standards (chemical and mechanical stability and biocompatibility) and output energy requirements by controlled energy release systems, according to their application.

19.4.1 *In Vitro* BFCs

Development of feasible BFC designs that cope with the mass transport criteria for implantable applications is important for the realization of efficient BFCs. In order to improve mass transport properties, BFCs have been incorporated into microfluidic systems, where the enzymes were dissolved in the electrolyte and membranes were employed to avoid cross-reactions [101,105,109,110]. However, since no mixing of the substrate was observed due to low flow rates, the separator could be excluded. In 2009, Zebda et al. developed a Y-shaped microfluidic system with a flow rate of $1000\,\mu l\,min^{-1}$ of $10\,mM$ glucose with GOx and laccase in the solution, reporting a power density of $110\,\mu W\,cm^{-2}$ [111,112]. Two years later, Rincón et al. introduced a 3D flow-through microfluidic system using dehydrogenase immobilized in a chitosan/ CNT polymer within a foam-form reticulated vitreous carbon electrode and NADH in the solution, combined with a laccase-based biocathode [113]. For all these systems, the fuel needs to be pumped through an external device, which is naturally unrealistic for IMDs. In 2012, Ciniciato et al. showed the feasibility of employing a "fan-shaped" quasi-2D system [114]. The catalytic layer was fed with fuel by capillary action of microfluidic channels made of cellulose material, which reduced the mass transport limitation of the BFC.

In 2010, in order to design miniature BFCs, Pan et al. used a proton-conductive nanowire to design a glucose/O_2 EFC [115]. Laccase was used to design the biocathode and GOx was used in the bioanode, where the nanowires served as a substrate carrier and H^+ conductor. The whole biodevice could be realized at a micrometer scale, and the BFC was able to generate a maximum power of $30\,\mu W\,cm^{-2}$

in buffer with an OCV of 0.23 V. In blood, however, the maximum power output was reduced to $5.6\,\mu W\,cm^{-2}$ with an OCV of 0.12 V. The problem associated with the design, when implantation is considered, is the fact that laccase was used to design the biocathode, making operation under physiological conditions rather inefficient.

Another example of a miniature EFC was reported in 2012 by Falk et al. [116], using a rather simple design with the enzymes immobilized on AuNP-modified microwires. The EFC used CDH at the bioanode and BOx at the biocathode, and the biodevice was investigated in human tear fluid *in vitro* in a microscaled cell with a total volume of $10\,\mu l$. When harvesting glucose from tear fluid, the EFC generated a power density of $1\,\mu W\,cm^{-2}$ at a cell voltage of about 0.5 V. This type of design has been proposed for noninvasive *ex vivo* situations, where the EFC could be incorporated into a contact lens and used to power sensors or other electronic devices. The main drawbacks of this design, as the authors pointed out, are the stability of enzymes for long-term operation, as well as the inefficient oxidation of glucose, which is present at quite low concentration in tears, by CDH. The operational voltage of the device was decreased by cross-reactions occurring at the biocathode; however, incorporating a gas diffusion biocathode could increase it. Modeling of the system and calculations for feasibility demonstrated that power output for the cell, at glucose concentration of $50\,\mu M$ and a rate of tear production of $3.4\,\mu l\,min^{-1}$, should be $0.3\,\mu W$. Albeit low, it would suffice to power a $1\,mm^3$ wireless autonomous system, requiring only $5.3\,nW$ at $500\,mV$. This study demonstrated the feasibility of the design to be employed for powering contact lens-based bionic devices.

An interesting concept for BFCs is not only the use of fuel cells as pure power sources, but also their integration with sensing systems to either power or provide a trigger mechanism for other implants. This could allow effective "communication" between complex physiological processes and electronic systems, providing autonomous, individual, and on-demand medical care [117]. Starting in 2009, the Katz research group designed several switchable BFCs logically controlled by the change in pH solution [117–119], changing the BFC operation between ON and OFF and, thus, the outgoing signal. The devices were created using different enzymes, where the presence of a specific chemical or a combination of chemicals triggers or inhibits particular enzymatic reactions, which leads to pH changes in the electrolyte. By influencing the pH of the solution, the power output of the BFC is also drastically changed. The first prototype of such a BFC, controlled by biochemical reactions to deliver power on demand for an appropriate IMD, according to physiological needs, was published in 2009 by Katz and coworkers [119]. The cathode was modified with a special pH-switchable Os polymer and laccase was dissolved in solution, making the electrode active toward O_2 electroreduction at pH below 4.5 and inactive above pH 5.5. As illustrated in Figure 19.2, by combining a laccase-based biocathode with a GOx-based bioanode employing enzymatic logic systems, the EFC could be switched off by increasing the pH above 5.5 and switched on by decreasing the pH below pH 4.5, controlled by the enzyme logic operations. Other designs have instead used aptamer- [120], antigen–antibody- [121], and DNAzyme-controlled BFC logic systems [122]. Controlled energy releases by logically processed biochemical signals have great promise in implantable applications.

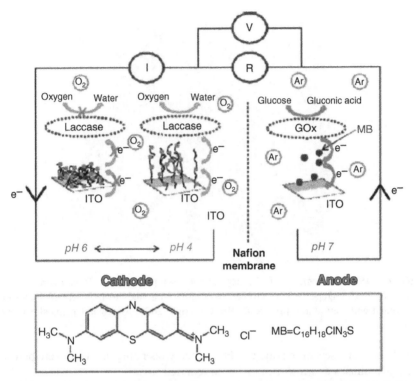

FIGURE 19.2 Glucose/O$_2$ EFC composed of a bioanode and a pH-switchable logically controlled biocathode. (Reprinted with permission from Ref. [119]. Copyright 2009, American Chemical Society.) (Please see the color version of this figure in Color Plates section.)

In 2011, Hanashi et al. described an example of using a BFC as a sensing device [3]. Specifically, they designed a self-powered, wireless glucose-sensing device for diabetes control by integrating an EFC together with a radio transmitter and a capacitor. The EFC was designed by using recombinant FAD-dependent GDH immobilized by glutaraldehyde to create the bioanode and BOx mixed with platinum/carbon/Nafion ink and further treated with glutaraldehyde to make the biocathode. The electrodes were based on KB-printed electrodes with sputtered platinum deposition. When glucose was present in the solution, the BFC generated an electric current that charged the capacitor. The charging/discharging cycle of the capacitor was used to determine the concentration of glucose. Simultaneously, a resonance frequency was generated and detected by a radio receiver. The designed EFC could detect glucose in the range between 1.5 and 6.6 mM.

19.4.2 *In Vivo* Operating BFCs

Many examples of potentially implantable BFCs exist nowadays; nonetheless, few records of actually implanted devices have been reported. However, in just the last

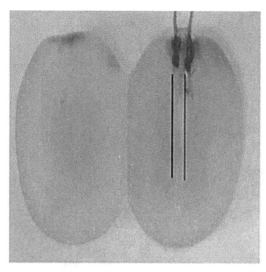

FIGURE 19.3 Photograph of the experimental setup of a GOx/BOx microscale EFC implanted into a grape (two electrodes are drawn with lines to highlight their position). (Reprinted with permission from Ref. [49]. Copyright 2003, American Chemical Society.)

couple of years, several reports of EFCs actually operating *in vivo* in vertebrates, as well as mollusks and insects, have been reported.

The very first report of an implanted EFC was published in 2003 by Heller and coworkers [49]. The work was a result of earlier efforts by the group in wiring enzymes to electrode surfaces [39,41], and it led to the first EFC actually being implanted into a living organism, a grape. Grapes contain rather high amounts of glucose, more than 30 mM, and have an acidic pH (roughly between pH 2.8 and 4.5) [123]. A photograph of the implanted EFC is shown in Figure 19.3, displaying a truly miniature cofactorless and membraneless EFC. The electrodes were made from 2 cm long and 7 μm wide carbon fibers, where BOx was used to create the biocathode and GOx was used to design the bioanode. The enzymes were incorporated into different Os-containing redox polymers to facilitate the electronic coupling. The power output generated by the EFC *in vivo* was heavily dependent on the position of the biocathode; close to the skin a maximum power of $240\,\mu\text{W}\,\text{cm}^{-2}$ was registered, whereas near the center a maximum power of $47\,\mu\text{W}\,\text{cm}^{-2}$ was obtained, both at a cell potential of about 0.52 V. The power output was clearly limited by diffusion of O_2, and after 24 h operation, 78% of the initial power output remained.

In 2010, the first EFC implanted in an animal was reported by Cosnier and coworkers, who demonstrated the successful function of an implanted biodevice in a living organism that could be accepted as a model organism with physiology close to the human body [110]. A glucose/O_2 EFC was implanted in the retroperitoneal space of a rat and could provide a peak power of $24.4\,\mu\text{W}\,\text{cm}^{-3}$ at a cell voltage of 0.13 V. No significant inflammation was registered after 3 months of implantation, although formation of a vascular network was observed around the implant. The recorded

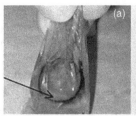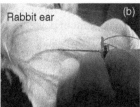

FIGURE 19.4   (a) Photograph of a GDH/PPO EFC based on soluble mediators encased by membranes implanted in a rat. (Reprinted with permission from Ref. [110]. Copyright 2010, PLoS.) (b) Photographs of the assembled BFC for power generation from a rabbit vein. (Reprinted with permission from Ref. [124]. Copyright 2011, Royal Society of Chemistry.) (c) Photograph of the experimental setup during *in vivo* tests of a CDH/BOx microscale EFC operating in the brain tissue of a living rat [125]. (Please see the color version of this figure in Color Plates section.)

power output was significantly lower than what could be achieved in grapes; however, the conditions in which the EFC operated were also markedly different (i.e., vertebrate vs. plant). Similar to Heller's EFC implanted in grapes, Cosnier's approach was also based on mediators. However, instead of being immobilized on the electrodes directly, the mediators and enzymes were mechanically confined within dialysis bags, as shown in Figure 19.4a, allowing for the diffusion of fuel and oxidant but also confining the enzymes and mediators to the electrodes. The graphite anode was based on ubiquinone and GOx with the addition of catalase to eliminate the H_2O_2 produced by GOx during operation, whereas the graphite cathode was based on quinone combined with polyphenol oxidase (PPO). PPO is just like laccase and BOx, a member of the BMCO family, and the enzyme was chosen because of its tolerance of chloride ions and urate present in extracellular fluids. Due to the design, significant miniaturization of the EFC was not possible, limiting device applications to larger scaled implantable devices requiring insignificant power, where the size of the power supply is not the crucial factor.

To circumvent the limited amount of O_2 usually available in living organisms, in 2011, Miyake et al. designed a needle bioanode together with a gas diffusion biocathode [124]. Due to the omission of any separator membranes, significant simplification in the design could be made, enabling biofuels to be accessed from living organisms by the bioanode, whereas the biocathode avoided contact with potential inhibitors present in physiological fluids using the abundant O_2 available in the air (Figure 19.4b). To harness power from grapes, FDH was adsorbed on carbon nanoparticles (KB)-modified electrodes, whereas a similar electrode modified with GDH mediated with poly-L-lysine (PLL)-modified vitamin K_3, and PLL-modified NAD^+ was used as bioanode for an EFC operating in blood. The gas diffusion biocathode was constructed by adsorbing BOx on KB-modified carbon paper, coated with carbon particles to create a hydrophobic layer. The anode and the cathode were assembled together using an ion-conducting agarose hydrogel as inner matrix, and by inserting the FDH-based bioanode into a grape, a maximum power of 115 µW cm^{-2} at

0.34 V was generated. When the GDH-based bioanode was inserted in a vein on the ear of a rabbit, the EFC generated a maximum power of 131 $\mu W\,cm^{-2}$ at a cell voltage of 0.56 V. The authors suggested that improvement of these devices would allow the fabrication of microscale bioanodes for patchable EFCs.

Recently, Shleev and coworkers reported a microscale EFC implanted in the brain of a living rat, where microelectrodes attached to micromanipulators were inserted directly into the brain tissue, as shown in Figure 19.4c [125]. The EFC design was similar to earlier investigations by the same group on tear fluid based on 100 μm diameter AuNP-modified gold wire electrodes and a simple DET-based approach without the use of any membranes, immobilizing CDH as glucose-oxidizing enzyme on the bioanode and BOx as an O_2-reducing enzyme on the biocathode. The use of such a simple design in the construction of the EFC allowed for miniaturization and easy fabrication of the biodevice.

When the EFC was implanted in the rat's brain, a maximum power of 2 $\mu W\,cm^{-2}$ at a cell voltage of 0.4 V was registered. Under a constant load, the voltage was roughly halved in 2 h; however, a small manipulation of the electrodes brought back the voltage to the initial level. Thus, the implanted EFC operating in the living organism was limited by the diffusional properties of the surrounding tissue [125].

Apart from vertebrates and plants, insects and mollusks also provide interesting opportunities for BFCs to be used, mainly for military or environmental applications. The circulatory system of invertebrates consists of hemolymph instead of blood, which supplies the organisms with nutrients and O_2 in an open system [126]. The hemolymph does not have nearly as sophisticated an immune system as the blood of vertebrates, simplifying long-term implantation. In 2006, Heller and coworkers worked on developing electrodes capable of oxidizing trehalose at a concentration of around 30 mM, although no functional EFC was fabricated [127,128]. However, in 2012, Rasmussen et al. reported on a trehalose–glucose/O_2 EFC implanted in a living cockroach, based on splitting the sugar into glucose units, which then could be used as a fuel [67]. The bioanode was created by co-immobilizing trehalose oxidase (TOx), capable of dissociating trehalose into glucose monomers, with GOx as the glucose-oxidizing enzyme, by covalently attaching the enzymes on Os-containing redox hydrogel-modified carbon rod electrodes. The biocathode was designed in a similar way, however, using BOx to reduce oxygen instead of the trehalose/GOx bienzyme system. The electrodes were encased within a glass capillary tube and carbon fibers were used as wiring contacts. A photograph of the EFC setup is shown in Figure 19.5a, and when operating *in vivo*, the EFC yielded a maximum power of 55 μW cm^{-2} at a cell voltage of 0.2 V, only decreasing by 5% after 2.5 h. The same EFC was also implanted in a shiitake mushroom, where 1.21 $\mu W\,cm^{-2}$ at 0.2 V was generated [67]. The lower power output can be explained by the lower amount of fuel available in the animal compared with the fungus.

The Katz research group published two consecutive reports in 2012 of glucose/O_2 EFCs operating in mollusks [129,130]. The electrodes were fabricated using a DET-based design based on CNT paper (buckypaper), with laccase and PQQ-dependent GDH as cathodic and anodic enzymes, respectively. The enzymes were immobilized on nanostructured electrodes by cross-linking. Such a DET-based approach simplifies

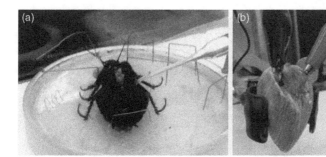

FIGURE 19.5 (a) Photograph of the experimental setup of a TOx–GOx/BOx EFC implanted into a cockroach. (Reprinted with permission from Ref. [67]. Copyright 2012, American Chemical Society.) (b) Photograph of the experimental setup of a clam with implanted (PQQ)-GDH/laccase EFC. (Reprinted with permission from Ref. [130]. Copyright 2012, Royal Society of Chemistry.) (Please see the color version of this figure in Color Plates section.)

the construction of EFCs and excludes possible toxic mediator compounds. When the EFC was implanted in a snail, a maximum power of $30\,\mu W\,cm^{-2}$ was obtained at a cell voltage of 0.39 V [129]. When an EFC was implanted in clams, a maximum power of $40\,\mu W\,cm^{-2}$ at a voltage of 0.17 V was registered. They also demonstrated a significant increase in voltage and the current, when several EFCs implanted into different organisms were connected in series and in parallel, respectively [130]. The setup used for investigating the power output generated from living clams is shown in Figure 19.5b. The EFC was also implanted in both a snail and a clam and a rather fast voltage drop was observed when the EFC was under load, reducing the voltage to half in less than 15 min. The drop can be attributed to inefficient mass transport in the biological tissue, similar to what was observed by Shleev and coworkers in the brain tissue (see below) [125]. When the electrodes were disconnected and the organism was allowed a certain relaxation time, similar power as obtained initially could yet again be generated (Figure 19.6).

A summary of all of the reported implanted BFCs in the literature and described herein is presented in Table 19.1. A direct comparison, however, is difficult due to the vastly different conditions in which they were operated, that is, grape juice, hemolymph, and blood. Moreover, the BFCs were also designed with different applications in mind.

The power an implanted BFC can generate highly depends on where in the body it is implanted. As concluded in 2004 by Calabrese Barton et al., implanted devices are expected to provide power on the microwatt scale in the tissue and milliwatt to watt scale in the blood vessel, where high velocities can be present [2]. The power output generated from a BFC implanted in tissue is limited by diffusion of a fuel and/or an oxidant to the electrode surface, which was first observed *in vivo* in 2003 by Heller and coworkers, when implanting a BFC into a grape [49]. Almost five times more power output was obtained when the biocathode was placed close to the grape's surface, being more freely supplied with O_2, compared with when it was placed at the center. Similarly, diffusion limitation of bioelectrodes was observed by both Katz's

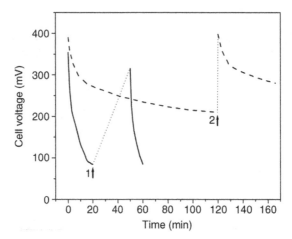

FIGURE 19.6 Drop in a cell voltage under constant load of EFCs implanted in the tissue of a clam (solid line) and rat brain (dashed line) [125,130]. At the times marked with arrows, (1) the load was switched off or (2) a slight manipulation of the position of the electrodes was performed, returning the voltage to initial values. (Reprinted with permission from Ref. [130]. Copyright 2012, Royal Society of Chemistry.)

and Shleev's research groups, as illustrated in Figure 19.6, with a rapid drop in power output over time. However, the generated power was completely restored when the organism was allowed to recover or slight manipulation of the electrode placements was performed, respectively. When comparing the actual power that can be generated from an implanted BFC (see Table 19.1), the output is high enough for the powering of certain modern bioelectronics systems to be fully feasible. The critical parameter would rather be the voltage, since the connection of several BFCs in series within the same organism will not increase the voltage [125]. Although significantly higher power output could be expected for BFCs operating in blood vessels, since the diffusion would not be a very crucial issue, great care would have to be taken with such biodevices regarding biocompatibility to avoid thrombosis (also during insertion to avoid damaging the blood vessel cell walls) [2].

19.5 OUTLOOK

The main advantage of fuel cells being used as an electric power source for IMDs compared with batteries is the theoretical potential for the source to function as long as the individual is alive. Devices can use substrates that already exist in the body; that is, no artificial fuel and oxidant are required for proper performance of the implanted fuel cells. Among different types of fuel cells, BFCs are viewed as very promising devices with huge potential to power miniature IMDs.

Many decades ago, the development of implantable fuel cells for biomedical application was demonstrated to be feasible in theory, and it was tested

TABLE 19.1 Summary of the Characteristics of Reported Implanted Enzymatic Fuel Cells

Fuel/Oxidant	Bioelements (Anode/Cathode)	Membrane/ Mediator	Cell Voltage	Power (μW cm^{-2})	Activity Loss in 2.5 h	Electrode Description	Organism	Reference
Glucose/O$_2$	GOx/BOx	No/yes	0.52	240	$\ll$ (22% after 24 h)	Carbon fibers modified with Os redox polymer	Grape	[49]
Glucose/O$_2$	GOx/PPO	Yes/yes	0.13	24.4^a	25%	Graphite electrodes with soluble mediators, encased in membranes	Rat	[110]
Glucose/O$_2$	GDH/BOx	No/yes	0.56	131	—	KB on carbon modified with PLL-NAD, gas diffusion cathode^b	Rabbit	[124]
Glucose/O$_2$	FDH/BOx	No/yes	0.34	115	—	KB on carbon, gas diffusion cathode^b	Grape	[124]
Trehalose–glucose/O$_2$	TOx–GOx/BOx	No/yes	0.20	55	5%	Carbon rods modified with Os redox polymer	Cockroach	[67]
Trehalose–glucose/O$_2$	TOx–GOx/BOx	No/yes	0.20	1.21	—	Carbon rods modified with Os redox polymer	Mushroom	[67]
Glucose/O$_2$	(PQQ)-GDH/laccase	No/no	0.39	30	0 ($\gg$)	CNT-based electrodes	Snail	[129]
Glucose/O$_2$	(PQQ)-GDH/laccase	No/no	0.17	40	0 ($\gg$)	CNT-based electrodes	Clam	[130]
Glucose/O$_2$	CDH/BOx	No/no	0.4	2	0 (40%)	AuNP-modified Au wire	Rat	[125]

^{a}Power is given in μW cm^{-3}.
^{b}Cathode was positioned outside the organism.

441

experimentally. However, after many years of research, the technology has brought us to the stage where biodevices are applied to *in vivo* trials. The very first report about implanted BFCs was published in 2003 by Heller and coworkers when a miniature biodevice was tested in a plant. It took almost a decade for researchers to colonize the animal kingdom; beginning in 2010, several reports appeared, in which the performance of different types of BFCs in mollusks, insects, and mammals have been presented.

Although most of the systems described in the chapter cannot be used as implantable devices in human hosts and do not represent a model for trials toward biomedical implantable application, there are some facts that can be useful toward real biomedical application of EFCs. Those studies demonstrate the feasibility of using enzymatic systems for *in vivo* power production.

Researchers in the field have overcome many of the limitations for real-life applications, but there are many others that are still to be overcome, specifically the stability of the biocatalyst, the biocompatibility of the materials and the products, the chemical and mechanical stability of the electrode materials, the electrolytic solution, and the packaging of the device. Recent advances in nanotechnology have brought great improvements to the development of composite nanomaterials that are feasible to be integrated into BFC designs. Better understanding of the enzymatic systems and processes certainly helped the development of biosensors and made us envision close future applications of BFCs for biomedical purposes. Furthermore, new alternatives for *ex vivo* applications of biodevices may be implemented in a shorter period of time.

LIST OF ABBREVIATIONS

AOx	ascorbate oxidase
AuNP	gold nanoparticle
BFC	biological fuel cell
BMCO	blue multicopper oxidase
BOx	bilirubin oxidase
CDH	cellobiose dehydrogenase
CNT	carbon nanotube
DET	direct electron transfer
EFC	enzymatic fuel cell
ET	electron transfer
FAD	flavin adenine dinucleotide
FDH	fructose dehydrogenase
GDH	glucose dehydrogenase
GOx	glucose oxidase
IMD	implantable medical device
KB	Ketjenblack
MET	mediated electron transfer
MFC	microbial fuel cell

MWCNT	multiwalled carbon nanotube
NAD^+	nicotinamide adenine dinucleotide
NADH	nicotinamide adenine dinucleotide, reduced form
NHE	normal hydrogen electrode
OCV	open-circuit voltage
PLL	poly-L-lysine
PPO	polyphenol oxidase
PQQ	pyrroloquinoline quinone
SEM	scanning electron microscopy
SWCNT	single-walled carbon nanotube
T1	type 1
T2	type 2
T3	type 3
TOx	trehalose oxidase

REFERENCES

1. Soykan O. Power sources for implantable medical devices. In: Cooper E (ed.), *Business Briefing: Medical Device Manufacturing and Technology*. Business Briefings Ltd., London, 2002, pp. 76–79.

2. Calabrese Barton S, Gallaway J, Atanassov P. Enzymatic biofuel cells for implantable and microscale devices. *Chem Rev* 2004;104:4867–4886.

3. Hanashi T, Yamazaki T, Tsugawa W, Ikebukuro K, Sode K. BioRadioTransmitter: a self-powered wireless glucose-sensing system. *J Diabetes Sci Technol* 2011;5:1030–1035.

4. Heller A. Miniature biofuel cells. *Phys Chem Chem Phys* 2004;6:209–216.

5. Hu Y, Zhang Y, Xu C, Lin L, Snyder RL, Wang ZL. Self-powered system with wireless data transmission. *Nano Lett* 2011;11:2572–2577.

6. Drake RF, Kusserow BK, Messinger S, Matsuda S. A tissue implantable fuel cell power supply. *Trans Am Soc Artif Intern Organs* 1970;16:199 205.

7. Kannan AM, Renugopalakrishnan V, Filipek S, Li P, Audette GF, Munukutla L. Bio-batteries and bio-fuel cells: leveraging on electronic charge transfer proteins. *J Nanosci Nanotechnol* 2009;9:1665–1678.

8. Wolfson SK, Gofberg SL, Prusiner P, Nanis L. The bioautofuel cell: a device for pacemaker power from direct energy conversion consuming autogenous fuel. *Trans Am Soc Artif Intern Organs* 1968;14:198–203.

9. Appleby AJ, Ng DYC, Weinstein H. Parametric study of the anode of an implantable biological fuel cell. *J Appl Electrochem* 1971;1:79–90.

10. Jacobson B (ed.), *Digest of the Seventh International Conference on Medical and Biological Engineering*. Royal Academy of Engineering Sciences, Stockholm, 1967, p. 520.

11. Bockris JO'M, Piersma BJ, Gileadi E. Anodic oxidation of cellulose and lower carbohydrates. *Electrochim Acta* 1964;9:1329–1332.

12. Talaat ME, Kraft JH, Cowley RA, Khazei AH. Biological electrical power extraction from blood to power cardiac pacemakers. *IEEE Trans Biomed Eng* 1967;14:263–265.

13. Malachesky P, Holleck G, McGovern F, Devarakonda R. *Parametric studies of implantable fuel cell.* 7th Intersociety Energy Conversion Engineering Conference Proceedings, American Chemical Society, Washington, DC, 1972, pp. 727–732.

14. Rao JR, Richter G Von Sturm F, Weidlich E, Wenzel M. Metal–oxygen and glucose–oxygen cells for implantable devices. *Biomed Eng* 1974;9:98–103.

15. Rao MLB, Drake RF. Studies of electrooxidation of dextrose in neutral media. *J Electrochem Soc* 1969;116:334–337.

16. Ohm OJ, Danilovic D. Improvements in pacemaker energy consumption and functional capability: four decades of progress. *Pacing Clin Electrophysiol* 1997;20:2–9.

17. Rubino RS, Gan H, Takeuchi ES. *Implantable medical applications of lithium-ion technology.* 17th Annual Battery Conference on Applications and Advances, 2002, pp. 123–127.

18. Kerzenmacher S, Ducrée J, Zengerle R, von Stetten F. Energy harvesting by implantable abiotically catalyzed glucose fuel cells. *J Power Sources* 2008;182:1–17.

19. von Stetten F, Kerzenmacher S, Sumbharaju R, Zengerle R, Ducrée J. *Biofuel cells as power generator for implantable devices.* Proc Eurosensors XX, 2006, pp. 222–225.

20. Palmore GTR, Whitesides GM. Microbial and enzymatic biofuel cells. *ACS Symp Ser* 1994;566:271–290.

21. Kerzenmacher S, Mutschler K, Kraeling U, Baumer H, Ducrée J, Zengerle R, von Stetten F. A complete testing environment for the automated parallel performance characterization of biofuel cells: design, validation, and application. *J Appl Electrochem* 2009;39:1477–1485.

22. Nishio K, Kouzuma A, Kato S, Watanabe K. *Microbial Biofilms: Current Research and Applications.* Caister Academic Press, Norfolk, UK, 2012, pp. 175–191.

23. Willner I, Yan YM, Willner B, Tel-Vered R. Integrated enzyme-based biofuel cells—a review. *Fuel Cells* 2009;9:7–24.

24. Han Y, Yu C, Liu H. A microbial fuel cell as power supply for implantable medical devices. *Biosens Bioelectron* 2010;25:2156–2160.

25. Justin GA, Zhang Y, Cui XT, Bradberry CW, Sun M, Sclabassi RJ. A metabolic biofuel cell: conversion of human leukocyte metabolic activity to electrical currents. *J Biol Eng* 2011;5:5.

26. Sun M, Justin GA, Roche PA, Zhao J, Wessel BL, Zhang Y, Sclabassi RJ. Passing data and supplying power to neural implants. *IEEE Eng Med Biol Mag* 2006;25:39–46.

27. Bhatnagar D, Xu S, Fischer C, Arechederra RL, Minteer SD. Mitochondrial biofuel cells: expanding fuel diversity to amino acids. *Phys Chem Chem Phys* 2011;13:86–92.

28. Arechederra R, Minteer SD. Organelle-based biofuel cells: immobilized mitochondria on carbon paper electrodes. *Electrochim Acta* 2008;53:6698–6703.

29. Zebda A, Gondran C, Le Goff A, Holzinger M, Cinquin P, Cosnier S. Mediatorless high-power glucose biofuel cells based on compressed carbon nanotube–enzyme electrodes. *Nat Commun* 2011;2:1–6.

30. Cracknell JA, Vincent KA, Armstrong FA. Enzymes as working or inspirational electrocatalysts for fuel cells and electrolysis. *Chem Rev* 2008;108:2439–2461.

31. Habermüller K, Mosbach M, Schuhmann W. Electron-transfer mechanisms in amperometric biosensors. *Fresenius J Anal Chem* 2000;366:560–568.

32. Katz E, Shipway AN, Willner I. Mediated electron transfer between redox-enzymes and electrode supports. In: *Encyclopedia of Electrochemistry*. Wiley-VCH Verlag GmbH, Weinheim, 2002, pp. 559–626.

33. Heller A. Electron-conducting redox hydrogels: design, characteristics, and synthesis. *Curr Opin Chem Biol* 2006;10:664–672.

34. Schuhmann W, Ohara TJ, Schmidt HL, Heller A. Electron transfer between glucose oxidase and electrodes via redox mediators bound with flexible chains to the enzyme surface. *J Am Chem Soc* 1991;113:1394–1397.

35. Heller A, Feldman BJ, Say J, Vreeke MS, Tomasco MF. *Small volume* in vitro *analyte sensor*. WIPO Pub No. WO/1998/035225, 1998 (to E. Heller & Co.).

36. Gao Z, Binyamin G, Kim HH, Calabrese Barton S, Zhang Y, Heller A. Electrodeposition of redox polymers and co-electrodeposition of enzymes by coordinative crosslinking. *Angew Chem Int Ed* 2002;41:810–813.

37. Mao F, Mano N, Heller A. Long tethers binding redox centers to polymer backbones enhance electron transport in enzyme "wiring" hydrogels. *J Am Chem Soc* 2003;125:4951–4957.

38. Heller A, Gao Z, Dequaire M. *Electrodeposition of redox polymers and co-electrodeposition of enzymes by coordinative crosslinking*. WIPO Pub No. 20030168338, 2003 (to TheraSense, Inc.).

39. Chen T, Calabrese Barton S, Binyamin G, Gao Z, Zhang Y, Kim HH, Heller A. A miniature biofuel cell. *J Am Chem Soc* 2001;123:8630–8631.

40. Liu Z, Feldman BJ, Mao F, Heller A. *Redox polymers for use in analyte monitoring*. WIPO Pub No. 20121232525, 2012 (to Abbott Diabetes Care Inc.).

41. Heller A, Mano N, Kim H-H, Zhang Y, Mao F, Chen T, Calabrese Barton S. *Miniature biological fuel cell that is operational under physiological conditions, and associated devices and methods*. WIPO Pub No. WO/2003/106966, 2003 (to TheraSense, Inc.).

42. Falk M, Blum Z, Shleev S. Direct electron transfer based enzymatic fuel cells. *Electrochim Acta* 2012;82:191–202.

43. Ramanavicius A, Ramanaviciene A. Hemoproteins in design of biofuel cells. *Fuel Cells* 2009;9:25–36.

44. Shleev S, Tkac J, Christenson A, Ruzgas T, Yaropolov AI, Whittaker JW, Gorton L. Direct electron transfer between copper-containing proteins and electrodes. *Biosens Bioelectron* 2005;20:2517–2554.

45. Tsujimura S, Kuriyama A, Fujieda N, Kano K, Ikeda T. Mediated spectroelectrochemical titration of proteins for redox potential measurements by a separator-less one-compartment bulk electrolysis method. *Anal Biochem* 2005;337:325–331.

46. Pardo-Yissar V, Katz E, Willner I, Kotlyar AB, Sanders C, Lill H. Biomaterial engineered electrodes for bioelectronics. *Faraday Discuss* 2000;116:119–134.

47. Smolander M, Boer H, Alkiainen M, Roozeman R, Bergelin M, Eriksson JE, Zhang XC, Koivula A, Viikari L. Development of a printable laccase-based biocathode for fuel cell applications. *Enzyme Microb Technol* 2008;43:93–102.

48. Calabrese Barton S, Kim HH, Binyamin G, Zhang Y, Heller A. The "wired" laccase cathode: high current density electroreduction of O_2 to water at +0.7 V (NHE) at pH 5. *J Am Chem Soc* 2001;123:5802–5803.

49. Mano N, Mao F, Heller A. Characteristics of a miniature compartment-less glucose–O_2 biofuel cell and its operation in a living plant. *J Am Chem Soc* 2003;125:6588–6594.

50. Christenson A, Shleev S, Mano N, Heller A, Gorton L. Redox potentials of the blue copper sites of bilirubin oxidases. *Biochim Biophys Acta* 2006;1757:1634–1641.

51. Kang C, Shin H, Heller A. On the stability of the "wired" bilirubin oxidase oxygen cathode in serum. *Bioelectrochemistry* 2006;68:22–26.

52. Shin H, Kang C, Heller A. Irreversible and reversible deactivation of bilirubin oxidase by urate. *Electroanalysis* 2007;19:638–643.

53. Wong CM, Wong KH, Chen XD. Glucose oxidase: natural occurrence, function, properties, and industrial applications. *Appl Microbiol Biotechnol* 2008;78:927–938.

54. Hu F, Chen S, Wang C, Yuan R, Chai Y, Xiang Y, Wang C. ZnO nanoparticle and multiwalled carbon nanotubes for glucose oxidase direct electron transfer and electrocatalytic activity investigation. *J Mol Catal B: Enzym* 2011;72:298–304.

55. Wang Y, Liu L, Li M, Xu S, Gao F. Multifunctional carbon nanotubes for direct electrochemistry of glucose oxidase and glucose bioassay. *Biosens Bioelectron* 2011;30:107–111.

56. Courjean O, Gao F, Mano N. Deglycosylation of glucose oxidase for direct and efficient glucose electrooxidation on a glassy carbon electrode. *Angew Chem Int Ed* 2009;48:5897–5899.

57. Holland JT, Lau C, Brozik S, Atanassov P, Banta S. Engineering of glucose oxidase for direct electron transfer via site-specific gold nanoparticles conjugation. *J Am Chem Soc* 2011;133:19262–19265.

58. Ferri S, Kojima K, Sode K. Review of glucose oxidases and glucose dehydrogenases: a bird's eye view of glucose sensing enzymes. *J Diabetes Sci Technol* 2011;5:1068–1076.

59. Svoboda V, Cooney MJ, Rippolz C, Liaw BY. *In situ* characterization of electrochemical polymerization of methylene green on platinum electrodes. *J Electrochem Soc* 2007;154: D113–D116.

60. Zhou D, Fang H, Chen H, Ju H, Wang Y. The electrochemical polymerization of methylene green and its electrocatalysis for the oxidation of NADH. *Anal Chim Acta* 1996;329:41–48.

61. Karyakin AA, Karyakina EE, Schuhmann W, Schmidt HL. Electropolymerized azines. Part II. In a search of the best electrocatalyst of NADH oxidation. *Electroanalysis* 1999;11:553–557.

62. Tanne C, Goebel G, Lisdat F. Development of a (PQQ)-GDH-anode based on MWCNT-modified gold and its application in a glucose/O$_2$-biofuel cell. *Biosens Bioelectron* 2010;26:530–535.

63. Ivnitski D, Branch B, Atanassov P, Apblett C. Glucose oxidase anode for biofuel cell based on direct electron transfer. *Electrochem Commun* 2006;8:1204–1210.

64. Ivnitski D, Atanassov P, Apblett C. Direct bioelectrocatalysis of PQQ-dependent glucose dehydrogenase. *Electroanalysis* 2007;19:1562–1568.

65. Ludwig R, Harreither W, Tasca F, Gorton L. Cellobiose dehydrogenase: a versatile catalyst for electrochemical applications. *ChemPhysChem* 2010;11:2674–2697.

66. Ikeda T, Matsushita F, Senda M. Amperometric fructose sensor based on direct bioelectrocatalysis. *Biosens Bioelectron* 1991;6:299–304.

67. Rasmussen M, Ritzmann RE, Lee I, Pollack AJ, Scherson D. An implantable biofuel cell for a live insect. *J Am Chem Soc* 2012;134:1458–1460.

68. Ha S, Wee Y, Kim J. Nanobiocatalysis for enzymatic biofuel cells. *Top Catal* 2012;55:1181–1200.

69. Minteer SD. Nanobioelectrocatalysis and its applications in biosensors, biofuel cells, and bioprocessing. *Top Catal* 2012;55:1157–1161.

70. Fei G, Ma G-H, Wang P, Su Z-G. Enzyme immobilization, biocatalyst featured with nanoscale structure. *Encyc Ind Biotechnol* 2010;3:2086–2094.

71. Atanassov P, Apblett C, Banta S, Brozik S, Calabrese Barton S, Cooney M, Liaw BY, Mukerjee S, Minteer SD. Enzymatic fuel cells. *Electrochem Soc Interface* 2007;16(2): 28–31.

72. Narváez Villarrubia CW, Rincón RA, Radhakrishnan VK, Davis V, Atanassov P. Methylene green electrodeposited on SWNTs-based "bucky" papers for NADH and L-malate oxidation. *ACS Appl Mater Interfaces* 2011;3:2402–2409.

73. Jaegfeldt H, Kuwana T, Johansson G. Electrochemical stability of catechols with a pyrene side chain strongly adsorbed on graphite electrodes for catalytic oxidation of dihydro-nicotinamide adenine dinucleotide. *J Am Chem Soc* 1983;105:1805–1814.

74. Yan Y, Zheng W, Su L, Mao L. Carbon-nanotube-based glucose/O_2 biofuel cells. *Adv Mater* 2006;18:2639–2643.

75. Joshi PP, Merchant SA, Wang Y, Schmidtke DW. Amperometric biosensors based on redox polymer–carbon nanotube–enzyme composites. *Anal Chem* 2005;77:3183–3188.

76. Le Goff A, Moggia F, Debou N, Jegou P, Artero V, Fontecave M, Jousselme B, Palacin S. Facile and tunable functionalization of carbon nanotube electrodes with ferrocene by covalent coupling and π-stacking interactions and their relevance to glucose biosensing. *J Electroanal Chem* 2010;641:57–63.

77. Guiseppi-Elie A, Lei C, Baughman RH. Direct electron transfer of glucose oxidase on carbon nanotubes. *Nanotechnology* 2002;13:559–564.

78. Whitby RLD, Fukuda T, Maekawa T, James SL, Mikhalovsky SV. Geometric control and tuneable pore size distribution of buckypaper and buckydiscs. *Carbon* 2008;46:949–956.

79. Wang D, Song P, Liu C, Wu W, Fan S. Highly oriented carbon nanotube papers made of aligned carbon nanotubes. *Nanotechnology* 2008;19:075609.

80. Ivnitski D, Artyushkova K, Rincón RA, Atanassov P, Luckarift HR, Johnson GR. Entrapment of enzymes and carbon nanotubes in biologically synthesized silica: glucose oxidase-catalyzed direct electron transfer. *Small* 2008;4:357–364.

81. Weidlich E, Richter G, von Sturm F, Rao JR, Thoren A, Lagergren H. Animal experiments with biogalvanic and biofuel cells. *Biomater Med Devices Artif Organs* 1976;4:277–306.

82. Heller A. Potentially implantable miniature batteries. *Anal Bioanal Chem* 2006;385:469–473.

83. Sharma T, Hu Y, Stoller M, Feldman M, Ruoff RS, Ferrari M, Zhang X. Mesoporous silica as a membrane for ultra-thin implantable direct glucose fuel cells. *Lab Chip* 2011;11:2460–2465.

84. Ratner BD, Hoffman AS, Schoen FJ, Lemons JE. *An Introduction to Materials in Medicine*, 2nd edition. Elsevier Academic Press, San Diego, CA, 2004.

85. Guyton AC, Hall JE. *Textbook of Medical Physiology*, 11th edition. Elsevier Saunders, Philadelphia, PA, 2006.

86. Palmer JK, Brandes WB. Determination of sucrose, glucose, and fructose by liquid chromatography. *J Agric Food Chem* 1974;22:709–712.

87. Sokic-Lazic D, Minteer SD. Citric acid cycle biomimic on a carbon electrode. *Biosens Bioelectron* 2008;24:939–944.

88. Moehlenbrock MJ, Toby TK, Waheed A, Minteer SD. Metabolon catalyzed pyruvate/air biofuel cell. *J Am Chem Soc* 2010;132:6288–6289.

89. Moehlenbrock MJ, Toby TK, Pelster LN, Minteer SD. Metabolon catalysts: an efficient model for multi-enzyme cascades at electrode surfaces. *ChemCatChem* 2011;3:561–570.

90. Sokic-Lazic S, Rodrigues de Andrade A, Minteer SD. Utilization of enzyme cascades for complete oxidation of lactate in an enzymatic biofuel cell. *Electrochim Acta* 2011;56:10772–10775.

91. Moehlenbrock MJ, Meredith MT, Minteer SD. Bioelectrocatalytic oxidation of glucose in CNT impregnated hydrogels: advantages of synthetic enzymatic metabolon formation. *ACS Catal* 2012;2:17–25.

92. Xu S, Minteer SD. Enzymatic biofuel cell for oxidation of glucose to CO_2. *ACS Catal* 2012;2:91–94.

93. Shao M, Nadeem Zafar M, Sygmund C, Guschin DA, Ludwig R, Peterbauer CK, Schuhmann W, Gorton L. Mutual enhancement of the current density and the coulombic efficiency for a bioanode by entrapping bi-enzymes with Os-complex modified electro-deposition paints. *Biosens Bioelectron* 2013;40:308–314.

94. Shleev D, Shumakovich G, Morozova O, Yaropolov A. Stable "floating" air diffusion biocathode based on direct electron transfer reactions between carbon particles and high redox potential laccase. *Fuel Cells* 2010;10:726–733.

95. Gupta G, Lau C, Branch B, Rajendran V, Ivnitski D, Atanassov P. Direct bio-electro-catalysis by multi-copper oxidases: gas-diffusion laccase-catalyzed cathodes for biofuel cells. *Electrochim Acta* 2011;56:10767–10771.

96. Gupta G, Lau C, Rajendran V, Colon F, Branch B, Ivnitski D, Atanassov P. Direct electron transfer catalyzed by bilirubin oxidase for air breathing gas-diffusion electrodes. *Electrochem Commun* 2011;13:247–249.

97. Brocato S, Lau C, Atanassov P. Mechanistic study of direct electron transfer in bilirubin oxidase. *Electrochim Acta* 2008;61:44–49.

98. Calabrese Barton S. Oxygen transport in composite biocathodes. *Proc Electrochem Soc* 2005; 2002–31: 324–335.

99. Gutierrez-Sanchez C, Pita M, Vaz-Dominguez C, Shleev S, De Lacey AL. Gold nano-particles as electronic bridges for laccase-based biocathodes. *J Am Chem Soc* 2012;134:17212–17220.

100. Ivnitski DM, Khripin C, Luckarift HR, Johnson GR, Atanassov P. Surface characteriza-tion and direct bioelectrocatalysis of multicopper oxidases. *Electrochim Acta* 2010;55:7385–7393.

101. Klotzbach T, Watt M, Ansari Y, Minteer SD. Effects of hydrophobic modification of chitosan and Nafion on transport properties, ion-exchange capacities, and enzyme immobilization. *J Membr Sci* 2006;282:276–283.

102. Ghanem A, Ghaly A. Immobilization of glucose oxidase in chitosan gel beads. *J Appl Polym Sci* 2004;91:861–866.

103. Liu Y, Qu X, Guo H, Chen H, Liu B, Dong S. Facile preparation of amperometric laccase biosensor with multifunction based on the matrix of carbon nanotubes–chitosan com-posite. *Biosens Bioelectron* 2006;21:2195–2201.

104. Siso MIG, Lang E, Carreno-Gomez B, Becerra M, Espinar FO, Mendez JB. Enzyme encapsulation on chitosan microbeads. *Process Biochem* 1997;32:211–216.

105. Akers NL, Moore CM, Minteer SD. Development of alcohol/O_2 biofuel cells using salt-extracted tetrabutylammonium bromide/Nafion membranes to immobilize dehydrogenase enzyme. *Electrochim Acta* 2005;50:2521–2525.

106. Chen RJ, Zhang Y, Wang D, Dai H. Noncovalent sidewall functionalization of single-walled carbon nanotubes for protein immobilization. *J Am Chem Soc* 2001;123:3838–3839.

107. Ramasamy RP, Luckarift HR, Ivnitski DM, Atanassov PB, Johnson GR. High electro-catalytic activity of tethered multicopper oxidase–carbon nanotube conjugates. *Chem Commun* 2010;46:6045–6047.

108. Moore CM, Akers NL, Hill AD, Johnson ZC, Minteer SD. Improving the environment for immobilized dehydrogenase enzymes by modifying Nafion with tetraalkylammonium bromides. *Biomacromolecules* 2004;5:1241–1247.

109. Moore CM, Minteer SD, Martin RS. Microchip-based ethanol/oxygen biofuel cell. *Lab Chip* 2005;5:218–225.

110. Cinquin P, Gondran C, Giroud F, Mazabrard S, Pellissier A, Boucher F, Alcaraz J-P, Gorgy K, Lenouvel F, Mathé S, Porcu P, Cosnier S. A glucose biofuel cell implanted in rats. *PLoS One* 2010;5:e10476.

111. Zebda A, Renaud L, Cretin M, Innocent C, Pichot F, Ferrigno R, Tingry S. Electro-chemical performance of a glucose/oxygen microfluidic biofuel cell. *J Power Sources* 2009;193:602–606.

112. Zebda A, Renaud L, Tingry S, Cretin M, Pichot F, Ferrigno R, Innocent C. Microfluidic biofuel cell for energy production. *Sens Lett* 2009;7:824–828.

113. Rincón RA, Lau C, Luckarift HR, Garcia KE, Adkins E, Johnson GR, Atanassov P. Enzymatic fuel cells: integrating flow-through anode and air-breathing cathode into a membrane-less biofuel cell design. *Biosens Bioelectron* 2011;27:132–136.

114. Ciniciato GPMK, Lau C, Cochrane A, Sibbett SS, Gonzalez ER, Atanassov A. Development of paper based electrodes: from air-breathing to paintable enzymatic cathodes. *Electrochim Acta* 2012;82:208–213.

115. Pan C, Fang Y, Wu H, Ahmad M, Luo Z, Li Q, Xie J, Yan X, Wu L, Wang ZL, Zhu J. Generating electricity from biofluid with a nanowire-based biofuel cell for self-powered nanodevices. *Adv Mater* 2010;22:5388–5392.

116. Falk M, Andoralov V, Blum Z, Sotres J, Suyatin DB, Ruzgas T, Arnebrant T, Shleev S. Biofuel cell as a power source for electronic contact lenses. *Biosens Bioelectron* 2012;37:38–45.

117. Katz E. *Bioelectronic systems logically controlled by biochemical signals: towards biologically regulated electronics*. Laboratory of Bioelectronics and Bionanotechnology, Clarkson University, Potsdam, 2012 (http://people.clarkson.edu/~ekatz).

118. Tam TK, Pita M, Ornatska M, Katz E. Biofuel cell controlled by enzyme logic network—approaching physiologically regulated devices. *Bioelectrochemistry* 2009;76:4–9.

119. Amir L, Tam TK, Pita M, Meijler MM, Alfonta L, Katz E. Biofuel cell controlled by enzyme logic systems. *J Am Chem Soc* 2009;131:826–832.

120. Zhou M, Du Y, Chen C, Li B, Wen D, Dong S, Wang E. Aptamer-controlled biofuel cells in logic systems and used as self-powered and intelligent logic aptasensors. *J Am Chem Soc* 2010;132:2172–2174.

121. Tam TK, Strack G, Pita G, Katz E. Biofuel cell logically controlled by antigen–antibody recognition: towards immune-regulated bioelectronic devices. *J Am Chem Soc* 2009;131:11670–11671.

122. Zhou M, Kuralay F, Windmiller JR, Wang J. DNAzyme logic-controlled biofuel cells for self-powered biosensors. *Chem Commun* 2012;48:3815–3817.

123. Kliewer WM. Concentration of tartrates, malates, glucose, and fructose in the fruits of the genus *Vitis*. *Am J Enol Vitic* 1967;18:87–96.

124. Miyake T, Haneda K, Nagai N, Yatagawa Y, Onami H, Yoshino S, Abe T, Nishizawa M. Enzymatic biofuel cells designed for direct power generation from biofluids in living organisms. *Energy Environ Sci* 2011;4:5008–5012.

125. Andoralov V, Falk M, Suyatin D, Granmo M, Schouenborg J, Sotres J, Ludwig R, Popov V, Blum Z, Shleev S. Biofuel cell based on microscale nanostructured electrodes with inductive coupling to rat brain neurons. *Sci Rep* 2013;3:3270.

126. Gilbert LI. *Insect Molecular Biology and Biochemistry*, 1st edition. Elsevier Academic Press, San Diego, CA, 2012.

127. Lee Y, Keeley LL. Intracellular transduction of trehalose synthesis by hypertrehalosemic hormone in the fat body of the tropical cockroach *Blaberus discoidalis*. *Insect Biochem Mol Biol* 1994;24:473–480.

128. Pothukuchy A, Mano N, Georgiou G, Heller A. A potentially insect-implantable trehalose electrooxidizing anode. *Biosens Bioelectron* 2006;22:678–684.

129. Halámkova L, Halámek J, Bocharova V, Szczupak A, Alfonta L, Katz E. Implanted biofuel cell operating in a living snail. *J Am Chem Soc* 2012;134:5040–5043.

130. Szczupak A, Halámek J, Halámkova L, Bocharova V, Alfonta L, Katz E. Living battery—biofuel cells operating *in vivo* in clams. *Energy Environ Sci* 2012;5:8891–8895.

20

CONCLUDING REMARKS AND OUTLOOK

GLENN R. JOHNSON

Airbase Sciences Branch, Air Force Research Laboratory, Tyndall Air Force Base, FL, USA

HEATHER R. LUCKARIFT

Universal Technology Corporation, Dayton, OH, USA; Airbase Sciences Branch, Air Force Research Laboratory, Tyndall Air Force Base, FL, USA

PLAMEN ATANASSOV

Department of Chemical and Nuclear Engineering and Center for Emerging Energy Technologies, University of New Mexico, Albuquerque, NM, USA

20.1 INTRODUCTION

The transition of enzymatic fuel cells (EFCs) from model bioelectrochemical systems and laboratory prototypes to a viable technology hinges on numerous scientific, design, and economic factors. EFC development truly requires multidisciplinary contributions, principally in the areas of biochemistry, electrochemistry, and materials science. Engineering expertise that translates advances in the fundamental science to practical use is the next critical piece of the transition. System design depends not only on effective material choices and assembly but also on conventional electronic connections and circuitry. Since most applications require conditioned power, appropriate energy storage must also be coupled with EFCs for effective devices. Taking advantage of advances in rechargeable battery and capacitor technologies in concert with EFCs will provide compact,

Enzymatic Fuel Cells: From Fundamentals to Applications, First Edition. Edited by Heather R. Luckarift, Plamen Atanassov, and Glenn R. Johnson.
© 2014 John Wiley & Sons, Inc. Published 2014 by John Wiley & Sons, Inc.

efficient power sources. The economic factors to overcome are challenging. In many proposed applications, EFCs would be competing with relatively mature battery technologies. On a strict cost basis, it is unlikely that EFCs can displace chemical batteries in most contemporary applications. The key to market penetration is to identify and then promote how EFCs overcome performance limitations and life cycle costs of the standard approaches.

Bioelectrochemistry and the EFC field continue to grow as research topics, and this advance is evident from published literature as well as related intellectual property filings. A survey of the broader field is provided in Figure 20.1; the search results include many reports focused on microbial fuel cells, biosensors, and redox enzyme-based diagnostics, not EFCs alone. Nevertheless, the community strongly supports advances in the relevant fields of research; novel applications and devices are likely to enter the marketplace as a result. It is the responsibility of everyone in the industry to use and promote the new products to their best (and realistic) potential, and not allow them to falter due to underperformance of overreaching end goals.

The preceding chapters provide background on the state of the science underpinning EFC development, tools and techniques applicable to development, and some examples of emerging applications. The materials and characteristics of EFCs offer specific design advantages compared with conventional fuel cells (Chapter 10). The fuels that may be used in EFCs have high energy densities and are readily available (Chapters 4, 5, and 16). Moreover, the exceptional catabolic range of biology allows numerous fuels to be considered. Various sugars, alcohols, and polyols have been well investigated to date, but more complex fuels (e.g., polysaccharides, mixed waste streams, hydrocarbons) might be considered (Chapters 5, 16, and 19). Ultimately, the

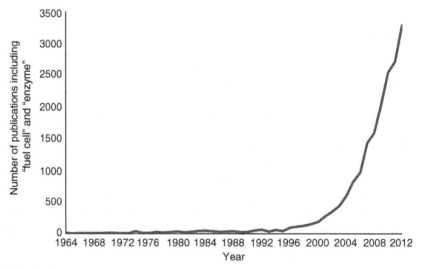

FIGURE 20.1 Trend in bioelectrochemistry publications. Cumulative research articles including terms "fuel cell" and "enzyme". The search did not include patent literature or citations within document bibliographies. Search completed May 2013 using Google Scholar.

redox enzymes used as catalysts can be less expensive than traditional heavy metal or platinum catalysts; the enzymes are harvested from bacterial or fungal strains and their production benefits readily from economy of scale (Chapters 8 and 17). Additionally, the electrochemistry of redox enzymes provides distinct performance advantages in a fuel cell. The enzyme-catalyzed fuel oxidation and oxygen reduction is achieved with low overpotentials compared with metal catalysts, providing high operating voltage from an EFC or biobattery (Chapters 3 and 4). Moreover, the operating conditions are favorable for many applications. Enzymes function well at physiological conditions (room temperature and neutral pH) compared with conditions for conventional fuel cells (high temperatures and highly alkaline or acidic pH); accordingly, both operational and disposal hazards are minimal for EFCs. The specificity of enzyme catalysis leads to a series of design and performance advantages. An exceptional benefit of this selectivity is the potential for miniaturization and conforming (flexible) design options. Conventional electrochemical power sources require robust (and relatively expensive) separator membranes to isolate half-cell reactions and typically rigid casings to contain the chemistry of the overall device. EFCs do not suffer those limitations, which provide great opportunity for implantable and portable electronic devices. If used appropriately, the broad range of potential advantages can promote EFC applications on the basis of performance. Additionally, the sustainability of the fuels and catalysts is a considerable attribute, but viable implementation of any technology is undoubtedly tied to performance and cost, in particular with power sources.

20.2 PRIMARY SYSTEM ENGINEERING: DESIGN DETERMINANTS

The high energy density but relatively low power output of EFCs provides specific design opportunities for application space and also dictates certain aspects of the device duty cycle (i.e., output, charge and discharge rates). There are countless low power demand (micro- to milliwatts) devices presently in use that might be directly driven by EFCs, such as radio frequency identification (RFID) chips, environmental sensors, and medical implants (monitors, drug delivery, etc.) (Chapter 19). Devices that have high power demand but intermittent use could also be sustained with EFCs. By using the complex duty cycle and appropriate energy storage (e.g., battery or capacitor), an EFC could charge the storage component, which provides high power discharge on demand, and then the system would return to charging mode. The recharging approach opens the technology for an exceptionally broad range of applications. Advances in energy storage and control electronics coupled with efficient EFCs may provide means for powering larger systems and other high power/short duty cycle applications, such as personal electronics, surveillance and monitoring equipment, and autonomous devices. The next phases of work in EFCs (and biological fuel cells overall) must emphasize integration of the power source in realistic (and important) applications. Effective demonstration in these settings will greatly contribute to the transition of EFCs from a novelty item or "potential application" to self-sustaining technology.

20.3   FUNDAMENTAL ADVANCES IN BIOELECTROCATALYSIS

Applications and efficiency of bioelectrocatalysis have improved dramatically in recent years by taking advantage of advances in enzymology, molecular biology, and spectroscopy. The advances reveal insight toward modification of reactor conditions (design and electrolyte) that optimize catalysis and electrochemical response. By using molecular biology and general biochemistry, researchers have improved aspects of the anodic and cathodic biocatalysts (Chapters 3 and 4). Normally disparate enzymes have been combined in engineered enzyme cascades that support complete fuel oxidation and have considerably improved power output (Chapter 5). Similarly, defining the mechanisms of inter- and intramolecular electron transfer processes of multicopper oxidases provides insight and improves their application as oxygen reduction catalysts (Chapters 8, 10, and 15). In addition, molecular biology provides efficient means to produce and isolate catalysts by using recombinant bacterial and yeast strains for elevated synthesis of select proteins (Chapters 7 and 8). In other work, genetic modification is used to enhance electrocatalysis by altering protein structure to interact differently with electrode material surfaces (Chapters 7 and 10) [1]. Predicting and defining those interfacial relationships can improve direct electron transfer processes between redox enzymes and electrodes, improving EFC performance.

20.4   DESIGN OPPORTUNITIES FROM EFC OPERATION

The mild operating conditions and sustainable materials that may be used for EFCs allow for design options that were previously unavailable or incompatible with conventional electrochemical power sources. For example, chemical fuel cells operate at temperatures ranging from 80 °C for polymer exchange membrane fuel cells (PEMFCs) to as high as 1000 °C using solid oxide fuel cells. Even at the relatively modest temperature of PEMFCs, operating safety would be a problem for personal electronics or medically implanted devices. Because EFCs typically function well at ambient conditions, these safety concerns are effectively addressed. The operation characteristics of EFCs can have other benefits; environmental sensors or surveillance devices powered by EFCs would be more discrete because any thermal operating signature is eliminated. Along those lines, for distributed environmental applications, systems can be designed that are entirely disposable; the materials, catalysts, and reagents pose no harmful residue or hazard. The fuels and catalysts are readily biodegradable and the mild reaction conditions may be contained in minimal, nonhazardous, even biodegradable polymers and materials [2–4]. The nontoxic, biocompatible EFC components may be effectively incorporated with personal products, such as cosmetic applicators, powered appliques, and contact lenses [5,6]. Electronic devices for food safety, antitamper sensors, or any small powered items that are shipped in restricted conditions are opportunities for EFC integration.

20.5 FUNDAMENTAL DRIVERS FOR EFC MINIATURIZATION

The classic "lock-and-key" specificity of enzymes affords a number of opportunities and benefits for fuel cell design. One direct outcome often noted is that the selective nature eliminates the "fouling" problems often associated with platinum or other metal catalysts [7]. The broader benefit is development of an ultimately "packageless" design that can optimize energy density by minimizing peripheral materials; that is, with effective design, the bulk of the final device's weight is the fuel itself. The electrode materials and enzymes applied as catalysts in the anode and cathode are not affected by fuel–oxidant crossover like conventional chemical fuel cells. Consequently, membrane-free fuel cells are feasible. Enzyme specificity and mild reaction conditions help eliminate restrictive packaging or casing materials. The minimal materials requirement simplifies potential designs and decreases bulk of the power source (Chapters 2, 16, and 17). The elimination of casing materials allows fully flexible and conforming design elements to be incorporated. In that approach, the penalty paid due to load and restrictive space for the power source is lessened. One may even take advantage of a system-of-systems design approach, in which the conformable power source could serve additional purposes, such as a structural component in the device.

Although some of the packaging minimization advantages can be realized with mediated electron transfer (Chapter 9), robust direct electron transfer (DET) (Chapters 3, 10, and 18) is necessary to fully exploit EFCs. DET is particularly important in order to minimize bulk design and maximize EFC output. DET reduces the need for separator materials because problems such as electrolyte containment and crossover between half-cells would be eliminated. Additionally, service lifetime of the EFC can be extended, since the typically labile mediator molecules are eliminated. Furthermore, the high operating potentials that may be attained with DET decrease the number of EFCs that must be stacked in series to achieve the required operating potentials compared with numbers needed for mediated systems.

20.6 COMMERCIALIZATION OF EFCs: STRATEGIES AND OPPORTUNITIES

As EFC performance and reliability matures, it must yield a commercially viable application and product. The outcome will validate the significant financial and intellectual investments made in EFC development and will serve to sustain the emerging field of bioelectrochemistry. To date, the customers for EFCs have essentially been funding agencies supporting the fundamental and developmental research, and, to a lesser degree, venture capital investors endorsing the vision with their checkbooks. If all concerned are keenly committed to a return on each category of investment, the best opportunity for payoff should follow as consumers of the technology adopt the EFC as a power solution.

EFCs have an exceptional range of application areas such as military, medical, and environmental. The military has extensive needs for portable power. A recent report described that a U.S. soldier may carry as many as 70 batteries (as much as 20 lb) to power electronic instruments for a 72 h patrol [8]. Further military uses include developing microscale autonomous systems. Robotic devices, in particular, maneuverable, flapping-wing flight systems, have limited operation times using rechargeable and even primary batteries as power sources. The high energy density afforded by EFCs would be an enormous advantage for these and other mobile autonomous devices. A reasonable approach may have an EFC and its high-density fuel continually charging a small battery to reliably power the autonomous device. An even more ambitious vision has the autonomous device seeking and consuming fuel from the environment, the natural biofuels or waste found in its operating environment could recharge the system, fully mimicking biology. In the medical field, various implanted devices and sensors are ideally suited for EFC power (Chapter 19). Typically, these devices' applications have low power requirements and extended operation times. Conventional batteries can provide the necessary power in many instances, but design factors (rigidity, size) can be a drawback for batteries, and safety problems can arise from leaking caustic chemicals or electrochemical burns from uncontrolled discharge. EFCs do not suffer the same safety problems, operating around neutral pH with no risk of large and rapid discharge. EFC-controlled implanted systems are likewise suited for scavenging fuel from the operating environment; the implant is literally bathed in buffered, biocompatible fuel that may be transformed for electrical power. Environmental sensors and monitors are also application-rich areas. The low-power, energy-dense, green material characteristics of EFCs are desirable attributes of distributed systems and are well matched to requirements for various networks. Additionally, the harmless materials could allow for easy disposal after service, as all components could decompose readily and without consequence [9].

These general areas only introduce the opportunity for EFC applications. Through effective development, EFCs can be a truly revolutionary technology. The unique operating characteristics that have been described throughout this book enable service in "niche" environments that cannot be satisfied by any extant power source. Again, the nearly limitless capability to scale the EFC *down*, coupled with ability to consume ubiquitously available fuels, cannot be achieved with any other developing technology. Accordingly, the next generation of microdevices may be very well served through a bottom-up, system-of-systems design strategy focused on a device that operates indefinitely by seeking and scavenging fuel to sustain operation. EFCs, however, are not bound to such niche applications. In the nearer term, EFCs should integrate and be adopted with current requirements for initial market penetration. The annual multibillion dollar (and growing) personal electronics and communication area is an obvious target. Versatile charging stations (e.g., USB port) for a range of devices are an attractive application for mobile and emergency service in developed markets and a valuable commodity in emerging markets where the conventional electrical grid is unreliable (e.g., rural China and India) or nonexistent (e.g., sub-Saharan Africa). Another effective near-term demonstration for commercialization may be toys and educational kits. This scope of application is more modest than that for personal electronics, but it has already been an important starter market for portable PEMFCs [10].

In short, the EFC field is reaching a point where fundamental principles and understanding are in hand to legitimately engineer and fabricate devices for testing the practical value of EFCs. Although the transition from legitimate demonstration to sustained marketable product is tremendously challenging, there are many encouraging factors. EFCs are a provocative and engaging concept. Effectively using natural and benign processes for portions of the portable consumer power market and to power the next generation of autonomous and distributed devices offers an extensive and tremendous payoff. Truly, there are aspects of the EFC operation that cannot be matched by other technology solutions. The flexible form and "packageless" design opportunities are unique to EFC and are being realized in demonstration prototypes. Another highlight is the recent maturation of this interdisciplinary and competitive field largely via multilaboratory collaborations. This spirit is demonstrated through completion of this book; its organization is intended to summarize the field and to inspire the next leg of advances for EFCs and bioelectrochemistry. We, the editors, enthusiastically look forward to further development and achievements in the technology, and ultimately to the success of EFCs in the marketplace.

ACKNOWLEDGMENT

This research was prepared as an account of work sponsored by the United States Air Force Research Laboratory, Materials and Manufacturing Directorate, Airbase Technologies Division (AFRL/RXQ), but the views of authors expressed herein do not necessarily reflect those of the United States Air Force.

LIST OF ABBREVIATIONS

DET	direct electron transfer
EFC	enzymatic fuel cell
PEMFC	polymer exchange membrane fuel cell
RFID	radio frequency identification

REFERENCES

1. Trohalaki S, Pachter R, Luckarift H, Johnson G. Immobilization of the laccases from *Trametes versicolor* and *Streptomyces coelicolor* on single-wall carbon nanotube electrodes: a molecular dynamics study. *Fuel Cells* 2012;12:656–664.
2. Ciniciato GPMK, Lau C, Cochrane A, Sibbett SS, Gonzalez ER, Atanassov P. Development of paper based electrodes: from air-breathing to paintable enzymatic cathodes. *Electrochim Acta* 2012;82:208–213.

3. Minteer SD, Atanassov P, Luckarift HR, Johnson GR. New materials for biological fuel cells. *Mater Today* 2012;15:166–173.

4. Wu XE, Guo YZ, Chen MY, Chen XD. Fabrication of flexible and disposable enzymatic biofuel cells. *Electrochim Acta* 2013;98:20–24.

5. Falk M, Andoralov V, Blum Z, Sotres J, Suyatin DB, Ruzgas T, Arnebrant T, Shleev S. Biofuel cell as a power source for electronic contact lenses. *Biosens Bioelectron* 2012;37:38–45.

6. Falk M, Blum Z, Shleev S. Direct electron transfer based enzymatic fuel cells. *Electrochim Acta* 2012;82:191–202.

7. Leech D, Kavanagh P, Schuhmann W. Enzymatic fuel cells: recent progress. *Electrochim Acta* 2012;84:223.

8. Parsons D. Effort to reduce battery weight may soon hit brick wall. *Natl Defense* 2012; 96(6):22–24.

9. Gellet W, Kesmez M, Schumacher J, Akers N, Minteer SD. Biofuel cells for portable power. *Electroanalysis* 2010;22:727–731.

10. Carter D, Ryan M, Wing J. *The fuel cell industry review 2012*. Fuel Cell Today, Herts, UK, 2012.

INDEX

Enzymatic Fuel Cells: From Fundamentals to Applications, First Edition. Edited by Heather R. Luckarift, Plamen Atanassov, and Glenn R. Johnson.
© 2014 John Wiley & Sons, Inc. Published 2014 by John Wiley & Sons, Inc.